Basic Animal Nutrition and Feeding

by

D.C. Church, Ph.D.
Professor of Ruminant Nutrition
Department of Animal Science
Oregon State University
Corvallis, Oregon 97331

and

W.G. Pond, Ph.D.
Professor of Animal Science
Department of Animal Science
Cornell University
Ithaca, New York 14850

ISBN Number 0-9601586-2-6

Fifth printing, April, 1978

Published and Distributed by **O & B BOOKS**
1215 NW Kline Place
Corvallis, Oregon 97330
United States of America
Telephone 503-752-2178

Printed by Oxford Press
1427 SE Stark
Portland, Oregon 97214

TABLE OF CONTENTS

PART I — INTRODUCTION

PART II — THE REQUIRED NUTRIENTS

PART III — APPLIED ANIMAL NUTRITION

Preface

This book is intended for students or other readers interested in the principles as well as the application of animal nutrition. As such, it is assumed that the reader will have a minimal knowledge of general chemistry, preferably some exposure to organic and biochemistry, and some understanding of animal biology and husbandry.

Advances in animal nutrition have been quite rapid in recent decades, and the tremendous amount of data published in research journals, field day reports, bulletins and trade publications make a formidable task in keeping up with the current literature in a speciality area to say nothing about the total field of animal nutrition and closely related subject matter. We have now reached the point that it seems likely that few required nutrients remain to be identified. Perhaps some minor nutrients remain to be discovered for some species as it is only within the past few years that silicon has been shown to be a required element for chicks, not a surprising finding in view of the fact that feathers contain appreciable amounts of silicon. Although our knowledge of nutrient needs and functions of the various nutrients is continually expanding, much remains to be learned for a complete understanding and application of the findings to the farm and ranch, particularly with respect to nutrient needs in very exact situations and with regard to nutrient interactions and the effect of many different stresses and diseases on nutrient requirements and metabolism.

With a book of this type, there is always a question of how much and what type of detail to present to the reader. The author's preconception of the audience for the book may be correct and then, again, it may not. Consequently, it is a matter of picking and choosing what to include or exclude. For this reason historical coverage has been omitted. More space could have been allotted on any of the subjects. For some readers, there may be more information on nutrient metabolism than they might desire. Others will probably wish for more complete coverage in Part III on applied nutrition, especially with respect to more complete discussion when dealing with the various species and classes of animals. Regardless of the approach to this subject, some area or subject must be slighted in order to keep the size of the book within bounds and at a reasonable cost. Whatever the deficiencies of this book may be, it is hoped that it will serve a useful purpose by covering in broad scope a complicated and voluminous subject, and that it will serve to guide the student through the important areas of basic and applied animal nutrition.

D.C. Church

Preface

Chapter 1 – Nutrition and its Importance in our Modern Agriculture

Nutritional science, as practiced today by competent nutritionists, is partly the outgrowth of observations by animal husbandrymen and livestock feeders over many centuries and, more recently, by teachers, scientists, and veterinarians. The quantitative aspects — the ability to describe nutrient requirements quantitatively, deficiency symptoms, and metabolism of nutrients — are the result of countless experiments carried out by scientists throughout the world, primarily with domestic and laboratory animals, but also with various animal tissues or microorganisms and, to a lesser extent, with man. As this information has become available, science has gradually replaced the "art" of animal feeding and care in modern agriculture. Much still remains to be learned, however, and a skilled and observant husbandryman often can make up for lack of specific knowledge of animal needs through observation and trial.

Nutrition, as with most other biological sciences, does not have the precision that is possible in a physical science. This is partly because biological organisms are quite variable and, in higher animals, probably no two are exactly alike; also, the environment that any two animals are exposed to is nearly always different, thus nutrient needs are apt to be different.

Nutrition is a subject that interests many readers simply because they must eat to provide for body needs; in addition, for most people, eating and drinking are pleasurable social experiences as well. If one needs added incentives to become partially familiar with the subject, there is the profit motive if one is dealing with domestic animals, or simply the desire to know more about the animal body, its functions, and its needs. In animal agriculture, adequate feeding of animals for production of meat, milk, eggs or fiber is an essential component of the total enterprise. It can be amply demonstrated with man that nutrition may affect his health and welfare, emotions, physical capabilites, and susceptibility to and recovery from disease; there is also the likelihood that the infirmites of old age may be delayed by adequate nutrition.

What is nutrition? A dictionary such as Webster's may define nutrition as "being nourished" or "The series of processes by which an organism takes in and assimilates food for promoting growth and replacing worn or injured tissue." These are obviously, very simplified definitions. Nutrition today, as practiced by competent professionals, requires that the nutritionist be knowledgeable with respect not only to the nutrients, their function, occurrence and so forth, but also with animal behavior and management, digestive physiology, and some aspects of biochemistry and analytical chemistry. In addition, knowledge is required in the fields of crop and soil science, endocrinology, bacteriology, genetics, and disease as related to nutrient needs and dietary requirements.

Although a very wide base of knowledge may be necessary for a thorough understanding of the subject, this is not to say that most people will not benefit from some knowledge of the basic fundamentals of animal nutrition; for example, one does not need to be a complete nutritionist to appreciate that vitamins are important to the animal body and that they may be required in different amounts during growth, maintenance, or lactation.

It might be noted that most people feel that they may have some degree of expertise in nutrition. To a certain extent this may be true, just as it is true that the layman may have a degree of knowledge of law and medicine. It is not a common practice for nutritionists to be licensed, thus anyone can call himself a nutritionist. The result has been, generally, that a great deal of bad nutritional advice has been passed out in the guise of accurate information. The old adage, "a little knowledge is dangerous," still applies to the study of nutrition.

In the author's (Church) opinion, a competent nutritionist must be able to formulate diets, rations, or supplemental feeds that are sufficiently appetizing to ensure an intake (not necessarily maximal) adequate for the purposes desired. He must nearly always take into account the cost of the supplemental or total mixture, and formulated rations should supply adequate nutrients without detrimental imbalances for the desired level of production and take into account the need and required level of growth promotants, medicants, or other nonnutritive additives. In addition, the rations so formulated must have adequate milling, mixing, handling, and storage properties. A nutritionist may be called on to know or to do numerous other things, but these functions would seem to be the minimum that should be expected.

The Role of Animals in Meeting the World Food Needs

One of the greatest challenges in the years ahead is to produce the amount of food needed to feed a human population which is predicted to be approximately double the present level by the year 2000 A.D. (Table 1-1). The amounts of energy and protein needed to feed the human population in 1970, 1985 and 2000 A.D. have been projected as shown in Table 1-2. Animals and animal products seem certain to have a major role in meeting these increased needs. The task of developing the knowledge and technology in animal nutrition and feeding required to meet this challenge in the immediate future is a monumental one, but should be attainable by the students of today who are the scientists and livestock producers of tomorrow.

Factors Affecting Nutritional Needs

A variety of factors operate in today's agricultural businesses that make it imperative for the nutrition and feeding of animals to be at least reasonably adequate. In the USA, as well as in many other countries, the overriding factor is the cost-price squeeze on producers.

The cost of labor, land, and capital have generally increased at a more rapid rate than agricultural products, with the result that many small or inefficient operators have been forced out of the business or suffer from a reduced income.

The typical producer response to low prices in an agricultural market is to increase production or improve the efficiency of production. Thus, the livestock producer is forced to take a reduced income, to accept a reduced income/unit marketed, or to become more efficient if he is to remain in business. Gross efficiency may usually be improved by achieving a higher turnover rate, but net efficiency will depend upon managerial capabilities.

Several changes in American agriculture have occurred in recent years in an effort to improve productivity of animals or their feed supply. The relative importance of each may vary from place to place or from species to species, but all may have some bearing on nutrient requirements or nutrient utilization. Most of these factors are related to more rapid gain or higher production by the animals involved or to higher plant yields or such factors as greater stresses placed on the animals. The reader should bear in mind that a low level of productivity — gain, milk production, egg production, and so on — requires a lower level of nutrient intake than does a high level of production. Thus, unless rations are carefully formulated for high-producing, animals, some nutrients are much more apt to be limiting, with the result that productivity usually will not be sustained at levels at which a well-nourished animal should perform, or outright deficiencies may develop which, at the very least, will reduce productivity.

Some of the developments that affect nutrient needs are discussed in subsequent paragraphs. Note that the first two items may be expected to increase animal productivity, but that the others are more apt to depress it.

Genetic Improvement

Improvement in the genetic potential of animals in recent decades by selection, crossbreeding, development of hybrid lines of chickens, and the use of high quality sires in bull studs has brought about a general improvement in the genetic capabilites of domestic animals and, where the environment is suitable, an increase in productivity. The increased rate of growth one might expect from a well-bred broiler chick, however, requires a higher quality diet to supply the needed nutrients.

Improved Health and Management

The availability of antibiotics and other drugs in recent years has allowed many diseases to be kept under adequate control. Likewise, vaccines of various types have been developed with which animals may be immunized against many diseases. Accompanying these changes has been a gradual development and appreciation for management techniques and procedures which may result in

Table 1-1. Projected human population (millions) in 1985 and 2000 A.D.[a]

	1965	% of Total	1985	% change from 1965	2000	% change from 1965
World	3,281	100	4,746	45	5,965	82
Asia	1,828	56	2,710	48	3,402	86
Europe	440	13	492	12	527	20
Africa	306	9	513	68	646	111
Latin America	245	7	436	78	624	255
USSR	231	7	297	28	380	65
North America	213	7	283	33	354	66

[a] United Nations Dept. Social Affairs, Population Division, 1966. Population Studies No. 41 and the World Food Problem. 1967, a report of the President's Science Advisory Committee.

Table 1-2. Energy and proteins needed to feed population in 1970, 1985 and 2000 A.D.[a]

	1970	1985	2000
Protein, metric tons[b]	127,922	168,139	211,325
Energy, Kcal, billions	7,787	11,118	13,974

[a] From: The World Food Problem (for 1970, 1985 only). A report of the President's Science Advisory Committee, Vol. II.
[b] FAO Reference Protein.

more efficient means of handling and managing animals. The overall result is that a greater productivity may be expected along with more critical nutrient needs.

Development of Large Specialized Operations

The broiler and egg industries are two excellent examples where a tremendous growth in large, specialized operations has occurred in which thousands or even millions of birds are managed in one large operation. These large units depend upon a relatively high productivity/bird, and efficient utilization of feedstuffs is a must when the product sold is very cheap. Furthermore, the birds are usually completely confined so that all essential nutrients must be available and the feeds must be palatable and readily consumed to achieve rapid growth rates or high egg production.

The beef industry in the USA provides another good example. Over the past 20 years, there has been a tremendous increase in the numbers of large feedlots for finishing cattle. These lots supply an increasingly larger and larger share of the total market. Feedlots are currently in operation that have a capacity of around 100,000 head of cattle, and a great many in operation have a capacity of 25-50,000. With the small producer, particularly a farmer-feeder, a slight reduction in efficiency of feed utilization or a reduced gain or a slightly higher cost of feed may not seem important. For the large feedlot, however, a small change can mean a tremendous difference in operating expense and profit in a year's time. Many of the large lots use highly specialized nutrition and management consultants to advise on feeding practices, nutrition, and overall management of the animals they feed. In a well-run operation, the result is a rapid turnover at relatively low profit per animal.

Other examples of this type could be cited for swine, dairy cattle, and sheep. What is important to remember is that each of the situations mentioned — where large numbers of animals are confined and pushed to gain or produce rapidly — may be expected to make the nutrition of the animal a more critical factor than if production were maintained at a lower level in less crowded conditions.

Greater Stresses on Animals

Several factors may contribute to greater stressing effects on animals in today's agriculture. Of these, the movement of animals throughout the world by ships, air, and ground transportation may result in very rapid dissemination of disease organisms and their vectors. Close confinement of large numbers of animals is more apt to result in disease outbreaks. In addition, the close confinement must, in many instances, result in psychological and sociological stresses that have unknown effects on animal performance and nutrient requirements.

The process of assembling large numbers of animals from diverse backgrounds (this applies particularly to cattle and sheep) must result in a variety of stresses. Animals may be transported over long distances in adverse weather. Their previous exposure to disease and parasites differs, and sudden confinement with strangers is another adverse factor. Many range cattle or sheep have never consumed any feed but forage, and suddenly they are confronted with strange surroundings, strange feed in bunks, water in containers, and are subjected to loud and totally unfamiliar noises. The stress of weaning may also be imposed on top of these other stresses.

In many of the large beef feedlots it is the practice to dehorn and castrate where required, brand, inoculate for a variety of diseases, administer vitamins of various kinds and, perhaps, treat for stomach worms or other parasites in one quick trip through the squeeze chute — all this being done to an animal that probably has not recovered from stresses to which it has previously been exposed. If nothing else, the ability of cattle to withstand such a combination of stresses convinces one that they are extremely hardy animals. Furthermore, most of them recover and do surprisingly well in subsequent months in the feedlot. Death losses, as a whole, are remarkably low in well-run feedlots. Thus, the important role of adequate nutrition for animals subjected to such stresses in current commercial practice seems clear.

Economic Pressure to Increase Crop Production

Agricultural production of various crops faces the same type of cost-price squeeze that applies to the animal industries. One solution to this problem is the application of increasing amounts of fertilizers and of cultural or management practices that serve to increase yields. One possible result of the increased yields is the production of plant tissues that may be deficient in one or more of the mineral elements, particularly the trace elements. Long-time irrigation of sandy soils may result in the same thing in the western parts of the USA, resulting sooner or later in malnutrition of animals that consume the plant tissue.

Fertilization under some conditions may result in toxicities, as well. One example is in the heavy use of N fertilizers, particularly on soils borderline or deficient in sulfur. The plant tends to accumulate much higher levels than normal of nonprotein amino acid N as well as nitrate-N. Plants grown under these conditions can easily accumulate enough nitrate-N to be toxic. A second illustration applies to acid soils where either Mo or Ca will result in increased production of legumes. Mo applications may be used without markedly increasing plant tissue levels or at least not to an objectionable level. When both Ca and Mo are applied, however, the plant may accumulate toxic quantities of Mo. This would, of course, reduce performance of animals consuming such forage in quantity.

These examples clearly illustrate some of the problems that can occur as a result of nutrient imbalance in the soil, although the frequency of occurrence is poorly documented. Undoubtedly, many other analogous but unknown situations occur from time to time but are not recognized at this time. The incidence of such situations will increase with time unless remedial measures are taken.

Increasing Use of Waste and By-Product Feedstuffs

Many animal producers are increasingly turning to the use of any apparently useful product which is competitively priced that the animal will consume without apparent harm in an effort to hold down the cost of production.

Materials having potential value for feed include garbage, sewage, poultry litters, manures, papers, wood, and bark products. Currently, several experiment stations are vigorously working on the utilization of poultry and ruminant manures which are beginning to present disposal problems in some heavily populated areas. Numerous papers have appeared in recent years dealing with use of wood products and paper, and others on garbage, sewage sludge, and crop residues. Currently there is a great deal of interest worldwide in improving the utilization of low-quality roughage and straw. More of this type of information may be expected in the future.

Increasing Use of Synthetic and Purified Products

There has been a tremendous increase in the usage of urea over the past 20 years in ruminant rations. Although this has often resulted in a reduced cost of crude protein in the formula, it has not always resulted in efficient utilization of the needed N because of limiting factors such as readily available carbohydrates or to any of several required mineral elements. Inappropriate use has often resulted in toxicity, as well. There is no reason to believe , however, that usage will not increase, but this will require, again, competent nutritional wisdom to take advantage of the lower cost of N from such products. There has been much interest recently, in the use of methionine hydroxy analogue in ruminant rations because a limited amount of work indicates that methionine (an amino acid) may be limiting, particularly for high-producing dairy cows. Other examples of this type may be expected to come up from time to time in the future.

Individual Variation in Nutrient Requirements

Examples of variations in nutrient requirements that may be seen in laboratory animals and man have been well documented and discussed by Williams (1, 2). These examples clearly show that a wide range in requirements for a variety of different nutrients may be expected, sometimes even in closely inbred strains of laboratory rats. Obviously, this variable presents great problems in the field of human nutrition, requiring the development and dissemination of needed information on this subject and more training in nutrition of people in the medical field. With domestic animals, except for pets, the common practice would be to cull animals that do not perform adequately, whether the cause be an excessive requirement for nutrients or for other reasons. The incidence of occurrence of domestic animals with very high (or low) nutrient requirements is poorly documented at this time, but it is a subject that is certainly worthy of further study.

References Cited

1. Williams, R.J. 1956. Biochemical Individuality: The Basis for the Genetotrophic Concept. Wiley Pub. Co., New York.
2. Williams, R.J. 1971. Nutrition Against Disease. Pitman Pub. Corp., New York.

Chapter 2 – Common Methods of Analysis for Nutrients and Feedstuffs

The science of nutrition has progressed rapidly in recent decades, partly because of the large amount of effort that has been expended to learn more about nutrition. Expansion of knowledge and a better understanding of nutritional needs and nutrient metabolism are possible, in part, because methods are continually being developed and improved to quantitatively evaluate the nutrient content of foods, feeds, and animal tissues. Thus, to have a moderately good understanding of nutrition, the reader should have at least a minimal understanding of laboratory analyses that are commonly utilized and at least a limited knowledge of chemistry. Although a detailed knowledge is not required, some understanding of analytical procedures will make for easier reading and some knowledge of organic structures will facilitate understanding of those chapters dealing with specific nutrients.

At this point it is logical to define some terms often used in nutrition. Nourish means to feed or sustain (an animal or plant) with substances necessary to life and growth. Thus, a nutrient may be defined as something that nourishes an animal or, more specifically, an element or compound that is required in the diet of a given animal to permit normal functioning of the life processes. It is difficult to give a short, precise definition of a nutrient and still be accurate. Some compounds, such as starch, are readily utilized by most species as a source of energy (and thus provide nourishment), yet starch is not specifically required as a source of energy or for any other purposes. Food is generally used to mean an edible material that will provide nourishment; feed means the same thing, but is more commonly applied to animal food than to human food. A foodstuff or feedstuff is any material made into or used as food or feed, respectively. A ration, on the other hand, is either a daily supply of feed or a mixture of feedstuffs used to supply nourishment to an animal.

Simple Classification of Nutrients

At this point it is necessary to give a simple classification of nutrients in order to facilitate discussions that follow on various analytical methods. In a diet-conscious country such as the USA, most school children probably know that nutrients, foods and feeds may be broadly divided into water, proteins, carbohydrates, fats (lipids), minerals, and vitamins. Although food may contain many other chemical compounds, these broad groups include all of the nutrients known to be required, as well as those (such as starch) which may be metabolized as a source of energy but for which the body has no specific need. A simple flow diagram of nutrient classification is shown in Fig. 2-1. The various components that may be found within each main nutrient group are discussed in detail in the chapters dealing with these specific groups.

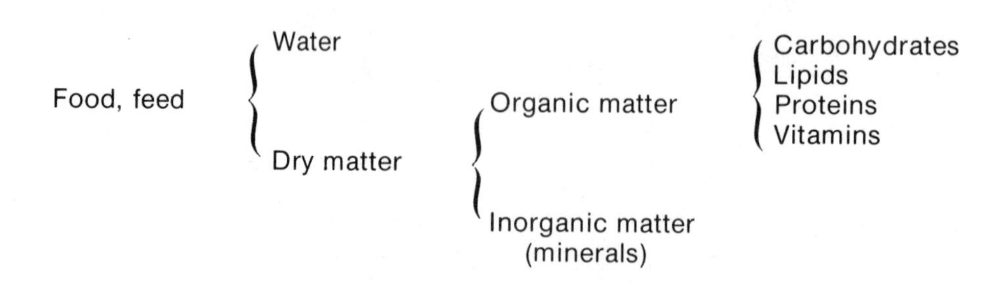

Figure 2-1. A simple diagrammatic outline of nutrient classification.

Analytical Methods

Most of the analytical methods in common use depend upon various chemical procedures which are specific for a given element, compound, or group of compounds. Quantitative data may be obtained by gravimetric procedures, but more often are obtained by other methods which involve the use of acid or base titration, colorimetry, chromatography, and so on. A relatively common characteristic of chemical methods is that they often involve drastic degradation of feeds with reagents such as concentrated acids or bases, extraction with concentrated solvents or other treatments that are biologically harsh. As a result, one of the big problems in nutrient analysis is that a chemical procedure may be quantitative in terms of finding out how much of a given nutrient or compound is in the feed, but such analyses are often difficult to relate to animal utilization. For example, we might analyze forage for its Ca content, but the data provide no information on how much of the Ca is available to the animal.

Because chemical methods often leave questions regarding the availability of nutrients from feeds, biological procedures are sometimes used, although they are usually more tedious and expensive. Such methods may, however, give a much more accurate estimate of animal utilization; in other words, biological methods tell us how much of a nutrient the animal 'sees' in the feed. Chicks or rats are often used in the tests. If, for example, we want to determine the effect of heating on utilization of proteins from soybean meal, we can feed one or more groups of chicks unheated meal along with other necessary nutrients and feed other chicks meal that has been heated to different temperatures or for different lengths of time and, thus, get a pretty good biological estimate of the effect of heat treatments on the soy proteins.

Microbiological methods may be used in a manner similar to the biological procedures just described for chicks. Bacteria have been isolated that have specific requirements for one or more of the essential amino acids or for specific water-soluble vitamins. These organisms can be used to determine how much of a given amino acid or vitamin is available in a given product or mixture. The information may or may not be applicable to rats, chicks or swine, but it is more likely to be applicable than that from a chemical method.

Sampling for Analysis

In the early days of nutrition research, it was not uncommon to analyze the whole animal body. Today, such practices are less feasible because equipment is not adapted to such methods and the cost would be tremendous. It is seldom feasible today to analyze the whole bodies even of small animals such as rats or chicks. Modern chemical methods are geared to procedures that require small amounts of material which must be collected and prepared in a manner that gives us the best reasonable estimate of the total batch in which we are interested. For example, if we are interested in the protein content of hay produced from a field, where do we begin? We certainly can't grind up all of the hay produced; even one bale would tax the facilities of most laboratories. Consequently, we resort to the use of core samples taken from as many different bales as is reasonable. Perhaps as many as 25-50 core samples may be taken from one stack of bales which represents the hay from the field in question. The assumption is that each core will correspond reasonably well to the total composition of the bale from which it came and that, if we sample enough bales, our composite sample will be representative of the total hay crop. This is an assumption that may not always be valid, but it is the appropriate statistical approach.

The core samples are then brought to the laboratory, ground, mixed well, and small subsamples taken for analyses. For the common Kjeldahl analysis which is used for crude protein (see section on proximate analyses), a common sample size is 2 g of material. A micro-Kjeldahl procedure now in use allows for the use of a sample that contains only about 1 mg of N, or a sample of about 100 mg of the hay in question. Thus, we may base our estimate of protein content of the total field on a very small amount of material. Consequently, the material being analyzed must be representative if results are to be meaningful.

Similar procedures are used for other commodities. One small sample of grain may be used to evaluate a carload. With liquid samples, we assume that liquids are more homogeneous than solids, but this is not always true and errors may creep in if care is not taken in sampling. With respect to the beef carcass, the 9-10-11 rib cut has been shown to give a relatively accurate estimate of the total carcass for fat, protein, water, and ash (minerals). As a result, we can obtain this cut from one side of the carcass, grind it up, and analyze for the constituents of interest.

Some of the more sophisticated equipment developed in recent years requires samples of only microliter or microgram size; sometimes even nanogram size (10^{9} g) or picogram (10^{12} g) amounts can be detected in plant or animal tissues. For example, current procedures can detect diethylstilbestrol, a synthetic hormone used to stimulate weight gain in cattle, in tissues at levels of 1-2 ppb (parts/billion).

Specific Methods of Analysis

Dry Matter

The determination of dry matter is probably the most common procedure carried out in nutrition laboratories. The reason for this is that natural feedstuffs, animal tissues, and so on, may be quite variable in water content, and we must know the amount of water if analytical data are to be compared for different feeds. When grain is bought or fed, obviously its value with 14% moisture is not the same as with 10% moisture. After analysis, nutrient composition can be expressed on a dry basis or a normal as-fed basis, which would be about 90% dry matter for most grains.

The simplest means of determining dry matter is to place the test material in an oven and leave it until all of the free water is evaporated. Temperatures used are usually 100-105°C. Moisture can also be estimated with moisture meters, devices that give immediate results by means of a probe inserted into the test material. These devices depend upon electrical conductivity; they are useful for quick answers, but results are not as precise as those obtained by actually drying the test material.

The determination of dry matter, as with most procedures, is not always as simple as the previous discussion indicated. This statement applies to any material that has a relatively high content of volatile compounds. Most fresh plant tissue contains some volatile compounds, but the amount is low enough that the volatiles can usually be ignored. Some plants, however, contain large amounts of essential oils, terpenes, and so on, which may be lost in drying and, thus, give an erroneous answer with the usual procedures. Of the common feedstuffs, silages or other fermented products may have large amounts of easily vaporized compounds such as the volatile fatty acids (acetic, propionic, butyric) and ammonia. In addition, some sugars may decompose at temperatures above 70°C and many proteins become partially insoluble at temperatures above 70°C.

There are several means of avoiding excessive losses of volatiles. Drying in vacuum ovens, freeze drying, oven drying at 70°C or less, and distillation with toluene have been used. One example of the effect of these different procedures on different silages is shown (1): Oven drying at 100°C, 44.4% dry matter; oven drying at 70°C, 46.8%; freeze drying, 47.2%; toluene distillation, 47.7%; and toluene distillation corrected for total acids, ethanol and ammonia, 48.2%. These data show that very substantial losses may occur if silage is dried in the usual manner. Hood et al (3) have devised a procedure in which an ethanolic silage extract is mixed with a water-sensitive reagent, the mixture is incubated and then titrated with HCl. It is said to be rapid, simple, specific for water, and suitable for routine laboratory analyses.

Proximate Analysis

The proximate analysis is a combination of analytical procedures developed in Germany over a century ago. It is intended for the routine description of feedstuffs and, although from a nutritional point of view, it has many faults, it is still widely used. In some instances, its use has been encouraged and prolonged because of laws that require listing of minimum and maximum amounts of components that may be present in commercial feed mixtures. The different fractions that result from the proximate analysis include: Water, crude protein, ether extract, ash, crude fiber, and nitrogen-free-extract. We have already discussed water (or dry matter); more detailed discussion of the other factors follows.

Crude Protein. The procedure used is known as the Kjeldahl procedure. Material to be analyzed is first digested in concentrated H_2SO_4 which converts the N to NH_4SO_4. This mixture is then cooled, diluted with water, and neutralized with NaOH, which puts the N into the form of ionized ammonium. The sample is then distilled, and the distillate is titrated with acid. This analysis is accurate and repeatable, but it is relatively time-consuming and involves the use of hazardous chemicals. A micro apparatus is shown in Fig. 2-2.

From a nutritional point of view, the data are applicable to ruminant species which can efficiently utilize almost all forms of N, but the

Figure 2-2. A micro-Kjeldahl apparatus which is designed for nitrogen analyses of small samples. The digestion apparatus is shown on the left and the distillation equipment immediately to the right. Photo by R.W. Henderson.

information may be of little value for monogastric species (such as man, swine, poultry). Monogastric species have specific requirements for various amino acids (see Ch. 6) and do not efficiently utilize non-protein-nitrogen compounds such as amides, ammonium salts, and urea. The crude protein analysis does not distinguish one form of N from another, thus we cannot tell if a feed mixture has urea or the highest quality of protein such as casein (from milk). In addition nitrate N is not converted to ammonium salts by this procedure so N in this form is not included.

Ether Extract. This procedure requires that ground samples be extracted with ether for a period of 4 hr or more. Ether-soluble materials include quite a variety of organic compounds (see Ch. 8), only a few of which have much nutritional significance. Those of quantitative importance include the true fats and fatty acid esters, some of the compound lipids, and fat-soluble vitamins or provitamins such as the carotenoids. The primary reason for obtaining ether extract data is an attempt to isolate a fraction of feedstuffs that has a high caloric value. Provided the ether extract is made up primarily of fats and fatty acid esters, this may be a valid approach. If the extract contains large percentages of plant waxes, essential oils, resins, or similar compounds, however, it has little meaning, as compounds such as these are of little value to animals.

Ash. Ash is the residue remaining after all the combustible material has been burned off in a furnace heated to 500-600°C. Nutritionally, ash values have little importance, although excessively high values may indicate contamination with soil or dilution of feedstuffs with such substances as salt and limestone. In the proximate analysis, data on ash are required to obtain other values. It should be noted that some mineral elements, such as iodine and selenium, may be volatile and are lost on ashing. Normally, these elements represent only very small percentages of the total, so little error is involved.

Crude Fiber. Crude fiber is determined by using an ether-extracted sample, boiling in dilute acid, then in dilute base, drying, and burning in a furnace. The difference in weight before and after burning is the crude fiber fraction. This is a tedious laboratory procedure that is not highly repeatable. It is an attempt to simulate digestion that occurs first in the gastric stomach and then in the small intestine of animals. Crude fiber is made up primarily of plant structural carbohydrates such as cellulose and hemicellulose (see Ch. 7), but it also contains some lignin, a highly indigestible material associated with the fibrous portion of plant tissues. For the monogastric animal, crude fiber is of a variable but low value; for ruminants, it is of variable value, but is much more highly utilized than by monogastrics.

Nitrogen-Free-Extract [NFE]. This term is a misnomer in that no extract is involved. It is determined by difference; that is, NFE is the difference between the original sample weight and the sum of weights of water, ether extract, crude protein, crude fiber, and ash. It is called nitrogen-free because it ordinarily would contain no N. NFE is made up primarily of readily available carbohydrates, such as the sugars and starches (see Ch. 7), but it may also contain some hemicellulose and lignin, particularly in such feedstuffs as forages. A more appropriate analysis would be one specifically for readily available carbohydrates — one in which starches are hydrolyzed to sugars and then an analysis is done for all sugars present. Nutritionally, the NFE fraction of grains is well utilized by nearly all species, but NFE from forages and other roughages are less well utilized.

A diagram of the proximate analysis scheme illustrating the sequence of procedures as well as the major fractions that are isolated is shown in Fig. 2-3.

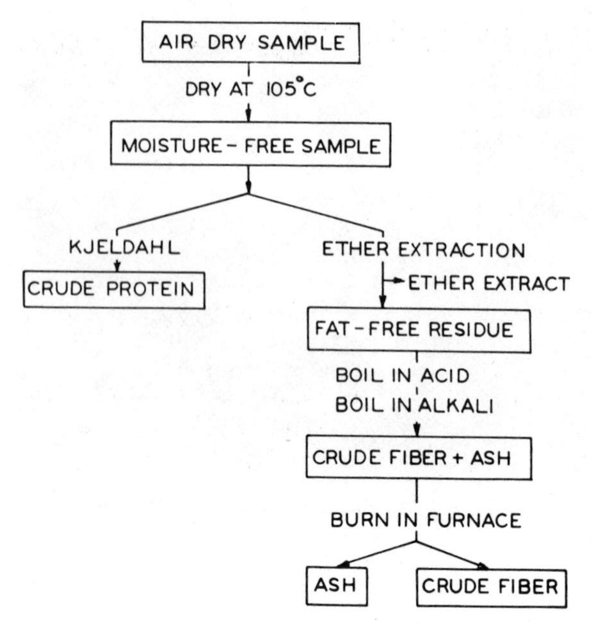

Figure 2-3. Flow diagram for the proximate analysis.

Van Soest Scheme for Forage Analysis

Analytical methods, primarily intended for forages, have been developed at the USDA laboratories at Beltsville by Van Soest and coworkers (2, 4). Micro methods have also been developed (5). These analyses divide the nutrients in plant tissues into a group (cell contents) that are generally highly available to animals and into a second group (cell walls) which are much less available. This is effected by extracting the test material with neutral detergent solutions. The resulting fractionation is shown in Table 2-l.

Table 2-1. Classification of forage fractions by Van Soest method.[a]

Fraction	Components included	Nutritional availability	
		Ruminant	Nonruminant
Cell contents	Sugars, soluble carbohydrates, starch	complete	complete
	Pectin	complete	high
	Nonprotein N	high	moderate-high
	Protein	high	high
	Lipids	high	high
	Other solubles	high	high
Cells walls	Hemicellulose	partial	low
	Cellulose	partial	low
	Heat-damaged protein	indigestible	indigestible
	Lignin	indigestible	indigestible

[a] From Van Soest (4)

pH of Feedstuffs

The pH of feedstuffs is rarely used to evaluate materials except for fermented products such as silage, cannery residues, or mixtures such as potato slurp (fermented cull potatoes and other feedstuffs). It should be pointed out that pH of mineral supplements may be of importance with respect to palatability or metabolism by the animal. With respect to silage, pH may be determined by taking a sample of 100 g, mixing with 100 ml of water, expressing the juice, and measuring with a pH meter. Good quality silages should have a pH between 3.8 and 5.0.

Specialized Analytical Methods

A wide variety of analytical methods find some use in nutrition from time to time. Such methods may be used for feedstuffs and rations, animal tissues, or with urine and fecal samples, depending on the situation. The list of such methods is far too long to discuss in a book of this type; however, there are several methods involving specialized equipment that are used extensively and which deserve some discussion.

Bomb Calorimetry

The bomb calorimeter (Fig. 2-4) is an instrument used to determine energy values of solids, liquids, or gases. The energy value of a given sample is determined by burning it in an atmosphere of oxygen. When the sample is burned, the heat produced raises the temperature of water surrounding the container in which the sample is enclosed, and the temperature increase provides the basis for calculating the energy value. Bomb calorimetry finds extensive use for evaluating fuels such as natural gas and coal. In nutrition, its most useful application is in determining the digestible energy of feedstuffs or rations. The gross energy value (that obtained by burning) of feedstuffs has little or no application of its own, however, as it is almost impossible to distinguish between constituents that are well utilized by animals and those that are poorly utilized (see Ch. 9).

Figure 2-4. A modern bomb calorimeter used for energy determinations of feed and other combustible materials. Photo by R.W. Henderson.

Amino Acid Analysis

Chemical methods for amino acid analysis have been around for a good many years, but it is only in the last 10-15 years that simi-automated equipment such as that shown in Fig. 2-5 has been available. This type of equipment is capable of fractionating protein preparations that have been hydrolyzed into the constituent amino acids. The preparations are placed on chromatographic columns, and various solutions are passed through the columns, resulting in separation and evolution of the individual amino acids in a relatively short time (a few hours). This type of equipment has greatly facilitated collection of data on amino acid composition of foods and feeds as well as on metabolism and requirements of amino acids.

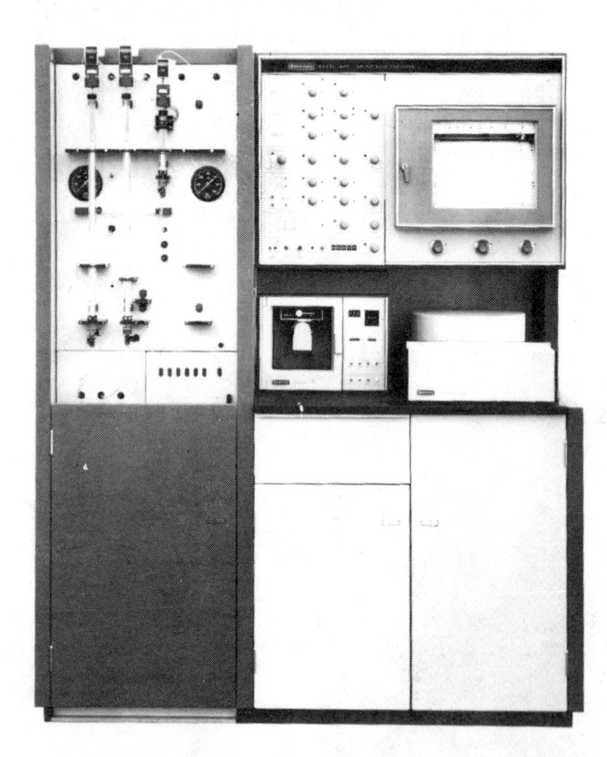

Figure 2-5. A modern automated amino acid analyzer used for quantitatively determining individual amino acids. Courtesy of Beckman Instrument Co.

Atomic Absorption

Atomic absorption spectrophotometric instruments (Fig. 2-6) have greatly facilitated analyses for most mineral elements (cations). In the operation of these instruments, liquid or solid materials are ashed and resuspended in liquid solution which may be put directly into the instrument. Body fluids such as blood plasma and urine may be used directly. The solution passes through a flame which serves to disperse the molecules into individual atoms. Radiation from a cathode lamp is passed through the flame, and the atoms absorb some of this radiation at specific wavelengths. With instruments such as this, vast numbers of samples can be analyzed in a short time.

Figure 2-6. A modern atomic absorption instrument which is used for analyses of most mineral [cations] elements.

Gas Chromatography

The forerunner of the gas chromatograph was developed to analyze for rumen volatile fatty acids. Since that time (early 1950's), a tremendous development has occurred in this technique and in the available instrumentation (Fig. 2-7). Such instruments are capable of

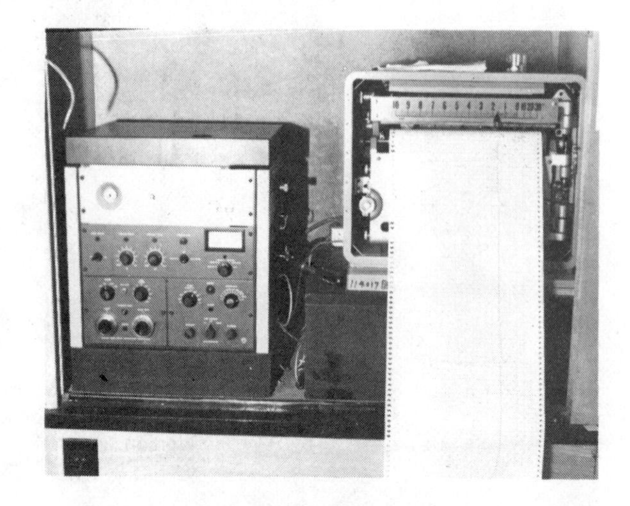

Figure 2-7. A simple gas chromatograph instrument used for many different analyses, but which is especially useful for nutritional studies of lipids. Photo by R.W. Henderson.

handling almost any compound that can be vaporized or those that are in gas form. The sample to be analyzed is placed in the instrument and it is moved through a heated chromatographic column by means of gas. This process allows the quantitative separation of closely related chemical compounds (such as acetic and propionic acid) quite rapidly. This process requires only very small samples. In nutrition, gas chromatographs have been particularly useful for fatty acid analyses, but are capable of handling many other organic compounds.

Automated Analytical Equipment

The gradually increasing cost of labor has stimulated the development of instrumentation designed to do a number of simultaneous repetitive analyses. Such equipment has found widespread use in the medical field, particularly, but has application as well in the nutrition laboratory. For example, it is possible to obtain simultaneous data on blood serum for glucose, total lipids, cholesterol, Ca, P, Mg, urea, and total protein. This is just an example of the type of information that may be obtained on one tissue. The speed of analysis and the fact that such equipment is highly automated have greatly increased the volume of information that may be obtained at a given cost, even though the equipment itself is expensive.

References Cited

1. Brahmakshatriya, R.D. and J.D. Donker. 1971 J. Dairy Sci. 54:1470.
2. Goering, H.K. and P.J. Van Soest. 1970. Forage fiber analyses. ARS Agr. Handbook No. 379.
3. Hood, R.L., C.E. Allen, R.D. Goodrich and J.C. Meiske. 1971. J. Animal Sci. 33:1310.
4. Maynard, L.A. and J.K. Loosli. 1969. Animal Nutrition. 6th ed. McGraw-Hill Book Co., New York.
5. Van Soest, P.J. 1967. J. Animal Sci. 26:119.
6. Waldern, D.E. 1971. Can J. Animal Sci. 51:67.

Chapter 3 – The Gastro-Intestinal Tract and Nutrition

Some knowledge of the gastro-intestinal tract (GIT) is important to those who study nutrition because it is so intimately concerned with the utilization of food and nutrients. The various organs, glands, and other structures involved are concerned with procuring, chewing and swallowing food and with the digestion and absorption of nutrients as well as some excretory functions.

Digestion and absorption are terms which we will make frequent reference to in this chapter. **Digestion** has been defined simply as the preparation of food for absorption. As such, it may include mechanical forces (chewing or mastication; muscular contractions of the GIT), chemical action (HCl in the stomach; bile in the small intestine), or enzymic activity from enzymes produced in the GIT or from microorganisms in various sites in the tract. The overall function of the various digestive processes is to reduce food particles to a size or solubility that will allow for absorbtion. **Absorption** includes various processes that allow small molecules to pass through the membranes of the GIT into the blood or lymph system.

Anatomy and Function of the Gastro-Intestinal Tract

The GIT of simple-stomached mammalian species includes the mouth and associated structures and glands, esophagus, stomach, small and large intestines, and the pancreas and liver. For convenience, this type of GIT is often termed **monogastric.** In avian species, the tract is somewhat different in anatomy than in typical monogastrics, but overall function is believed to be similar. **Ruminant** and **pseudo-ruminant** species have a much more complex stomach than monogastric species. These various types of tracts are described briefly in subsequent sections, but sufficient space to thoroughly describe them cannot be justified in a book of this type. For more detail than is given here the reader is referred to other books such as Sisson and Grossman (7) and Swenson (10) on various domestic animal species or Church (3) on ruminant species.

Monogastric Species

The mouth and associated structures — tongue, lips, teeth — are used for grasping and masticating food; however, the degree of use of any organ depends on the species of animal and the nature of its food. In omnivorous species — those that consume both plant and animal food — such as humans or swine, the incisor teeth are used primarily to bite off pieces of the food and the molar teeth are adapted to mastication of nonfibrous materials. The tongue is used relatively little. In carnivorous species the canine teeth are adapted to tearing and rending, while the molars are pointed and adapted to only partial mastication and the crushing of bones. Herbivorous species (plant eaters), such as the horse, have incisor teeth adapted to nipping off plant material, and the molars have relatively flat surfaces that are used to grind plant fibers. The jaws are used in both vertical and lateral movements which efficiently shred plant fibers. Rodents have incisor teeth which continue to grow during the animal's lifetime, allowing the animal to use its teeth extensively for gnawing on hard material such as nut shells. Their incisor teeth would not withstand such rugged use without the continual growth and would be worn down greatly.

In the process of mastication, saliva is added, primarily from three bilateral pairs of glands — submaxillary, at the base of the tongue; sublingual, underneath the tongue, and the parotids, below the ear. Saliva aids in forming food into a bolus, which may be swallowed easily, and has other functions such as keeping the mouth moist, aiding in the taste mechanisms, and providing a source of enzymes (see later section) for initiating enzymic digestive processes.

Portions of the GIT of the rat, rabbit, and pig are shown in Fig. 3-1. These pictures indicate the relative differences in size of the stomach as compared to the intestines. The pig, for example, has a relatively large stomach, with the capacity in the adult said to be 6-8 liters. The shape of the stomach of different species varies as does the relative size. Mucosal tissues lining the interior of the stomach are divided into different areas which are supplied with different types of glands as illustrated in Fig. 3-2. In the cardiac region, the cells produce primarily mucus, probably as a means of protecting the stomach lining from gastric secretions. In the peptic gland region, the lining is covered with gastric pits (Fig. 3-3) which open into gastric glands. These produce a mixed secretion of acid, enzymes, and mucas.

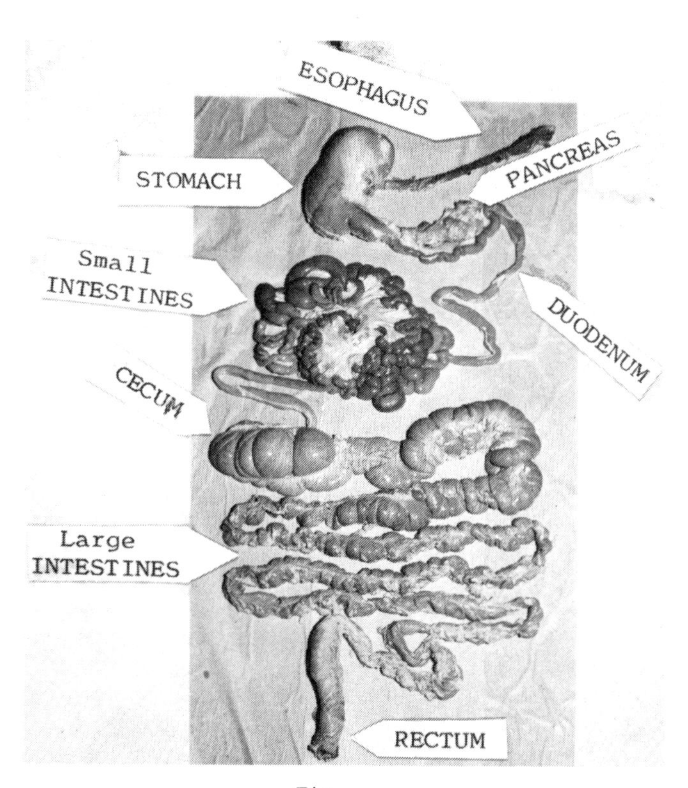

Pig

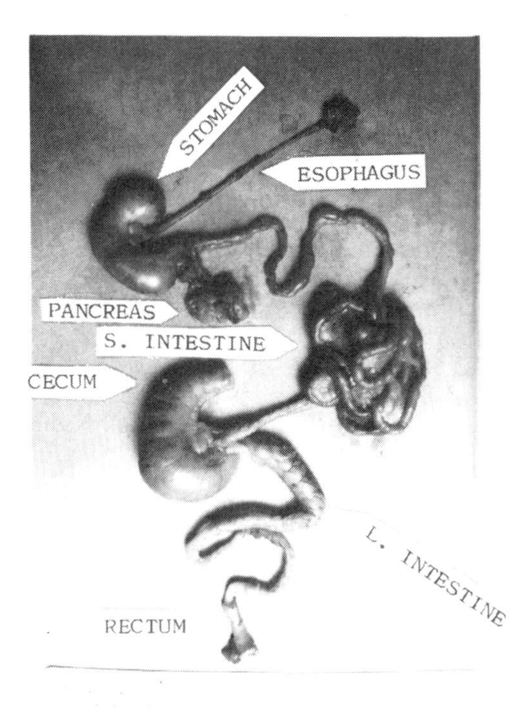

Rat

Rabbit

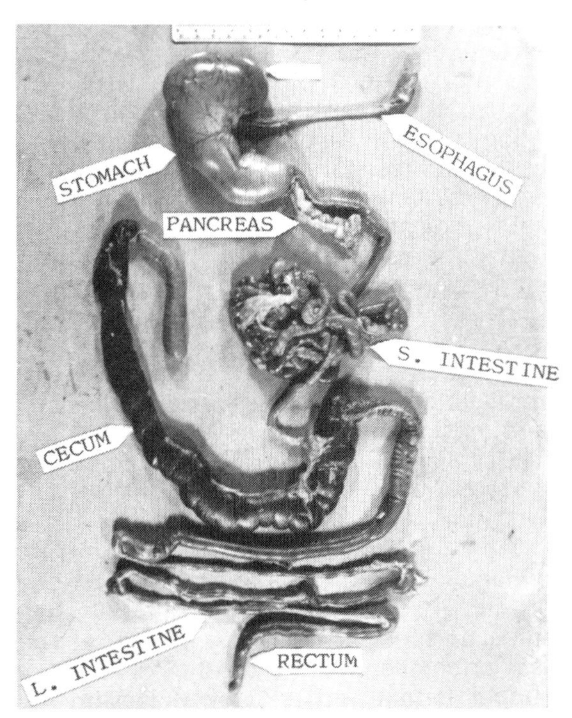

Figure 3-1. Photographs showing the principal parts of the gastro-intestinal tract [GIT] of the pig, rabbit and rat. Note the differences in relative size of the various parts. Approximate weights of the donors were: Pig, 100 kg; rabbit, 2 kg; and rat, 150 g.

The glands consist of two main types of cells: The body chief or peptic cells, which produce proteolytic enzymes, and the parietal or oxyntic cells, which secrete HCl. In the pyloric region, mucus-producing cells again are found. The enzymes produced and their function are discussed in a later section.

The relative length of the small intestine of various animal species varies greatly. In the pig, it is relatively long (15-20 m), but in the dog it is relatively short (ca. 4 m). The duodenum, the first short section, is the site of production of various digestive juices. Furthermore, a variety of digestive juices from the pancreas as well as bile come into the duodenum within a short distance of the pylorus of the stomach. Ducts from the liver and pancreas join to form a common bile duct which empties into the duodenum in some species, and other species have separate ducts. An appreciable amount of absorption may occur in the duodenum (see later section on absorption). The small intestine, in general, accounts for most of the absorption in the GIT, and it is lined with a series of finger-like projections, the villi, which serve to increase the absorption area. Each villus contains an arteriole and venule, together with a drainage tube of the lymphatic system, a lacteal. The venules ultimately drain into the portal blood system which goes directly to the liver; the lymph system empties via the thoracic duct into the vena cava, a large vein.

The large intestine is made up of the cecum, colon, and rectum. Relative sizes of these organs vary considerably in different animal species. In the pig, the large intestine is about 4-4.5 m in length and is considerably larger in diameter than the small intestine. The length and diameter of the cecum varies considerably, generally being much larger in herbivorous species than in omnivorous or carnivorous species. Note the very large size (Fig. 3-1) of the cecum of the rabbit as compared to the pig and rat. In general, the large intestine acts as an area for absorption of water and secretion of some mineral elements such as Ca. An appreciable amount of bacterial fermentation takes place in the cecum and colon. Recent data on horses indicate that the volatile fatty acids (acetic, propionic, butyric) may be absorbed from the cecum as are some peptides and other small molecules. This area may be vital for synthesis of some of the water-soluble vitamins and, perhaps, proteins in species such as the horse and rabbit. Because proteins and other large molecules originating in the cecum and colon are not subject to action of the digestive juices, they are assumed to be of little use to the host. Further data are required to clearly understand overall function of the cecum and large intestine.

The pancreas and liver are vital to digestive processes because of the digestive secretions produced (see later section). Bile, from the liver, has many important functions, and the liver is an extremely active site of synthesis and detoxification. The liver is also an important storage site for most vitamins and trace minerals.

Avian Species

The GIT of avian species (Fig. 3-4) differs considerably in anatomy from typical monogastric species. Birds have no teeth, for example, although some prehistoric forms did have teeth. Thus, the beak and/or claws serve to partially reduce food to a size that may be swallowed. Although insect-eating species have no crops, other types have crops of variable sizes. Ingested food goes directly to the crop where fermentation probably occurs in some species. The proventriculus of birds is the site of production of gastric juices and the gizzard serves some of the functions of teeth in mammalian species, acting to physically reduce particle size of food. Current data indicate that little proteolytic digestion occurs in either the proventriculus or gizzard and that removal of the gizzard has little effect on digestion if the food is ground. In the small intestine, most of the enzymes found in mammalian species are present, with the exception of lactase. The pH of the small intestine is slightly acid, and protein digestion is assumed to result from a combination of the common proteolytic enzymes. Data on absorption indicate that it is similiar to mammalian species except that the hormone, enterogastrone, which affects fat absorbtion, is not present in birds. The ceca and large intestines are sites for water resorption. Some fiber digestion occurs in the ceca because of bacterial fermentation, but at much lower levels than in most mammals. Total digestibility by birds is similar to that occurring in mammals for nonfibrous diets. Further information on avian species may be found in Sturkie (8, 9).

Ruminants

The mouths of ruminants differ from other mammalian species in that they have no upper incisor or canine teeth. Thus, they depend on an upper dental pad and lower incisors in conjunction with lips and tongue for pre-

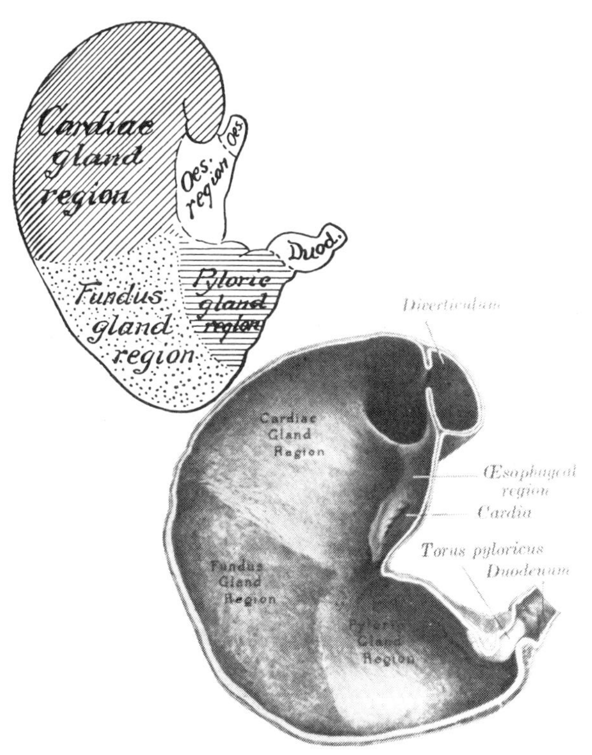

Figure 3-2. Diagrams of the stomach of the pig illustrating the various zones and types of mucosal areas found in the stomach. From Sisson and Grossman [1953] by permission of W.B. Saunders Co.

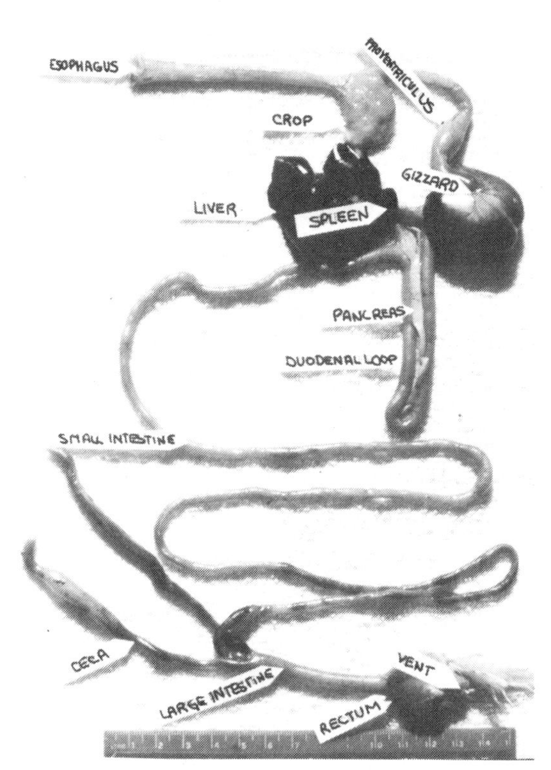

Figure 3-4. Digestive tract of the chicken. Photo by Don Helfer, Oregon State University Diagnostic Laboratory.

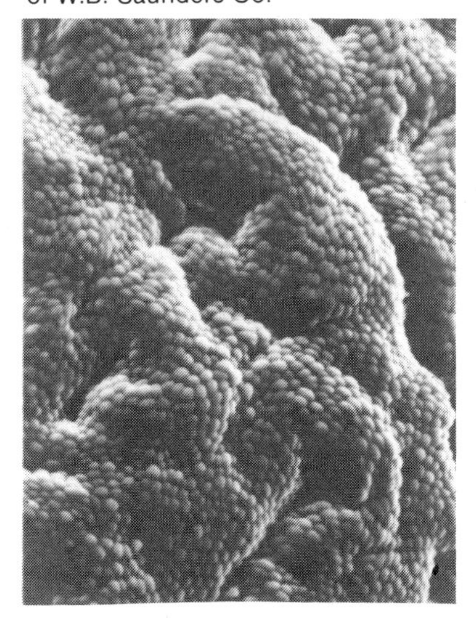

Figure 3-3. Surface of human stomach's inner lining [glandular mucosa] is seen enlarged some 280 diameters in this scanning electron micrograph. The view shows the tops of epithelial cells, the gastric pits, and characteristic folds of a normal stomach. Courtesy of Jeanne M. Riddle, Wayne State University School of Medicine.

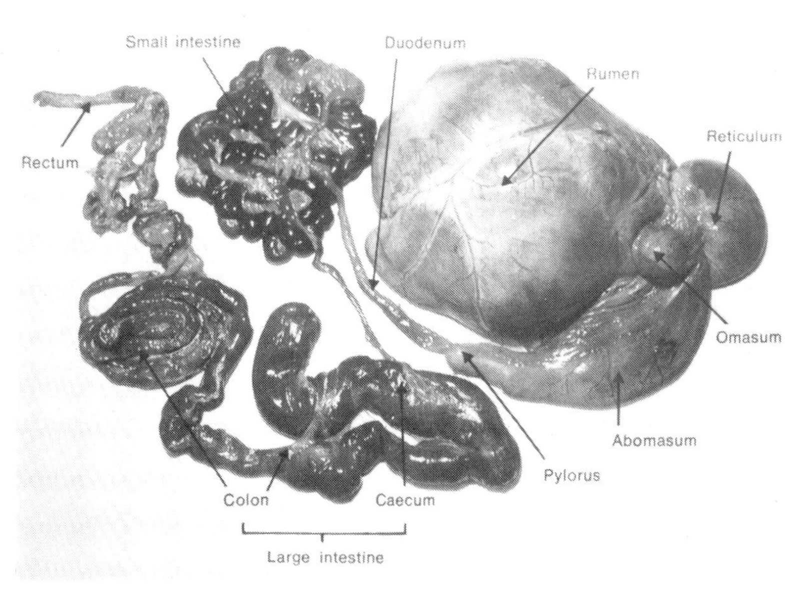

Figure 3-5. The stomach and intestines of the sheep. From Anonymous [1972]. Courtesy of C.S.I.R.O.

hension of food. Ruminant species may be divided into roughage eaters, selective eaters, and transitional types (5), and these various types utilize differences in tongue mobility and in lip structure, particularly, to facilitate selection and consumption of feedstuffs. With respect to mastication, ruminant species have molar teeth so shaped and spaced that the animal can chew only on one side of the jaw at one time. Lateral jaw movements aid in shredding tough plant fibers.

Saliva production in ruminants is very copious, reaching amounts of 150+ l./day in adult cows and 10 l. or more in sheep. Production of saliva is relatively continuous, although greater quantites are produced when eating and ruminating than when resting. Saliva provides a source of N (urea and mucoproteins), P, and Na, which are utilized by rumen microorganisms. It is also highly buffered and aids in maintaining an appropriate pH in the rumen, in addition to other functions common to monogastric species.

The stomach of the ruminant (Fig. 3-5) is divided into 4 compartments — **reticulum, rumen, omasum,** and **abomasum**. The reticulum and rumen are not completely separated, by any means, but have different functional purposes. The reticulum functions in moving ingested food into the rumen or into the omasum and in regurgitation of ingesta during rumination. The rumen acts as a large fermentation vat and has a very high population of microorganisms (see later section). Function of the omasum is not clearly understood although it appears to aid in reducing particle size of ingested food, and some absorption occurs there. The abomasum is believed to be comparable to the gastric stomach of monogastric species.

The stomach of ruminant species makes up a greater percentage of the total GIT than is the case for other species. In adults, the stomach may contain 65-80% of total digesta in the entire GIT. The intestinal tract is relatively long, typical values for cattle and sheep, respectively, are: Small intestine, 40 and 24-25 m; cecum, 0.7 and 0.25 m; colon, 10 and 4-5 m (7).

In the young ruminant, the reticulum, rumen and omasum are relatively under developed, because the suckling animal depends primarily on the abomasum and intestine for digestive functions. As soon as the animal starts to consume solid food, the other compartments rapidly develop, reaching relative mature size by about 8 wk in lambs and goats, 3-4 mo. in black-tailed deer, and 6-9 mo. or more in domestic bovines.

Another anatomical preculiarity of ruminant species is that they have a structure called the **esophageal** or **reticular groove**. This structure begins at the lower end of the esophagus and, when closed, forms a tube from the esophagus into the omasum. Its function is to allow milk from the suckling animal to bypass the reticulorumen and escape bacterial fermentation. Closure of this groove is stimulated by the normal sucking reflexes, by certain ions, and by solids in suspension in liquid. It does not appear to remain functional in older animals unless they continue to suckle liquid diets.

There is a well-developed pattern of rhythmic contractions of the various stomach compartments that act to circulate ingesta into and throughout the rumen, into and through the omasum, and on to the abomasum. Of importance also are contractions that aid in regurgitation during **rumination**. This is a phenomenon peculiar to ruminants. In effect, it is a controlled form of vomiting, allowing semiliquid materials to be regurgitated up the esophagus, swallowing of the liquids, and a deliberate remastication of and reswallowing of the solids. Ruminants may spend 8 hr/day or more in rumination, depending upon the nature of their diet. Coarse, fibrous diets result in more time ruminating. The origin of ruminating is not clear; perhaps it was an evolutionary development allowing animals to hastily consume feed and then retire in relative safety from predators to rechew their food at leisure.

Eructation (belching of gas) is another mechanism which is quite important to ruminants. Microbial fermentation in the rumen results in production of large amounts of gases (primarily CO_2 and methane) which must be eliminated. This is accomplished by contractions of the upper sacs of the rumen which force the gas forward and down; the esophagus then dilates and allows the gas to escape. During this process much of the gas penetrates into the trachea and lungs. A common problem in ruminants is bloat, a condition which results, for the most part, formation of froth in the rumen. Froth, if found in the area where the esophagus enters the rumen, inhibits eructation, a safety mechanism preventing inhalation of froth into the lungs.

Rumen Metabolism

In the GIT of the ruminant, as opposed to other types of animals, ingested food is exposed to very extensive pre-gastric fermentation. In other words, most of the ingesta

ingesta is fermentated by microbes before it is exposed to typical gastric and enteric digestive enzymes and chemicals. It might be noted that a limited amount of fermentation occurs in the stomach of some monogastric species, but much, much less than in ruminants.

The reticulo-rumen provides a very favorable environment for microbial survival and activity — it is moist and warm and there is an irregular introduction of new digesta and a more or less continual removal of fermented digesta and end-products of digestion. A vast number of bacterial types may be found in the rumen, typical counts approaching numbers of 25-50 billion/ml. Characteristics of species that have been studied are too detailed to discuss here except to note that a wide variety exists in cell size, shape, structure, and in metabolism. In addition to bacteria, some 30+species of ciliate protozoa have been identified from the rumens of animals in different situations, although the variety that may be found in any one animal is considerably less. Protozoal counts vary widely, but typical values to be expected are on the order of 200,000-500,000/ml. Most, if not all, protozoal species ingest rumen bacteria and many also ingest food particles. Considerably lower concentrations of flagellated protozoa are sometimes found. In addition, relatively large counts of phages (bacterial viruses) have been noted in recent years, and almost any organism found on feed or in water may be recovered, although many of them may not be natural inhabitants of the rumen. Yeasts sometimes occur in large numbers, but not with great regularity. The fate of rumen microorganisms is that, eventually, they will pass into the abomasum and intestines where they are digested by the host animal.

Comparative Capacity of the Gastro-Intestinal Tract

The relative capacities of the GIT from different species varies greatly as shown in Table 3-1. Assuming that these values are truly representative of these different animal species, it is evident that there are tremendous differences. Man has a very small GIT as compared to any of the other species. The pig has a very large stomach capacity for a monogastric animal, and the horse shows the adaptation of a herbivorous animal to handle the large amounts of roughage naturally consumed. Ruminant species have, by far, the largest stomach. It might be pointed out that volume is subject to change depending upon the amount or bulkiness of the diet, so these values are not fixed.

Table 3-1. Estimated capacity (liters) of the digestive tracts of some different species.[a]

	Animal species				
	Man	Pig	Horse	Sheep	Cattle
Body wt, kg	75	190	450	80	575
Reticulo-rumen				17	125
Omasum				1	20
Abomasum				2	15
Gastric stomach	1	8	8		
Total stomach	1	8	8	20	160
Small intestine	4	9	27	6	65
Cecum		1	14	1	10
Large intestine	1	9	41	3	25
Total digestive tract	6	27	90	30	260
Stomach, % of body wt	1.3	4.2	1.8	25.0	27.8
Total GIT, % of body wt	8.0	14.2	20.0	37.5	45.2

[a] After Maynard and Loosli (6)

The Role of Digestive Juices in Digestion

Digestive juices have very important roles in the overall digestive processes. In ruminant species, the digestive juices are supplementary to the digestion that first occurs in the rumen; in monogastric and avian species the digestive juices attack the food before it is subjected to microbial action in the cecum and large gut. A list of the various enzymes involved is shown in Table 3-2 together with information on their origin, substrate acted on, and end products produced. The enzymes are listed according to the general type of compound hydrolyzed — amylolytic (carbohydrates), lipolytic (lipids) or proteolytic (proteins). The information shown in the table should be adequate except for a few miscellaneous comments.

With respect to the proteolytic enzymes, it might be pointed out that pepsin is relatively inactive except at rather low pH. Thus, in the young suckling animal, where stomach pH is apt to be on the order of 4-4.5, pepsin is relatively inactive. In fact, some information indicates very little pepsin secretion in calves until they start to consume solid food. Note also that the activity of the various proteolytic enzymes is quite specific, as is the case for most enzymes. Pepsin, for example, tends to attack peptide bonds involving an aromatic amino acid (phenylalanine, tryptophan, or tyrosine), and it also has significant action on peptide bonds involving leucine and acidic residues. However, it liberates only a few free amino acids. Trypsin acts on peptide linkages involving the carboxyl group of arginine and lysine, and chymotrypsins are most active on peptide bonds involving phenylalanine, tyrosine, and tryptophan. The action of trypsin and chymotrypsin is additive, resulting in more complete degradation of proteins to small peptides. One type of carboxypeptidase rapidly liberates carboxyl terminal amino acids, but the other type acts only on peptides with terminal arginine or lysine residues. Similar comments apply to the other peptidases.

In addition to gastric enzymes, HCl has an important effect on gastric digestion. It activates both pepsin and rennin and provides a pH that is more or less optimal for pepsin activity. HCl also has the property of coagulating milk proteins and has some hydrolytic activity in addition.

In the small intestine, bile from the liver has several important roles. Note that not all species have gall bladders, examples being the rat, horse, deer, elk, moose, and camel. In other species including man, swine, chickens, cattle and sheep, the gall bladder serves as a reservoir for temporary storage of bile. Bile contains various bile salts which are important in providing an alkaline pH in the small intestine and in emulsifying fats. These salts are easily absorbed in the small intestine and rapidly recirculated back to the liver. In addition to salts, bile pigments are present which are responsible for its color and, ultimately, for most of the color in feces and urine. Bilirubin or oxidation products of it account for much of the pigment. These pigments are waste products, being derived from the porphyrin nucleus of hemoglobin which is metabolized in the liver. Bile also serves as a route of excretion for many different metallic elements, inactivated hormones, and various harmful substances.

Control of Gastro-Intestinal Secretions

The reader is assumed to understand that motility and secretory activities of the GIT are under hormonal and nervous control, the result of which is a coordinated series of secretory and motor activities that provide for an orderly sequence of action. The coordinated muscle activity serves to mix and transport digesta through the total GIT. Gastro-intestinal hormones have been studied extensively and several have been isolated and identified, but their precise site of production is not known. Some of the hormones important in digestive processes are listed in Table 3-3. Perusal of this table will clearly indicate the importance of these hormones.

Role of the Intestinal Tract in Transport of Nutrients

Most of the absorption of nutrients takes place in the upper intestinal tract, including the duodenum and the jejunum — and to a lesser extent, the ileum. The rate of passage of nutrients through the digestive tract is only a matter of a few hours in most species, so it is clear that opportunity for processing and absorption of nutrients is limited. The degree of absorption of nutrients from the intestinal tract is increased enormously by an increase in the absorptive surface. For example, in the human adult, the surface area of the intestine, if in the form of a simple cylinder would be

Table 3-2. Principal digestive enzymes secreted by the gastrointestinal tract.

Type, name	Origin	Substrate, action	End products	Comments
Amylolytic				
Salivary amylase	saliva	starch, dextrins	dextrins, maltose	None in ruminants; of minor importance in other species
Pancreatic amylase	pancreas	starch, dextrins	maltose	Low in ruminants
Maltase	s. intestine	maltose	glucose	Low in ruminants
Lactase	s. intestine	lactose	glucose, galactose	High in young mammals
Sucrase	s. intestine	sucrose	glucose, fructose	Very low in ruminants
Oligoglucosidase	s. intestine	oligosaccharides	misc. monosaccharides	
Lipolytic				
Salivary Lipase	saliva	triglycerides	free fatty acids, mono- and diglycerides	of minor importance in young mammals
Lipase	pancreas	triglycerides	free fatty acids, mono- and diglycerides	
Lecithinase	pancreas, s. intestine	lecithin	lysolecithin, free fatty acids	
Proteolytic				
Pepsin*	gastric juice	clots milk; hydrolyzes native proteins in acid pH	polypeptides	
Rennin*	abomasum	clots milk (casein)	Ca caseinate	Important in young ruminants
Trypsin*	pancreas	native proteins, polypeptides	poly- and dipeptides	
Chymotrypsin* (two types)	pancreas	polypeptides	small peptides	
Carboxypeptidase* (two types)	pancreas	peptides	amino acids	
Aminopeptidases*	s. intestine	peptides	amino acids	
Dipeptidases	s. intestine	dipeptides	amino acids	
Nucleases (several types)	pancreas, s. intestine	nucleic acids	nucleotides	
Nucleotidases	s. intestine	nucleotides	purine & pyrimidine bases, phosphoric acid, pentose sugars	

*These enzymes are given off in inactive forms, probably a means of protecting the tissues. Pepsinogen is activated by HCl to the active form, pepsin; rennin is activated by HCl; trypsinogen by an intestinal enzyme, enterokinase, and by trypsin. The proforms of chymotrypsin, carboxypeptidases, and amino peptidases are activated by trypsin.

Table 3-3. Hormones important in gastro-intestinal function.[a]

Hormone	Origin	Releasing mechanism	Function
Gastrin	Pylorus	Distension and movement of stomach	Stimulation of acid secretion by gastric glands
Enterogastrone	Duodenum	Fat and fatty acids plus bile in duodenum	Inhibition of gastric secretion and motility
Secretin	Duodenum	Acid and peptones in duodenum	Stimulation of pancreatic secreation (water and electrolytes)
Pancreozymin	Duodenum	Acid and peptones in duodenum	Stimulation of pancreatic secreation (enzymes)
Cholecystokinin	Duodenum	Fat in duodenum	Contraction of gallbladder and relaxation of sphincter of Oddi
Enterocrinin	Jejunum	Food digestion products (?)	Stimulation of intestinal secretion

[a] After Hill (4)

approximately 3300 cm^2, is increased to about 2 million cm^2 by virtue of folds, villi, and microvilli, each of which increases surface area by several times. A diagram of nutrient absorption from the GIT showing epithelial cells along the surface of the villi and the microvilli associated with each individual cell is shown in Fig. 3-6. Figure 3-7 shows a diagram of a larger segment of the GIT illustrating the arrangement of the single layer of epithelial cells lining the intestine and the villi which increase the absorptive surface. In Fig. 3-8 are photographs of a jejunal intestinal epithelial cell from newborn and 4-day-old pigs. The microvilli and subcellular components of a single cell are visible at a magnification of l800X (2). The rate of metabolism of the intestinal mucosa (the intestinal epithelium) is one of the fastest of any tissue in the body. In the human adult, there is a turn-over of about 250 g of dry matter/day, an amount which contributes significantly to the maintenance requirements of the animal. The passage of individual nutrients from the intestinal lumen into the intestinal epithelial cell and then into the blood or lymph may occur by passive diffusion (pore size of cells in jejunum is 7.5 angstroms or .0000075 mm), by active transport, or by pinocytosis (phagocytosis). Pinocytosis involves engulfment of large particles or large ions in a manner similar to the way an amoeba surrounds its food. This occurs in newborn animals to allow absorption of immune globulins from colostrum.

The passage of a nutrient across the intestinal mucosal cell membrane and into the blood or lymph, whether by diffusion or by active transport, involves (a) penetration of the microvillus and of plasma membrane, which encapsulates the epithelial cell, (b) migration through the cell interior, (c) possible metabolism within the cell, (d) extrusion from lateral and basal aspects of the cell, (3) passage through the basement membrane, and (f) penetration through the vascular or lymphatic epithelium into blood or lymph. The exact means of transfer of each individual nutrient will be discussed in subsequent chapters on individual nutrient requirements and metabolism.

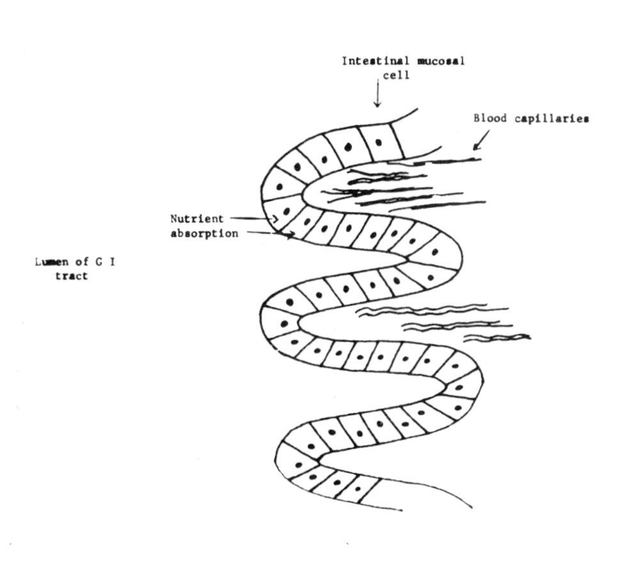

Figure 3-6. Diagram of epithelial cells lining the intestinal tract and villi which increase the absorptive surface.

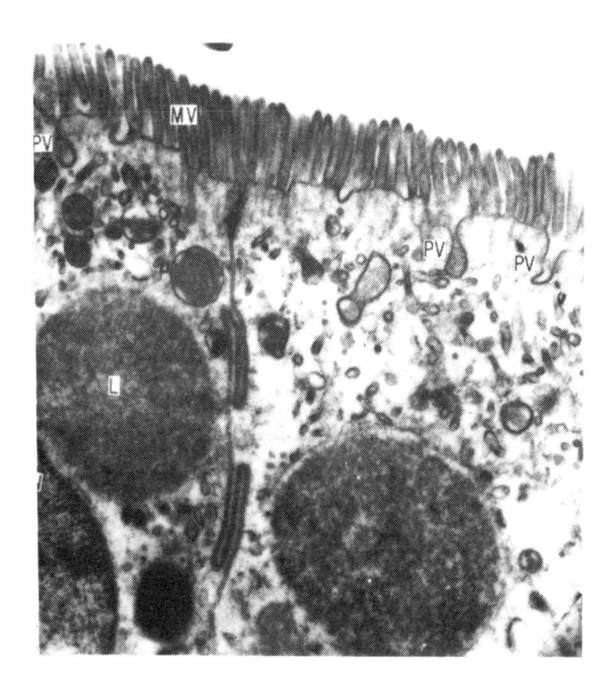

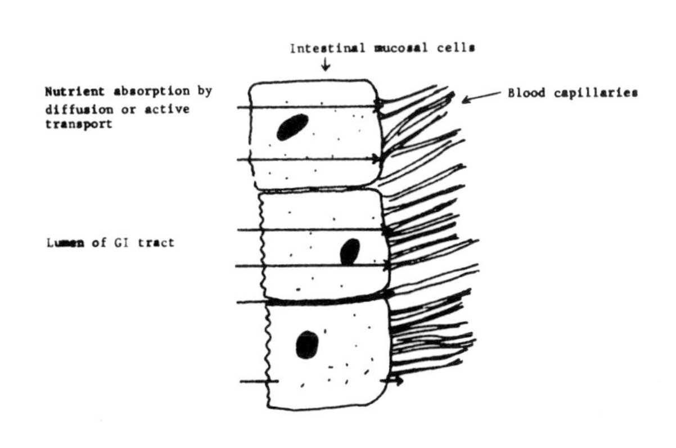

Figure 3-7. Diagram of arrangement of intestinal mucosal cells.

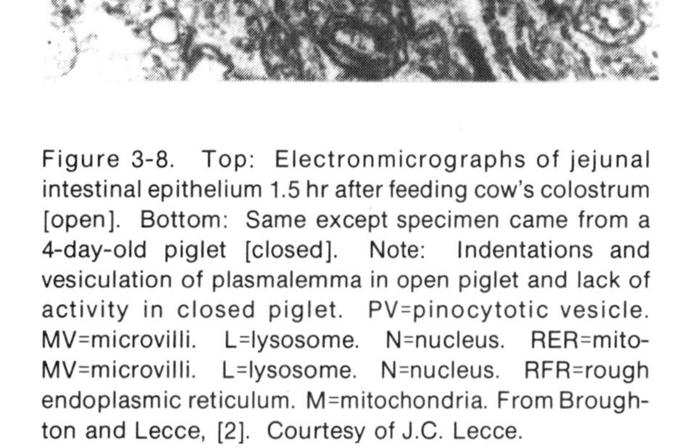

Figure 3-8. Top: Electronmicrographs of jejunal intestinal epithelium 1.5 hr after feeding cow's colostrum [open]. Bottom: Same except specimen came from a 4-day-old piglet [closed]. Note: Indentations and vesiculation of plasmalemma in open piglet and lack of activity in closed piglet. PV=pinocytotic vesicle. MV=microvilli. L=lysosome. N=nucleus. RER=mito- MV=microvilli. L=lysosome. N=nucleus. RFR=rough endoplasmic reticulum. M=mitochondria. From Broughton and Lecce, [2]. Courtesy of J.C. Lecce.

Role of Blood and Lymph in Nutrient Transport

A broad concept of nutrient flow through the body is illustrated in Fig. 3-9. Briefly, a nutrient passes across the epithelial cell and enters either blood capillaries or the lymph system and is carried through the portal vein to the liver or, when materials enter the lymph, through the thoracic duct to the heart. The liver acts as a central organ in metabolism because many complex and vital reactions occur there. After transversing the liver, venous blood reaches the right atrium of the heart and passes into the right ventricle, from which it is pumped through the lung for oxygenation. From the lung, it returns to the left atrium of the heart and passes into the left ventricle, the largest and most muscular chamber of the heart. From the left ventricle, it is pumped through the aorta to enter all tissues of the body, carrying oxygenated blood and nutrients from the GIT and those synthesized in other tissues, including the liver. The nutrients enter the capillaries in all tissues of the body and, in this way, nourish every cell. The same capillary system carries waste products from the cell into veins for transport to sites of further metabolism or for excretion. The kidney plays an important role in acting as a filter to dispose of waste materials. In addition to waste product excretion through the kidney, other waste products can be disposed of by re-excretion back into the intestinal lumen. This takes place by accumulation in the liver and excretion via the bile. In addition, excretory products pass through the blood capillaries into the intestinal epithelium. Thus, two-way traffic moves across the epithelial cells — both absorption of nutrients from the intestinal lumen and excretion of metabolites and waste materials across the intestinal wall — but in opposite directions. In addition to excretory products, there is considerable loss from the body of nutrients that have already served a productive function. For example, the loss via feces of cells sloughed off from the intestinal mucosa represents an obligatory loss of protein, minerals, and other nutrients, even though they have been utilized in normal body function.

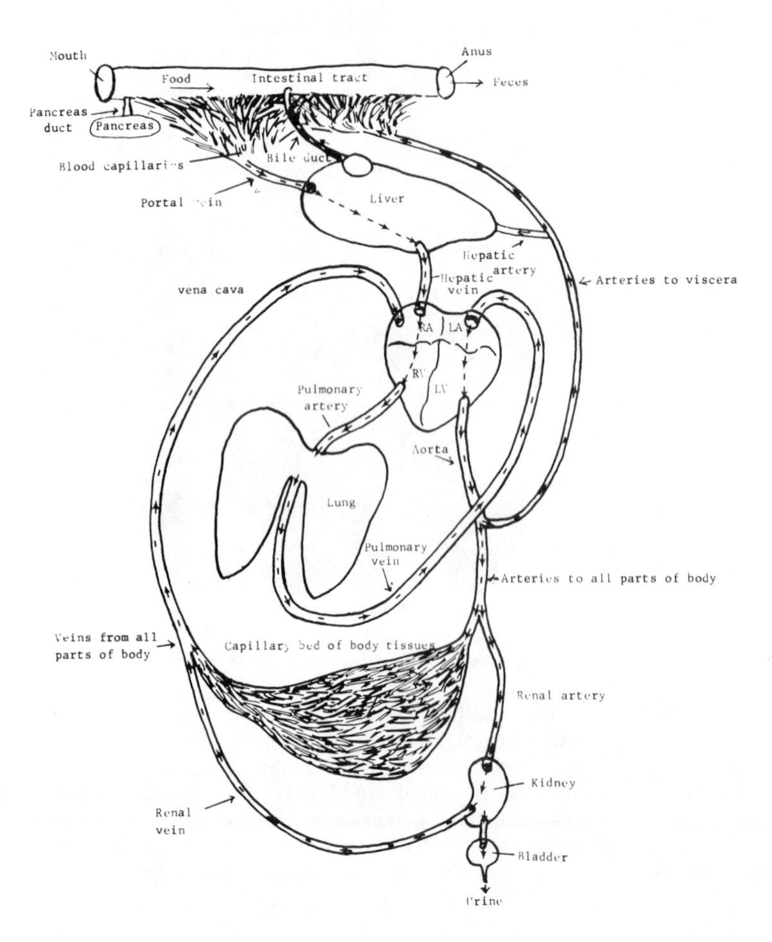

Figure 3-9. Diagram of routes of nutrient absorption, flow of nutrients and fluids in the body, and routes of excretion.

Digestibility and Partial Digestion

Digestion, as indicated previously, is defined as the preparation of food for absorption by the GIT. Most authors tend to use an implied definition meaning disappearance of food from the GIT, however; this broad definition would include absorption as well as digestion.

Digestibility data are used extensively in animal nutrition to evaluate feedstuffs or study nutrient utilization (see Ch. 4 for more detail). The reader should be aware that digestibility is variable; in other words, the same feedstuff given to the same animal is not always digested to the same extent. Several factors may alter the extent of digestion. These include level of feed intake, digestive disturbances, nutrient deficiencies, frequency of feeding, feed processing, and associative effects of feedstuffs (nonadditive effects of combining different feedstuffs). Marked differences also exist in the ability of different animal species to digest a particular feedstuff, particularly roughages. Note in Table 3-4, where 7 different species have been fed the same alfalfa hay, that there are appreciable differences in digestion of the different fractions listed. This information reflects the ability of the GIT to utilize fibrous feeds, clearly indicating an advantage of ruminant species and the horse over swine, rabbits and guinea pigs.

The term, **partial digestion,** implies that total digestion in the GIT can be subdivided into fractions that are digested in different parts of the tract. In ruminant species, enough research has been done in recent years to provide a considerable amount of information on this subject. Less information is available on other domestic animals. Data on sheep and cattle (see Ch. 9 in Church, 3) indicate that about 2/3 of digestible organic matter and energy disappear from the forestomach (reticulo-rumen, omasum), the remainder being digested in the intestines. The more fibrous portions (crude fiber, cell walls), as well as readily available carbohydrates, are digested to a somewhat greater extent (65-90% of total digestion) in the forestomach. Digestibility of N is quite variable, but is relatively more important in the small intestine; this is a reflection of passage of large amounts of microbial proteins into the small intestine.

Fecal and Urinary Excretion

Fecal material excreted by animals is comprised of undigested residues of food material; residues of gastric juices, bile, pancreatic and enteric juices; cellular debris from the mucosa of the gut; excretory products excreted into the gut; and cellular debris and metabolites of microorganisms that grow in the large intestine with some, probably coming from the forestomach in ruminants.

Undigested food residues are largely dependent upon the type of food consumed and the type of GIT; thus, in roughage eaters, undigested residues will usually account for more of the total than would be the case with monogastric species consuming a diet low in fiber. Sloughed cellular debris may reach substantial amounts. Using data from rats, it has been estimated that the dairy cow may

Table 3-4. Digestibility of alfalfa hay by different species.[a]

Species	Digestion coefficients, %					
	Organic matter	Crude protein	Ether extract	Crude fiber	NFE	TDN
Cattle	61	70	35	44	71	48.3
Sheep	61	72	31	45	69	48.1
Goat	59	74	19	41	69	46.9
Horse	59	75	10	41	68	46.0
Pig	37	47	14	22	49	30.5
Rabbit	39	57	21	14	51	30.9
Guinea pig	52	58	12	33	65	40.5

[a] From Maynard and Loosli (6). The alfalfa hay in question contained 86.1% dry matter, 16.2% crude protein, 1.6% ether extract, and 26.9% crude fiber.

slough about 2500 g of cells daily from the gut wall.

The color of feces is from plant pigments and stercobilinogen produced by bacterial reduction of bile pigments. The odor is from aromatic substances, primarily indole and skatole, which are derived from deamination and decarboxylation of tryptophan in the large intestine.

Urine represents the main route of excretion of nitrogenous and sulfurous metabolites of body tissues. In addition, it is usually the principal route for excretion of some of the mineral elements, particularly Cl, K, Na, and P. Urine is, essentially, an aqueous solution of these various components with minor amounts of pigments and sloughed cells from the urogenital tract.

The color of urine is primarily due to urochrome, a metabolite of bile pigments complexed with a peptide. The urine from ruminants (other than suckling animals and those on high-grain feed) is usually basic, being in the pH range of 7.4 to 8.4. The basic reaction is characteristic of herbivorous animals because of the relatively large amounts of Na and K ions found in vegetation. Urine is usually in the acidic range in monogastric species. In mammals, urea is the principal N-containing compound excreted with lesser amounts of other compounds such as ammonia, allantoin, creatine, and creatinine. In birds, the major N-containing compound in urine is uric acid.

Only traces of protein are normally found in urine. The presence in quantity of proteins such as albumin and globulins is indicative of kidney disease in adults, although it may occur in young animals under normal conditions, particularly within a few days after normal consumption of colostrum. Carbohydrates such as glucose or fructose may sometimes be found in mammalian urine following ingestion of a meal high in soluble carbohydrates. In adult ruminants, however, carbohydrates in the urine are indicative of disease. Ketones are found at times, particularly in animals suffering from starvation or ketosis.

References Cited

1. Anonymous. 1972. Rural Research in C.S.I.R.O. p. 4, June issue.
2. Broughton, C.W. and J.G. Lecce. 1970. J. Nutr. 100:445.
3. Church, D.C. 1969. Digestive Physiology and Nutrition of Ruminants. Vol. 1. O & B Books, 1215 NW Kline Pl., Corvallis, Ore.
4. Hill, K.J. 1970. In: Dukes' Physiology of Domestic Animals. 8th ed. Comstock Publishers Associates, Ithaca, New York.
5. Hoffman, R.R. 1968. In: Comparative Nutrition of Wild Animals. Academic Press. New York.
6. Maynard, L.A. and J.K. Loosli. 1969. Animal Nutrition. 6th ed. McGraw-Hill Book Co.
7. Sisson, S. and J.D. Grossman. 1953. The Anatomy of the Domestic Animals. 4th ed. W.B. Saunders Co., Philadelphia.
8. Sturkie, P.D. 1965. Avian Physiology. 2nd ed. Comstock Publishers Associates, Ithaca, New York.
9. Sturkie, P.D. 1970. In: Duke's Physiology of Domestic Animals. 8th ed. Comstock Publishers Associates, Ithaca, New York.
10. Swenson, M.J. (ed.). 1970. Dukes' Physiology of Domestic Animals. 8th ed. Comstock Publishers Associates, Ithaca, New York.

Chapter 4 – Measurement of Nutrient Utilization and Requirements of Animals

The utilization of nutrients after absorption is remarkably similar among animal species despite the wide differences in anatomy of the digestive system as described in Ch. 3. Methods applicable to determination of requirements of individual nutrients are, for the most part, common to all species. This chapter describes the methods commonly used to measure nutrient utilization by animals and the means by which qualitative and quantitative nutrient requirements are determined. Subsequent chapters will cover the metabolism of individual nutrients in more detail within each broad class — proteins, lipids, carbohydrates, minerals, and vitamins.

Protein, energy, and minerals constitute the bulk of body storage of nutrients during growth and pregnancy and of body losses of nutrients during lactation. Of course, water makes up a large part of milk and of tissues formed during growth, but, because it does not contribute energy, it can be considered independently. The changes in proportions of water, fat, protein, and ash that occur from youth to adulthood are illustrated in Fig. 4-1, using data from pigs. Vitamins and trace elements are required in minute amounts and contribute insignificant amounts of mass to the growing or adult animal and so need not be considered in the same context as protein, energy, and the major mineral elements.

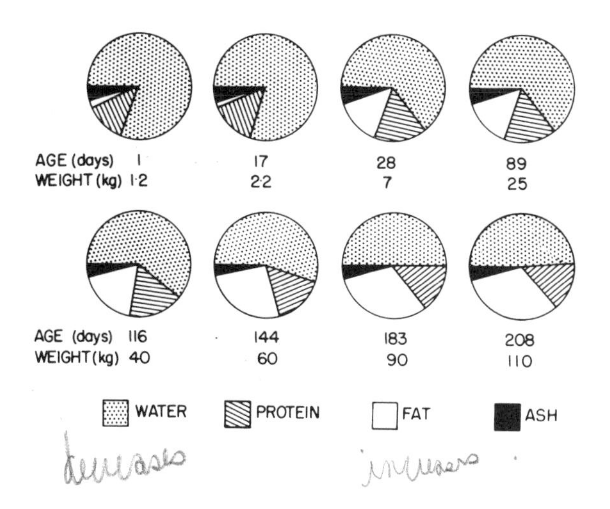

Figure 4-1. Changes in body composition with increased age and weight of pigs. From Oslage [11].

Variation in Biological Availability of Nutrients

Biological availability of all nutrients varies depending on a number of factors; a major one is the efficiency of absorption from the intestinal tract. Increased rate of passage of feed residues down the GIT depresses digestibility (absorbability); increased feed intake above maintenance tends to depress digestibility, especially in ruminants. Part of this variation results from inefficient release of nutrients from feedstuffs. Aside from differences related to source of feedstuffs, variation exists in absorptive capacity based on the specificity of a wide variety of specialized transport systems present in the intestinal epithelium. This specialization is illustrated by comparing the maximum absorptive capacity of man for vitamin B_{12} which has been estimated at 1 mcg/day with that for glucose which is 3600 g/day (13). The recognition of existence of such specific absorption pathways of widely differing capacities has resulted in the development of a number of research techniques for absorption studies. These include permanent fistulas of stomach and intestine, cannulation of the portal vein or thoracic duct, whole-body counting of radioactivity after administration of radio-isotopes, everted intestinal tract techniques, and a variety of other techniques. Many of these will be discussed in depth in later chapters in conjunction with metabolism of individual nutrients. The essential concept here is to envision the utilization of energy and protein as the net difference between the amount ingested and the amount lost in the feces.

Methods of Measuring Nutrient Utilization

Growth Trials

Growth, as defined by Brody (2), is "the constructive or assimilatory synthesis of one substance at the expense of another (nutrient) which undergoes dissimilation." In the broadest sense, growth of an animal consists of an increase in body weight resulting from assimilation by body tissues from ingested nutrients. The increase in weight is composed of the sum of the increases in weight of individual components making up the body —

namely, water, fat, protein, carbohydrate, and ash. It can be expressed as absolute gain in a given period of time or as relative gain (usually expressed as a %). Growth trials usually include the measurement of absolute gain in body weight during a period of feeding a test diet. Rate of gain is then expressed as average daily or weekly gain (absolute gain) or in terms of final weight as a % of initial weight (relative gain). The greater the rate of gain in this period, generally the more efficient is the diet in meeting the animal's requirement for weight gain (see Table 4-1). Animals used in a growth experiment are usually feed the test diet concurrently with similar animals fed a standard (or basal) diet of known nutritive quality which allows normal growth. In this way, direct comparisons can be made among various feed or nutrient sources and they can be ranked in order of their ability to promote weight gain or efficiency of feed utilization. Growth is often used interchangeably with weight gain; strictly speaking, the two terms are different because an equal increase in body weight between animals does not necessarily indicate equal growth of body tissues. That is, weight gain does not identify changes in body composition.

Normally, a growth trial involves ad libitum feeding of a diet. Knowing rate of gain and total feed consumption, feed/unit of gain can be computed, which is also a meaningful estimate of nutrient adequacy of the diet. Diets that promote a maximum weight gain will usually promote maximum efficiency of feed utilization. Of course, physical factors or palatability factors may affect total feed intake (see Ch. 17). To rule out this type of variation, paired feeding experiments can be conducted, in which the test diets are fed to animals of comparable size and at equalized intake based on the voluntary consumption by the member of the pair eating the least. Paired feeding eliminates differences in animal performance related to palatability of the feed. It penalizes the animal consuming the more adequate diet and tends to reduce the magnitude of difference in growth of animals fed the test diets.

Digestion Trials

Digestion trials are used to estimate the proportion of a feed that is available to the animal for absorption from the GIT. Animals are fed a diet of known composition over a time period of several days during which the feces

Table 4-1. Inverse relationship between daily gain and feed per unit gain.

Diet	Daily gain, kg	Feed/gain
	Pigs[a]	
Experiment 1		
Basal diet	0.23	5.97
Basal diet + Zn	0.57	3.81
Experiment 2		
Basal diet	0.36	5.14
Basal diet + Zn	0.71	3.75
	Lambs[b]	
Normal phosphorus	0.246	5.26
High phosphorus	0.220	5.60

[a] Dahmer et al (3)
[b] Emerick et al (4)

are collected and analyzed for the components of interest. Maintaining a constant daily feed intake over several days is advisable, to minimize day-to-day variation in fecal output. Time required for feed residues to traverse the GIT is 1 to 3 days for nonruminants and 5-10 days for ruminants. Therefore, a preliminary period of 4 to 10 days is needed to void the GIT of residues of pre-test feed and to allow adaptation of the animal to the test diet. A collection period of 4-10 days follows the preliminary adjustment period. Values can be obtained for apparent digestibility of any desired nutrient, but data may be rather meaningless for some nutrients such as vitamins and minerals whose passage both from the lumen of the GIT into the body and from the body to the lumen of the GIT is quite variable and subject to change. There are two general means by which digestibility of a feed or its component can be estimated. One is by total collection of feed and feces, which allows a direct measure of apparent digestibility. It is computed as follows:

$$\text{Apparent digestibility (\%)} = \frac{\text{Nutrient intake} - \text{Nutrient in feces}}{\text{Nutrient intake}} \times 100$$

A second method, and the one of choice when it is impossible or inconvenient to measure total feed intake or to collect total feces, is the indicator method. This method depends on the use of an inert reference substance. Internal indicators are those such as lignin that are present in the feed. External indicators, such as chromic oxide, are added to the feed or given to the animal. An indicator must be nontoxic, palatable, and easily measured; it should be insoluble and should pass down the GIT at a uniform rate. The calculation for digestibility of a particular nutrient using the indicator method is as follows:

$$\text{Apparent digestibility} = 100 - \left(100\ \frac{\text{\% indicator in feed}}{\text{\% indicator in feces}} \times \frac{\text{\% nutrient in feces}}{\text{\% nutrient in feed}}\right)$$

Such a ratio provides an estimate of digestibility of a particular nutrient without knowing either the total intake of feed or the total excretion of feces. Indicators lend themselves to use in group-fed animals by "grab sampling" of feces and in pasture and range studies where measurement of both feed intake and fecal output is difficult. Consumption of feed on pasture can be estimated by a calculation as follows:

$$\text{Dry matter intake (units per day)} = \frac{\text{(units of dry matter in feces per day)} \times \text{(amount of indicator per unit of dry feces)}}{\text{amount of indicator per unit of dry matter in feed}}$$

The problem here is that total feces collection is required. A bag attached to the animal by straps can be used in this way (see Fig. 4-2). A problem in pasture studies arises from the animal's selecting certain plants and refusing others. Equating actual consumption with assumed consumption based on analysis of clippings from a given area may be misleading.

Apparent digestibility and total intake of feed from pasture forage under grazing conditions can be estimated using two indicators simultaneously. The external indicator (such as chromic oxide) can be administered to the animal in known amounts and the internal indicator (such as lignin), since it is undigested by the animal, can be measured as a % of the diet and used as previously described to calculate apparent digestibility.

Apparent Digestibility vs. True Digestibility

The apparent digestibility of N is calculated from the difference between feed intake and fecal output. The feces, however, include not only nutrients but other things such as sloughed intestinal cells and digestive enzymes. The true digestibility of a nutrient is that proportion which is absorbed from the lumen of the GIT as a nutrient passes down its length, excluding contributions from endogenous (body tissue) sources. Fecal N derived directly from ingested food is called exogenous N (not from body tissues); that derived from body tissues is termed fecal metabolic N (endogenous). Thus, for protein, true digestibility can be estimated by subtracting the amount of N appearing in the feces of animals fed a protein-free diet from the amount of N appearing in the feces of animals fed the test diet. The apparent digestibility of protein in a feed is influenced by the level of protein in the feed. This is because the amount of endogenous protein in the feces is relatively constant at differing protein intakes.

This relationship is illustrated as follows:

	High protein intake	Low protein intake
Daily N intake, g	20	10
Daily fecal N, g	5	3
Apparent N absorption, g	15	7
Apparent N digestibility, %	15/20=75	7/10=70
Daily fecal N on N-free diet, g (fecal metabolic nitrogen)	1	1
Daily fecal N on test diet	5 minus 1	3 minus 1
Minus daily fecal N on N-free diet, g	=4	=2
True N absorption, g	16	8
True N digestibility, %	16/20=80	8/10=80

These data show that the apparent digestibility of a protein is reduced as protein intake is decreased because of the greater contribution of fecal metabolic N (endogenous), even though true N digestibility remains unchanged. Fecal metabolic N is that derived from sloughed intestinal cells, digestive enzymes and other endogenous sources. Although true digestibility is theoretically a better means of assessing protein quality, it is difficult to determine because it requires a separate feeding trial of questionable validity (animals cannot survive on a protein-free diet). Therefore, apparent digestibility is the most easily obtained and commonly used measure for all animals.

Balance Trials

Closely aligned to the digestion trial is the balance trial. Balance trials account for total losses of a particular nutrient or nutrients from the body after ingestion of a know quantity of the nutrient. Thus, the balance of N, energy, lipids, or minerals can be studied. A mineral or N balance trial requires provisions for collecting both feces and urine. An energy balance trial requires, in addition, (especially in ruminant animals) measurement of gaseous losses resulting from the fermentation processes (see Ch. 9). As with digestion trials, several days must be included in the balance trial to cancel out short-term variations in fecal and urinary output. Because of rate of passage differences between ruminants and nonruminants, balance trials for ruminants are run conventionally for a collection period of about 10 days (following a preliminary period), whereas trials for nonruminants can usually be completed over a 4- to 7-day collection period.

Because N metabolism is so closely linked with growth, the N balance trial is perhaps the most common in animal nutrition. A growing animal is in positive N balance. Short term liabilities, as for lactation, involve negative N balance if the animal is unable to consume as much N as is being secreted in the milk and lost in feces and urine. An adult in a nonproductive capacity would normally be in N equilibrium; that is, the average amount of N taken in/day would equal the amount of N lost in the urine and feces/day. In addition to these major losses, slight N losses occur through shedding of hair, sloughing of skin, and sweating, which can be accounted for if we are interested in very precise measurements. Thus, to study the protein adequacy of a particular diet or feedstuff for growth, measurement of N balance allows us to compare a test diet with a standard diet of known adequacy. The protein level of the diet must be kept low or marginal; when the protein requirement is exceeded, the extra protein is deaminated and the carbon skeleton used for energy, while the N from the protein appears in the urine as urea and gives an unrealistically low estimate of the amount of N used for productive purposes. The N balance method is often criticized as being erroneous in that, if one extrapolates the amount of N retained by an animal on a good diet during a typical collection period to the lifetime of the animal, the projected amount of N becomes excessively high — so high as to suggest the absurdity of an animal the size of a pig having N in the body equal to that of a mature elephant. Such errors result from incomplete collection of excreta and losses in the process of collection. The criticisms have some validity, but it must be remembered that we are usually interested in a relative comparison of different protein sources and not with absolute amounts of N retained in a given short period of time. The same criticisms can be made for other nutrient balance experiments.

Balance methods of all kinds, as described, are expensive and tedious to carry out, but they are of vital importance in evaluating feedstuffs or animal requirements. Considerable additional biochemical work may be necessary to describe the metabolic activities that occur between consumption and excretion of a nutrient. It is the final overall picture that is important from a productive standpoint in nutrition, and this is the basis on which balance data have retained their value.

Rumen Digestion Techniques

The high cost of digestion trials, especially with cattle, has prompted the development of in vitro techniques that allow simulation of rumen fermentation under controlled conditions. Quite a variety of methods have been developed, depending upon the objective in mind. In practice, a small amount of rumen fluid is obtained from a rumen-fistulated animal. This material is placed in a container along with some buffer (to simulate saliva) and the test sample. The combination is then fermented at rumen temperature (39°C) for a period of time. Where the object is to predict (or correlate) live-animal digestion from in vitro digestion, adequate methods (5) have been developed which give a better estimate of animal digestion of roughage and forage than can be obtained by using chemical analyses. These methods are not as reliable for evaluating animal utilization of grains and other concentrates. Rumen fermentation procedures have proved to be useful for screening feedstuffs or generally characterizing them for use by ruminants, especially when only small samples are available. These methods are also useful for studying rumen function and metabolism of specific nutrients. For example, determination of how much urea can be utilized by the rumen microorganisms, production of volatile fatty acids from a given ration, and so forth.

The nylon bag technique (1, 10) is also an efficient means of evaluating rumen digestion. In this procedure the feedstuff in question is placed in a nylon bag which is suspended in the rumen of an animal with a rumen fistula. The bags are then removed after a period of time and loss of material (from fermentation) in the bag is determined. Such methods are useful in evaluating relative differences between feedstuffs, but do not give values similar to live animal digestion. This method is, perhaps, more useful for studying rumen digestion of concentrates (grains) than the in vitro method where relative digestion is of interest.

Figure 4-2. One example of a harness and plastic bag used for fecal collections with sheep. Courtesy of G. fishwick, Glasgow University Vet School.

Surgical Procedures for Studying Nutrient Absorption and Utilization

A wide variety of surgical techniques have been developed to aid in study of nutrient absorption. Fistulation of the rumen of cattle and sheep is a common procedure that allows both sampling of contents and infusion of known quantities of substances at known rates directly into the rumen. Stomach fistulas of simple-stomached animals and abomasal fistulas of ruminants are also commonly used as are cecal fistulas and cannulas in a variety of animal species. Studies of rates of absorption of individual radioisotopes can be effectively measured by a surgical technique that consists of tying off a segment of the intestinal tract and measuring disappearance over time of known amounts of a substance injected directly into the lumen of the GIT between the two ligatures. An extension of this in vivo procedure involves a related in vitro technique whereby a segment of intestinal tract is removed, everted so that the serosal side is exterior, and the ends ligated as above. The substance whose transport across the intestinal lining is being studied is then injected into the lumen of the everted intestinal segment which has been placed in a beaker containing physiological saline. The appearance of the injected substance in the beaker contents or its disappearance from the everted intestinal segment is then used to estimate its absorption. Such in vivo and in vitro techniques based on surgical procedures add another dimension to the study of nutrient absorption and utilization and, although they do not substitute for more conventional methods, they can frequently add significant knowledge.

Estimation of Nutrient Requirements of Animals

The establishment of a substance as an essential nutrient for any animal depends on the demonstration of adverse effects on the animal in the absence of that substance from the diet. The array of such nutrients depends on the animal species and, in some instances, on the stage of the life cycle. For example, only primates (including man), guinea pigs, the red vented bulbal bird, and the fruit-eating bat (both native to India) require dietary vitamin C (12). All other species apparently are able to synthesize vitamin C in sufficient quantities to prevent scurvy. Adult birds and man can synthesize the amino acid, arginine, in sufficient quantities to meet body needs, but adult swine cannot. Also, genetic differences exist, even within species, in quantitative requirements for individual nutrients. Nesheim (9) has developed strains of chickens requiring either high or low levels of arginine in the diet. Much remains to be learned concerning variations in quantitative nutrient requirements within species and between different species. Williams (14) has discussed the importance of this phenomenon. It is of importance in both human and animal nutrition.

Sequence of Events in Nutrient Deficiency

The discovery of most of the nutrients as essential dietary constituents has been accomplished largely with farm and laboratory animals. Regardless of the nutrient deficiency, the same sequence of events prevails:

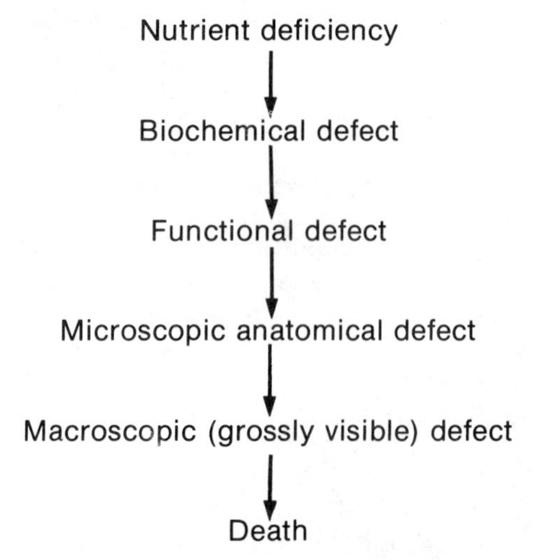

The above sequence of events is well illustrated in the deficiency of the B-vitamin, thiamin. The biochemical defect is a failure to produce the coenzyme, cocarboxylase (thiamin pyrophosphate) which is responsible for removing one carbon from pyruvic acid in the formation of acetyl coenzyme A. Thus, pyruvic acid accumulates in the tissues, resulting in poor appetite and reduced growth (functional defect). Microscopic lesions of nervous tissue follow and, later, grossly visible signs of deficiency develop, including emaciation and tremors (polyneuritis in birds). Death is the final result if the deficiency is not alleviated; if alleviated after microscopic or macroscopic lesions have developed, damage to the animal may be permanent.

The biochemical, functional and structural defects associated with nutrient deficiencies are usually specific for each nutrient. Details of these changes are given in subsequent chapters dealing with each class of nutrient. For some nutrients, biochemical changes in blood or tissues are the best indices of dietary adequacy, but, for others, growth or balance trials provide the best information.

Examples of Experimental Data Obtained with Growth and Balance Trials

The requirements of structural nutrients such as protein, individual amino acids, Ca, and P can be determined effectively by growth and balance experiments. An example of the type of experimental data from a growth trial which is used to estimate nutrient requirements is shown in Fig. 4-3 (6). The quantitative lysine requirement of the weanling pig was determined in a series of 3 different growth experiments. Ad libitum feeding was used throughout. Precautions in experimental work are important to minimize the variability from factors other than the dietary variable of interest, in this instance, lysine. Thus, standardized environmental conditions (pen size, temperature, number of animals/pen, and so on) are important, and animals should be assigned to experimental diets with minimum bias introduced by differences in sex, breeding, body weight, disease level, and other similar potential sources of error. Increments of the nutrient in question are then added to the basal diet, and body weight gain and feed consumption over the experimental period are recorded. Thus, Fig. 4-3 clearly shows that in experiment 1 the weight gain was directly related to level of lysine added to the diet up to and including the highest level (0.4% L-lysine added to the basal diet which contained 0.33% lysine). From this single experiment, the minimum lysine requirement for growth could not be judged because a plateau was not reached. In subsequent experiments, higher levels of lysine were added to the basal diet and a plateau in weight gain could be demonstrated beyond which no further gain was achieved by an additional increment of L-lysine. By applying appropriate statistical procedures, the quantitative lysine requirement of the weanling pig could be stated with reasonable assurance under the condition of the experiment (0.71% lysine at 12.8% protein; 0.95% lysine at 21.7% protein). Simply by observing the shape of the curve formed by connecting points within experiments, a reasonable estimate of the requirement can be made even without resorting to mathematical procedures more complex than calculating the mean for each group. In Fig. 4-3, experiments 1 and 2 each employed two protein levels, 12.8% (A) and 21.7% (B). Thus, the data further show that the response to increments of lysine is similar at both protein levels, although weight gain is higher at the 21.7% level. Clearly, by careful attention to experimental design, much valuable information can be obtained on nutrient requirements using only the growth trial, as illustrated here. The quantitative requirements of a single nutrient often depend on age as well as on the levels of other nutrients in the diet. An example of this phenomenon is illustrated in Fig. 4-4 which shows the effects of age and dietary P on Ca and P retention by young pigs (7). This illustration shows clearly that one cannot consider the requirement for a single nutrient without attention to dietary levels of others. Figure 4-5 illustrates this in another way, by considering the partitioning of Ca and P in feces, urine, and body tissues as affected by level of dietary P.

Usually, two or more types of data can be combined appropriately to arrive at a more valid estimate of nutrient requirement. For example, although a growth trial may not reveal differences between groups of animals fed 2 levels of a nutrient, blood concentration of the nutrient may indeed show differences of importance. In general, the larger the number of appropriate types of measurement used in establishing a nutrient requirement, the greater the confidence justified in the estimate. Cost is often the limiting factor in the size and scope of a particular experiment, so data must be synthesized from a number of experiments in arriving at a reasonable conclusion.

Extrapolating experimental results too far, is always tempting and a common limitation of carefully controlled laboratory experiments is the possibility of a different result when the same dietary variables are applied under more practical field conditions. For this reason, research results obtained under laboratory conditions must be put to the test under field conditions before broad inferences are drawn.

Laboratory Animals as Models for Farm Animals and Man

Laboratory animals play an important role in contributing to knowledge of nutritional requirements and interrelationships of farm animals and man. The numbers of animals and housing facilities necessary to obtain meaning-

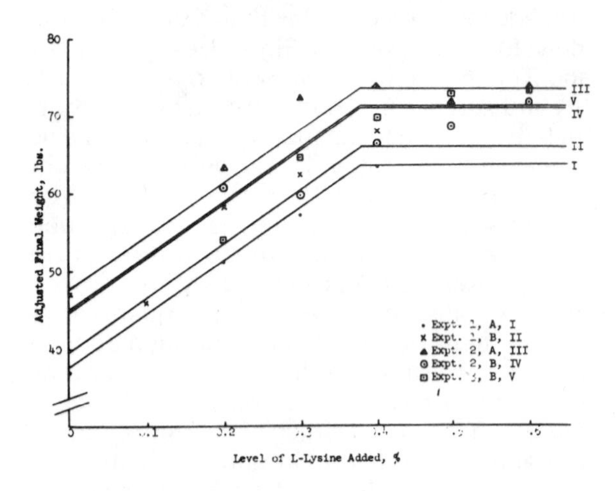

Figure 4-3. Relation of final weight [adjusted to equal initial weight] to lysine level.

ful data with farm animals are often so large as to be economically prohibitive. Also, the life cycle of smaller animal species is generally shorter than that of farm animals, so data can be obtained over several generations/year. Each year thousands of mice, rats, hamsters, guinea pigs, rabbits, and other mammalian species — as well as avian species, such as Japanese quail, and a variety of reptiles and fishes — are used in nutritional research. Of course, the results obtained provide information applicable directly to the species used, but often the information can be extrapolated to farm animals or to man. Among the farm animals, the pig is especially useful as a model for human nutrition studies because its digestive system is similar anatomically and functionally to that of man. The National Research Council has published a series on the nutrient requirements of laboratory and farm animals and of man. One of these (8) provides both narrative and tabular data on the nutrient requirements of cats, guinea pigs, hamsters, monkeys, mice, and rats. Similar publications are available for other individual species of larger animals including dogs, rabbits, horses, swine, sheep, mink and foxes, beef cattle, dairy cattle, and poultry. The science of nutrition knows no limits in terms of the kind of animal life selected for obtaining new information of use in the feeding of man and animals, as numerous microorganisms are utilized in addition to the various animals mentioned.

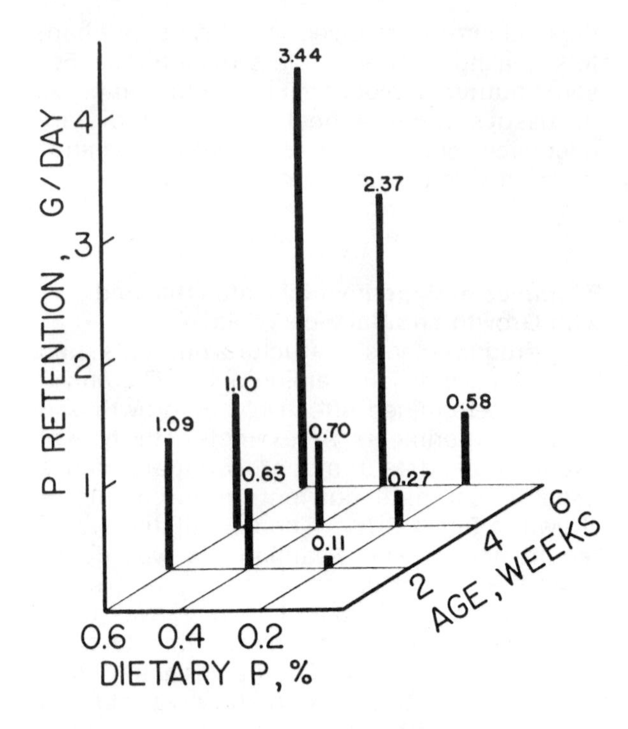

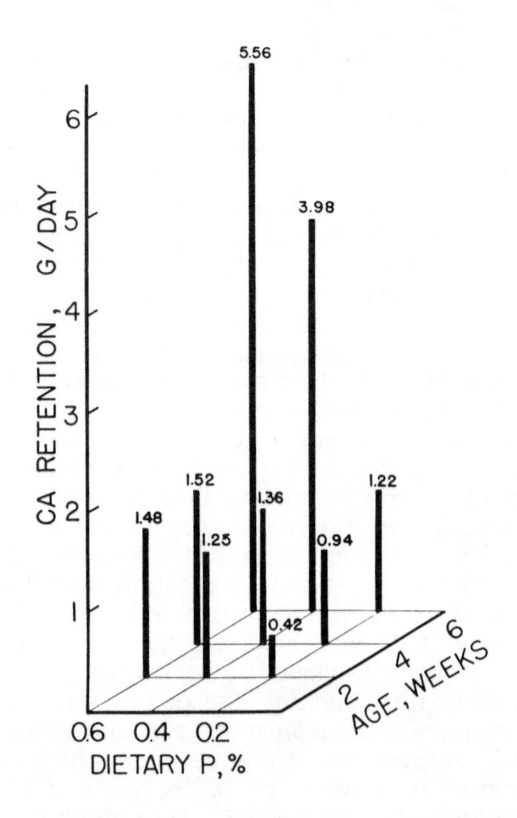

Figure 4-4. Top. Phosphorus retention of baby pigs as affected by age and phosphorus level [synthetic diet]. Bottom. Calcium retention of baby pigs as affected by age and phosphorus level. From Miller et al [7].

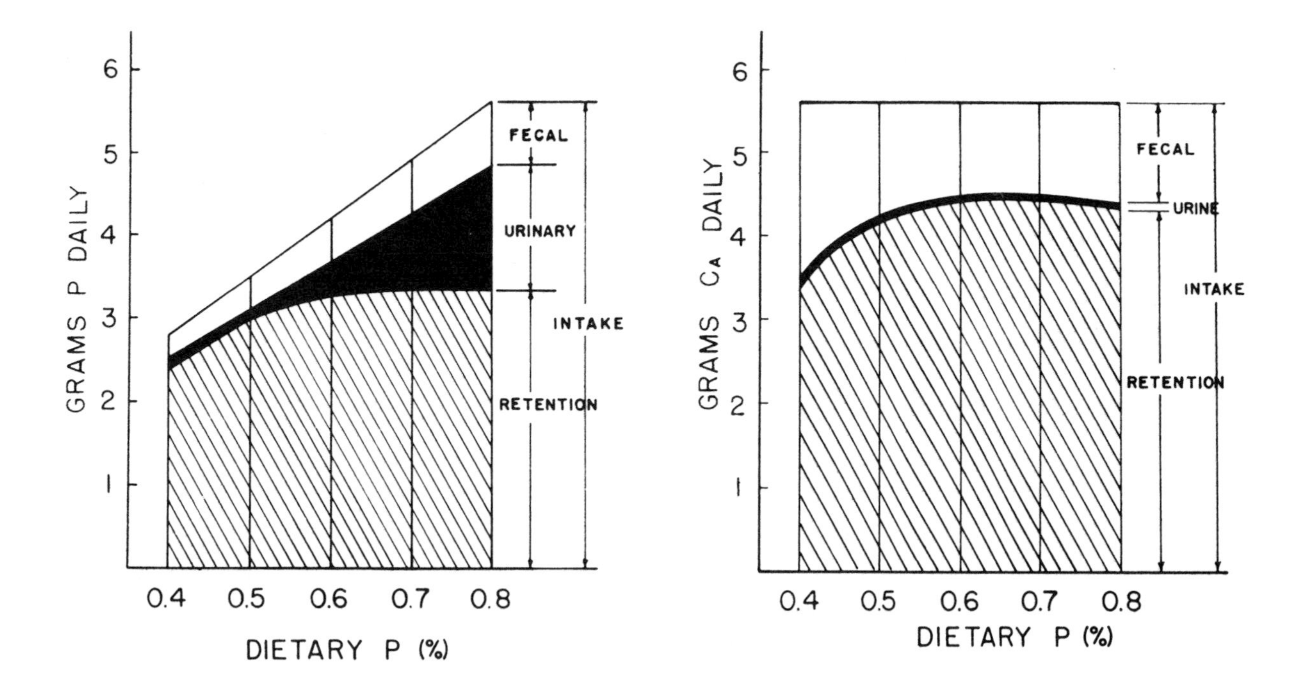

Figure 4-5. Left. Phosphorus balance as affected by dietary P level. Right. Ca balance as affected by dietary P level. From Miller et al [7].

References Cited

1. Barton, J.S., L.S. Bull, and R.W. Hemken. 1962. J. Animal Sci. 21:340
2. Brody, S. 1945. Bioenergetics and Growth. Reinhold Pub. Corp., New York.
3. Dahmer, E.J., R.H. Grummer, and W.G. Hoekstra. 1972. J. Animal Sci. 34:1176.
4. Emerick, R.J., H.R. King and L.B. Embry. 1972. J. Animal Sci. 35:901.
5. McLeod, M.N. and D.J. Minson. 1972. J. Br. Grassland Soc. 27:23.

06. McWard, G.W. et al. 1959. J. Animal Sci. 18:1059.

7. Miller, E.R. et al. 1964. J. Nutr. 82:64.
8. NRC. 1972. Nutrient Requirements of Laboratory Animals. NRC Pub. #10. Washington, D.C.
9. Nesheim, M.C. 1966. Proc. Cornell Nutr. Conf., pp 16-23.
10. Neatherly, M.W. 1972. J. Animal Sci. 34:1075.
11. Oslage, H.J. 1963. Z. Tierphysiol. Tierernahr. 17:357.
12. Roy, R.N. and B.C. Guha. 1958. Nature 182:319; 1689.
13. Van Campen, D. 1969. Digestive System. McGraw-Hill Encyclopedia of Sci. and Tech. McGraw-Hill Book Inc., New York.
14. Williams, R.J. 1971. Nutrition Against Disease. Pitman Pub. Co., New York.

Chapter 5 – Water

Water is seldom, if ever, classed as a nutrient even though it makes up about 1/2 to 2/3 of the body mass of adult animals and up to 90% of that of newborn animals. The importance of an adequate supply of potable water for livestock is well recognized, however, and it is currently receiving more emphasis in the quest to clean up polluted environments by improving the quality of water supplies so that most resemble that shown in Fig. 5-1.

From a functional point of view, water is extremely important to any biological organism, a fact easily substantiated by the sudden termination of productive functions and of life when insufficient water is available, as contrasted to relatively long-term life when the supply of other nutrients is restricted. Water has a number of functions that might be mentioned briefly. Many of the biological functions of water are dependent upon the property of water acting as a solvent for an extremely wide variety of compounds and also because many compounds ionize readily in water. Solvent properties are extremely important because most protoplasm is a mixture of colloids and crystalloids in water. In addition, water serves as a medium for transportation of slurries and semi-solid digesta in the GIT, for various solutes in blood and tissue fluids and cells, and in excretions such as urine and sweat. Water provides for dilution of cell contents and body fluids so that relatively free movement of chemicals may occur within the cells and in the fluids and GIT. Water is extremely important in regulation of body temperature because of its high specific heat, high thermal conductivity, and high latent heat of vaporization — properties that allow accumulation of heat, ready transfer of heat, and loss of large amounts of heat on vaporization. Lubrication of joints and cushioning of joints, organs in the body cavity, and of the central nervous system by cerebral-spinal fluids are important functions. In addition, water provides the base media for conduction of sound in the middle ear as well as aiding in transmission of the other special senses.

Figure 5-1. An example of a clear, unpolluted stream which will supply an adequate quantity of good quality water for livestock. Courtesy of Dow Chemical Co.

Body Water

Water content of the animal body varies consiuerably; it is influenced over the long term by age of the animal and the amount of fat in the tissues. Water content is highest in fetuses and in newborn animals, declines rapidly at first, and then slowly declines to adult levels. When body water is expressed on the basis of the fat-free body, the water content is relatively constant for many different animal species, ranging from 70-75% of fat-free weight, with an average of about 73%. As a result of this constant relationship, the composition of the animal body can be estimated with reasonable accuracy if either the fat or water content is known. Body water can be estimated with various dyes or tritium, by administering them intravenously and determining the amount of dilution of the tritium or dye; fat content of the tissues may be calculated by the formula:

$$\text{percent fat} = 100 - \frac{\text{percent water}}{0.732}$$

The greatest amount of water in the body tissues will be found as intracellular fluids, perhaps accounting for 40% or more of total body weight. Most of the intracellular water is found in muscle and skin. Extracellular water is found in interstitial fluids, which occupy spaces between cells, blood plasma, and miscellaneous other fluids such as lymph, synovial, and cerebrospinal fluids. Extracellular water accounts for the second largest water "compartment," or roughly 1/3 of the total body water of which about 6% is blood plasma water.

Most of the remaining body water will be found in the contents of the GIT and urinary tract. The amount present in the GIT is quite variable, even within species, and is greatly affected by type and amount of feed consumed. As indicated, body water tends to decrease with age and has an inverse relationship with body fat. Body water is apt to be higher in lactating cows than in dry cows (probably less body fat also), and extracellular water is greater in young male calves than in female calves of the same age. It might be noted that water easily passes through most cell membranes and from one fluid compartment to another. Passage from compartment to compartment is controlled primarily by differences in osmotic or hydrostatic pressure gradients.

Physiological abnormalities or disease (fever, diarrhea) may result in body dehydration or in retention of excess water in the tissues (faulty blood circulation or adrenal gland activity). The result is, of course, that normal values for body water may not be applicable to these situations.

Water Turnover

Water turnover is a term used to express the rate at which body water is excreted and replaced in the tissues. Tritium-labeled water has been used to estimate turnover time in a variety of different species. In cattle, a typical half-life value (time for 1/2 of tritium to be lost) is about 3.5 days. Nonruminant species probably have a more rapid turnover because they have less water in the GIT. Those species which tolerate greater water restriction (camel, sheep) also have lower turnover rates than species which are less tolerant to water restriction (horse, European cattle). Water turnover is greatly affected by climatic factors such as temperature and humidity or by ingestion of material, such as NaCl, which increases urinary excretion.

Water Sources

Water available to an animal's tissues is derived from (a) drinking water, (b) water contained in or on feed, (c) metabolic water produced by oxidation of organic nutrients, (d) water liberated from polymerization reactions such as condensation of amino acids to peptides, and (e) preformed water associated with body tissues which are catabolized during a period of negative energy balance. The importance of these different sources differs from animal species to species, depending upon their diet and habitat and on their ability to conserve body water. Some species of desert rodents and antelopes are said not to require drinking water except in rare situations. The ability to survive on such limited water intake does not apply to most avian or mammalian species, however. Information on lower forms of animals is very sparse.

The water content of feedstuffs consumed by animals is highly variable. For example, in forage it may range from a low of 5-7% to as high as 90% or more in lush young grass. Likewise, precipitation and dew on ingested feed may, at times, account for a very substantial amount of water consumption.

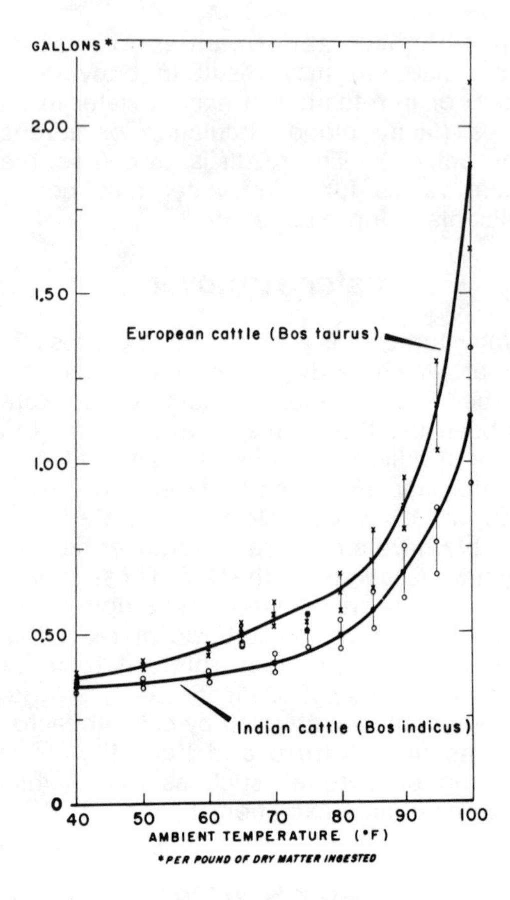

Figure 5-2. Water requirements of European and Indian cattle as affected by increasing temperatures. From Winchester and Morris [7].

Metabolic water from oxidation of organic nutrients in the tissues may also be quite variable, depending upon the diet consumed. Oxidation of carbohydrates produces about 60 g of water/100 g of carbohydrate; fat yields about 108 g and protein about 42 g/100 g. Studies with domestic cattle and horses indicate that metabolic water may account for about 5-10% of total water intake. It should be noted that ingestion of these organic nutrients results in increased respiration, urinary excretion, and heat dissipation. In the case of protein, the end product of catabolism is urea in most mammalian species. Large amounts of water are required for dilution and excretion via the kidney and the amount of water derived from oxidation is not sufficient to meet the excretory demands. Consequently, ingestion of protein really has a negative effect on conservation of water. With respect to fats, Schmidt-Nielsen (5) has shown, in dry climates at least, that ingestion of fat results in less net metabolic water than ingestion of carbohydrates. This results because fats are less oxidized compounds than carbohydrates, so more respiration is required to take in oxygen, and water loss from respiration is greater when fats are consumed. The overall result is that carbohydrates supply more net metabolic water than either proteins or fats.

Water Losses

Losses of water from the animal body occur by way of urine, feces, so-called insensible water — lost via vaporization from the lungs and dissipation through the skin, and by sweat from the sweat glands in the skin during warm or hot weather.

Water excreted via urine acts as a solvent for excretory products excreted from the kidney. Some species have greater ability to concentrate urine than others. Sometimes this is related to the type of compound excreted. For example, avian species excrete primarily uric acid rather than urea as an end produ tt of protein metabolism. These species excrete urine in semi-solid form with only small amounts of water. Mammalian species, however, cannot concentrate urine to nearly such an extent.

The kidney of most species has great flexibility in the amount of water that may be excreted. Minimal amounts required for excretion (called osligatory water) are usually gre rtly exceeded, except when water intake is restricted. Consumption of excess water during periods of heat stress or consumption of diuretics such as caffeine and alcohol may greatly increase kidney excretion of water.

Fecal losses of water in man are usually about 7-10% of urinary water. In ruminant species, such as cattle, fecal water loss usually exceeds urinary losses. Other species tend to be intermediate. Animals that consume fibrous diets usually excrete a higher percentage of total water via feces, and those that form fecal pellets usually excrete drier feces and, presumably, are more adapted to drier climates and more severe water restriction.

Insensible water losses account for a relatively large amount of total loss, particularly at temperatures that do not induce sweating or in species which do not sweat. Air inhaled into the lungs may be very dry, but is about 90% saturated with water when exhaled.

During periods of hyperventilation that occur with hot temperatures, loss from the lungs is greatly increased. Insensible water loss through the skin is relatively low.

Loss of water via sweat may be very large in species such as man and horses whose sweat glands are distributed over a large portion of the body surface. Sweating is used as a means of dissipating body heat and is said to have an efficiency of about 400% as compared to respiratory heat loss. Heat-tolerant species generally have well-developed sweat glands. This is one explanation why *Bos indicus* cattle are more heat tolerant than *Bos taurus* breeds.

Regulation of Drinking

The regulation of drinking is a highly complex physiological process (1, 4). At times, it is induced as a result of dehydration of body tissues. Drinking may also occur, however, when there is no apparent need to rehydrate tissues. When an animal is thirsty, salivary flow is usually reduced and dryness of mouth and throat may stimulate drinking — a relationship that may, indirectly, be traced to a decrease in plasma volume. Other information indicates that salivary flow is not a critical factor in initiation of drinking. Oral sensations apparently are involved that may be influenced by osmotic receptors in the mouth. For example, dogs with esophageal fistulas (which do not allow water to enter the stomach) will stop drinking after shamdrinking a more or less normal amount of water. However, the shamdrinking will be repeated again in a few minutes. There is ample evidence to indicate that passage of water through the mouth is required for a feeling of satiety, because placing water in the stomach by a tube leaves animals restless and unsatisfied.

Water Requirements

Water requirements for any class or species of healthy animals are exceedingly difficult to specify except in very specific situations. This is so because numerous dietary and environmental factors affect water excretion and because water is so important in regulation of body temperature. Other factors, such as ability to conserve water, differences in activity, gestation, lactation, and so on, compound the problem when different classes or species of animals are compared. As a result of these different factors relatively little effort has been made to quantify water needs, except in a few specific situations.

It is well recognized that water consumption is related to heat production and, sometimes, to energy consumption. At environmental temperatures that do not result in heat stress, there is a good linear relationship between dry matter consumption and water consumption. However, when the temperature reaches stressing levels, feed consumption is apt to decrease while water consumption is greatly increased. One example of this is shown in Fig. 5-2. Note in the figure that water consumption/unit of feed consumption by cattle goes from about 0.35 gal (3.0 lb)/lb of dry matter consumed at 40°F to about 1.9 gal (16.3 lb) at 100°F for *Bos taurus* cattle. When expressed as a % of body weight, non-heatstressed non-lactating cattle may drink on the order of 5-6% of body wt/day. Water consumption may increase to 12% + of body wt/day when heat stressed. Seasonal differences in confined cattle primarily reflect temperature stress. Recent data on feedlot steers (3) show average daily winter consumption to be 19 liters of water as opposed to 31 liters in the summer.

Dietary and Environmental Factors

Dietary Factors

As mentioned, dry matter intake is highly correlated to water intake at moderate temperatures. There is also evidence to show that water content of feed consumed affects total water intake. This is, however, primarily a factor in comsumption of excess water when forage is very lush with a high water content. Other dietary factors affecting water intake include consumption of high levels of protein, but here the effect is because of greater required urinary excretion. An increased intake of fat may also increase water intake, and consumption of silages tends to increase intake and urinary excretion (Table 5-1).

There is ample evidence that consumption of NaCl or other salts will greatly increase consumption and excretion of water by different species. Some salts may cause diarrhea and greater fecal excretion of water, but those, such as NaCl, that are almost completely absorbed, result in much greater urinary excretion, and tissue dehydration occurs if water is not available.

Environmental Factors

High temperature, as mentioned previously is the major factor causing increased consumption of water. Associated with heat stress is high humidity, which also increases the need for water as high humidity at a given temperature puts more heat stress on the animal because heat losses — resulting from evaporation of water from the body surface and lungs — are reduced.

In confined animals, design and accessibility of watering containers influences intake as does cleanness of the containers. In range animals, the distance that must be traveled between water and forage affects the frequency of drinking and the amount consumed, i.e., the greater the distance the less frequently the animal drinks and the less it consumes.

Water Quality

Water quality may be an important factor in obtaining adequate consumption of water and feed. Herrick (2) points out that most domestic animals can tolerate a total dissolved solid concentration of 15,000-17,000 mg/l., but production is apt to be reduced at this concentration. Water classified as good should have less than 2,500 mg/l. of dissolved solids. Cattle and sheep can tolerate 1% NaCl in water, but much more than this is apt to be detrimental, especially with warm summer temperatures. With sulfates, a level over 1 g/l. may cause diarrhea and, in the case of nitrates, levels of 100-200 ppm are potentially toxic. More information is needed on water pollutants such as sewage and industrial wastes with respect to animal utilization and consumption.

Water Restriction

Fluid intake by animals is intermittent, even more so than food, but the loss of water from the body is continuous, although variable. Thus, the body must be able to compensate in order to maintain its physiological functions. The most noticeable effect of moderate restriction is reduced feed intake and reduced productivity. With more severe restriction, weight loss is rapid as the body dehydrates. The dehydration is accompanied by increased excretion of N and electrolytes such as Na^+ and K^+. Water restriction causes more severe or quicker responses when temperatures are stressing. Animals vary greatly in the amount of dehydration which they can withstand; the camel is an example of one that can withstand weight loss of about 30% or more. Most mammals cannot survive such severe dehydration. With moderate restriction most species show some adaptation and can partially compensate for it by reducing excretion.

Water intoxication may occur in some species as a result of sudden ingestion of large amounts. As an example, calves are susceptible and death may result from a slow adaptation of the kidney to the high water load.

Table 5-1. Effect of rations and level of feeding on water intake of Holstein heifers.[a]

Item[b]	Roughage fed and level of feeding			
	Hay		Silage	
	Ad lib.	Maintenance	Ad lib.	Maintenance
Dry matter intake, kg/100 kg BW	2.06	1.24	1.70	1.15
Feed water, kg/kg feed DM	0.11	0.12	3.38	3.38
Water drunk, kg/kg feed DM	3.36	3.66	1.55	1.38
Total water, kg/kg feed DM	3.48	3.79	4.93	4.76
Urine, kg/kg feed DM	0.93	1.14	1.85	1.68

[a] From Waldo et al (6)

[b] BW=body weight; DM=dry matter

References Cited

1. Chew, R.M. 1965. In: Physiological Mammalogy. Vol. 2. Academic Press, New York.
2. Herrick, J.B. 1971. Feedstuffs 43(8):28.
3. Hoffman, M.P. and H.L. Self. 1972. J. Animal Sci. 35:871.
4. Roubicek, C.B. 1969. In: Animal Nutrition and Growth. Lea & Febiger, Philadelphia.
5. Schmidt-Neilsen, K. 1964. Desert Animals, Physiological Problems of Heat and Water. Oxford Univ. Press.
6. Waldo, D.R., R.W. Miller, M. Okamoto and L.A. Moore. 1965. J. Dairy Sci. 48:1473.
7. Winchester, C.F. and M.J. Morris. 1956. J. Animal Sci. 15:722.

Chapter 6 – Protein

Protein is a most important component of animal tissue in that it is the nutrient in highest concentration in muscle tissues of animals. All cells synthesize protein for part or all of their life cycle and without protein synthesis life could not exist. Except in animals whose intestinal microflora can synthesize protein from non-protein N sources, protein must be provided in the diet to allow normal growth and other productive functions. All cells contain protein and rapid cell turnover occurs, especially for those such as the epithelial cells of the intestinal tract. Consequently, providing replacement protein from the diet is essential to meet these turnover requirements in addition to those needed for growth and other productive functions. The % of protein required in the diet of young growing animals is highest, but declines gradually to maturity when only enough protein to maintain body tissues is required. Productive functions such as pregnancy and lactation increase the requirement because of increased output of protein in products of conception and in milk and because of an increased metabolic rate associated with the productive function.

Structure

Proteins vary widely in chemical composition, physical properties, size, shape, solubility, and biological functions. All proteins have one common property: Their basic structure is made up of simple units, amino acids. More than 20 amino acids occur naturally of which up to 10 are required in the diet of animals (12 for broiler chicks), because tissue synthesis is not adequate to meet metabolic needs. The basic structure of amino acids is illustrated by glycine, the simplest amino acid:

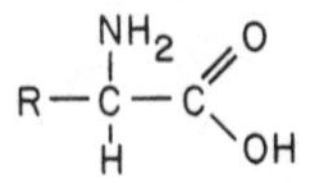

General structure for amino acids

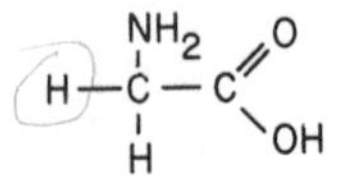

Glycine

The essential components are a carboxyl group (-COOH) and an amino group (NH_2) on the C atom adjacent to the carboxyl group. The general structure representing all amino acids is shown on the left above where R is the remainder of the molecule attached to the C atom associated with the α-amino group of the amino acid. The chemical structure of 20 of the amino acids is given in Fig. 6-1. Amino acids that are either not synthesized in animal tissues or not in sufficient amounts to meet metabolic needs are termed **essential** or **indispensable**, and those generally not needed in the diet because of adequate tissue synthisis are termed **nonessential** or **dispensable**.

The synthesis of protein from amino acids involves joining of individual amino acids to form long chains. The length of the chain and the arrangement of amino acids in the chain are basically the two factors determining the composition of the protein.

The linkage of the amino group of one amino acid to the carboxyl group of another is called a peptide bond, illustrated as follows:

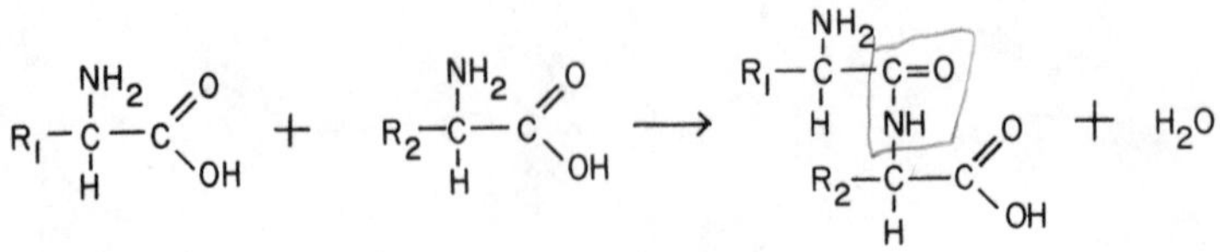

Figure 6-1

Chemical structure of amino acids

Aliphatic

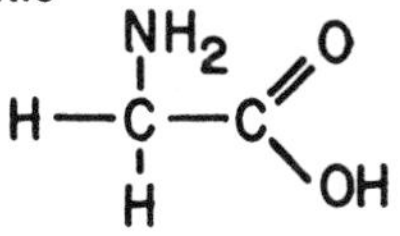

Glycine
[amino acetic acid]

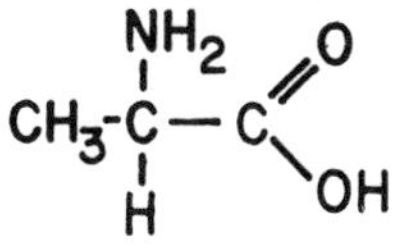

Alanine
[α-amino propionic acid]

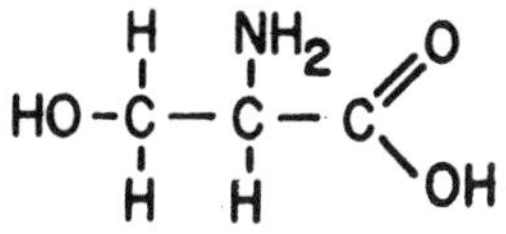

Serine
[β-hydroxy α-amino propionic acid]

Threonine
[α-amino β-hydroxy-N-butyric acid]

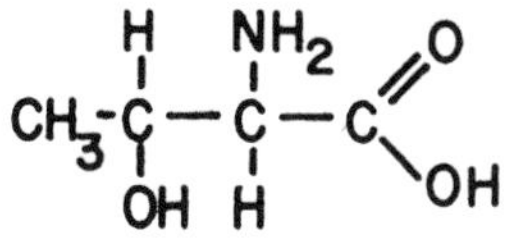

Valine
[α-amino isovaleric acid]

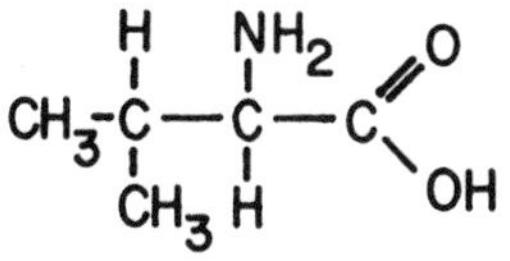

Leucine
[α-amino isocaproic acid]

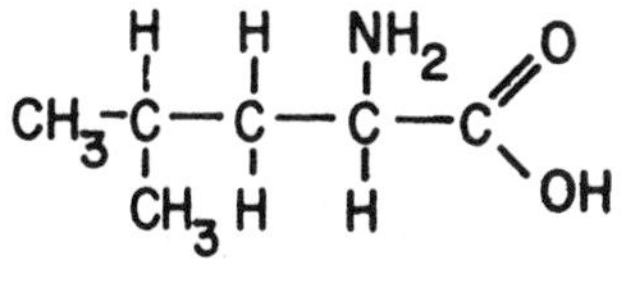

Aromatic

Isoleucine
[β-methyl α-amino valeric acid]

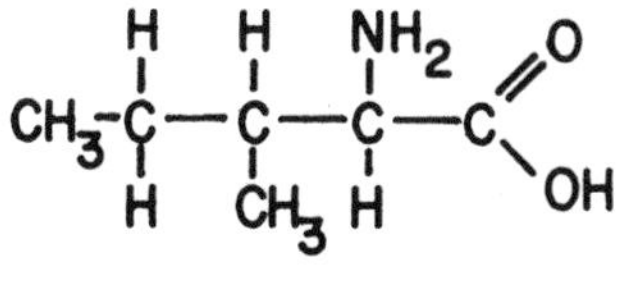

Phenylalanine
[β-phenyl α-amino propionic acid]

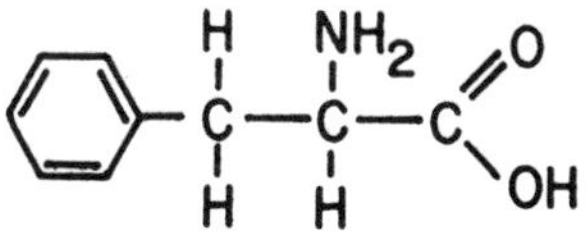

Tyrosine
[β-para hydroxy phenyl α amino propionic acid]

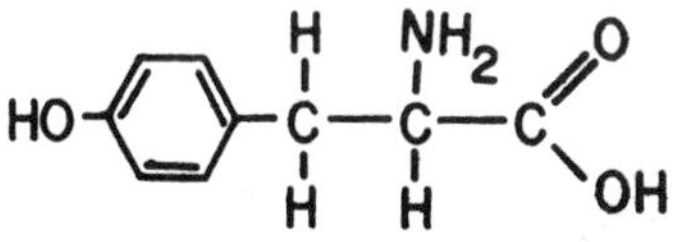

Sulfur-containing

Cysteine
[β-thiol α-amino propionic acid]

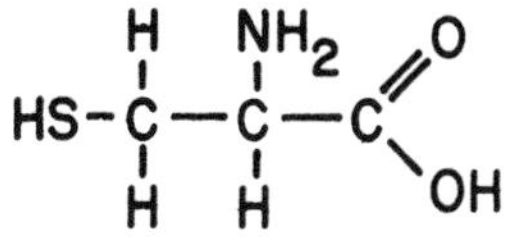

Cystine
di-[β-thiol α-amino propionic acid]

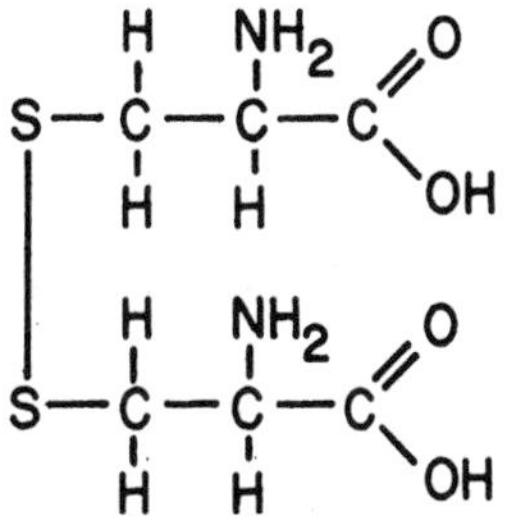

Methionine
[gamma methyl thiol α-amine-N-butyric acid]

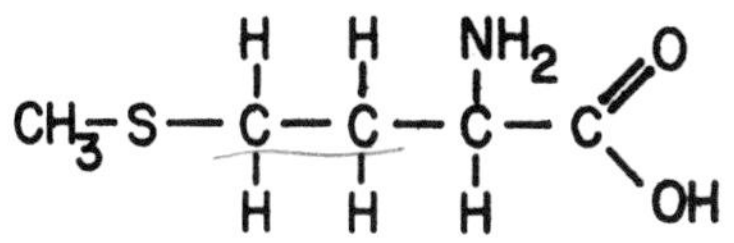

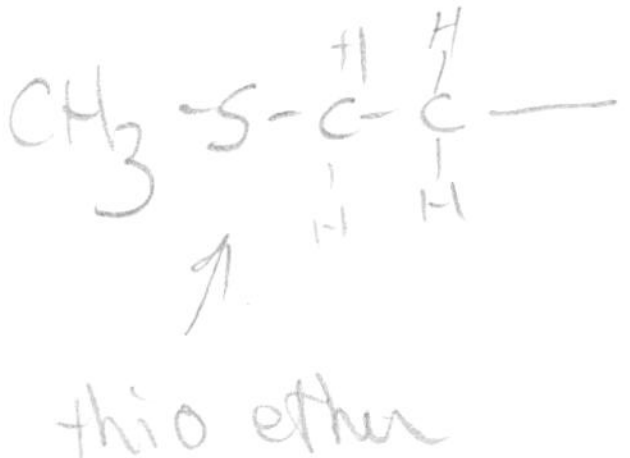

thio ether

Heterocyclic

Tryptophan
[β-3-indole-α-amino propionic acid]

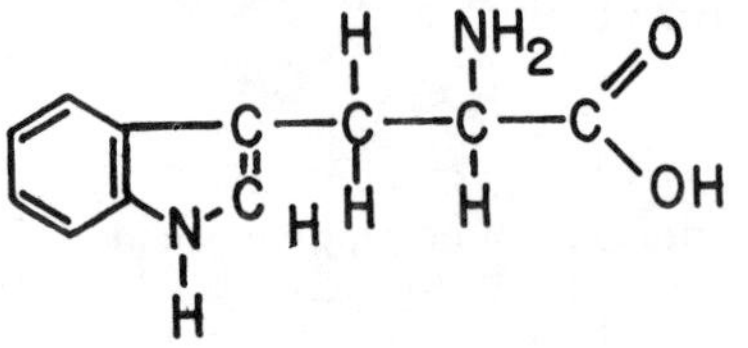

Proline
[pyrolidine-2-carboxylic acid]

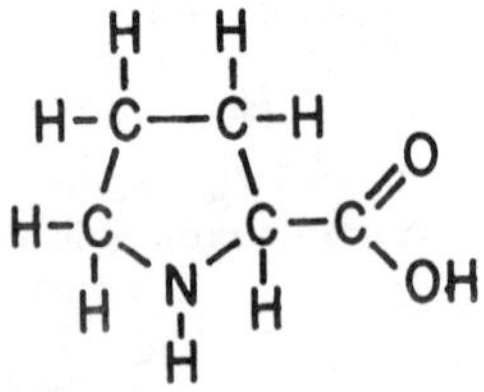

Hydroxy proline
[4-hydroxy pyrolidine-2-carboxylic acid]

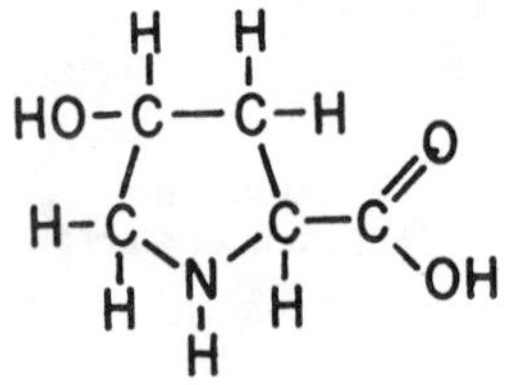

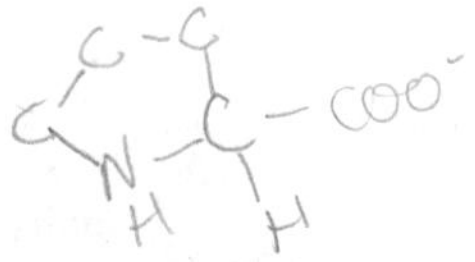

Acidic

Aspartic acid
[α-amino succinic acid]

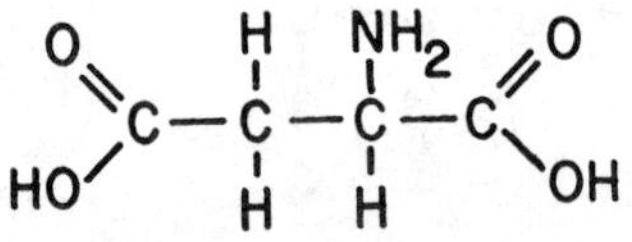

Asparagine
[α-amino succinamic acid]

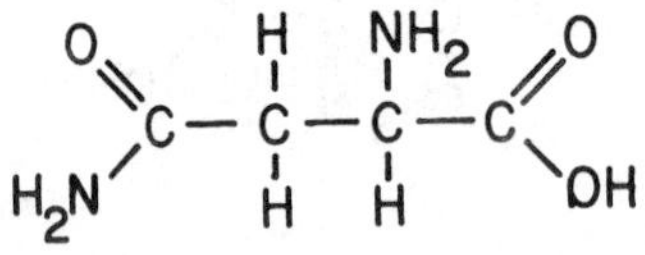

Glutamic acid
[α-amino glutaric acid]

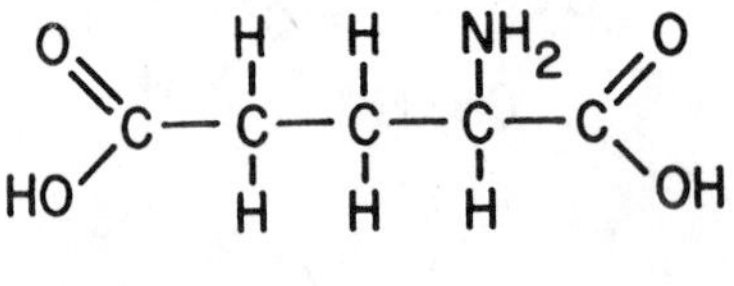

Glutamine
[α-amino glutaramic acid]

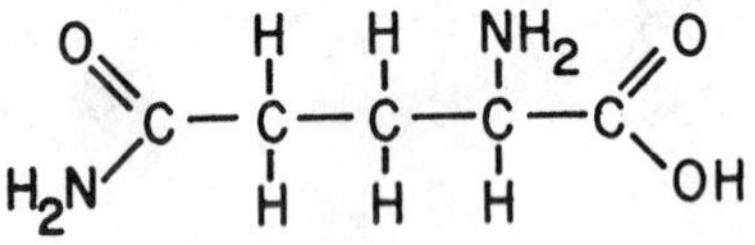

Basic

Arginine
[β-guanidino α-amino valeric acid]

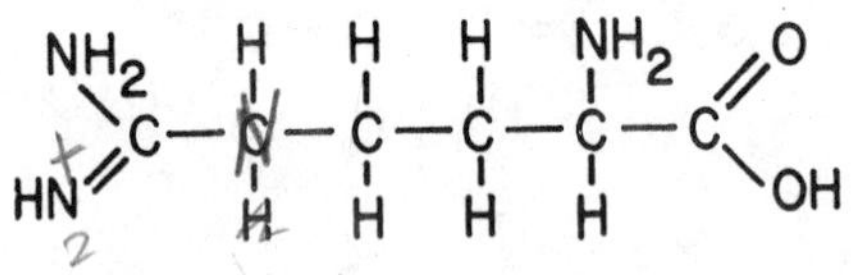

Histidine
[β-imidazol α-amino propionic acid]

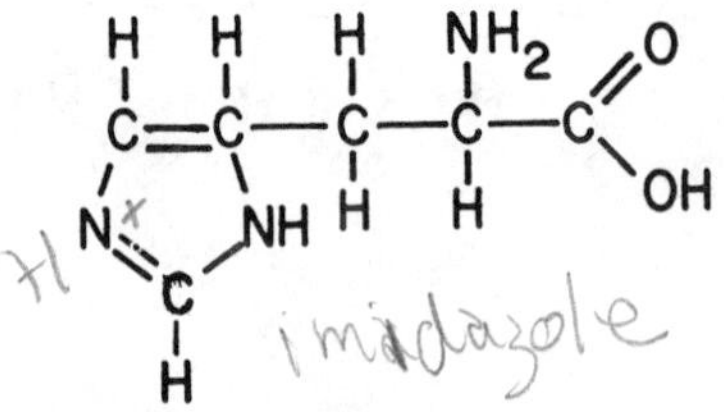

Lysine
[α, epsilon diamino caproic acid]

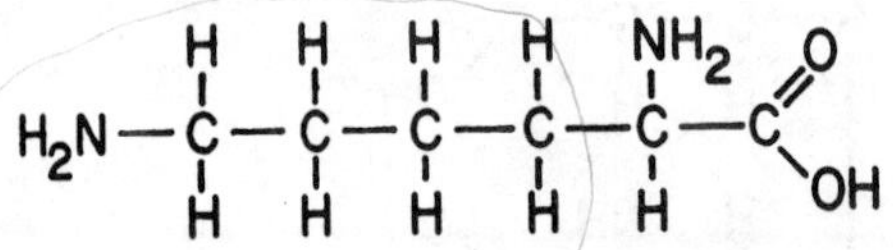

The dipeptide alanyl-glycine would be formed as follows:

L-alanine + Glycine → Alanyl-glycine

All naturally occurring amino acids are in the L-configuration. Synthetic amino acids are usually found as a racemic mixture of L- and D-isomers. The L-form is biologically the more active, with only a few exceptions.

The schematic representation of L- and D-amino acids is as follows:

L-form D-form

Elongation of the chain by addition of amino acids proceeds from tripeptides to polypeptides and, eventually, to a complete protein molecule of specific amino-acid content and sequence. Biochemists have been able to isolate and determine the amino acid composition of hundreds of proteins and, in several cases, the exact structure of the protein is known. For example, the structure of a ribonuclease, an enzyme that digests RNA, is diagrammed in 2 dimensions in Fig. 6-2 (34). This protein has 587 C atoms, 909 H atoms, 197 O atoms, 171 N atoms and 12 S atoms. It is a long protein with 4 S-S bridges (cystine-cystine) in the molecule. The molecule contains 19 different amino acids and a total of 124 amino acids. This protein is a very small one compared to most natural proteins. The molecular weight of proteins ranges from 35,000 to 500,000 with the number of amino acid residues from 350 to 5,000.

The amino acid compositions of some common proteins of plant and animal origin are shown in Table 6-1. Egg albumin is considered the most nearly perfect protein for meeting animal needs because of its amino acid composition and high digestibility. Some fractions of corn endosperm are listed to illustrate the principle that a single feedstuff may be composed of several distinct proteins

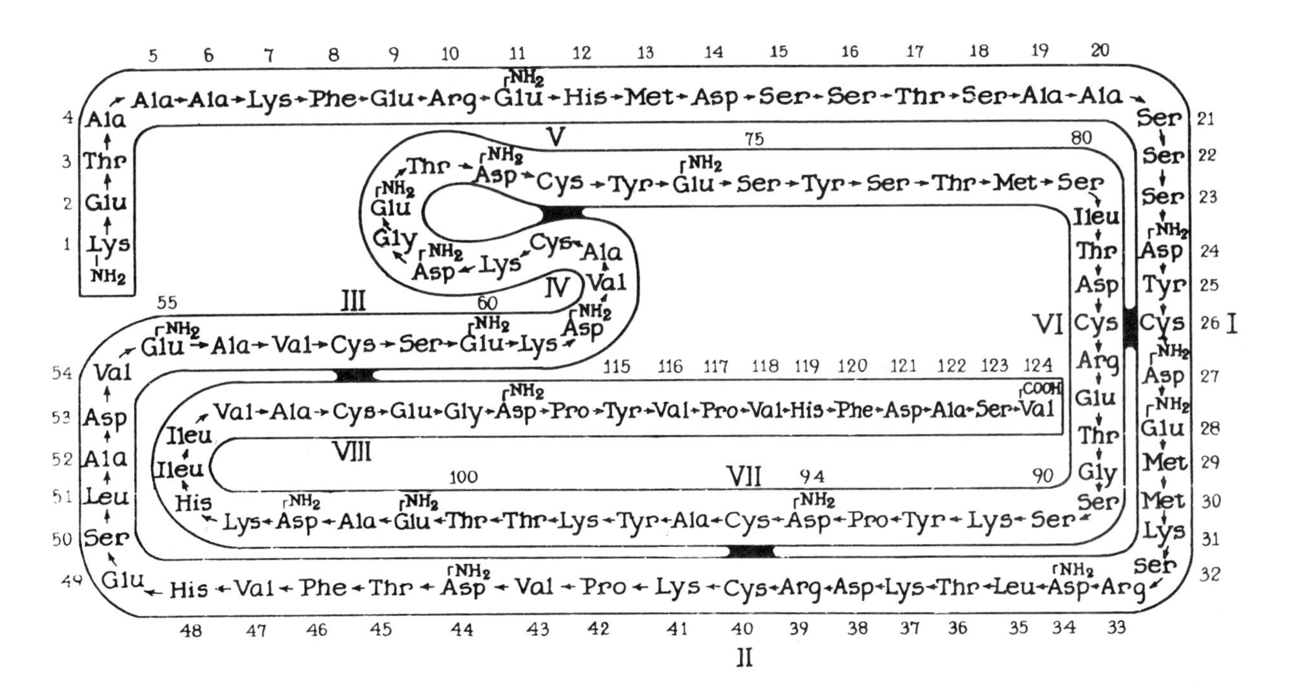

Figure 6-2. Structure of ribonuclease A from bovine pancreas. Courtesy of W.H. Stein. From Smyth, D.G., W.H. Stein and S. Moore [34].

and the adequacy of its protein will depend on the total amino acid mixture supplied by the individual fractions. Zein is notably low in lysine and tryptophan. A genetic mutation discovered by Purdue University scientists (20) involves a gene for high lysine and tryptophan (referred to as the "opaque-2" gene). The proportion of zein in opaque-2 corn is lower and that of glutelin and other protein fractions is higher than in common corn, resulting in a corn with a more favorable balance of amino acids.

Some amino acids are present in animal tissues in higher concentrations than in plant tissues. For example, hydroxyproline is absent from the plant proteins listed in Table 6-1, but it is abundant in animal connective tissue, especially collagen. The ratio of essential to nonessential amino acids in a protein may be indicative of its biological value for animals, but a more sensitive measure is the balance among the essential amino acids. The essential amino acid composition of proteins is as variable as the number of proteins present in nature.

All proteins can be classified on the basis of their shape and solubilities in water, salt, acids, bases and alcohol. Such a broad classification includes the following:

A. Globular proteins
   - Albumins - Soluble in water and coagulable by heat.
   - Globulins - Soluble in dilute neutral solutions of salts of bases and acids and coagulable by heat.
   - Glutelins - Soluble in dilute acids or bases.
   - Prolamins (gliadins) - Soluble in 70-80% ethanol.
   - Histones - Soluble in water; characterized by a large excess of basic amino acids; most are combined with nucleic acids in animal cells.
   - Protamines - Soluble in water, not coagulable by heat; resemble histones in having excess of basic amino acids, but of lower molecular weight.

B. Fibrous proteins
   - Collagens - Insoluble in water, resistant to digestive enzymes, soluble in dilute acids or bases to become gelatin.
   - Elastins - Similar to collagens but cannot be converted to gelatin.
   - Keratins - Insoluble in water, resistant to digestive enzymes, contain up to 15% cystine.

C. Conjugated proteins

In addition to the unconjugated proteins described above, a wide array of compounds containing protein plus other components exists naturally.

Protein-Lipid Complexes

The existence of an egg yolk phospholipid-protein complex was recognized more than a century ago; now hundreds of examples of lipid-protein complexes are known. Electrostatic bonds, hydrogen bonds, hydrophobic bonds and other forces, apparently contribute to the stability of lipoproteins (8). Myelin is a lipoprotein abundant in the central and peripheral nervous system as a sheath around nerve fibers. Peripheral and central nervous tissue contain about 80 and 35% myelin, respectively. Erythrocyte membranes contain mucolipids, phospholipids and loosely bound proteins. Mitochondria contain structural protein, a smaller amount of soluble matrix protein and about 30% lipids, mainly phospholipids. Soluble lipoproteins are a major component of blood lipids (31). Fatty acids and other lipids are absorbed to the surface of proteins. The amino acid composition of the serum lipoproteins is similar to that of other serum proteins. The protein content of plasma lipoproteins ranges from 2% in chylomicrons to about 50% in high-density lipoproteins (31).

A very important type of protein-lipid complex is represented by the membrane proteins of animal cells. Biological membranes act as a permeability barrier, transport substances across the boundary between the interior and exterior of the cell, act as supports for catalytic functions, and probably perform other important, though less well-defined functions (11). These membranes are composed of proteins, lipids and carbohydrates in various proportions. The chemical composition of some important cell membranes is summarized in Table 6-2 (11).

Protein-Carbohydrate Complexes [glycoproteins]

Proteins can complex with carbohyrate either loosely or tightly. The formation of these glycoproteins arises from the acceptance of sugars by amino acid residues in the polypeptide chain (17). Protein complexes of sulfated polysaccharides occur in 3 types designated chondroitin sulfate A, B and C (11). Cartilage, tendon and skin are high in chondroitin sulfates complexed with protein. Mucoproteins are complexes of protein with

Table 6-1. Amino acid composition of some plant and animal products.

	Egg	Corn endosperm[a]											Bovine Intra-muscular
		Albumin		Globulin		Zein		Glutelin					
	albumin	N[b]	O[c]	N	O	N	O	N	O	Beef[d]	Pork[d]	Lamb[d]	collagen[e]
Alanine	5.7	9.8	7.9	8.4	7.3	10.9	10.5	4.5	4.7	6.4	6.3	6.3	10.4
Arginine	5.9	11.6	11.7	9.6	8.9	2.1	2.3	5.8	5.3	6.6	6.4	6.9	4.5
Aspartic acid	9.2	11.1	10.8	8.2	8.0	5.9	5.7	6.8	6.9	8.8	8.9	8.5	3.8
Cystine/2	3.0		5.3	1.8	1.8	1.0	1.4			1.4	1.3	1.3	
Glutamic acid	15.7	16.5	13.2	19.2	18.2	27.4	24.9	12.9	14.6	14.4	14.5	14.4	7.7
Glycine	3.2	8.8	7.9	5.3	5.5	1.6	2.0	3.3	3.8	7.1	6.1	6.7	32.8
Histidine	2.4	3.2	2.6	3.4	3.8	1.4	1.4	3.8	4.2	2.9	3.2	2.7	0.6
Hydroxyproline	0	0	0	0	0	0	0	0	0				10.5
Isoleucine	7.1	4.6	3.9	4.2	4.3	4.2	4.6	3.4	3.4	5.1	4.9	4.8	1.3
Leucine	9.9	6.3	6.3	13.2	10.7	22.4	21.2	8.1	8.6	8.4	7.5	7.4	2.5
Lysine	6.4	6.4	5.5	4.4	4.6	0.1	0.2	3.7	3.6	8.4	7.8	7.6	2.3
Methionine	5.4		1.6	2.0	1.9	1.7	1.3	1.4	1.1	2.3	2.5	2.3	0.6
Phenylalanine	7.5	2.9	4.6	5.8	6.3	7.6	8.0	3.8	3.8	4.0	4.1	3.9	1.4
Proline	3.8	6.6	5.7	7.2	7.6	10.7	10.5	8.7	10.1	5.4	4.6	4.8	11.8
Serine	8.5	7.0	5.4	5.9	5.4	6.2	5.6	3.7	3.7	3.8	4.0	3.9	4.0
Threonine	4.0	5.4	4.7	4.0	4.1	3.2	2.9	3.3	3.4	4.0	5.1	4.9	1.8
Tryptophan	1.2	1.1	3.0	0.3	1.3		0.2		0.5	1.1	1.4	1.3	
Tyrosine	3.8	3.2	5.0	4.7	4.9	5.6	6.0	2.9	3.1	3.2	3.0	3.2	0.4
Valine	8.8	12.6	6.0	5.8	7.1	4.5	3.4	5.4	5.6	5.7	5.0	5.0	2.3

[a] Concon (7). Expressed as g/100 g protein. [b]Normal maize. [c]Opaque-2 maize. [d]Schweigert and Payne (32). Expressed as % of crude protein (6.25). [e]McClain et al (18). Expressed as residues/1000 residues.

the amino-sugars, glucosamine and galactosamine; the hexoses, mannose and galactose; the methyl pentose, fucose and sialic acid. Serum cholinesterase, an enzyme, is a mucoprotein as is the hormone, gonadotrophin. Mucus secretions contain abundant amounts of mucoproteins. Ovalbumin is a mucoprotein containing 3 glucosamine and 5 mannose residues. Aspartic acid is apparently the amino acid immediately concerned in linkage to the carbohydrate part of the complex. Ovomucoid, the trypsin inhibitor in egg white, is distinguished by its high heat stability and high carbohydrate content. It contains about 14% hexosamine and 7% hexose.

Functions

Proteins perform many different functions in the animal body. Quantitatively, those that are most important function as components of cell membranes, in muscle and in other supportive capacities such as in skin, hair and hooves. In addition, blood serum proteins, enzymes, hormones and immune antibodies all serve important specialized functions in the body even though they may not contribute significantly to the total protein content of the body.

Tissue Proteins

Collagen. The molecule consists of a triple-helix, 2800 Å long and 15 Å diameter; molecules are arranged in parallel, quarter-staggered to give the characteristic banded appearance of collagen (2). It has a compact structure and great strength. Collagen content increases with aging of the animal and, due to its characteristic shrinkage on heating, the toughness of cooked meat from older animals is a well-known phenomenon.

Table 6-2. Chemical composition of some cell membranes.[a]

Membrane	Protein %	Lipid %	Carbohydrate %	Ratio of protein to lipid
Myelin	18	79	3	0.23
Blood platelets	33-42	51-58	7.5	0.7
Human red blood cell	49	43	8	1.1
Rat liver cells	58	42	5-10	1.4
Nuclear membrane, rat liver cell	59	35	2.9	1.6
Mitochondrial outer membrane	52	48	2.4	1.1
Mitochondrial inner membrane	76	24	1-2	3.2
Retinal rods, cattle	51	49	4	1.0

[a] From Guidotti (11)

Elastin. The molecule resembles denatured collagen and consists of long, randomly ordered polypeptide chains. It is rubber-like in its response to stretching and, when stretched to the elastic limit, it breaks more easily than collagen. Because of this, it is always found in combination with a large proportion of collagen, even in places where it is most effective in restoring a tissue to original shape or position as in ligaments and in artery walls. It is only a minor component of musculature.

Myofibrilar proteins. These are proteins of sarcoplasm, which is the material extracted from finely homogenized muscle with dilute salts. Most of the protein in this extract is in solution and contains more than 20 enzymes involved in muscle metabolism as well as mitochondrial fragments and particles of sarcoplasmic reticulum.

Contractile proteins. Three proteins — actin, tropomycin B, and myosin — take part in muscle contraction. Tropomycin B has no enzyme properties and does not combine in solution with either actin or myosin. Myosin is the major component of the thick filaments of striated muscle. Its most important feature is its enzyme activity as an ATPase. Actin, when extracted from myosin is in globular form. For it to remain in this form, ATP must be present. The globular form polymerizes on the addition of neutral salts to give chain-like molecules of fibrous actin with the simultaneous change of the bound ATP to ADP and liberation of a molecule of inorganic P. The reverse process occurs in the absence of ATP. The complicated interactions between the contractile proteins are beyond the scope of this discussion. Suffice it to say that the unique chemical structure of each of these proteins is vital to normal muscle metabolism.

Keratins. Proteins of hair, feathers, hoofs, claws, beaks, and horns are similar in composition in that they are resistant to acid, alkali and heat treatment, and especially, to breakdown by digestive enzymes.

Blood proteins. Albumin and a series of globulins(α, β, γ) along with thromboplastin, fibrinogen and hemoglobin, a conjugated protein, are the chief proteins of the blood. Typical composition of the serum proteins of the normal pig is shown in Table 6-3. In addition, blood contains a large array of conjugated proteins, including lipoproteins, enzymes and hormones. Each protein component of blood has a distinct composition and structure and specific physicochemical properties.

Enzymes. Animal cells contain innumerable enzymes each with a specific structure and distinct reactive group. Sometimes referred to as organic catalysts, all known enzymes are protein in nature and are relatively specific in their reactions. Some are hydrolytic (as the digestive enzymes), some are involved in other degradative metabolic reactions, and others are involved in synthetic processes.

Hormones. Hormones, like enzymes are produced in minute quantities by cells and have profound effects on metabolism, but unlike enzymes whose action is usually restricted within the same cell or in close proximity to the site of elaboration, many hormones characteristically are carried by the blood from the site of release to a target organ far removed from site

Table 6-3. Approximate serum protein concentrations of pigs of various ages.

	Newborn	6 weeks	3 months	1 year
Total serum protein, g/100 ml serum	2 to 3	4 to 5	5 to 6	7 to 8
Albumin, g/100 g serum protein	18	47	45	52
α-globulin, g/100 g serum protein	60	25	29	18
β-globulin, g/100 g serum protein	16	19	16	13
γ-globulin, g/100 g serum protein	6[a]	9	10	16
Albumin: Globulin	0.2	0.90	0.82	1.1

[a] At 24 hr (after sucking) the value is 45-48 g/100 g serum protein because of colostrum ingestion.

of release. Not all hormones are proteins, as, for example, the sex hormones, estrogen, testosterone and progesterone. Important protein hormones include insulin, growth hormone, gonadotrophic hormone, calcitonin and adrenal corticotrophic hormone.

Immune antibodies. Antibodies against specific infections can be obtained passively by placental transfer to the fetus from the blood of the dam, by ingestion and absorption from colostrum by the newborn, or by injection into the body of purified sources from other animals. Such antibody protection is termed passive immunity. Exposure of a susceptible animal to an antigen stimulates antibody protection. This type of antibody protection is termed active immunity. Antibodies acquired either passively or actively are protein and, as such, perform a vital role in protecting the animal against specific infections.

Metabolism

Protein metabolism must be considered in two phases: Catabolism (degradation) and anabolism (formation). Individual amino acids, which are the basic units required in metabolism by the animal, are present in the diet as intact proteins or polypeptides which must be hydrolyzed to their component amino acids before absorption. Thus, the conversion of dietary protein to tissue protein involves hydrolysis to amino acids in the GIT, absorption, and resynthesis into tissue proteins.

Hydrolysis of dietary proteins is accomplished by proteolytic enzymes elaborated by epithelial cells lining the lumen of the GIT and by the pancreas (see Ch. 3). The efficiency with which hydrolysis occurs determines the degree of absorption of individual amino acids and, hence, contributes to the nutritional value of the dietary protein. The other important factor contributing to nutritional value is the balance of essential amino acids. Even proteins easily hydrolyzed in the GIT do not have a high nutritional value if they have a deficiency or an imbalance of one or more amino acids. Proteins can be characterized nutritionally on the basis of both digestibility and utilization of amino acids after absorption.

Digestible protein refers to that disappearing from the feed as the feed passes through the GIT and ultimately appears in the feces. In the same way that apparent indigestible energy includes both unabsorbed residues and endogenous energy (that derived from body tissues), so also indigestible protein appears with the endogenous protein fraction of the feed to contribute to total fecal N. Thus, **apparent protein digestibility** of feed represents the difference between what is present in the feed and what appears in the feces and includes both unabsorbed and endogenous N. After absorption, amino acids are subject to further metabolism in the liver and other tissues which eventually results in deamination. This loss of N occurs in the mammal mainly as urinary urea N and in birds as uric acid. Because the degree of utilization of a feed protein depends not only on its digestibility but also on its utilization after absorption, protein sources can be characterized on the basis of their overall value, often termed biological value. **Biological value** (BV) is defined as that % of N absorbed from the GIT that is available for productive body functions. BV is derived from experiments involving measurement of total N intake and N losses in urine and feces. Formulas for the Thomas-

Mitchell method and a simpler method of determining BV are shown:

Thomas-Mitchell formula for BV [as %] =

$$\frac{\text{N intake - [fecal N-metabolic N] - [urinary N-endogenous N]}}{\text{N intake - [fecal N-metabolic N]}} \times 100$$

Simplified formula for BV [as %] =

$$\frac{\text{N intake - fecal N - urinary N}}{\text{N intake - fecal N}} \times 100$$

Because some N is lost in the feces as a result of endogenous losses (metabolic fecal N) and the urinary loss of N involves both excess dietary N and end-products of metabolism involving obligatory losses (endogenous N), the Thomas -Mitchell method of determining BV takes these metabolic and endogenous N losses into account. It provides an estimate of the efficiency of use of the absorbed protein for combined maintenance and growth. The BV of several plant and animal proteins for growing and adult rats are shown in Table 6-4. Egg protein is considered to have the highest BV of natural sources. Proteins that individually have very poor BV, when combined in correct proportions with other proteins, may yield a BV similar to that of a single high quality protein. This is exemplified in the use of a mixture of corn and soybean meal to supply the total amino acid requirement of growing pigs. Corn alone or soybean meal alone will not promote maximum growth.

Table 6-4. Biological value of proteins for growing and adult rats.

Protein	State of life	
	Growing	Adult
Egg albumin	97	94
Beef muscle	76	69
Meat meal	72-79	—
Casein	69	51
Peanut meat	54	46
Wheat gluten	40	65

[a] From Mitchell and Beadles (22)

Other measures of protein adequacy are the protein efficiency ratio (PER) and net protein utilization (NPU) or net protein value (NPV). PER is by definition the number of grams of weight gain produced by an animal/unit of protein consumed. Thus,

$$\text{PER} = \frac{\text{body wt gain, g}}{\text{protein consumed, g}}$$

Conventionally, this index is obtained by feeding laboratory rats, but the same calculations could be made for any animal species fed a particular protein mixture. It is important in any measure of protein utilization to maintain a low or marginal level of dietary protein, because even poor quality protein may allow reasonable growth and productive performance when fed at a higher level of total N than required by the animal. NPU (21) measures efficiency of growth by comparing body N content resulting from feeding a test protein with that resulting from feeding a comparable group of animals a protein-free diet for the same length of time. Thus,

$$\text{NPU} = \frac{\text{[Body N with test protein] - [body N with protein-free diet]}}{\text{Total N intake}}$$

A large number of values can be obtained over a brief test period of 1-2 wk.

When digestibility as well as BV data are used, a net protein value (NPV) can be computed, which is simply the product of the BV x the digestion coefficient. Thus, NPV = BV x digestion coefficient. This value, then, is corrected for either very low or very high digestibility and would seem to be a more useful value than BV.

With the development of automated methods for assay of free amino acid concentrations of blood plasma, using changes in amino acid patterns after ingestion of the test protein to assess protein quality is a possibility. Most studies of this type have yielded disappointing results, but the technique may be useful with improvement. Changes in ratio of nonessential to essential amino acids in the plasma in protein deficiency also offer encouragement for further refinements in this concept of protein evaluation.

All of the estimates of protein utilization described have their limitations and no one estimate is superior to all others under all conditions. If one is interested in simplicity and a minimum of facilities and analytical work, perhaps PER provides the best estimate of protein value for growth, because only weight gain and protein consumed in a particular time period of 2-3 wk are needed.

Absorption of Amino Acids

The intestinal epithelium is an effective barrier to diffusion of a variety of substances. Except in early postnatal life of mammals (in most species during the first 24 to 48 hr), when intact protein is absorbed by pinocytosis, little or no transfer of proteins, polypeptides or even dipeptides occurs across the intestinal epithelium. Overwhelming experimental evidence shows that amino acid absorption takes place by active transport. That is, the amino acid moves across the intestinal cell membrane against a concentration gradient requiring energy supplied by cellular metabolism. The naturally occurring L-forms of amino acids are absorbed preferentially to the D-forms, probably as a result of specificity of active transport systems. Neutral amino acids may compete with each other for transport. For example, a high concentration of leucine in the diet increases the requirement for isoleucine. Removal of the carboxyl group by formation of an ester, removal of the charge on the amino group by acetylation, or introduction of a charge into the side chain of a variety of amino acids destroys active transport, emphasizing the highly specific nature of the carrier system (39). The 3 basic amino acids — ornithine, arginine and lysine — share the same transport system along with cystine. Arginine, cystine and ornithine inhibit lysine transport, and arginine, lysine and ornithine inhibit cystine transport. Some neutral amino acids inhibit basic amino acid transport; for example, methionine inhibits lysine transport. Apparently, basic amino acids do not inhibit neutral amino acid transport. Proline and hydroxyproline appear to share a common transport system along with sarcosine and betaine, and have a high affinity for the neutral amino acid transport system as well as for their own. It is unknown if a common transport system for glycine, alanine and serine exists in intestinal epithelium as it does for many microorganisms (26).

Biological Availability of Amino Acids

Several methods exist for the estimation of amino acid availability to the animal from a variety of feedstuffs. Meade (19) categorized these methods as follows: Microbiological assay; fecal analysis; growth assay; and plasma free amino acids.

In the microbiological assay the test material is digested by enzymatic, acid, or alkaline hydrolysis, and the amino acid composition of the hydrolysate is evaluated by growth of microorganisms which have specific amino acid requirements. The rate and degree of release of amino acids from the protein is taken as an index of the availability to the animal.

The fecal analysis method is a balance trial in which % amino acid availability is estimated by the following formula (3).

$$\%\ \text{amino-acid [AA] availability} = \frac{\text{Total AA intake} - \left(\text{Total fecal AA, protein diet}\right) - \left(\text{Total fecal AA, protein - free diet}\right)}{\text{Total AA intake}} \times 100$$

In the chick, in which urinary and fecal N are excreted together, this method combines urinary and fecal losses (3). However, the same formula can be applied to mammals by collecting feces — if the assumption is made that urinary loss of free amino acids is negligible. This method allows calculation of availability of individual amino acids. Values greater than 90% availability are commonly obtained. The absorbability of amino acids from different protein sources by rats and pigs appears to be remarkably constant (10, 15).

The growth assay can be used to study availability of one or a series of amino acids from a test protein. The procedure is to feed the test diet alongside diets of known amino acid availability such as a crystalline amino acid diet. By comparing the growth curve of animals fed the test protein with that of animals fed the amino acid diet with the amino acid of concern supplied at several increment levels, it is possible to estimate the proportion of the amino acid in the test protein utilized for growth.

Measurement of free amino acids in the plasma of animals at intervals following a meal of the test protein allows a means of estimating the availability of one or more amino acids. Stockland and Meade (35) found differences in the availability to the rat of isoleucine, threonine and phenylalanine from several sources of meat and bone meal. Using the same method, they found isoleucine, methionine and threonine to vary in availability from different samples of meat and bone meal for the pig. The basis for judging availability is the change in relative concentrations of amino acids in the plasma after a meal of the test protein. The duration of fasting and the selection of appropriate intervals for blood sampling are of importance and must be established for each species and for the conditions of a particular experiment.

A chemical method for estimating lysine availability from animal protein feeds was developed by Carpenter (6). The method suffers because of lysine destruction during prolonged hydrolysis of the protein and because of the reaction of lysine with the carbohydrates of many commonly fed feedstuffs. The applicability of this and modifications of this method to commercial feed formulation remains in question.

Probably the greatest single factor affecting amino acid availability from feedstuffs is proper heating of feedstuffs during processing. Overheating depresses availability of all amino acids, especially lysine. An ideal method for quick, accurate prediction of amino-acid availability in a particular feedstuff applicable for animals under a variety of conditions still awaits development.

Synthesis of Amino Acids and of Other Metabolites from Them

The ability of animal tissues to synthesize amino acids from other compounds is the basis for their classification as essential or nonessential. Of the 10 amino acids listed as essential for most simple-stomached animals, several can be replaced by their corresponding α-hydroxy or α-keto analogs, illustrating that the carbon skeleton is what the animal is unable to synthesize. For example, the α-keto analogs of arginine, histidine, isoleucine, leucine, methionine, phenylalanine, tryptophan and valine promote normal growth of rats fed diets devoid of the corresponding amino acid. The α-hydroxy analogs of isoleucine and tryptophan can partially replace the amino acid, but those of threonine and lysine are not utilized for growth by rats (27). Several amino acids, though not themselves dietary essentials, are synthesized from essential amino acids. Cystine is produced from methionine and can replace about 1/2 of dietary methionine; tyrosine is produced from phenylalanine and can replace about 1/3 of dietary phenylalanine. Many amino acids are precursors or supply part of the structure of other metabolites. For example, methionine supplies methyl groups for creatine and choline and is a precursor of homocysteine, cystine, and cysteine; arginine, when urea is removed in metabolism, yields ornithine; histidine is decarboxylated to form histamine; tyrosine is iodinated in the thyroid gland to form the hormone, thyroxin, and is used in the formation of adrenaline and noradrenaline and for melanin pigments; tryptophan is a precursor of serotonin (5-hydroxy-tryptamine) and the vitamin, niacin.

The removal of the α-amino group from an amino acid to form an a-keto acid is termed deamination and is illustrated as follows:

$$R-\underset{H}{\overset{NH_2}{C}}-C\overset{\displaystyle /\!\!/ O}{\underset{\displaystyle \backslash OH}{}} + O \longrightarrow R-\overset{O}{\overset{\|}{C}}-C\overset{\displaystyle /\!\!/ O}{\underset{\displaystyle \backslash OH}{}} + NH_3$$

The interchange of amino groups from an amino acid to an α-keto acid is termed transamination as illustrated.

$$R{-}CH(NH_2){-}COOH + R'{-}CO{-}COOH \rightleftharpoons R{-}CO{-}COOH + R'{-}CH(NH_2){-}COOH$$

Synthesis of Protein

Protein synthesis in animal tissues requires the presence of a series of nucleic acids. All living cells contain many different nucleic acids which play a number of vital roles. Deoxyribonucleic acid (DNA), a chromosomal component of cells, carries the genetic information in the cell and transmits inherited characteristics from one generation to the next. DNA controls the development of the cell and the organism by controlling the formation of ribonucleic acid (RNA). There are 3 different kinds of RNA in cells as follows: Ribosomal RNA, transfer RNA, and messenger RNA. All 3 types are involved in the synthesis of proteins. Ribosomal RNA is part of the structure of the ribosome which is the site of formation of proteins in the cell. Transfer RNA carries specific amino acids to the ribosome where they interact with messenger RNA. Messenger RNA determines the sequence of amino acids in the protein being formed. Thus, each enzyme or other protein that is being synthesized is controlled by a different messenger RNA. This basic information on protein synthesis is needed for appreciation of the overall role of proteins in normal growth and development of animals.

DNA is the carrier of genetic information and is the blueprint of protein synthesis. It is composed of phosphate-linked deoxyribose and 4 nitrogenous bases — adenine, cytosine, guanine and thymine (Fig. 6-3). The molecule is in the form of a long, double-helix chain of nucleotides-phosphate-linked deoxyribose sugar groups to each of which is attached 1 of the 4 bases. The bases are always paired, adenine with thymine and guanine with cytosine. The sequence of pairs can vary infinitely and this sequence determines the exact protein to be synthesized by the cell. The paired arrangement is diagrammatically illustrated in Fig. 6-4 (1). Changes in DNA occur in mutation and subsequent selection.

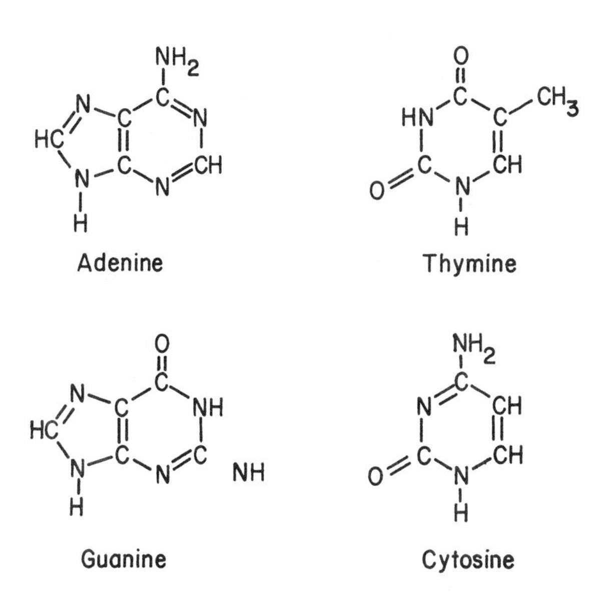

Figure 6-3. Structure of the four nitrogenous bases found in DNA.

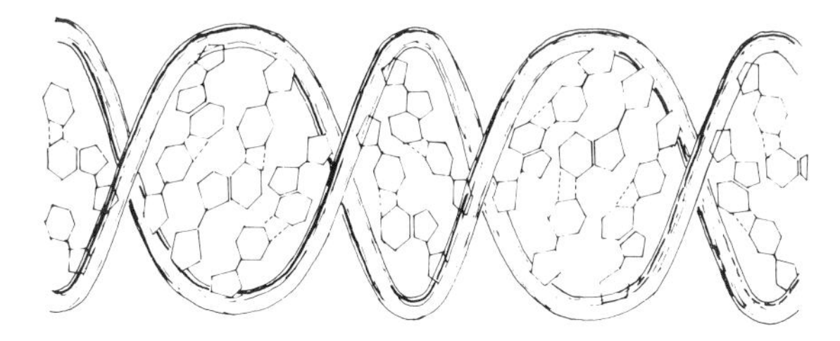

Figure 6-4. DNA molecule, shown in the figure is depicted as a twisted double helix, according to the generally accepted Watson-Crick model. In this drawing the phosphate groups are not shown. The sugar and base molecules are shown diagrammatically; in the actual three-dimensional model the base pairs that make up the crosslinks all lie in parallel planes. Courtesy of Allfrey and Mirsky [1].

The transfer of information from DNA in the nucleus to the site of protein synthesis in the cytoplasm is accomplished by RNA. Its composition is similar to that of DNA, except that ribose is the sugar instead of deoxyribose and uracil is 1 of the 4 nitrogenous bases instead of thymine. The nucleotides of RNA are linked through their phosphate groups to form long chains as in DNA.

Protein synthesis occurs by transfer of amino acids to ribosomes, particles that are attached to membrane surfaces and to which amino acids are linked in sequence predetermined by the sequence of nitrogenous bases in DNA and, in turn, in RNA. Ribosomes have molecular weights approximating 4 million, 40% of which is RNA, the remainder of which is protein. The ribosomes act as templates for the orderly array of amino acid linkage during protein synthesis. The addition of amino acids to a growing polypeptide chain in this way occurs very rapidly. For example, hemoglobin synthesis occurs at the rate of 2 amino acid additions to the chain/second so that a molecule of hemoglobin, which contains 150 amino acid residues, is synthesized in about 1½ minutes. Fig. 6-5 illustrates symbolically the nucleotides of DNA and RNA and the RNA template on which the amino acids are positioned in forming the polypeptide chain during protein synthesis (1).

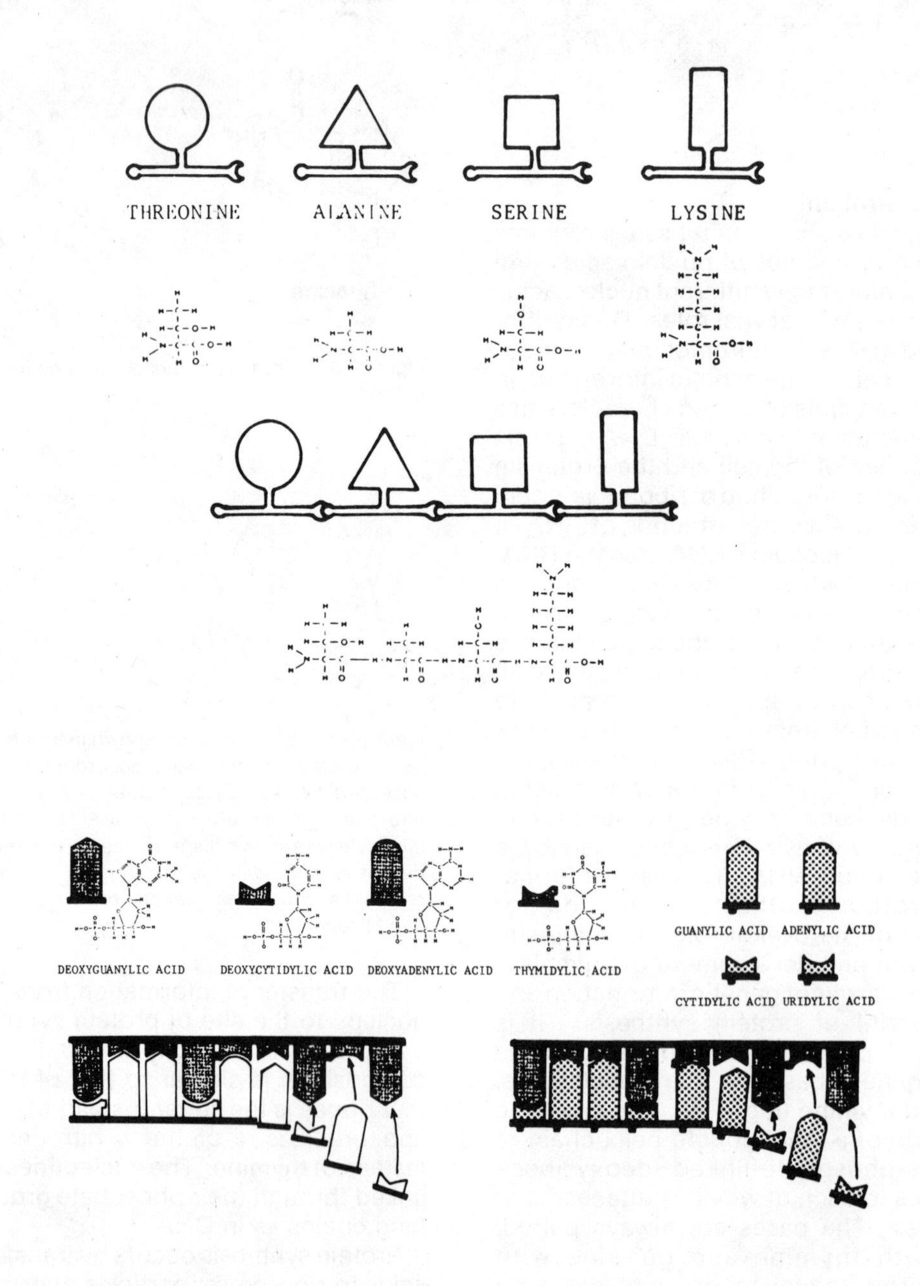

Figure 6-5. Amino acids are the constituents of proteins. Here four amino acids [of 20-odd that are know] are represented by building blocks. In the top row the symbols and formulas of the four are shown separately; in the second row they have been linked by peptide bonds, which H_2O is dropped from adjacent COOH and NH_2 groups to form a fragment of protein. The bottom rows symbolize the nucleotides of DNA and RNA and the RNA template on which amino acids are positioned during protein synthesis. By permission of Allfrey and Mirsky [1].

Urea Cycle

The fate of amino acids after absorption can be divided broadly into 3 categories: (a) tissue protein synthesis; (b) synthesis of enzymes, hormones and other metabolites; and (c) deamination and use of the carbon skeleton for energy.

The first 2 of these (protein synthesis) have been discussed. The third involves release of the amino group from the carbon skeleton of the amino acid (deamination) and entrance of the amino group into the urea (ornithine) cycle. The metabolism of the carbon skeleton for energy is discussed in Ch 7. Ammonia resulting from deamination is joined by CO_2 and a phosphate from ATP to form carbamyl phosphate which in turn combines with ornithine to form citrulline. Through a series of reactions involving formation, successively, of citrulline, arginosuccinate, and arginine, the formation of urea takes place as diagrammed in Fig. 6-6. The breakdown of arginine produces urea and ornithine, making ornithine available to repeat the cycle. The urea cycle is the chief route of N excretion in mammals. In birds, because ornithine synthesis does not occur, the main route of N excretion is as uric acid whose structural formula is:

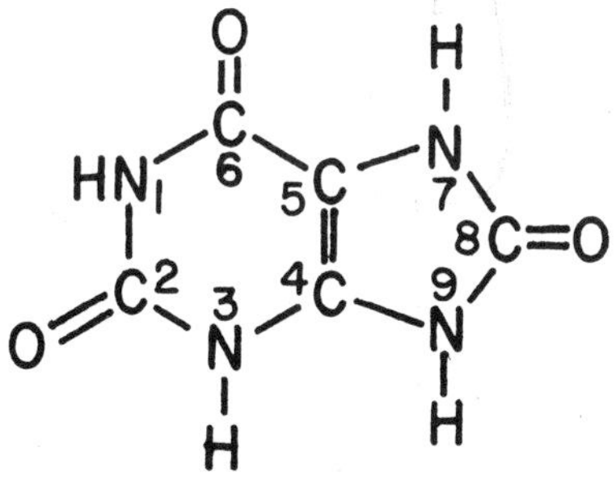

The N in position 1 comes from aspartate, that in positions 3 and 9 from glutamine, and that in position 7 from glycine. This high metabolic requirement for glycine probably accounts for the fact that during periods of rapid growth the chick may require dietary glycine even though some tissue synthesis occurs.

Uric acid is also the principal end product of purine metabolism in man and other primates, but in other mammals it is the oxidation product of uric acid, allantoin.

Hippuric acid is a common constituent of the urine of herbivorous animals, where it acts as a detoxification product of benzoic acid which is often present in high quantities in plant diets of herbivores.

Its structural formula is:

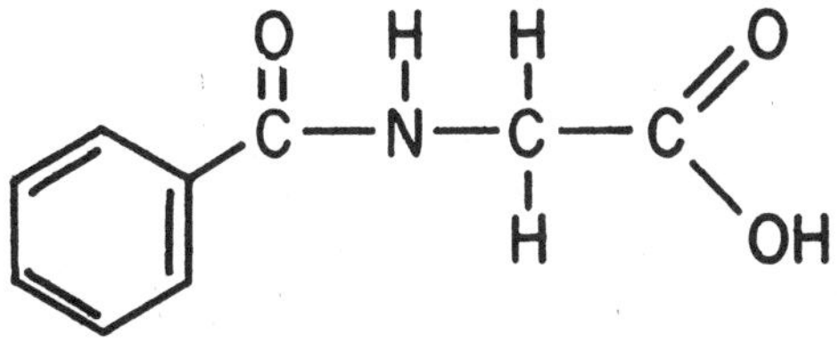

Protein and Amino Acid Requirements and Deficiencies

There is little evidence for a metabolic requirement for dietary protein per se, only for the amino acids which are the breakdown products of protein. Some evidence shows that polypeptides added to crystalline amino-acid diets may stimulate weight gain. Quantitative requirements for N and amino acids for different species performing various productive functions are discussed in more detail in Ch. 16. The array of amino acids which cannot be synthesized in sufficient amounts by animal tissues and therefore must be supplied in the diet varies somewhat according to species and stage of the life cycle (growth vs. maintenance), but the original classification in 1938 by Rose (30) still serves as a good guide. According to his classification the list of dietary essential and nonessential amino acids for growth of rats is as follows:

Essential (Indispensable)	Nonessential (Dispensable)
Arginine	Alanine
Histidine	Aspartic acid
Isoleucine	Citrulline
Leucine	Cystine
Lysine	Glutamic acid
Methionine (can be replaced partly by cystine)	Glycine
Phenylalanine (can be replaced partly by tyrosine)	Hydroxyproline
Theonine	Proline
Tryptophan	Serine
Valine	Tyrosine

Arginine is required in the diet for some species for maximum growth, but not for maintenance. This may also be true for other amino acids in some species. Asparagine is required for maximum growth during the first few days of consumption of a crystalline amino acid diet; certain other dispensable amino acids may also be required for maximum growth under some conditions, for example, glycine for the growing chick. In animals whose gastro-intestinal microflora synthesize protein from NPN sources (ruminants and some other herbivores), the amino acid balance of the diet is of little or no nutritional consequence; only the quantities of N and readily available carbohydrate are important.

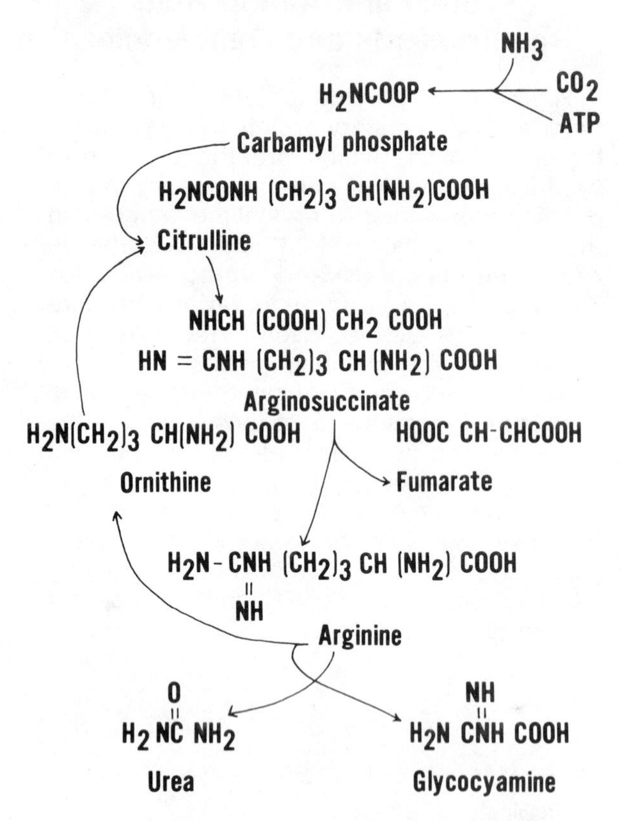

Figure 6-6. Diagram of the urea [ornithine] cycle which occurs in mammals.

Rumen Metabolism of Nitrogen

In the rumen, most ingested food proteins are hydrolyzed to peptides and amino acids, many of which will be degraded further to organic acids, ammonia, and CO_2. The majority of rumen bacteria utilize ammonia and many require it. A few organisms require or utilize amino acids or peptides, but the overall requirement for the total rumen population is for a readily available form of N that can be converted to ammonia. Part of the reason is that a considerable amount of recycling occurs in the rumen. Bacteria grow and eventually die, the cells lysing into the rumen contents. This material, containing preformed proteins, can be utilized by other rumen microorganisms. Protozoa consume bacteria or other protozoa. Their lysed cells also provide a source of preformed proteins, so the cycle can continue until the organisms pass out of the rumen. Even though a high quality protein is fed to the host, it is apt to be largely degraded and microbial protein resynthesized. The net result is that protein quality supplied to the lower GIT is remarkably constant, although sometimes of lower quality than ingested protein. The quality of protein from moderate to poor diets will usually be improved by rumen metabolism.

When protein is degraded, ammonia is given off. This may be rapidly absorbed through the rumen wall, particularly at a relatively high rumen pH or if the diet contains large amounts of protein. However, not all of this is lost to the rumen since blood urea may pass back through the rumen wall and, also, an appreciable amount of urea comes into the rumen via saliva.

Urea is only one of many NPN compounds that may come into the rumen. When it passes into the rumen from saliva or through the rumen wall, it is very rapidly hydrolyzed to ammonia and CO_2. Many forages, particularly young plants and legumes, have large amounts of NPN in the form of amino acids, peptides, nucleic acids, amines, and amides. These are all readily available to rumen microorganisms. The efficiency of utilization of ammonia and other NPN depends on relative solubility of the NPN (or protein) and on the availability of readily available carbohydrates such as the monosaccharides and starch. Starch allows the most efficient use of NPN such as urea, probably because it is not fermented quite so rapidly as sugars such as glucose or sucrose. Urea is now a common feed ingredient as cost often favors its use over typical plant protein concentrates. Although urea is well utilized by rumen microorganisms, if it is fed without care, particularly in the absence of enough readily available carbohydrate, absorption of rumen ammonia may be rapid enough to cause toxicity and death. Thus, care in its use is required.

The microbial protein produced in this way from NPN sources such as urea and biuret has a high biological value so that when hydrolyzed by digestive enzymes it promotes normal performance. Loosli et al (16) showed that lambs fed a nearly protein-free diet used urea N to form protein adequate for normal growth; Virtanen (37) showed that cows fed NPN and no protein can produce milk; Oltjen (25) obtained normal reproduction in beef cattle fed diets containing only NPN. Thus, tremendous potential exists for capitalizing on the protein synthetic abilities of GIT microflora in animal production as a means of supplying protein for man and other animals.

Protein Deficiency

Inadequate protein (N or amino acids) is probably the most common of all nutrient deficiencies, because most energy sources are low in protein and protein supplements are expensive. The quantitative protein requirement is greater for growth than for maintenance and is affected by sex (males tend to have a higher requirement) and species, and probably by genotype within species. The ratio of calories to protein in the diet is important. On a fixed adequate protein intake the energy level of the diet determines N balance, but on a fixed calorie intake, protein level is the determining factor. Proteins are not normally used to a great extent for energy except when calorie intake is insufficient; then protein is used for energy.

Signs of protein deficiency include: anorexia, reduced growth rate, reduced N balance, reduced efficiency of feed utilization, reduced serum protein concentration, anemia, fat accumulation in liver, edema (in severe cases), infertility, reduced birth weight of young, reduced milk production, and reduced synthesis of certain enzymes and hormones.

A young pig fed a diet deficient in lysine and tryptophan is shown in Fig. 6-7 alongside a littermate fed an adequate diet.

The small stature, low serum-protein concentration, anemia, and edema (pot belly appearance) of infants suffering from kwashiorkor are typical manifestations of protein deficiency in man.

Growth is completely arrested in severe protein deficiency as illustrated by histologic sections taken through the humerus of an 11-wk-old pig fed for 8 wk on a 3% protein diet (Fig. 6-8). The normal growth plate of a littermate pig fed an adequate diet is shown for comparison.

Deficiencies of individual essential amino acids generally produce the same symptoms because a single amino acid deficiency prevents protein synthesis in the same way that a shortage of a particular link in a chain prevents elongation of the chain. Thus, individual amino acid deficiencies result in deamination of the remaining amino acids, loss of the ammonia as urea, and use of the carbon chain for energy. Certain amino acid deficiencies produce specific lesions. For example, tryptophan deficiency produces eye cataracts; threonine or methionine deficiency produces fatty liver; and lysine deficiency in birds produces abnormal feathering (Fig. 6-9).

Figure 6-7. Amino acid deficiency in pigs. Littermates fed an adequate diet containing opaque-2 corn [pig A] as compared to a pig fed inadequate amounts of lysine and tryptophan. Photo by W.G. Pond.

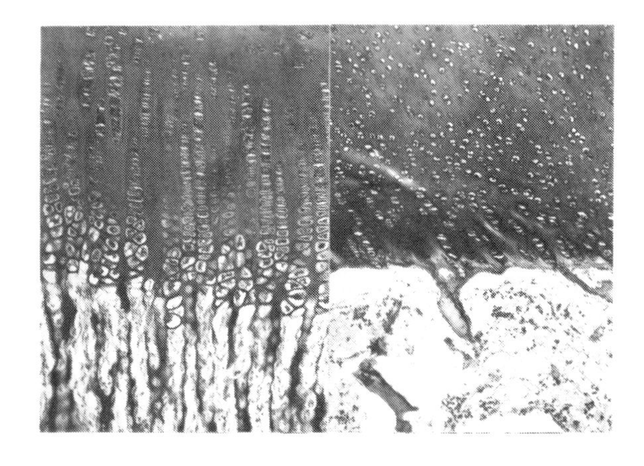

Figure 6-8. Left. Histological section from humerus of 11-wk-old pig fed a diet with normal protein content. Note normal growth plate. Right. Humerus of 11-wk-old pig fed a protein deficient diet [3% protein]. Note complete absence of differentiation of cartilage terminal plate sealing off the growth plate and presence of severe osteoporosis. H & E stain, 100X. Photos courtesy of L.E. Krook, Cornell University.

Figure 6-9. Lysine deficient turkey poult. Note white barring of the flight feathers of the deficient bird on the right. Both birds are the same age. By permission of S.J. Slinger, Univ. of Guelph, Ontario, Canada.

Use of D-Amino Acids and NPN

Supplementation of the diet with D-isomers of amino acids can serve as a source of nonspecific N for growth of rats (28), and perhaps of other simple-stomached animals. NPN compounds such as diammonium citrate can also serve this purpose. Urea utilization by simple-stomached animals for nonessential amino acid synthesis depends on its hydrolysis to ammonia. Current evidence indicates that urease is not produced by mammalian cells. Thus, any incorporation of N of urea into body tissues depends on microbial urease in the lumen of the GIT for hydrolysis of urea. Microorganisms can use ammonia for synthesis of amino acids and proteins, but the restriction of ureolytic activity to the area of the GIT past the site of maximal amino acid absorption would limit the availability to the host of amino acids which the microbes synthesize, unless coprophagy occurs. Therefore, in simple-stomached animals, a response to urea can be possible only if the basal diet meets essential amino acid requirements, but not total N requirement, and the ammonia release by bacterial urease is incorporated into nonessential amino acids by body tissues. Snyderman (33) has reported such an effect in humans, and reports on other species suggest that some urea N is incorporated into body tissues if the above conditions are met. In ruminants, where the rumen microbial activity is anatomically upstream from the sites of most active absorption of nutrients, NPN is the main dietary N requirement. Usually, however, performance of ruminant animals fed NPN is improved by supplementation with intact proteins.

Amino Acid Antagonisms, Toxicity and Imbalance

Amino acid antagonism is the term referring to growth depression which can be overcome by supplementation with an amino acid structurally similar to the antagonist. Excess lysine, for example, causes growth depression in chicks which can be reversed by additional arginine. Antagonism differs from imbalance in that the supplemented amino acids need not be limiting.

The term, amino acid toxicity, is used when the adverse effect of an amino acid in excess cannot be overcome by supplementation with another amino acid. Methionine, if added to the diet in excess, produces growth depression not overcome by supplementation with other amino acids.

Amino acid imbalance has been defined by Harper (12) as any change in the proportion of dietary amino acids that has an adverse effect preventable by a relatively small amount of the most limiting amino acid(s). The growth depression caused by the addition of amino acids to a mixture so as to create an imbalance depends on the growth rate supported by the original diet. Attempts to explain the increased need for the most limiting amino acid in the presence of amino acid imbalance have included suggestions such as: changes in proportions of fat, protein, and water deposited in the tissues; increased rate of catabolism of the limiting amino acid; increased anabolism of protein. A simple method for detecting amino-acid imbalances is to offer a choice of an imbalanced diet and a protein-free diet (13). Rats (29) and pigs (9) reject an imbalanced diet that supports growth and consume largely the protein-free diet that does not. The rejection of the imbalanced diet is probably caused by some biochemical or physiological disturbance. Nassett et al (23) suggested that it may be related to the change in plasma-free amino acid pattern which in turn alters the excitability of the satiety center in the hypothalamus. Whatever the mechanism, the amino acid balance of the diet has a profound effect on feed intake.

Of major practical concern in animal nutrition is provision of the minimum amount of protein compatible with normal performance, whether for maintenance, growth, pregnancy, lactation or egg production. Due to the relatively high cost of protein, the probability of feeding a diet deficient in protein and amino acids is far greater than that of feeding a higher-than-required protein level. Nevertheless, one must be aware of any adverse effects of feeding excesses of any nutrient, and protein is no exception. Sugahara et al (36) fed growing pigs 16, 32, or 48% protein and obtained a linear depression in weight gain with increasing protein. Feed intake was depressed and hair became dull and coarse. Other work (24) has shown that a high-protein diet reduces activity of several adipose tissue enzymes associated with fatty acid synthesis in swine. The extent to which the reduced feed intake and weight gain associated with high protein intake is due to increased blood and tissue ammonia concentration or to other metabolic changes is unknown. Ammonia toxicity is a practical problem in ruminants fed urea. Ammonia is absorbed from the rumen as well as from the omasum, small intestine and cecum of ruminants (4); its absorption is governed by both concentration gradient and pH; absorption is increased at higher pH. Toxic symptoms in ruminants include labored respiration, excessive salivation, muscle tremors, incoordination and death within 1 or 2 hr of onset of symptoms. Similar signs have been reported in ammonia intoxicated pigs (5). Peripheral blood contains 1-4 mg ammonia/100 ml at the height of toxic symptoms. Much higher levels would be found in portal blood before reaching the liver for metabolism. Blood glucose, lactate, pyruvate, pentoses and ketones all rise during ammonia toxicity, suggesting a drastic effect on energy metabolism, perhaps by inhibition of the citric acid cycle (5). Visek (38) has provided evidence that ammonia causes cell death and has suggested that excessive protein intake of man and animals may increase the incidence of cancer of the GIT by increasing cell turnover rate due to greater exposure to ammonia and, thereby, increasing the probability of mutations resulting in neoplastic cell formation.

Some or all of the above effects of ammonia may affect the animal as a result of greater-than-optimum protein intake. The growth depression associated with excessive levels of protein in the diet is the most visible sign of protein toxicity, whether or not the effects are produced through its catabolism to ammonia or by other means.

References Cited

1. Allfrey, V.G. and A.E. Mirsky. 1961. Scientific American, Sept.
2. Bendall, J.R. 1964. Meat Proteins. Symposium on Food-Proteins and Their Reactions. The AVI Publishing Co., Westport, Connecticut.
3. Bragg, D.B., C.A. Ivy and E.L. Stephenson. 1969. Poult. Sci. 48:2135.
4. Chalupa, W. 1968. J. Animal Sci. 27:207.
5. Chow, Kye-Wing, W.G. Pond and E.F. Walker. 1970. Proc. Soc. Expt'l Biol. Med. 134:122.
6. Carpenter, K.J. 1960. Biochem. J. 77:604.
7. Concon, J.M. 1966. The protein of opaque-2 maize. Proc. High Lysine Corn Conf., Purdue U., Lafayette, Indiana, p 67.
8. Cornwell, D.G. and L.A. Horrocks. 1964. Protein-Lipid Complexes. Symposium on Foods-Proteins and Their Reations. The AVI Publishing Co., Westport, Connecticut.
9. Devilat, J., W.G. Pond and P.D. Miller. 1970. J. Animal Sci. 30:536.
10. Eggum, B. 1973. Protein and amino acid utilization in animals. Ph.D. Thesis, Danish Agric. U.
11. Guidotti, G. 1972. Ann. Rev. Biochem. 41:731.
12. Harper, A.D. 1959. J. Nutr. 68:405.
13. Harper, A.E. 1964. Amino acid toxicities and imbalances. Mammalian Protein Metabolism. Academic Press, New York.
14. Jevons, F.R. 1964. Protein-Carbohydrate Complexes. Symposium on Food-Proteins and Their Reactions. The AVI Publishing Co., Westport, Connecticut.
15. Just-Nielsen, P. 1972. Energy and protein utilization in young pigs. Ph.D. Thesis. Danish Agric. U.
16. Loosli, J.K. et al. 1949. Science 110:144.
17. Marshall, R.D. 1972. Ann. Rev. Biochem. 41:802.
18. McClain, P.E. et al. 1965. Proc. Soc. Expt'l. Biol. Med. 119:493.
19. Meade, R.J. 1972. J. Animal Sci. 35:713.
20. Mertz, E.T., L.S. Bates and O.E. Nelson. 1964. Science 145:279.
21. Miller, D.S. and A.E. Bender. 1955. Br. J. Nutr. 9:382.
22. Mitchell, H.H. and J.R. Beadles. 1950. J. Nutr. 40:25.
23. Nasset, E.S., P.T. Ridley and E.A. Schenk. 1967. Amer. J. Physiol. 213:645.
24. O'hea, E.K. and G.A. Leveille. 1969. Fed. Proc. 28:687.
25. Oltjen, R.R. 1969. J. Animal Sci. 28:673.

26. Oxender, D.L. 1972. Ann. Rev. Biochem. 41:777.
27. Pond, W.G., L.H. Breuer, J.K. Loosli and R.G. Warner. 1964. J. Nutr. 83:85.
28. Rechcigal, M., J.K. Loosli and H.H. Williams. 1960. Proc. Soc. Expt'l Med. 104:448.
29. Rogers, Q.R., R.I. Thomas and A.E. Harper. 1967. J. Nutr. 91:561.
30. Rose, W.C. 1948. J. Biol. Chem. 176:753.
31. Scanu, A.M. and C. Wisdom. 1972. Ann. Rev. Biochem. 41:703.
32. Schweigert, P.S. and B.J. Payne. 1958. Amer. Meat Inst. Found. Bul. No. 30.
33. Snyderman, S.E. 1967. Urea as a source of unessential nitrogen for the human. In: Urea as a Protein Supplement Pergamon Press, Oxford, England.
34. Smyth, D.G., Stein, W.H. and S. Moore. 1963. J. Biol. Chem. 238:227.
35. Stockland, W.L. and R.J. Meade. 1970. J. Animal Sci. 31:1156.
36. Sugahara, M., D.H. Baker, B.G. Harmon and A.H. Jensen. 1969. J. Animal Sci. 29:598.
37. Virtanen, A.I. 1966. Science 153:1603.
38. Visek, W.J. 1969. Proc. Cornell Nutr. Conf., p 91.
39. Wilson, T.H. 1962. Intestinal Absorption. W.B. Saunders Co., Philadelphia, Pa.

Chapter 7 – Carbohydrates

Carbohydrates provide the chief source of energy contained in feedstuffs consumed by animals and man. Solar energy is utilized by plants for converting atmospheric CO_2 and H_2O to glucose by photosynthesis, illustrated in simplest form:

$$6\,CO_2 + 6\,H_2O + 673\ \text{calories} \xrightarrow{\text{Plant chlorophyll}} C_6H_{12}O_6 + 6\,O_2$$

carbon dioxide + water + solar energy → glucose + oxygen

Animal life could not exist without this transformation of energy by photosynthesis, and, of course, the free oxygen supplied as a by-product of the reaction is vital for animal life. Hundreds of different carbohydrates occur in nature. Only those most closely involved with animal nutrition and metabolism are discussed here.

Structure

The basic chemical structure of carbohydrates consists of carbon atoms arranged in chains, to which are attached oxygen and hydrogen as in glucose (shown below). Carbohydrates are characterized by having an aldehyde $-C\begin{smallmatrix}\diagup\!\!\!\!= O\\ \diagdown H\end{smallmatrix}$ or a ketone $(C - \overset{O}{\overset{\|}{C}} - C)$ group in their structure.

D-glucose (the naturally occurring isomer)

WAYS OF DEPICTING THE STRUCTURE OF GLUCOSE

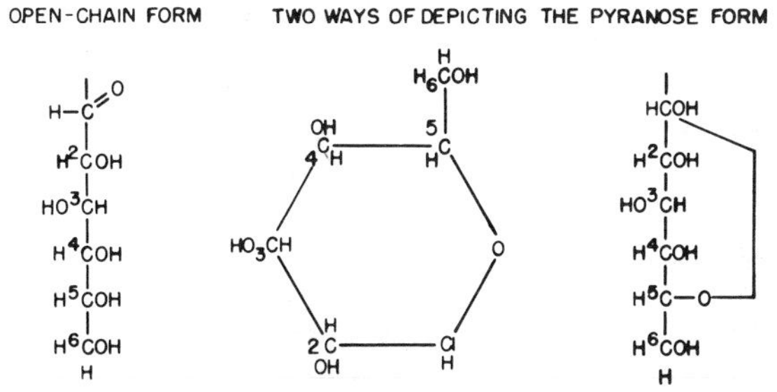

OPEN-CHAIN FORM

H–C=O
H²COH
HO³CH
H⁴COH
H⁵COH
H⁶COH
H

TWO WAYS OF DEPICTING THE PYRANOSE FORM

The broad classification of carbohydrates (Table 7-1) is based on their chemical structures (Table 7-2). Monosaccharides are the simplest units. Combinations of these in various ways, result in disaccharides (2 monosaccharide units), trisaccharides (3 units), and polysaccharides (many units).

Functions

The primary function of carbohydrates in animal nutrition is to serve as a source of energy for normal life processes, or, in plants, as structural components. Carbohydrates and lipids are the two major compounds used by the body for energy (see Ch. 9). The lipid content of most diets is $<5\%$, so that far the greatest proportion of energy comes from carbohydrates. In human diets, in which a greater amount of food of animal origin is consumed, the proportion of energy coming from carbohydrate is considerably less than in diets of domestic farm animals. The percentage of total energy intake coming from fat in human diets in the USA is usually >30 and may often be as high as 50 or more.

Although carbohydrates serve as a significant source of energy for body tissues, only limited evidence shows a dietary requirement for higher animals and man, even though certain insects may have a requirement (8, 10). This point is academic, however, because virtually all natural food sources contain some carbohydrate. Brambila and Hill (5, 6) showed essentially that chicks can grow normally on carbohydrate-free diets if the calorie to protein ratio is optimum and if triglycerides are included in the diet. The use of free fatty acids as the sole nonprotein energy source resulted in growth depression, as a result of failure of the body to produce glucose from amino acids in the absence of the glycerol from triglycerides.

The ultimate source of energy for most cells is glucose. This basic unit is made available to the cell either by ingestion of glucose or its precursors by the animal, or by conversion from other metabolites. The carbon skeleton of these substances and of products of fat metabolism (Ch. 8) provide the energy for maintaining normal life processes through enzyme-driven reactions.

Table 7-1. Classification of carbohydrates and their occurrence

		Occurance
Monosaccharides (simple sugars)		
Pentoses (5-C sugars) ($C_5H_{10}O_5$)		
Arabinose		pectin; polysaccharide, araban
Xylose		corn cobs, wood; polysaccharide, xylan
Ribose		nucleic acids
Hexoses (6-C sugars) ($C_6H_{12}O_6$)		
Glucose		disaccharides; polysaccharides
Fructose		disaccharides; (sucrose)
Galactose		milk (lactose)
Mannose		polysaccharides
Disaccharides ($C_{12}H_{22}O_{11}$)	monosaccharide content	
Sucrose	glucose-fructose	sugar cane, sugar beets
Maltose	glucose-glucose (glucose-4-α-glucoside)	starchy plants and roots
Lactose	glucose-galactose	milk
Cellobiose	glucose-glucose (glucose-4-β-glucoside)	fibrous plants
Trisaccharides ($C_{18}H_{32}O_{16}$)		
Raffinose	glucose-fructose-galactose	certain varieties of eucalyptus, cottonseed, sugar beets
Polysaccharides		
Pentosans ($C_5H_8O_4$)$_N$		
Araban	Ababinose	pectins
Xylan	Xylose	corn cobs, wood
Hexosans ($C_6H_{10}O_5$)$_N$		
Starch (a polyglucose) glucoside)	Glucose	grains, seeds, tubers
Dextrin	Glucose	partial hydrolytic product of starch
Cellulose	Glucose	cell wall of plants
Glycogen	Glucose	liver and muscle of animals
Inulin (a polyfructose fructoside)	Fructose	potatoes, tubers, artichokes
Mixed polysaccharides		
Hemicellulose	Mixtures of pentoses and hexoses	fibrous plants
Pectins	Pentoses and hexoses mixed with salts of complex acids	citrus fruits, apples
Gums (partly oxidized to acids)	Pentoses and hexoses	acacia trees and certain plants

Metabolism

Preparation for Absorption

Only monosaccharides can be absorbed from the GIT, except in newborn animals which are capable of absorbing larger molecules. Thus, poly-, tri- and disaccharides must be hydrolyzed by digestive enzymes elaborated by the host or by microflora inhabiting the GIT of the host for absorption to occur. Principal digestive enzymes elaborated by animals were discussed and listed in Ch. 3. The carbohydrate-splitting enzymes (carbohydrases) are effective in hydrolyzing most complex carbohydrates to monosaccharides except for those with a glucose-4-β-glucoside linkage (see Table 7-1), such as in cellulose. Microflora of the rumen of ruminants and the cecum of some nonruminants, such as the horse and rabbit, produce cellulase, so that these species can utilize large quantities of cellulose. Other monogastric species, including man and swine, utilize lesser amounts of cellulose due to digestion in the large intestine.

In ruminants and other species with large microfloral populations in the GIT, anaerobic fermentation of carbohydrates results in the production of large quantities of volatile fatty acids (VFA), mainly acetic, propionic, and butyric acids. These acids provide a large proportion of the total energy supply. The role of VFA in nutrition will be discussed in greater detail in Ch. 8 and the remainder of this discussion will deal with the metabolism of glucose and other monosaccharides.

Table 7-2.

Structure of Monosaccharides and Disaccharides

PENTOSES

D-arabinose	D-ribose	D-xylose
HC=O	HC=O	HC=O
HOCH	HCOH	HCOH
HC-OH	HCOH	HOCH
HCOH	HCOH	HCOH
HCOH	HCOH	HCOH
H	H	H

HEXOSES

D-glucose	D-mannose	D-galactose	D-fructose
HC=O	HC=O	HC=O	H HC-OH
HCOH	HOCH	HCOH	C=O
HOCH	HOCH	HOCH	HOCH
HCOH	HCOH	HO-CH	HCOH
HCOH	HCOH	HCOH	HCOH
HCOH	HCOH	HCOH	HCOH
H	H	H	H

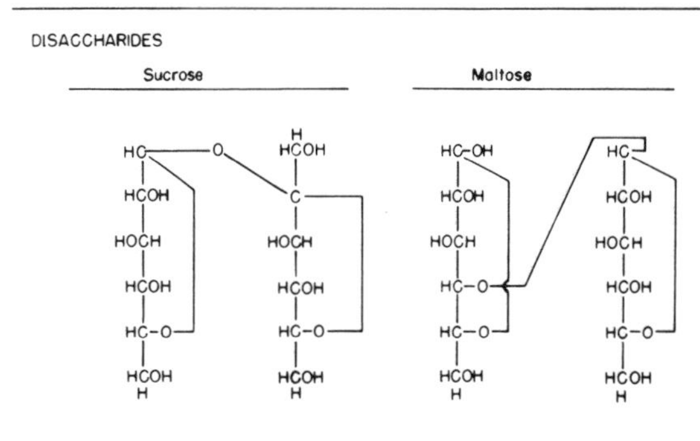

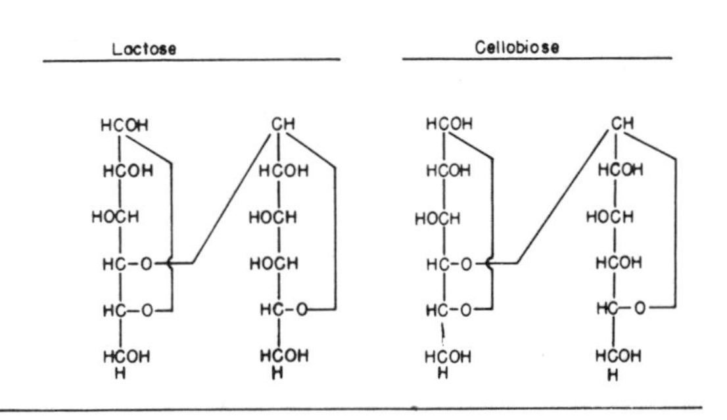

Absorption

The upper or cranial section of the small intestine, the duodenum and jejunum, has the greatest capacity to absorb monosaccharides. The lower small intestine (lower ileum) absorbs less, and the stomach and large intestine absorb little if any sugars. Cori (7) observed in 1925, that selective absorption of monosaccharides occurs from the GIT of the rat. He showed that galactose and glucose are absorbed very efficiently, but that mannose, which differs from glucose only in the configuration of the hydroxyl group at carbon 2 (see Table 7-2), is absorbed at only about 20% of the efficiency of glucose. Wilson (15) compiled a table showing the selective absorption of 6 different monosaccharides by a variety of animal species (Table 7-3). In general, glucose and galactose are absorbed at the highest rate and arabinose at the lowest rate among the 6 monosaccharides compared, regardless of the species tested. The exact mechanism of sugar absorption is still unknown. Active (energy-dependent) transport is established for glucose and galactose, but probably it does not operate for other sugars. Glucose and galactose appear unchanged in the portal vein after absorption. Failure to discover specific chemical alterations in the molecule during absorption suggests that, perhaps, transport involves absorption of the sugar to some type of membrane carrier located in the plasma membrane on the luminal border of the cell (15). Conversion of some monosaccharides to glucose occurs within the intestinal mucosal cell; conversion of fructose to glucose remains relatively constant over a wide range of fructose concentrations but the rate of movement of fructose into the cell is roughly proportional to the luminal concentration (9).

Sugars apparently share a common pathway of transport across the intestinal mucosal cell. This being true, competitive inhibition between glucose and galactose as well as between glucose and several derivatives of glucose is no surprise. Based on relationships of this kind, Wilson (15) proposed the minimal structural requirements for intestinal transport of sugars as shown:

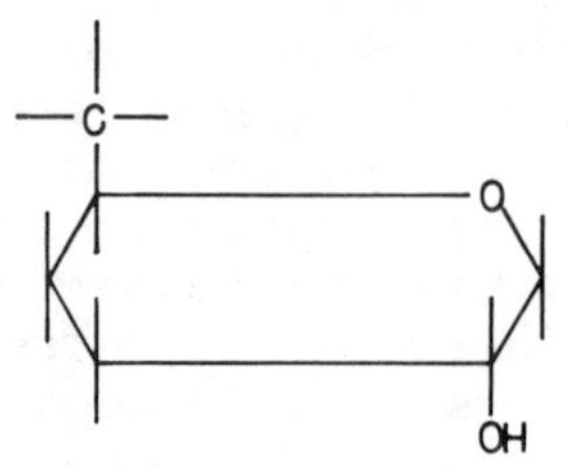

Table 7-3. Selective absorption of sugars by different animals.[a]

Animal	Rate of sugar absorption (glucose taken as 100) Galactose	Glucose	Fructose	Mannose	Xylose	Arabinose
Rat	109	100	42	21	20	12
Cat	90	100	35	—	—	—
Rabbit	82	100	—	—	—	60
Hamster	88	100	16	12	28	10
Man	122	100	67	—	—	—
Pigeon	115	100	55	33	37	16
Frog	107	100	46	51	29	
Fish	97	100	62	—	57	49

[a] Adapted from Wilston (15)

Fructose is converted to lactic acid by the intestine of some animal species (12); the rat converts up to 50% of fructose to lactic acid, as measured by mesenteric vein cannulation, but the guinea pig produces very little. Fructose in blood of mature animals is very low, but in fetal and newborn lambs (11, 13) and pigs (1) it is high. Aherne et al (2) found that fructose was absorbed with little or no intestinal conversion to glucose in 3-, 6- or 9-day-old pigs and suggested that fructokinase activity (enzyme neded for conversion of fructose to glucose) of the liver but not the intestine increases with age in the pig. Intravenous administration of sucrose, fructose, or lactose to hypoglycemic baby pigs fails to alleviate the hypoglycemia, and intravenous administration of disaccharides to animals generally results in their excretion in the urine.

Xylose is not actively transported against a concentration gradient and does not appreciably inhibit galactose transport, yet evidence suggests a common carrier even for such widely different sugars as glucose and xylose. Further details of the mechanism of transport of sugars across the mucosal cell to the blood are beyond the scope of this discussion.

Several important factors affect the absorption of glucose. It is reduced by short-term (24- or 48-hr) fasting, but increased by chronically restricted food intake (15). The basis for this apparent difference in functional capacity of the small intestine in fasting vs. underfed animals is not well understood. Diabetic animals absorb glucose more rapidly than normal; adrenalectomy results in a reduction in glucose absorption but has no effect on xylose absorption; thyroidectomy and ovariectomy reduce glucose absorption. Thus, a variety of endocrine factors clearly affect absorption of sugars.

Intravenous administration of glucose is a common means of reversing hypoglycemia. It results in a rise in blood glucose concentration followed by a gradual decline to normal concentration as uptake by tissues for energy occurs and as liver and muscle store glycogen, which is synthesized from excess glucose. No glucose appears in the urine of normal animals given a glucose load because the kidney functions to retain it in the body. Glucose is found in the urine of animals only with kidney damage or diabetes. In each instance, blood glucose concentration exceeds the kidney threshold, the level at which the kidney is no longer capable of preventing urinary loss.

Deficiency of specific disaccharidases (enzymes) in the GIT results in serious gastrointestinal upsets. Young mammals fed large amounts of sucrose develop severe diarrhea, and death may occur from an insufficiency of sucrase during the first few weeks of life. Ruminant species apparently produce no sucrase. Feeding appreciable amounts of sucrose to liquid-fed young animals results in severe diarrhea. In addition, these species have low levels of starch-splitting enzymes. Adult pigs fed lactose may develop diarrhea and gas discomfort because of a deficiency of lactase, and lactase deficiency is also prevalent in some human population groups and appears to have a genetic basis. Affected individuals are unable to tolerate milk products containing lactose.

Xylose feeding of young pigs results in depressed appetite and growth and causes eye cataracts (16). The mechanism whereby xylose causes these abnormalities in young animals is not known.

In the absence of diseases such as infectious diarrhea or of other pathological conditions affecting absorption, the absorption of soluble carbohydrates often exceeds 90%. Endogenous (fecal metabolic) losses result in a net (apparent) absorbability approximating 80% or more in nonruminant animals. Thus, except in abnormal situations such as those noted previously, the available energy is similar for a wide variety of carbohydrate sources.

Metabolic Conversions

Monosaccharides not converted to glucose in the intestinal mucosal cell during absorption may be converted to glucose by reactions in the liver. The animal body stores very little energy as carbohydrate, but some glucose is converted to glycogen which is stored in liver and muscle tissues. Glycogen is a starch-like compound and can be rapidly converted back to glucose. Thus, the level of blood sugar is maintained within a rather narrow range in normal animals by conversion of circulating blood glucose to glycogen and by reconversion to glucose by the process of **glycogenolysis** when the blood level declines. The blood glucose concentration increases after a meal, but returns to the fasting level within a few hr. This homeostasis is under endocrine control with insulin and glucagon from the pancreas playing an important role in maintaining the blood glucose concentration within normal limits for the species. Storage of glycogen after a carbohydrate meal prevents marked elevation in blood sugar **[hyperglycemia]**, and release of glucose by breakdown of glycogen during fasting prevents low blood sugar levels **[hypoglycemia]** although some variation occurs. In diabetes mellitus hyperglycemia occurs, and, with excess insulin release, hypoglycemia.

The formation of glycogen from glucose **[glycogenesis]** requires two molecules of adenosine triphosphate (ATP) for every molecule of glucose. Glucose molecules are added, one unit at a time, to form the long chain of glycogen. Uridine triphosphate (UTP) is also involved in the conversion of glucose to glycogen. The process is illustrated as follows:

Glucose + ATP ⟶ glucose-6-phosphate + ADP
Glucose-6-phosphate ⟶ glucose-1-phosphate
Glucose-1-phosphate + UTP ⟶ UDP-glucose + pyrophosphate
UDP-glucose ⟶ glycogen + UDT

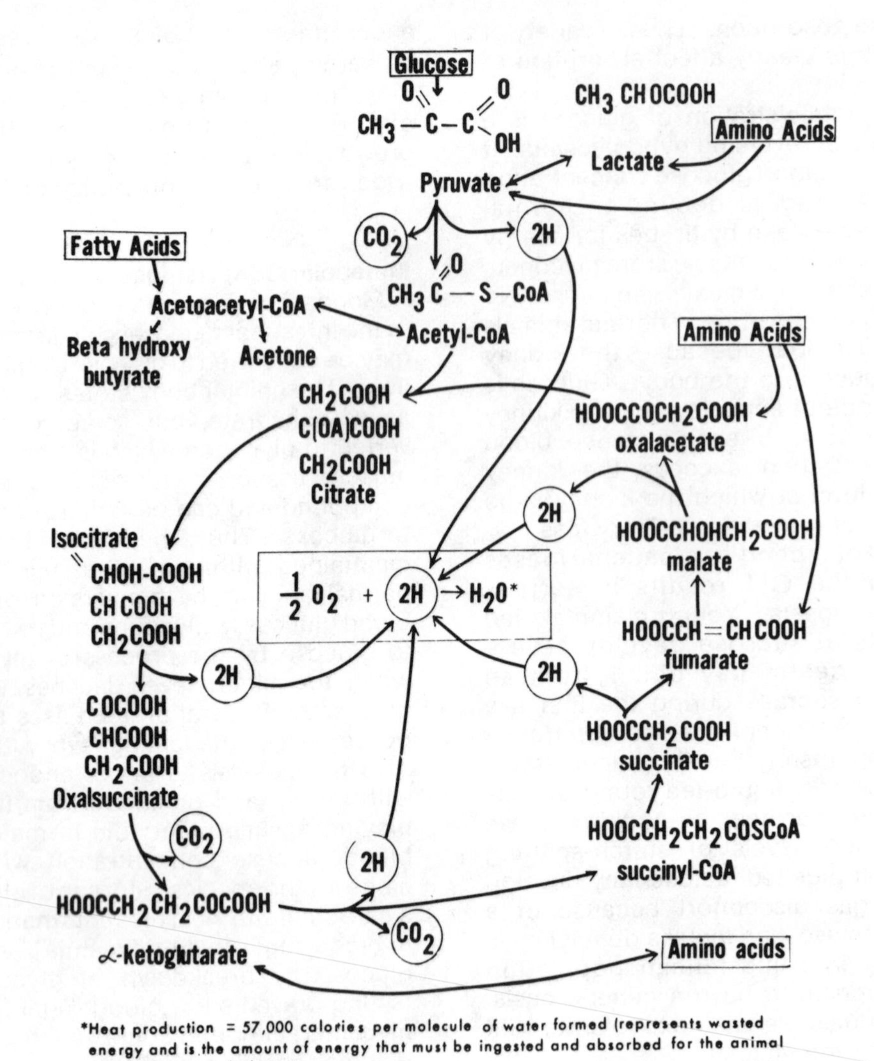

Figure 7-1. The citric acid [Krebs] cycle through which carbohydrates are oxidized to produce energy.

The breakdown of glycogen to form glucose is, in essence, the reverse of the process shown above.

Glucose can also be formed by body tissues from noncarbohydrate metabolites, including lipids and amino acids. This process is called **gluconeogenesis.** All of the nonessential amino acids along with several of the essential ones (arginine, methionine, cystine, histidine, threonine, tryptophan, and valine) are glucogenic. That is, when metabolized they can give rise to glucose. Some (isoleucine, lysine, phenylalanine, and tyrosine) are both **glucogenic** and **ketogenic;** they can give rise to glucose and acetone or other ketones. Leucine is strictly ketogenic.

The amino acids used for gluconeogenesis or for energy enter the citric acid cycle (Fig. 7-1) as acetate, pyruvate, or α-ketoglutarate as listed in Table 7-4. Thus, amino acids not used for protein synthesis enter the general pool of metabolites which provide energy for normal body maintenance and productive functions.

Table 7-4. Metabolism of the carbon chain of amino acids.

Citric acid cycle			
Acetate	Pyruvate	α-ketoglutarate	Nicotinic acid and serotinin
isoleucine	alanine	arginine	tryptophan
leucine	cystine	glutamic acid	
phenylalanine	glycine	hydroxy proline	
threonine	methionine	histidine	
tyrosine	serine	lysine	
valine		proline	

Glycogen storage is limited. Therefore, when ingestion of carbohydrate exceeds current needs for glycogen formation, glucose is converted to fat. This is accomplished by the breakdown of glucose to pyruvate, which then is available for fat synthesis. The conversion of glucose to pyruvate and then to lactate under anaerobic conditions in muscle **[glycolysis]** occurs through a series of transformations depicted below as the glycolytic or Embden-Meyerhoff pathway:

phosphate pathway has several important functions: allows synthesis of ribose-5-phosphate, an essential component of nucleic acids; allows a by-pass by certain tissues around the glycolytic pathway; and produces reduced nicotine adenine dinucleotide phosphate (NADPH), which is essential for synthesis of fatty acids and for hydroxylation reactions. The exact reactions in the pentose phosphate pathway are beyond the scope of this discussion.

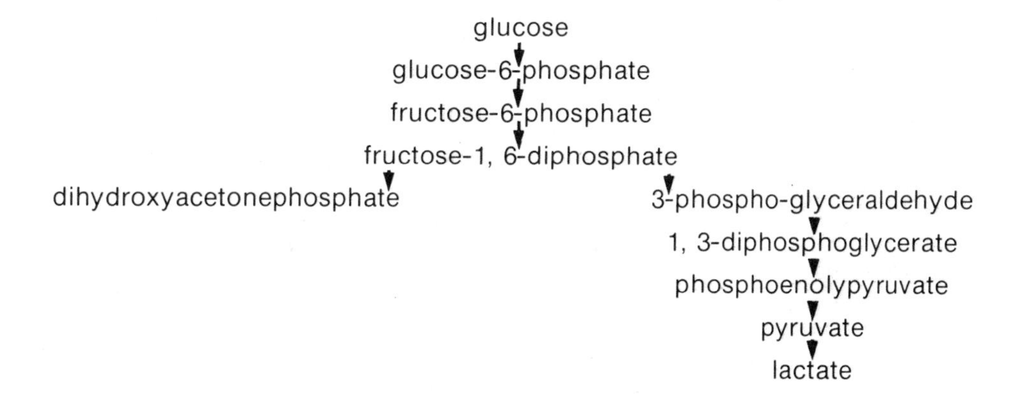

Pyruvate formed from glucose via the glycolytic pathway is available for entrance into the citric acid cycle, the final common pathway of energy metabolism (Fig. 7-1).

Another route of catabolism of glucose is via the phosphogluconic oxidative pathway (pentose-phosphate pathway; oxidative shunt; pentose shunt). The enzymes driving this pathway of glucose utilization are found in liver, adipose tissue, mammary gland cells and in the cornea and lens of the eye. The pentose

Energetics of Glucose Catabolism

The total energy released in the conversion of glucose to CO_2 and H_2O is 673 Kcal/mole. This can be illustrated as follows:

$$C_6H_{12}O_6 \longrightarrow 6\ CO_2 + 6\ H_20 + 673\ \text{Kcal}$$

The molecular weight of glucose is 180.2. Thus, the gross energy value of glucose is 673/180.2 = 3.74 Kcal/g. In the oxidation of metabolites via the citric acid cycle (Fig. 7-1),

57 Kcal/mole of water formed (total of 6 × 57 = 342 Kcal) represents heat production and is wasted energy, equivalent to the amount of energy that must be ingested and absorbed for the animal to stay in energy balance.

Catabolism of one mole of glucose by the glycolytic pathway is associated with the following amounts of adenosine triphosphate (ATP) trapped at each stage of oxidation to CO_2 and H_2O.

Glycolytic pathway	8 moles of ATP
2 pyruvate to 2 acetyl CoA	6 moles of ATP
2 acetate to CO_2 and H_2O	24 moles of ATP
Total	38 moles of ATP

ATP serves as a major form of high-energy phosphate bonds. One mole of ATP has a value of about 8 Kcal/mole. That is, ATP $\rightarrow$ ADP + 8 Kcal/mole. The conversion of the free energy of the oxidation of glucose has an efficiency of 40 to 65%, depending on the assumptions and calculations made.

Abnormal Carbohydrate Metabolism

Of the many problems in abnormal metabolism that occur in animals, diabetes and ketosis are primarily concerned with faulty carbohydrate metabolism. Although diabetes does occur in lower animals, adequate information is not at hand to evaluate its importance. In man, it is an important disease, affecting people of all ages. Ketosis, on the other hand, appears to be more of a problem with domestic animals.

Ketosis. This syndrome involves an excess of ketones (acetone, acetoacetate, and β-hydroxybutyrate) accumulating in body tissues due to a disorder in carbohydrate or lipid metabolism. Increased concentration in blood is termed **ketonemia** or **acetonemia**; if levels are high enough to spill into the urine the condition is called **ketonuria**. The disease is common in cattle at the peak of lactation and in sheep in late pregnancy; it is characterized by hypoglycemia, depleted liver glycogen, elevated mobilization of adipose tissue lipids, increased production of ketones, and lipemia. These changes in energy metabolism resemble diabetes mellitus of man and animals. The impaired utilization of energy results in increased breakdown of tissue proteins for energy, loss of body weight, decreased milk production in lactating animals and abortion in pregnant animals. Water consumption is increased due to excessive loss of body fluids in the urine in response to ketonuria. The ketones in urine are accompanied by excessive losses of electrolytes (K, Na); this triggers tissue dehydration and induces increased water intake. In ruminants, propionic acid is the major volatile FA (VFA) used for glucogenesis (4). The only pathway for fatty acids to be converted to glucose, except for propionic, is through acetyl CoA and the tricarboxylic acid cycle.

Because limited glucose is absorbed from the GIT in ruminants, liver synthesis of glucose is a major source for maintenance of blood glucose and tissue glycogen levels. Thus, during periods of great physiologic demand for glucose, such as in lactation or pregnancy, ketosis becomes a serious practical problem in ruminants. Metabolic vitamin B_{12} deficiency has been implicated in ketosis in dairy cattle; this possible relationship needs further study. In pregnant ewes, ketosis is partly precipitated by reduced feed intake caused by reduced stomach capacity as a result of increased uterine size.

Ketosis also occurs in swine and other nonruminants during starvation or chronic underfeeding, often at the onset of a sudden new energy demand such as at the beginning of lactation. Treatment of ketosis has generally centered on the restoration of normal blood glucose concentration. Thus, intravenous glucose injections are common. Hormones, such as ACTH and adrenal corticoid hormones, have also been used, but the mechanism of their action is outside the realm of this discussion.

Diabetes mellitus. This relatively common malady of man results from an insufficiency of insulin from the pancreas; it has been reported also in farm animals, dogs, and cats (3). The insulin deficiency has a genetic basis, but can also be induced by overfeeding and obesity or hyperactive anterior pituitary or adrenal glands. Because insulin is required for normal glucose utilization, its absence results in hyperglycemia, urinary loss of glucose (glucosuria), and other changes as described for ketosis. Man and animals with a tendency toward diabetes show an impaired glucose tolerance (an abnormally long time for clearance of an oral or injected dose of glucose from the blood). A glucose tolerance test indicates not only the ability of the pancreas to secrete insulin but also the ability of the liver to utilize glucose. Mild diabetes mellitus can be controlled by feeding low-carbohydrate, high-protein diets.

References Cited

1. Aherne, F.X., V.W. Hays, R.C. Ewan and V.C. Speer. 1969. J. Animal Sci. 29:906.
2. Aherne, F.X., V.W. Hays, R.C. Ewan and V.C. Speer. 1969. J. Animal Sci. 29:444.
3. Bergman, E.N. 1970. In: Duke's Physiology of Domestic Animals, Comstock Pub. Co., Ithaca, N.Y.
4. Bergman, E.N., R.S. Reid and K. Kon. 1966. Amer. J. Physiol. 211:793.
5. Brambila, S. and F.W. Hill. 1966. J. Nutr. 88:84.
6. Brambila, S. and F.W. Hill. 1967. J. Nutr. 91:261.
7. Cori, C.F. 1925. J. Biol. Chem. 66:691.
8. Dadd, R.H. 1963. Advances in Insect Physiology 1:47.
9. Fridhandler, L. and J.H. Quastel. 1955. Arch. Biochem. Biophys. 56:412.
10. Gilmour, D. 1961. The Biochemistry of Insects. Acad. Press, N.Y.
11. Hitchcock, M.W.S. 1949. J. Physiol. 108:117.
12. Kiyasu, J.Y. and I.L. Chaikoff. 1957. J. Biol. Chem. 224:935.
13. Newton, W.C. and J. Sampson. 1951. Cornell Vet. 41:377.
14. Shelley, H.G. and G.S. Dawes. 1962. Nature 194:296.
15. Wilson, T.H. 1962. Intestinal Absorption. W.B. Saunders Co., Philadelphia.
16. Wise, M.B., E.R. Barrick, G.H. Wise and J.C. Osborne. 1954. J. Animal Sci. 13:365.

Chapter 8 – Lipids

Lipids are organic compounds that are insoluble in water (but soluble in organic solvents) and serve important biochemical and physiological functions in plant and animal tissues. The lipids of importance in nutrition of man and animals can be classified as follows:

Simple lipids are esters of fatty acids with various alcohols. Fats and oils and waxes are simple lipids. Fats and oils are esters of fatty acids with glycerol, and waxes are esters of fatty acids with alcohols other than glycerol.

Compound lipids are esters of fatty acids containing groups in addition to an alcohol and fatty acid. They include phospholipids, glycolipids and lipoproteins. Phospholipids (phosphatides) are fats containing phosphoric acid and N. Glycolipids are fats containing carbohydrate and, often, N, and lipoproteins are lipids bound to proteins in blood and other tissues.

Derived lipids include substances derived from the above groups by hydrolysis; i.e., fatty acids, glycerol and other alcohols.

Sterols are lipids with complex phenanthrene-type ring structures (see later section), whereas terpenes are compounds which usually have isoprene-type structures.

Fats and oils quantitatively make up the largest fraction of lipids in most food materials and are characterized by their high energy value. One gram of a typical fat yields about 9.45 Kcal of heat when completely combusted, compared with about 4.1 Kcal (see Ch. 9) for a typical carbohydrate.

Structure

The most important lipid constituents in animal nutrition include: fatty acids; glycerol; mono-, di-, and triglycerides; and phospholipids. Glycolipids, lipoproteins, and sterols may be important in metabolism, but these lipids as well as waxes and terpenes are quantitatively unimportant, nutritionally, or are poorly utilized.

Fatty Acids

The fatty acids consist of chains of C atoms ranging from 2 to 24 or more C's in length and characterized by a carboxyl group on the end. The general structure is RCOOH, where R is a C chain of variable length. Acetic acid, a major product of microbial fermentation of glucose in ruminants, has 2 C's. Its formula is: CH_3COOH.

Myristic acid, a constituent of milk fat, has 14 C's. Its formula is: $CH_3(CH_2)_{12}COOH$.

These fatty acids are saturated. That is, each C atom in the chain (except the carboxyl group) has 2 H atoms attached to it (3 H's at the terminal C). Some fatty acids are unsaturated; one or more pairs of C atoms in their chain are attached by a double bond and H has been removed. Linoleic acid, a constituent of corn oil and other plant oils high in polyunsaturated fatty acids, has 18 C's and two double bonds. Its formula is:

$$CH_3(CH_2)_4CH{=}CHCH_2CH{=}CH(CH_2)_7COOH$$

Linoleic acid

Most fatty acids commonly found in animal tissues are straight chained and contain an even number of C's. Branched-chain fatty acids and those with an odd number of C's are more common in microorganisms; however, ruminant animals, particularly, have relatively large amounts of these acids as a result of rumen fermentation.

Fatty acids containing double bonds can occur as the *cis* or the *trans* isomer as illustrated below:

$$\begin{array}{l} H-C-(CH_2)_7-CH_3 \\ \quad \| \\ H-C-(CH_2)_7-COOH \end{array} \qquad \begin{array}{r} CH_3-(CH_2)_7-C-H \\ \| \quad \\ H-C-(CH_2)_7-COOH \end{array}$$

Oleic acid (*cis*) Elaidic acid (*trans*)

The names, number of C's and number of double bonds for fatty acids most common in plant and animal tissues are given in Table 8-1. Linoleic, linolenic and arachidonic acids cannot be synthesized by animals, although recent work (3, 4) suggests the pig may synthesize some linoleic and arachidonic. The position of the double bond in the C chain is critical to biological activity. Table 8-2 shows the position of the double bonds in each of the common unsaturated fatty acids.

Glycerol

The formula for glycerol is:

$$\begin{array}{c} HOCH_2 \\ | \\ HOCH \\ | \\ HOCH_2 \end{array}$$

Glycerol

It is the alcohol component of all triglycerides common in animal and plant tissues and is a component of the phosphatides — lecithin, cephalin, and sphingomyelin.

Mono-, Di-, and Triglycerides

Monoglycerides, diglycerides, and triglycerides are esters of glycerol and fatty acids. An ester is formed by reaction of an alcohol with an organic acid; the structure of an ester and the linkage between glycerol and fatty acids in glycerides is illustrated:

$$R-\overset{O}{\overset{\|}{C}}-OH + HOR' \rightleftharpoons R-\overset{O}{\overset{\|}{C}}-OR' + H_2O$$

A monoglyceride, diglyceride and triglyceride would have the following general structures, where R, R' and R'' represent 3 different fatty acids:

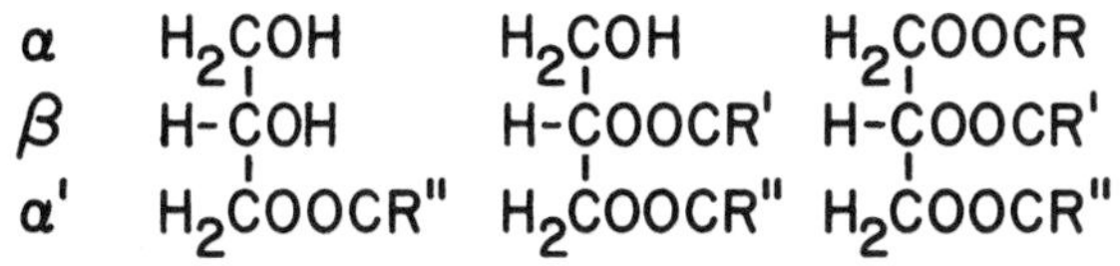

α	H_2COH	H_2COH	H_2COOCR
β	$H\text{-}COH$	$H\text{-}COOCR'$	$H\text{-}COOCR'$
α'	H_2COOCR''	H_2COOCR''	H_2COOCR''

The fatty acid composition of triglycerides is variable. The same or different fatty acids may be in all 3 positions; for example, if stearic acid occupied all 3 positions, the compound would be termed tristearin (a simple triglyceride), whereas if butyric, lauric, and palmitic acid each occupied one position, the compound would be called butyrolauropalmitin (glyceryl butyrolauropalmitate), a mixed triglyceride.

The chain length and degree of unsaturation of the individual fatty acids making up the triglyceride determines its physical and chemical properties. Simple triglycerides of saturated fatty acids containing ten or more C's are solid at room temperature, whereas those with less than ten C's are usually liquid. Triglycerides containing only long-chain saturated fatty acids are solids, whereas those containing a preponderance of unsaturated fatty acids are liquids.

Several constants commonly are used to characterize the chemical properties of fats. Constants of some common fats are given in Table 8-3. Each of these has some application in nutrition. **Saponification number** is the number of mg of KOH required for the saponification (hydrolysis) of 1 g of fat. The saponification number of a low molecular-weight fat (short-chain fatty acids) is large and becomes smaller as the molecular weight of the fat increases. Thus, the saponification number gives a measure of the average chain length of the 3 fatty acids in the fat. **Reichert-Meissl [RM] number** is the number of ml of 0.1N KOH solution required to neutralize the volatile water-soluble fatty acids (short-chain) obtained by hydrolysis of 5 g of fat. Beef tallow and other high molecular-weight fats contain practically no volatile acids and therefore have RM numbers of near zero, but butter contains a higher proportion of volatile acids and has a RM number of 17-35. **Iodine number** is the number of g of iodine that can be added to the unsaturated bonds in 100 g of fat. Iodine

Table 8-1. Fatty acids most common in plant and animal tissues.

Acid	No. carbons	No. double bonds	Abbreviated designation
Butyric (butanoic)	4	0	C 4:0
Caproic (hexanoic)	6	0	C 6:0
Caprylic (octanoic)	8	0	C 8:0
Capric (decanoic)	10	0	C10:0
Lauric (dodecanoic)	12	0	C12:0
Myristic (tetradecanoic)	14	0	C14:0
Palmitic (hexadecanoic)	16	0	C16:0
Palmitoleic (hexadecenoic)	16	1	C16:1
Stearic (octadecanoic)	18	0	C18:1
Oleic (octadecenoic)	18	1	C18:1
Linoleic (octadecadienoic)	18	2	C18:2
Linolenic (octadecatrienoic)	18	3	C18:3
Arachidic (eicosanoic)	20	0	C20:0
Arachidonic (eicosatetraenoic)	20	4	C20:4
Lignoceric (tetracosanoic)	24	0	C24:0

Table 8-2. Position of double bonds in unsaturated fatty acids.

Acid	Position of double bonds[a]	Precursor
Palmitoleic	9	Palmitic
Oleic	9	Stearic
Linoleic	9,12	None
Linolenic	9,12,15	None
Arachidonic	5,8,11,14	Linoleic

[a] C on carboxyl end of chain is number 1; thus a position of 9 means the double bond is between C9 and C10.

Table 8-3. Constants of some common fats.[a]

Fat	Saponification no.	Reichert-Meissl no.	Iodine no.
Beef	196-200	1	35-40
Butter	210-230	17-35	26-38
Coconut	253-262	6-8	6-10
Corn	187-193	4-5	111-128
Cottonseed	194-196	1	103-111
Lard	195-203	1	47-67
Linseed	188-195	1	175-202
Peanut	186-194	1	88-98
Soybean	189-194	0-3	122-134
Sunflower	188-193	0-5	129-136

[a] The constants for animal fats may vary outside the range of values listed because of unusual composition of dietary fats.

number is a measure of the degree of hydrogenation (saturation) of the fatty acids in the fat. A completely saturated fat such as tristearin has an iodine number of zero, whereas a liquid fat such as linseed oil has an iodine number of 175 to 202.

Phospholipids [phosphatides]

Phospholipids on hydrolysis yield fatty acids, phosphoric acid, and usually glycerol and a nitrogenous base. The general formula for lecithin is shown:

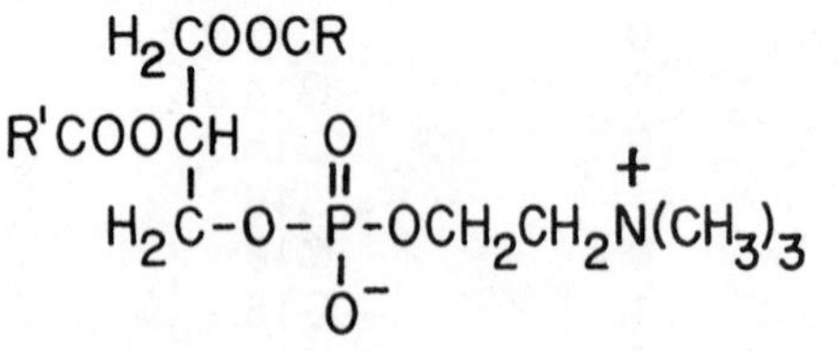

$$\begin{array}{l} \quad H_2COOCR \\ \qquad | \\ R'COOCH \quad O \\ \qquad | \qquad \| \qquad\qquad\qquad + \\ \quad H_2C-O-P-OCH_2CH_2N(CH_3)_3 \\ \qquad\qquad\;\; | \\ \qquad\qquad\;\; O^- \end{array}$$

L-α-Lecithin

Cephalins are similar to lecithins except that choline is replaced by hydroxyethyl amine in the molecule. Sphingomyelins do not contain glycerol, but contain fatty acids, choline, phosphoric acid, and the nitrogenous base, sphingosine.

These formulas are general representations of each group of compounds. The exact composition varies as to fatty acid composition and in other ways. Phospholipids of animal tissues are higher in unsaturated fatty acids than are the triglycerides of adipose tissue; phospholipids are more widely dispersed in body fluids than are neutral fats and have emulsifying properties that allow them to serve important functions in lipid transport.

Sterols

The most abundant sterol in animal tissue is cholesterol, shown below:

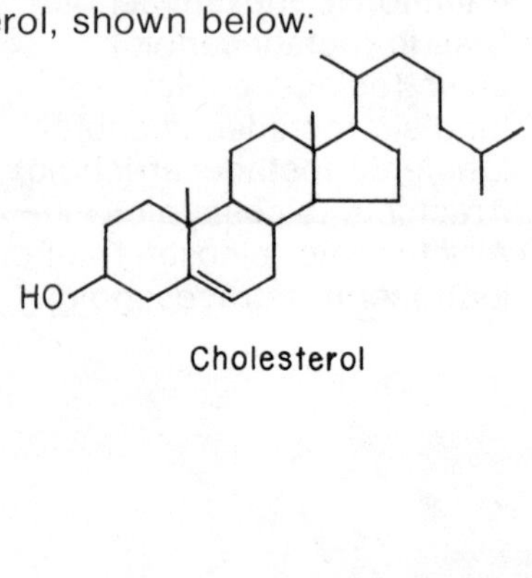

HO

Cholesterol

Other important sterols in animals are ergosterol (yields vitamin D_2 when irradiated); 7-dehydrocholesterol (yields vitamin D_3 when irradiated); bile acids; androgens (male sex hormones); and estrogens and progesterones (female sex hormones).

Functions

The functions of the lipids can be broadly listed as follows: to supply energy for normal maintenance and productive functions; to serve as a source of essential fatty acids; to serve as a carrier of the fat-soluble vitamins.

Energy supply

The hydrolysis of triglycerides provides glycerol and fatty acids which serve as concentrated sources of energy. Most of the variation among fat sources in the amount of utilizable energy they contain is related to their digestibility, but except in abnormal or special conditions of malabsorption, the true digestibility of fats exceeds 80%. When the total lipid content of the diet is low (<10%), as often occurs when animals are fed all-plant diets, apparent digestibility may be much less than this due to the higher proportion of metabolic fecal lipids on a low-fat diet. Also, a high proportion of waxes or sterols in the diet tends to reduce absorbability of the lipid, as these components are usually poorly digested and absorbed.

All of the energy in the diet except that present in essential fatty acids (see next section) may be provided by carbohydrate. Thus, no requirement exists for lipids as an energy source in the diet. Animals fed fat-free diets often develop fat-soluble vitamin deficiencies, however.

Figure 8-1. Skin lesions on a pig fed a fat-free diet [right] compared to a normal pig [left].

Essential Fatty Acids [EFA]

Linoleic acid (C18:2) and linolenic acid (C18:3) apparently cannot be synthesized by animal tissues, or at least not in sufficient amounts to prevent pathological changes, and so must be supplied in the diet. Arachidonic acid (C20:4) can be synthesized from C18:2, and therefore, is required in the diet only if C18:2 is not available.

The exact mechanisms by which EFA function in maintaining normal body functions are not known, but 2 probable vital areas are: they are an integral part of the lipid-protein structure of cell membranes (20); and they appear to play an important part in the structure of prostaglandins (6), hormone-like compounds widely distributed in reproductive organs and other tissues of man and animals. The functions and metabolism of prostaglandins are active areas of research (24). The prostaglandins are biosynthesized from arachidonic acid and have a wide variety of metabolic effects including the following: lower blood pressure, stimulate smooth muscle contraction, inhibit norepinephrine-induced release of fatty acids from adipose tissue and a variety of other tissue and species-specific effects.

Skin lesions and other abnormalities have been traced to deficiencies of certain fatty acids in monogastric species. The skin lesions which develop in pigs fed a fat-free diet are illustrated in Fig. 8-1. The following effects of a deficiency of EFA have been reported: scaly skin and necrosis of the tail; growth failure; reproductive failure; elevation of trienoic-tetraenoic ratio of tissue fatty acids; edema, subcutaneous hemorrhage and poor feathering in chicks.

Holman (26) suggested that in the rat a ratio of trienoic acids to tetraenoic acids of more than 0.4 in tissue lipids indicates a deficiency of EFA and that a level of linoleic acid at or exceeding 1% of the calories in the diet is sufficient to maintain a ratio of less than 0.4. Babatunde et al (3, 4) found increased triene-tetraene ratios in heart, liver, and adipose tissues of pigs fed diets containing no fat or 3% hydrogenated coconut oil compared to values obtained with 3% safflower oil, but found no skin lesions and no reduction in weight gain of pigs fed coconut oil, which aggravates EFA deficiency in the rat. They suggested the possibility that some linoleic acid synthesis occurs in the pig. Young ruminants (calves, kids, lambs) apparently require EFA in their diet, but no reports have appeared of EFA deficiency in adult ruminants. This is somewhat puzzling because rumen microflora hydrogenate most unsaturated fatty acids so that one would expect that EFA deficiency might be more likely than in other animals. Arachidonic acid has been found in high concentration in reproductive tissue of cattle and presumably is synthesized there as a precursor of prostaglandins. Additional studies clearly are needed to determine the degree of importance of linoleic and other fatty acids in the diet of pigs, cattle and other species, including man.

Carrier of the Fat-Soluble Vitamins

Absorption of the fat-soluble vitamins (A, D, E and K) is a function of digestion and absorption of fats. Fat-soluble vitamins are dispersed in micelles similar or identical to those formed in the absorption of fatty acids (see later section). Mixed micelles containing monoglycerides and free fatty acids take up fat-soluble vitamins more efficiently than micelles not containing them. A bile acid sequestrant, cholestyramine, has been shown (47) to reduce absorption of vitamin K when added to the diet, supporting the concept of an obligatory formation of bile salt-containing micelle formation. Because only a low level of dietary fat is needed for micelle formation, frank deficiencies of vitamin A, D, E, or K are unlikely to occur under normal dietary conditions.

Table 8-4. Effect of dietary fatty acid composition on fatty acid composition of the depot fat of growing pigs.[a]

Fatty acid designation	Dietary fat		Pig depot fat	
	Safflower	Hydrogenated coconut	Safflower	Hydrogenated coconut
	% of total fatty acids			
8:0		7.8		
10:0		5.7	0.2	0.2
12:0		44.5	0.2	0.6
14:0	trace	17.2	2.0	3.9
16:0	8.8	9.2	15.6	21.0
16:1			13.0	16.5
18:0	1.5	10.4	}50.7	}55.9
18:1	9.3	5.0		
18:2	80.4	0.2	17.2	0.9
18:3	trace		0.1	0.1

[a] From Babatunde et al (4)

Absorption

The preparation of lipids for absorption and the absorption process itself have been described (2, 27, 37). Digestion and absorption of lipids are illustrated schematically in Fig. 8-2.

The small intestine (duodenum) is the site of the major processes of preparation for absorption. Dietary lipids, mainly triglycerides, are discharged slowly from the stomach and are mixed with bile and pancreatic and intestinal secretions. Emulsification occurs here due to the detergent action of the bile salts and the churning action of the intestine, and the lipid particle size is reduced to spheres of 500-1000 mu in diameter. This smaller particle size allows for greater surface exposure to pancreatic and intestinal lipases which absorb on the particle surface and attack fatty acids in the 1 and 3 (α) positions, resulting in hydrolysis of triglycerides to β-monoglycerides and free fatty acids (FFA). The β-monoglycerides and FFA then combine with salt-phospholipid-cholesterol micelles (in about a 12.5:2.5:1 molar ratio) to form mixed micelles; these are essential for efficient absorption. Bile salts are detergent-like compounds that facilitate digestion and absorption of lipids. The presence of bile is necessary for efficient fat and fat-soluble vitamin absorption; in its absence, cholesterol absorption is reduced to near zero. Pancreatic lipase activity and resynthesis of triglycerides in the intestinal mucosal cells are promoted by bile salts.

The mixed micelles join with cholesterol and fat-soluble vitamins to form larger and more complex mixed micelles, each containing hundreds of molecules and having a diameter of 5-10 mu. These mixed micelles form microemulsions which then render the lipid ready for absorption as hydrolysis proceeds.

Bile salts are secreted in copious quantities (30 g/day in adult man). Bile salts are readily reabsorbed from the GIT in the lower jejunum and recycled to the liver; thus, the amount of daily secretion exceeds by several fold the amount present in the body at any one time as well as the amount of daily synthesis by the liver.

Several bile acids are common in animals, differing only in minor changes in the steroid portion of the molecule. Common bile acids are: cholic, deoxycholic, taurocholic and glycocholic acid. In the last 2, the amino acids, taurine and glycine, respectively, are part of the molecule. The structure of cholic acid is shown:

HO COOH HO OH

Cholic acid

The main site of absorption of lipids is the proximal (upper) jejunum, but some absorption occurs along the intestinal tract from the distal (lower) duodenum to the distal small intestine. Glycerol and short-chain FA (C2-C10) are absorbed by passive transport into the mesenteric veinous blood and then to the portal blood. Monoglycerides and long-chain FA enter the brush border (microvilli) and the apical core of the absorptive intestinal mucosal cells by diffusion. To a limited extent, some triglycerides may be absorbed intact as a fine emulsion of particles averaging 500 Angströms in diameter.

Most of the phospholipids in the intestinal lumen are partially hydrolyzed by pancreatic and intestinal lipases to yield FFA. The remainder of the molecule (lysophospholipid) is absorbed intact along with a small proportion of unhydrolyzed phospholipid. Although free cholesterol is readily absorbed, most other dietary sterols except vitamin D are poorly absorbed. Cholesterol esters must be hydrolyzed by pancreatic and intestinal lipases to form free cholesterol for absorption by displacement of the endogenous cholesterol of the microvilli lipoprotein. After entering the mucosal cell, free cholesterol is again esterified before transfer to the lymph system via the lacteals.

After entering the epithelial cell, long-chain FA are converted to derivatives of coenzyme A in the presence of ATP. This fatty acid-coenzyme A complex (termed fatty acyl coenzyme A) reacts with monoglyceride within

Symbols for Monomers
Bile salts
Phospholipids
Cholesterol
FFA
Monoglyceride
Fat-soluble vitamins

Micelle with FFA, monoglyceride, cholesterol, and fat-soluble vitamins added from emulsion particle. (5 to 10 mμ in diameter)

Cholesterol + Fat-soluble vitamins

Monoglyceride + FFA

Triglyceride

Emulsified lipid droplet containing mostly triglyceride but some cholesterol, fat-soluble vitamins, etc. (~750 mμ in diameter)

Pancreatic lipase

Micelle delivered to lumen in bile (<5 mμ in diameter)

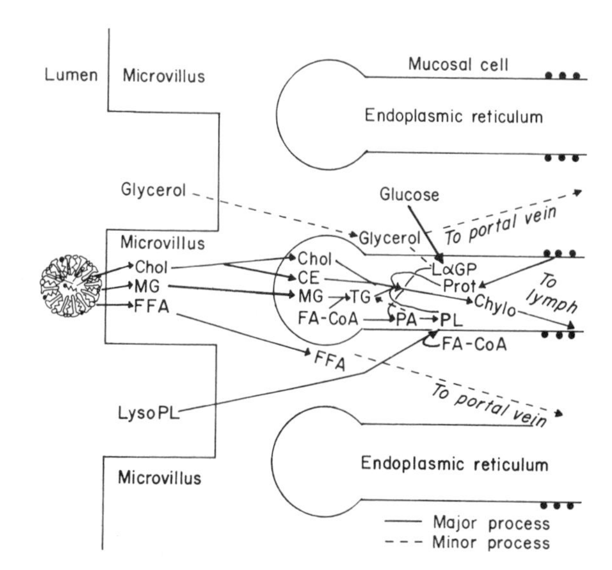

Figure 8-2. Digestion [left] and absorption [right] of lipids in the intestinal mucosa. Absorption. The symbols used in the micelle structure are the same as those used in digestion. In addition, the following symbols have been used: Chol, cholesterol; MG, monoglyceride; FFA, free fatty acid; CE, cholesterol ester; FA-CoA, fatty acyl-CoA; TG, triglyceride; PA, phosphate acid; PL, phospholipid; L-α-GP, L-α- glycerophosphate; Prot, protein; Chylo, chylomicron; ER, endoplasmic reticulum. From Masoro [37].

the cell to form di- and then triglycerides. The triglycerides thus formed contain only FA of C12 or greater chain length because shorter chain FA are absorbed directly into the portal system.

Before leaving the mucosal cell the mixed lipid droplets become coated with a thin layer of protein absorbed to the surface. These protein-coated lipid droplets are called chylomicrons and consist mainly of triglyceride with small quantities of phospholipids, cholesterol esters and protein. The chylomicrons leave the mucosal cell by reverse pinocytosis and enter the lacteals via the intercellular spaces. Lacteals lead to the lymphatic system which carries the chylomicrons to the blood via the thoracic duct. A summary of the major conversions that occur in transport of long-chain FA, phospholipids, cholesterol, and monoglycerides by the intestinal mucosal cell is diagrammed in Fig. 8-3.

Although mammals absorb most of these long-chain FA into the lymphatic system, the chicken apparently absorbs its dietary lipids directly into the portal blood (39, 50) which carries them directly to the liver. Nevertheless, the process of re-esterification of FA to triglycerides in the mucosal cell is similar in birds and mammals.

Transport and Deposition

Absorption of fat after a meal is associated with a large increase in lipid concentration of the blood referred to as lipemia. Blood lipids consist of chylomicrons formed within the intestinal mucosal cell during absorption, as well as lipids arising from mobilized depot stores and from synthesis in body tissues, especially the liver and adipose tissues. Blood lipids are transported as lipoproteins ranging from very low density (such as chylomicrons) to high density. The density is increased as the proportion of protein in the complex increases and the lipid decreases. Free FA (non-esterified FA) are transported as a complex with albumin. Very rapid removal of chylomicrons occurs from the blood by the liver, fat depots, and other tissues. For example, 1/2 of an injected dose of ^{14}C-labeled tripalmitin is removed from the blood plasma of dogs within 10 minutes (2).

The type and quantity of dietary lipid and the time after a meal are major determinants of the composition and concentration of lipids in blood at a particular time. In addition, such factors as species, age and endocrine status of the individual have an influence. Levels of cholesterol in the blood are affected by diet as well as by hepatic synthesis, but the ratio of free cholesterol to cholesterol esters and free cholesterol to phospholipid are rather constant in normal animals within a given species.

All tissues of the body store triglycerides. Adipose tissues (fat depots) are the most notable storage sites. Adipose tissue is capable of synthesizing fat from carbohydrate and of

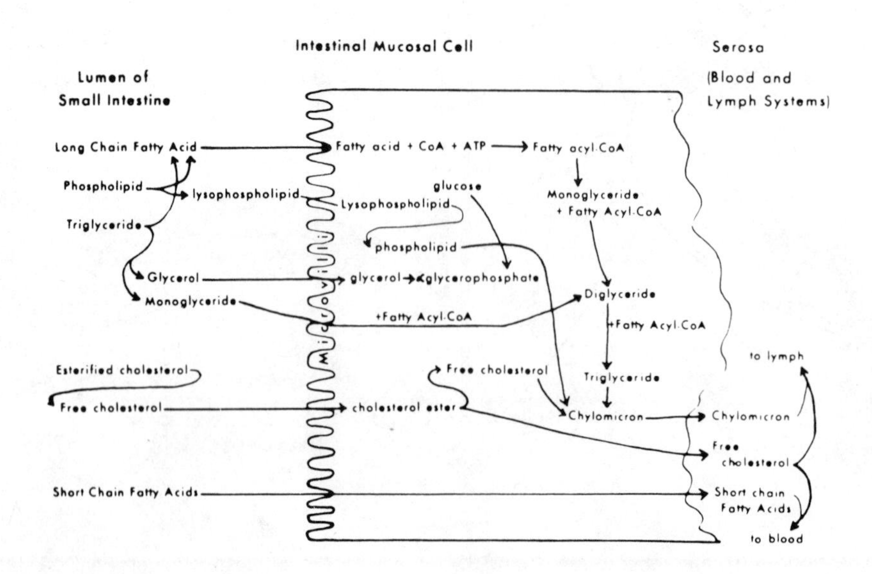

Figure 8-3. Schematic diagram of the major conversions that occur in transport of lipids across the intestinal mucosal cell during absorption.

oxidation of fatty acids. Because stored triglycerides are a ready source of energy, continuous deposition and mobilization clearly occurs in adipose tissue. Energy intake in excess of current needs results in a net deposition of triglycerides (fattening) and energy intake less than current needs (as in fasting) results in a net loss of triglycerides.

Triglycerides of depot fat tend to have a fatty acid composition characteristic for each animal species. In nonruminants, however, the fatty acid composition of the depot fat resembles that of the diet. This is illustrated in Table 8-4 for the pig fed semipurified diets containing various sources of fat. The depot fat of ruminant animals is less responsive to dietary fatty acid composition due to the action of the rumen microflora in metabolizing dietary fatty acids, although minor changes can be produced by dietary changes. Depot fat of ruminants can be changed to resemble dietary fat if the action of the rumen microflora is circumvented. Ogilvie et al (40) accomplished this by duodenal feeding of sheep, and others have been able to obtain softer depot fat in cattle and sheep by feeding very high levels of unsaturated FA, a portion of which presumably traversed the rumen without being metabolized by rumen microflora. Australian workers (14) showed that fatty acid composition of body fat can be modified by protecting dietary fat from rumen metabolism. The depot fat composition of ruminants can also be changed by altering the proportion of readily fermentable carbohydrates fed (35, 44). Ruminant fats are also characterized by odd-length and branched-chain fatty acids which are derived from volatile fatty acids and by the presence of *trans* isomers which result from metabolism of dietary unsaturated fatty acids.

Body lipids are clearly in a dynamic state of metabolism. The turnover rate of triglycerides in adipose tissue is extremely rapid. For example, the half-life in mice is 5 days; in rats, 8 days. The turnover rate of phospholipids and cholesterol is also rapid and may vary from 1 day for liver in some species to 200 days for brain in other species.

Fatty Acid and Triglyceride Metabolism

Liver, mammary gland and adipose tissue are the 3 major sites of biosynthesis of fatty acids and triglycerides (16). The liver is the central organ for lipid interconversion and metabolism and its role can be summarized as follows: synthesis of fatty acids from carbohydrates; synthesis of fatty acids from lipogenic amino acids; synthesis of cholesterol from acetyl CoA; synthesis of phospholipids; synthesis of lipoproteins; synthesis of ketone bodies (acetone, β-hydroxybutyric and acetoacetic acid); degradation of fatty acids; degradation of phospholipids; removal of phospholipids and cholesterol from blood; lengthening and shortening fatty acids; saturating and desaturating fatty acids; control of depot lipid storage; and storage of liver lipids.

Synthesis of fatty acids by liver and adipose tissue follows similar pathways, but the relative contribution made by each tissue differs greatly among the species studied. For example, in the mouse and rat about 1/2 of the synthesis occurs in the liver, but in the chicken (41) and pigeon (21) nearly all occurs in the liver; in the pig (41) nearly all occurs in adipose tissue and in the cow and sheep (5) both liver and adipose tissue are important, although the latter predominates (42).

Fatty Acid Biosynthesis

Synthesis begins with acetyl CoA (2 carbons) derived from carbohydrates, from certain amino acids or from degraded fats (see tricarboxylic acid cycle, Fig. 7-1). The fatty acid chain is assembled in 2-carbon units by joining of the carboxyl head of one fragment to the methyl tail of another. The details of fatty acid synthesis are illustrated in Fig. 8-4. Animal tissues synthesize carbon chains up to C16 in this way. Malonyl-CoA is, in effect, the source of the 2-C units. It combines with the even-numbered fatty acid-CoA esters. The synthesis of fatty acids occurs primarily in the microsomes of cells, but fatty acid oxidation and

Acetic acid + Coenzyme A → Acetyl CoA

Acetyl CoA + CO_2 → Malonyl CoA

Acetyl CoA + Malonyl CoA → Butyryl CoA

Butyryl CoA + Malonyl CoA → Caproyl CoA

Caproyl CoA + Malonyl CoA → etc.

Figure 8-4. Fatty acid synthesis. Fatty acids build up from acetic acid units which are made reactive by combining first with coenzyme A to form acetyl-CoA and then with carbon dioxide to form the CoA ester of malonic acid. Malonyl-CoA and acetyl-CoA can condense into an intermediate compound. Further reactions then reduce the intermediate to the CoA ester of a four-carbon fatty acid, butyryl-CoA. This, like acetyl-CoA, can condense with a molecule of malonyl-CoA, ultimately giving the ester of the 6-carbon fatt acid; the chain thus lengthens by successive steps. Adapted from Green [19].

CoA Ester of Caproic Acid (6C)

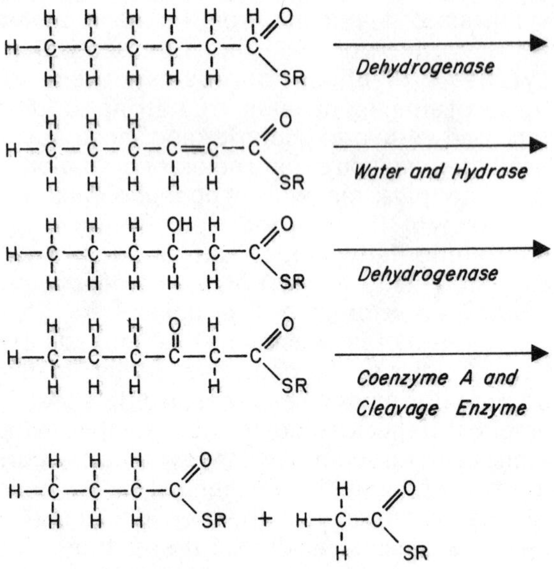

Dehydrogenase

Water and Hydrase

Dehydrogenase

Coenzyme A and Cleavage Enzyme

CoA Ester of Butyric Acid (4C) + Acetyl CoA

Figure 8-5. Breakdown of a fatty acid is an oxidative process, that is, hydrogens are removed by the actions of enzymes [dehydrogenases]. The fatty acid here [caproic acid, which has a 6-carbon chain] is not broken down in its free form but in the form of an ester of CoA [top]. After oxidation and hydration [first 3 steps], 2 carbon units split off from the chain in the form of acetyl-CoA [last step]. The remaining chain, still in the form of a CoA fatty-acid ester can go through the whole process again. Thus, a fatty acid of any length can be disassembled 2 units at a time until it is all reduced to acetyl-CoA or, if an odd-length acid, to propionyl-CoA. Adapted from Green [19].

synthesis of triglycerides occurs mainly in the mitochondria. Enzymes in mitochondria can use acetyl CoA to add C2 units to existing long-chain FA (C12, C14, C16), but formation of FA longer than C18 in this way is restricted to conversion of C18:2 to C20:4 and smaller amounts of other longer chain FA.

Desaturation of fatty acids occurs in animal tissues, but the extent is limited, as evidenced by the inability of most animals to synthesize C18:1, C18:3 and C20:4 from saturated FA of the same chain lengths. Desaturation of C18:0 to C18:1 and of C16:0 to C16:1 occurs at the 9, 10 position and subsequently moves 3 carbons toward the carboxyl end of the chain after elongation of the chain by 2 carbons at the carboxyl end. The mechanisms of desaturation of fatty acids in animal tissues are incompletely understood.

Triglyceride Biosynthesis

Synthesis occurs by fatty acyl CoA reacting with α-glycerol phosphate to form a phospholipid which is then converted to a diglyceride and thence to a triglyceride, or by fatty acyl CoA reacting with a monoglyceride to form a diglyceride and thence a triglyceride.

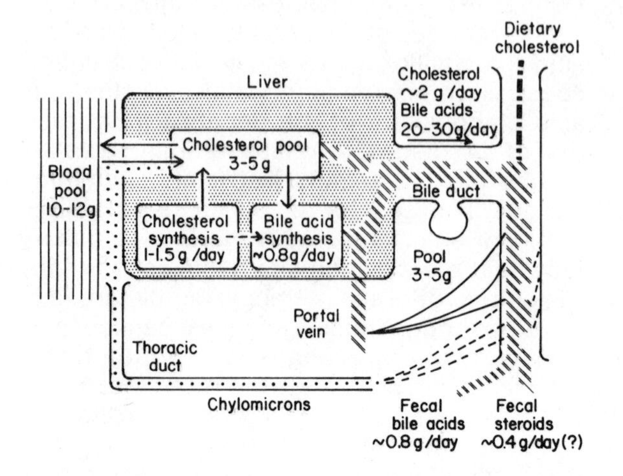

Figure 8-6. Summary of the turnover of cholesterol in man. From Bergstrom: In Ciba Foundation Symposium on the Biosynthesis of Terpenes and Sterols. Little, Brown and Co., 1959.

Fatty Acid Catabolism

Breakdown of long-chain FA proceeds by stepwise removal of two carbons at a time beginning at the carboxyl end (β-oxidation). The process is not exactly the reverse of synthesis, although acetyl CoA is the form in which the C2 fragments are removed. Before oxidation begins, the fatty acid is activated by esterification with CoA for form acyl CoA. At least 3 different enzymes are involved in oxidation of FA, one specific for short-chain (C2 and C3) and one for long-chain (C12 to C18) fatty acids. These enzymes (dehydrogenases) remove hydrogens and C2 units as acetyl CoA, as illustrated in Fig. 8-5. The acetyl CoA released in oxidation is available for resynthesis of fatty acids, for synthesis of steroids or ketones or for entry into the tricarboxylic acid cycle. The total energy produced by complete degradation of long-chain fatty acids comes partly from the β-oxidation sequence and partly from the oxidation of acetyl CoA in the tricarboxylic acid (TCA) cycle.

Although even-numbered carbon FA are by far the most prevalent in animal tissues, some odd-numbered carbon FA are present which are oxidized by a slightly different route.

Steroid Metabolism

Cholesterol is the most abundant steroid in the diet and the precursor of most other animal steroids. Biosynthesis is from acetyl CoA. Regulation of biosynthesis is partly by dietary intake; a high intake depresses synthesis by the liver, and low intake or reduced absorption results in increased synthesis. The turnover of cholesterol in man is summarized in Fig. 8-6. The liver of adult man contains about 3-5 g and the blood pool is 10-12 g. Daily synthesis is 1-1.5 g, of which about 1/2 is converted to bile acids. Secretion of cholesterol and bile acids into the intestinal lumen via the bile duct approximates 2 and 20-30 g/day, respectively, but due to reabsorption by the enterohepatic circulation, less than 1 g of each is lost in the feces. Thus, compounds that reduce absorption of cholesterol and bile acids may have a profound effect on the body pool of cholesterol and on its biosynthesis. One such compound is cholestyramine, a nonabsorbable resin which has been used to control hypercholesterolemia (13).

In addition to excretion of cholesterol in the bile and its conversion to bile acids, it can be used for steroid hormone synthesis (progesterone, adrenal cortical hormones, testosterone, and estrogen) or stored as a component of pathologic deposits in bile ducts (gall stones) and in arteries (atherosclerotic) plaques). Conjugation of bile acids with taurine or glycine results in the excretion of these conjugated bile acids in the bile.

Phospholipid Metabolism

The most abundant phospholipid in animal tissues is lecithin (phosphatidyl choline). It can be synthesized by 2 pathways, either by making use of choline directly or by methylation of phosphatidyl ethanolamine. The latter conversion occurs only in the liver, but the former can also occur in other tissues.

Degradation of phospholipids occurs in most mammalian tissues by hydrolysis of carboxy-esters and phosphate esters. Fatty acids and other metabolites released by hydrolysis can enter the tricarboxylic acid cycle. Glycerol can enter the glycolysis pathway or be used in triglyceride or phospholipid synthesis.

Ketones

Formation of ketones (ketogenesis) is a continuous process but may be excessive in certain disorders described in more detail later. The ketones are acetone, acetoacetic acid and β-hydroxybutyric acid. The ketones are rapidly removed from the blood by skeletal muscle and other peripheral tissues and provide a substantial supply of energy for use by these tissues. Their synthesis originates with acetyl CoA:

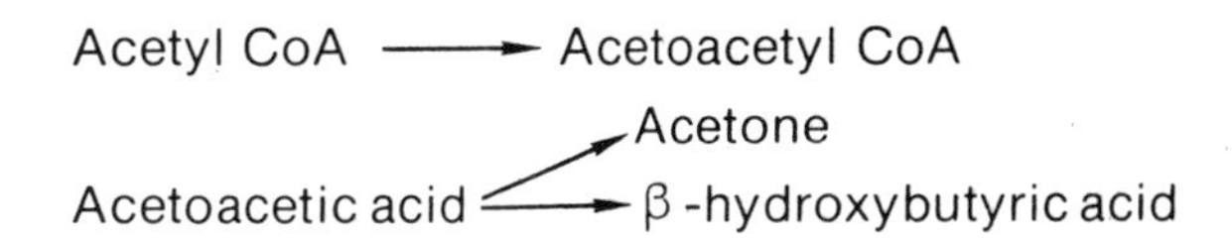

Table 8-5. Effect of frequency of feeding on pig adipose tissue enzymes, fatty acid synthesis, and and oxidation of plasma FFA.[a]

Feeding pattern	Malic enzyme[b]	Citrate cleavage[b] enzyme	Plasma FFA[c]	FA synthesis	FA oxidation[d]
Nibbler	118 + 10	33 + 5	190 + 15	188 + 36	89 + 16
Meal-fed (2 hr/24)	106 + 14	39 + 5	296 + 23	218 + 38	106 + 21
Meal-fed (2 hr/48)	219 + 15	36 + 4	628 + 94	248 + 26	118 + 18

[a] From Allee et al (1)
[b] Nanomoles substrates coverted to product/min./mg protein. Meq/l.
[c] Nanomoles glucose-5-^{14}C converted to fatty acid/100 mg tissue/2 hr.
[d] Nanomoles glucose-5-^{14}C converted to CO_2/100 mg tissue/2 hr.

Effects of Frequency of Feeding on Metabolism

It has generally been assumed that the composition of the diet and the level of dietary intake were the 2 factors responsible for controlling body composition and metabolism of energy. Now it is clear that the distribution of intake during a fixed period of time has an important effect on lipogenesis and body composition in some species (34). Such terms as meal eater vs. nibbler, single feeder vs.

multiple feeder and others have been used to describe the phenomenon of feeding frequency. Several reports have shown that bodies of meal-eating rats (those trained to consume the entire 24-hr ration in 2 hr) contain more fat and less protein and water than nibblers (11, 12, 22). Leveille (32) found that adipose tissue accounted for only 50-90% of total fatty acids synthesized in nibbling rats, but at least 95% in meal-fed rats. Meal-eating alters activities of enzymes involved in both carbohydrate and fat metabolism (10, 11). Metabolic adaptations developed in the rat by meal-eating persist for several weeks after feed is provided ad libitum. Meal-fed chickens have been shown to have elevated plasma cholesterol and much higher incidence of atherosclerosis than ad libitum-fed chickens (both groups on a high-cholesterol diet), even though daily feed intake was less in meal-fed chickens. Livers of meal-fed chickens incorporate more acetate into fatty acids than those of controls (31) and apparently are similar to the rat in their response to feeding frequency. In the pig, meal-eating also appears to influence weight gain and efficiency of feed utilization, but longer time of fasting (one 2-hr feeding every 48 hr) is needed than in the rat or bird to achieve increased adipose tissue lipogenesis associated with a longer post-absorptive state (1). The rise in malic enzyme, plasma FFA, increased FA synthesis, and oxidation are illustrated in Table 8-5. The capacity of the GIT also increases in meal-fed as compared with nibbling animals. Ruminants, whose rumen provides a reservoir of nutrients for absorption on a comparatively continuous basis, have not been shown to respond to frequency of feeding changes with appreciable changes in body composition (22).

Cohn (9) suggested, on the basis of data obtained with nonruminants, that man should increase frequency of food intake as a means of minimizing obesity, reducing serum lipid and cholesterol levels, and decreasing susceptibility to atherosclerosis and diabetes mellitus. More research is needed in man to evaluate this suggestion. Although data with humans are limited, Young et al (51) suggested that infrequent meals predispose humans to obesity, increase serum cholesterol and impair glucose tolerance.

Long-term fasting results first in depletion of liver and muscle glycogen and then oxidation of tissue lipids to meet energy requirements. Increased lipogenesis occurs in liver and less so in adipose tissue after refeeding of fasted animals. The rates of fatty acid and cholesterol synthesis are affected by prefasting diet (high fat diets inhibit the increase in lipogenesis), prefasting body weight (restricted-fed animals have greater fatty acid incorporation into depot fats than controls, but less turnover of cholesterol) and postfasting diet (high fat diets inhibit lipogenesis) (23).

Obesity in Man and Animals as Related to Lipid Metabolism

Mounting evidence (36, 45) shows that obesity has a strong genetic basis involving differences between lean and obese individuals in activities of tissue enzymes associated with lipid synthesis and oxidation. Fat cells from genetically obese rats convert more pyruvate or glucose to glyceride-glycerol than those from rats made obese by hypothalamic lesions which cause excessive feed intake. Adipose tissues, but not liver enzymes associated with lipogenesis, are several-fold higher in obese than in lean strains of swine (36), and gluconeogenic enzymes are higher in obese swine; the enzyme response to fasting and refeeding is greater in the lean pig.

The role of nutrition in these aspects of obesity and lipid metabolism is obscure at present, but nutritionists clearly should be aware of and concerned about possible interactions between genetics and nutrition, with respect to body composition and metabolism.

Effect of Dietary Carbohydrate Source on Lipid Metabolism

Considerable recent interest has focused on the effects of dietary carbohydrate on lipid metabolism (25), especially since reports linking high sucrose intake with atherosclerosis. Epidemiological studies with humans implicating such a relationship (49) stimulated research activity in this area and Brooks et al (8) have provided evidence that high intakes of sucrose can indeed produce atherosclerotic lesions in the heart of pigs. Plasma triglyceride and cholesterol tend to be elevated in animals fed high levels of sucrose and these plasma components have been implicated in human atherosclerosis. The effect of sucrose is assumed to be related to the metabolism of the fructose moiety, but no proof exists for that. It does seem clear that refined glucose or starch does not induce increased plasma triglycerides. Much more research is needed before the exact role of dietary carbohydrate source on lipid metabolism is understood.

Abnormalities in Metabolism of Lipids

Abnormal metabolism of lipids may occur in animals and man as a result of genetic factors or in response to alterations in environment, including diet.

Fatty Livers

Because the liver is a key organ in metabolism, it is not surprising that factors affecting liver function have far-reaching effects on the overall well-being of the animal. A common manifestation of abnormal liver function is accumulation of lipids in the liver. Normally, fat constitutes about 5% of the wet weight of the liver, but the value may be 30% or more in pathologic conditions. Fatty liver may arise from: high fat or high cholesterol diet; increased liver lipogenesis due to excessive carbohydrate intake or excessive intake of certain B-vitamins (biotin, riboflavin, thiamin); increased mobilization of lipids from adipose tissue due to diabetes mellitus, starvation, hypoglycemia, increased hormone output (growth hormone, adrenal corticotrophic hormone, adrenal corticosteroids); decreased transport of lipids from liver to other tissues due to deficiencies of choline, pantothenic acid, inositol, protein, or certain amino acids (methionine, threonine); cellular damage to the liver (cirrhosis, necrosis) because of infections, vitamin E-Se deficiency, or liver poisons such as chloroform and carbon tetrachloride.

Atherosclerosis

This is the name used to describe a disease characterized by progressive degenerative changes occurring in the blood vessels and heart of man and animals which ultimately are responsible for more deaths among humans in the USA than any other single cause. More than half (54.1%) of all deaths in the USA result from cardiovascular disease and about 2/3 of these are due to atherosclerotic heart disease and l/8 to cerebral accidents (strokes) resulting from atherosclerosis (38). Among the factors responsible for development of atherosclerosis, nutrition has received perhaps the most attention. The observation that serum cholesterol concentration seems to be correlated with incidence of atherosclerosis has led to numerous studies of the influence of diet on serum cholesterol. Rabbits and other animals fed cholesterol develop severe atherosclerosis. The cholesterol of blood is transported in an association with other lipids and with proteins. The composition of serum lipoproteins in man is summarized in Table 8-6. Each species of animals has its own peculiar serum lipoprotein composition, but the profile for normal man provides a useful guide. β-lipoproteins are highest in cholesterol, and it is this fraction that is elevated in atherosclerosis.

The effect of saturated animal fats (such as butter, tallow, lard) and eggs (high in cholesterol) on serum cholesterol of man has received much attention and publicity. Saturated fats tend to raise serum cholesterol and, of course, consumption of products high in cholesterol such as eggs add to the body burden of cholesterol in addition to that synthesized by liver. Despite the wide public attention to animals products as possible culprits in contributing to the prevalence of atherosclerosis in man, scientists still disagree as to their role. Some consider elevated serum triglycerides as a possible factor in the atherosclerotic process (dietary sucrose elevates serum triglycerides), but others believe other nutritional and metabolic factors are important. A series of genetic lipoprotein disorders asso-

Table 8-6. Composition of serum lipoproteins.[a]

Type	Density	Concentration	Protein	Triglyceride	Phospholipid	Cholesterol ester	Cholesterol free	Fatty acid
		mg/100 ml	%	%	%	%	%	%
Chylomicron	0.96	0.50	1	87	8	3	1	
β-lipoprotein	0.96 -1.006	150	7	52	19	14	7	1
	1.006-1.019	50	11	26	23	31	8	1
	1.019-1.063	350	21	10	22	38	8	1
α-lipoprotein	1.063-1.125	50	33	11	29	21	6	
	1.125-1.210	300	57	5	20	12	3	3

[a] From Kritchevsky (30)

ciated with early development of atherosclerosis has been described in man (34). In addition, Fredrickson et al (17, 18) have determined at least 5 genetically determined causes of hyperlipidemia in man (hypercholesterolemia and/or elevated triglycerides) which require dietary modifications to control. Individuals affected with any of these types of hyperlipidemia are treated by changing the dietary polyunsaturated to saturated fatty acid ratio to 2:1. Saturated fat should not exceed 10% and total fat 35% of total diet calories. Such drastic dietary changes as these have not been recommended for the general population.

Many animal models are being used to study the nutritional aspects of atherosclerosis. These include the rabbit, pig, chicken, monkey and many others; however, data obtained with animal models may be used for extrapolation to humans only with reservations.

The National Heart and Lung Institute Task Force on Atherosclerosis (38) has identified the following factors as increasing the risk from coronary heart disease: age; male sex; elevated serum lipids (hypercholesterolemia, hyperlipidemia); hypertension; cigarette smoking; impaired glucose tolerance (diabetes mellitus); and obesity. In addition, psychological stress and heredity are often mentioned as possible contributing factors.

References Cited

1. Allee, G.L., D.R. Romsos, G.A. Leveille and D.H. Baker. 1972. J. Nutr. 102:1115.
2. Allen, R.S. 1970. In: Duke's Physiology of Domestic Animals. 8th ed. Comstock Pub. Co.
3. Babatunde, G.M. et al. 1967. J. Nutr. 92:293.
4. Babtunde, G.M., W.G. Pond, E.F. Walker, Jr. and P. Chapman. 1968. J. Animal Sci. 27:1290.
5. Ballard, F.J., R.W. Hanson and D.S. Kronfeld. 1969. Fed. Proc. 28:218.
6. Bergstrom, S. and B. Samuelson. 1965. Ann. Rev. Biochem. 34:101.
7. Bloor, W.R. 1925-26. Chem. Rev. 2:243.
8. Brooks, C.C., A.Y. Miyahara, D.W. Huck and S.M. Ishizaki. 1972. J. Animal Sci. 35:31.
9. Cohn, C. 1964. Fed. Proc. 23:76.
10. Cohn, C. and D. Joseph. 1959. Amer. J. Physiol. 197:1347.
11. Cohn, C. and D. Joseph. 1960. Amer. J. Clin. Nutr. 8:682; Metabolism 9:492.
12. Cohn, C., D. Joseph, L. Bell and P. Allweiss. 1965. Ann. N.Y. Acad. Sci. 131:507.
13. Cook, D.A., L.M. Hagerman and D.L. Schneider. 1972. Proc. Soc. Exptl. Biol. Med. 139:70.
14. Hogan, J.P., P.J. Connell and S.C. Mills. 1972. Austral. J. Agr. Res. 23:87.
15. Elliot, J.M. 1966. Proc. Cornell Nutr. Conf., p 73.
16. Favarges, P. 1965. In: Handbook of Physiology, section 5: Adipose Tissue. Amer. Physiol. Soc., Washington, D.C.
17. Fredrickson, D.S. 1972. Amer. J. Clin. Nutr. 25:221.
18. Fredrickson, D.S., R.I. Levy and R.S. Lees. 1967. New Eng. J. Med. 276:34, 94, 148, 215, 273.
19. Green, D.E. 1960. Scientific American, Feb.
20. Green, D.E. and S. Fleischer. 1963. Biochem. Biophys. Acta 70:554.
21. Goodridge, A.G. and F.G. Ball. 1967. Amer. J. Physiol. 213:245.
22. Han, I.K. 1965. Ph.D. Thesis, Cornell Univ., Ithaca, N.Y.
23. Hill, R., J.M. Linazasoro, F. Chevallier and I.L. Chaikoff. 1958. J. Biol. Chem. 233:305.
24. Hinman, J.W. 1972. Ann. Rev. Biochem. 41:161.
25. Hodges, R.E. and W.A. Krehl. 1965. Amer. J. Clin. Nutr. 17:334.
26. Holman, R.T. 1960. J. Nutr. 70:405.
27. Isselbacher, K.J. 1965. Fed. Proc. 24:16.
28. Jackson, H.D. and V.W. Winkler. 1970. J. Nutr. 100:201.
29. Jansen, G.R., C.F. Hutchison and M.E. Zanetti. 1966. Biochem. J. 99:333.
30. Kretchevsky, D. 1967. J. Dairy Sci. 50:776.
31. Leveille, G.A. 1966. J. Nutr. 90:449.
32. Leveille, G.A. 1967. Proc. Soc. Exptl. Biol. Med. 125:85.
33. Leveille, G.A. 1972. Nutr. Rev. 30:151.
34. Lowe, C.U. 1972. Amer. J. Clin. Nutr. 25:245.
35. Luther, R. and A. Trenkle. 1967. J. Animal Sci. 26:590.
36. Martin, R.J., J.L. Bobble, T.H. Hartsock, H.B. Graves and J.H. Ziegler. 1973. Proc. Soc. Exptl. Biol. Med. 143:198.
37. Masoro, E.J. 1968. The Physiological Chemistry of Lipids in Mammals. W.B. Saunders Co., Philadelphia, Pa.
38. National Heart and Lung Institute Task Force on Arteriosclerosis. 1971. DHEW Publication No. (NIH) 72-219, Washington, D.C.

39. Noyan, A., W.J. Lossow, N. Brot and I.L. Chaikoff. 1964. J. Lipid Res. 5:538.
40. Ogilvie, B.M., G.L. McClymont and F.B. Shorland. 1961. Nature 190:725.
41. O'Hea, E.K. and G.A. Leveille. 1969a. Comp. Biochem. Physiol. 30:149.
42. Ingle, D.L., D.E. Bauman and U.S. Garrigus. 1972. J. Nutr. 102:617.
43. O'Hea, E.K. and G.A. Leveille. 1969b. J. Nutr. 99:338.
44. Trenkle, A. 1970. J. Nutr. 100:1323.
45. Trystad, O., I. Foss, E. Vold and N. Standal. 1972. FEBS Letters 26:311.
46. Warner, A.C.I. 1964. Nutr. Abst. Rev. 34:339.
47. Whiteside, C.H., R.W. Harkins, H.B. Fluckiger and A.P. Sarett. 1965. Amer. J. Clin. Nutr. 6:309.
48. Wilson, T.H. 1962. Intestinal Absorption. W.B. Saunders Co., Philadelphia, Pa.
49. Yudkin, J., Jr. 1964. Lancet 2:6.
50. Young, R.J. 1966. Proc. Cornell Nutr. Conf., p 107-110.
51. Young, C.M. et al. 1971. J. Amer. Dietetic Assoc. 59:473.

Chapter 9 – Energy Metabolism

The topic of energy and its metabolism is known as bioenergetics. In the overall subject of animal nutrition, it is important because energy is, quantitatively, the most important item in an animal's diet and all animal feeding standards (Ch. 16) are based on energy needs. As such, an appreciable amount of effort has been expended to study the metabolism of energy by animals. The reader may feel that nonruminant animals are unduly slighted in this chapter; however, considerably more work, particularly on net energy, has been done on ruminant species than on monogastric species. Those interested in more detail than is presented here are referred to Brody (5), Mitchell (11, 12), Blaxter (2) — strictly on ruminant species, and Ch. 4 and 15 in Hafez and Dyer (8).

With all other nutrients, modern laboratory procedures allow us to fractionate feedstuffs, animal tissues and so forth into their component parts, and we can isolate proteins, lipids, different minerals, and vitamins which we can weigh, see, smell, or taste. However, study of energy metabolism requires a different approach because energy may be derived from most organic compounds ingested by the animal. The animal derives energy by partial or complete oxidation of organic molecules ingested and absorbed from the diet or from metabolism of energy stored in the form of fat or protein.

The biochemical mechanisms by which biological organisms cope with energy transfer and oxidation are outside the scope of this chapter. Biochemists have, in general, pretty well defined the different compounds and enzyme systems that accomplish these reactions (see Ch. 7). Many readers are probably aware that energy transfer from one chemical reaction to another primarily occurs by means of high-energy bonds found in compounds such as ATP (adenosine triphosphate) and other related compounds. It is sufficient, for our purposes here, to say that all animal functions and biochemical processes require a source of energy to drive the various processes to completion. This applies to all life processes and animal activity such as chewing, digestion, and maintenance of body temperature, or liver metabolism of glucose, absorption from the GIT, storage of glycogen or fat, or protein synthesis.

In normal body metabolism, there is a tremendous transfer of energy from one type to another, for example, from chemical to heat (oxidation of fat, glucose or amino acids); from chemical to mechanical (any muscular activity), or from chemical to electrical (glucose oxidation to electrical activity of the brain). Based on biochemical laboratory data, the energy cost of many of these reactions can be estimated with reasonable precision, but other animal functions such as excretion and digestion have an energy input from so many different tissues and chemical reactions that it is difficult to evaluate their cost to the animal.

Energy Terminology

Energy may be defined as the capacity to do work where work is the product of a given force acting through a given distance. A broad definition such as this is not directly applicable to animals, however, as we are usually more concerned with the utilization of chemical energy. Chemical energy may be measured in terms of heat and expressed as calories (or B.T.U.'s), although, according to physicists, the joule is a more precise means of expression. In international usage, a calorie (cal) is the amount of heat required to raise the temperature of 1 g of water from 14.5° to 15.5° C and is equivalent to 4.1855 joules. A kilocalorie (Kcal) is equal to 1000 cal and a megacalorie (Mcal), or therm, is equal to 1000 Kcal or one million cal.

The manner in which energy is partitioned into various fractions in terms of animal utilization is shown schematically in Fig. 9-1. Detailed discussion on each of these fractions follows.

Gross Energy [GE]

Gross energy (GE) is the quantity of heat resulting from the complete oxidation of food, feed, or other substances. The term heat of combustion which is used in chemical terminology, means the same thing. GE is measured in apparatus called a bomb calorimeter (see Fig. 2-4, p. 11). GE values are often obtained on feedstuffs or rations in the process of arriving at energy utilization. Energy values of different feedstuffs or nutrients vary, but typical values are (Kcal/g): carbohydrates, 4.10; proteins, 5.65; and fat, 9.45. The differences here primarily reflect the state of oxidation of the initial compound. For example, a typical monosaccharide such as glucose has an empirical formula of $C_6H_{12}O_6$, or 1 atom of oxygen/atom of C; whereas in a fat molecule such as tristearin, we have 6 atoms of O and 57 atoms of C; thus the fat requires more oxygen for oxidation and gives off more heat in the process.

Examples of GE values of some selected tissues, nutrients, or compounds are shown in Table 9-1. Note, for example, that a poor quality feed such as oat straw has the same GE value as corn grain. This comparison clearly points out the fact that GE values, by themselves, are of little practical value in evaluating feeds for animal usage.

GROSS ENERGY (GE) of Feed (heat of combustion)

→ Fecal Energy
1. Undigested feed
2. Enteric microbes & their products
3. Excretions into the GIT
4. Cellular debris from the GIT

APPARENT DIGESTIBLE ENERGY (DE)

→ Urinary Energy
Gaseous Products of Digestion
(primarily methane)

METABOLIZABLE ENERGY (ME)

→ Heat Increment (heat of nutrient metabolism)
Heat of Fermentation
(from the rumen, cecum, large intestine)

NET ENERGY

Maintenance Energy (NE_m)
1. Basal Metabolism
2. Voluntary Activity
3. Energy to keep body warm or cool

Productive Energy (NE_p)
1. Growth
2. Fattening
3. Work
4. Milk
5. Wool, hair
6. Reproductive Energy Storage

Figure 9-1. Schematic diagram of energy utilization by animals.

Digestible Energy [DE]

The GE of food consumed minus fecal energy is called apparent DE. In practice, the GE intake of an animal is carefully measured over a period of time accompanied by collection of fecal excretion for a representative period. Both feed and feces are analyzed for energy content and this, then, allows for calculation of DE. It might be noted (see Fig. 9-2) that energy lost in the feces accounts for the single largest loss of ingested nutrients. Depending upon the species of animal and the diet, fecal losses may range from 10% or less in milk-fed animals to 60% or more in animals consuming poor quality roughage.

Apparent DE is not a true measure of the digestibility of a given diet or nutrient because the GIT of an animal is an active site for excretion of various products that end up in feces, and because there may be considerable sloughing of cellular debris from cells lining the GIT, neither factor having any direct relation to undigested food residues. In addition, undigested microbes and their metabolic byproducts may constitute a large portion of the feces of some species. Although some of these microbes might be digestible if passed through the stomach and small intestines, much of the growth occurs in the cecum and large intestine where there are no enteric enzymes and (apparently) relatively little absorption. Only in the case of fibrous plant components such as cellulose and xylan, which are foreign to the animal body, are values a measure of true digestibility.

True digestible energy is determined by measuring, in addition, the energy in fecal excretions (metabolic fecal energy) of an animal that is fasting or being fed a diet presumed to be completely absorbed, such as milk or eggs. This amount is then subtracted from total fecal excretion of the fed animal. This determination is not feasible with most herbivorous species and is seldom done in practice with any species.

Total Digestible Nutrients [TDN]

TDN is not shown in the scheme in Fig. 9-1, but it is probably the most common measure of energy used in ration formulation in the USA for ruminants and swine. TDN is roughly comparable to DE, but is expressed in units of weight or percent. When conversion of TDN to DE is desired, the values usually used are 2000 Kcal of DE/lb of TDN or 4.4 Kcal/g. TDN is determined by carrying out a digestion trial and summing the digestible protein and carbohydrates (NFE and crude fiber) plus 2.25 times digestible ether extract (crude fat). Fat is multiplied by 2.25 in an attempt to account for its higher caloric value. As compared to DE, TDN undervalues protein because protein is not completely oxidized by the body whereas it is in a bomb calorimeter. Multiplication of digestible protein by 1.25 would put TDN on a more comparable basis as compared to DE. The formula for TDN is: TDN = DCP + DNFE + DCF + 2.25(DEE).

Although most nutritionists recognize that TDN or DE tend to overvalue roughages as compared to some version of NE, the popularity of TDN is partly because of the relative ease of obtaining the necessary information and also because of the better understanding of its use by nonprofessionals. Current NRC publi-

Table 9-1. Gross energy values (dry basis) of various tissues, nutrients or feedstuffs.

Item	GE, Kcal/g	Item	GE, Kcal/g
Carbohydrates		Proteins, amino acids, urea	
Glucose	3.74	Beef muscle (ash-free)	5.3
Sucrose	3.94	Casein	5.9
Starch	4.18	Egg albumin	5.7
Cellulose	4.18	Gluten	6.0
Glycerol	4.31		
		Alanine	4.35
Fats, fatty acids		Tyrosine	5.91
Butterfat	9.1	Urea	2.52
Beef fat (ash-free)	9.4		
Corn oil	9.4	Ethyl alcohol	7.11
Coconut oil	8.9		
		Feeds	
Acetic acid	3.49	Corn grain	4.4
Propionic acid	4.96	Wheat bran	4.5
Butyric acid	5.95	Grass hay	4.5
Palmitic acid	9.35	Oat straw	4.4
Stearic acid	9.53	Soybean	5.5
		Linseed oil meal	5.1

cations tend to emphasize other energy values, but it should be pointed out that most of these values for ME or NE were derived from TDN values.

Metabolizable Energy [ME]

ME is defined as the GE of feed minus energy in feces, urine, and gaseous products of digestion. Values so obtained, thus, account for further losses as a result of digestion or metabolism of the ingested feed. Losses of combustible gases are negligible and are normally ignored in many monogastric species, although some losses occur as a result of fermentation in the cecum and large gut. ME is commonly used to evaluate feedstuffs for poultry and in establishing feeding standards, because feces and urine are excreted together. Thus, it is convenient to use ME values for these species. An appropriate formula for calculating ME for swine where DE is known is:

$$\text{ME (in Kcal/kg)} = \frac{\text{DE (in Kcal/kg)} \times 0.96 - (0.202 \times \text{protein \%})}{100}$$

Methane usually accounts for most of the combustible gases in ruminant species, and may range from 3 to 10% of GE, depending on the nature of the diet and level of feed intake. Low quality diets result in larger proportions of methane and, generally, the percentage of GE lost as methane declines as feed intake increases. Several formulas exist for calculating gaseous energy losses in ruminants. One given by Blaxter and Clapperton (3) is: methane = 1.30 + 0.112D-L (2.37 - 0.050D), where methane is expressed as Kcal/100 Kcal of GE of feed, D = digestibility of energy at a maintenance level of feeding, and L = the level of feeding as a multiple of maintenance. A second formula developed by Swift et al (19) is: methane = 2.41X + 9.80, where methane is in grams and X represents hundreds of grams of carbohydrate digested. On average, methane production is about 8% of GE at maintenance and it falls to 6-7% at higher levels of feeding.

Urinary energy losses are usually relatively stable in a given animal species, although they reflect differences in diet, particularly when excess protein is fed or when forages are consumed that may contain essential oils or detoxification products such as hippuric acid. For ruminants, a correction factor of 7.45 Kcal/g of N has been used to account for energy excreted from amino acids metabolized, and a factor of 8.22 has been used for poultry. Actual urinary energy losses run on the order of 3-5% of GE in ruminants, or 12-35 Kcal/g of N excreted.

ME for ruminants is often calculated by the formula: ME = DE × 0.82. Many of the NRC values given for ME of ruminant feeds are so calculated. However, as Flatt and Moe (7) point out, this is only an approximation as the ME/DE ratio may vary considerably, being affected by the nature of the diet and the level of feeding. ME values are seldom determined in practice, unless animals are used in calorimetric studies where apparatus is available to collect respiratory gases.

Net Energy [NE]

As indicated in Fig. 9-1, NE is equal to ME minus the heat increment (HI) and the heat of fermentation (HF). Many writers combine the HI and HF as it is very difficult to determine precise estimates of each in ruminants or some of the herbivores, such as horses, in which a considerable amount of fermentation may take place in the cecum and large intestine.

The NE of food is that portion that is available to the animal for maintenance or various productive purposes. The portion used for maintenance is used for muscular work, maintenance and repair of tissues, and for maintaining a stable body temperature; most of it will leave the animal body as heat. That used for productive purposes may be recovered as retained energy in the tissues, in some product such as milk, or used to perform work.

The HI (also called specific dynamic effect when referring to a specific nutrient) may be defined as the heat production associated with nutrient digestion and metabolism over and above that produced prior to food ingestion (or above basal metabolism). The resulting heat is produced by oxidative reactions which are (a) not coupled with energy transfer mechanisms (high energy bonds) or (b) the result of incomplete transfer of energy; (c) partly due to heat production resulting from work of excretion by the kidney, and (d) increased muscular activity of the GIT, respiratory, and circulatory systems resulting from nutrient metabolism.

Estimates made many years ago (5) indicate th 80%+ of the HI originates in the viscera. Short-term experiments with animals which have had their livers removed show very little additional heat production after food ingestion, indicating that metabolism in the liver accounts for most of the HI. When individual products are fed to rats or dogs, feeding lean meat results in a prolonged period of heat production that amounts to 30-40% of GE. The HI (dog) from other foodstuffs are: fat, 15%; sucrose, 6%; starch, 20-22%. In cattle, the HI is about 3% at maintenance and increases to 20+% at 2X maintenance and to higher levels in high-producing animals (see Fig. 9-2). Examples of values for different species are shown in Table 9-2.

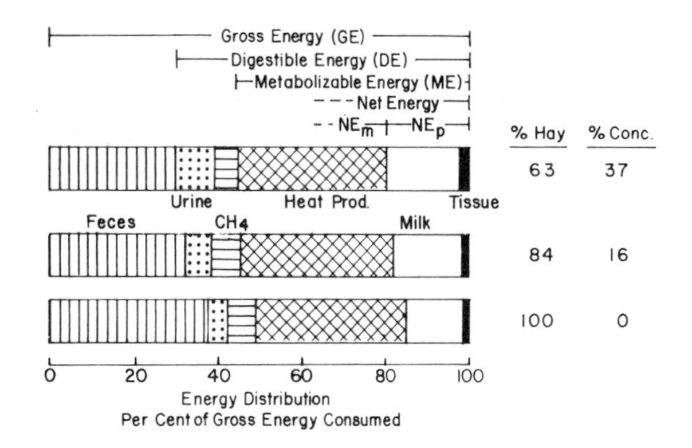

Figure 9-2. Illustration of energy terminology and the different systems of expressing the energy value of feeds. The bar chart shows relative energy losses when a mixed ration is fed to a lactating dairy cow. By permission of P.W. Moe, USDA, ARS, Beltsville, Md.

Table 9-2. Heat increment of feeding (Kcal/100 Kcal ME at maintenance).[a]

Nutrient	Species			
	Rat	Swine	Sheep	Cattle
Fat	17	9	29	35
Carbohydrate	23	17	32	37
Protein	31	26	54	52
Mixed rations	31	10-40	35-70	35-70

[a] From Armstrong and Blaxter (1)

It should be noted that the HI is not a constant for a given animal and a given foodstuff, but depends on how the nutrient is utilized. For example, if most of the material is absorbed and deposited in the tissues, the HI is very low. Incomplete proteins or amino acid mixtures fed to monogastric species result in oxidation of the amino acids and a high HI; a deficiency of an essential nutrient required in metabolic reactions, such as P or Mg, will result in a high HI. Frequent feeding results in a lower HI than infrequent feeding, and an increased feed intake results in a larger HI.

The HF is, as mentioned, poorly quantified. A nongrowing yeast culture comparable in size to that of an adult man has been estimated to produce 100X the heat production of the man. Blaxter (2) estimates the HF in ruminant animals to be 5-10% of GE. In monogastric species, some HF would originate from fermentation in the lower small intestine, cecum, and large gut, but quantitative information is not available.

Both the HI and HF may serve useful purposes to the animal in a cold environment. The heat resulting from the HI and HF may be used to warm the body just as well as that produced by more controlled metabolism of nutrients. At temperatures that result in heat stress to the animal, however, the HI is detrimental, requiring additional expenditure of energy to dissipate it by various means. In ruminant species, limited data indicate that feeding of urea in place of protein tends to reduce heat production, and that heat production is less when minimal amounts of protein are fed. Increasing the fat content of a ration appears to be helpful, because fat has a low HI, and reducing the fiber content may also be helpful in hot climates. Rumen fermentation of fiber results in substantial production of acetic acid which is used less efficiently than other volatile acids, resulting in a higher HI.

Other Methods of Measuring Feed Value

A method called Starch Equivalents (SE) is commonly in use in Europe at the present time. It was devised in Germany by Kellner many years ago. In effect, energy retention of fattening animals is measured by the C-N balance methods and feed values are expressed in relation to the value of starch, rather than in Kcal as with the NE method.

The Scandinavian Food Unit System is also used in Europe. With this method, feedstuffs are evaluated in feeding trials by replacing barley with the feed in question, and feed value is expressed relative to barley. Thus, this method is essentially the same as the SE method.

Methods of Measuring Heat Production and Energy Retention

If one wishes to study the utilization of ME, it is necessary to measure either (a) the animals heat production or (b) energy retained in the tissues, that used for productive work, or in a product such as milk. If one of these quantities (a or b) is known, then the other can be determined by subtracting the known one from ME. Heat production may be measured in various ways. Some of these methods are discussed briefly. For more detail see the references cited at the beginning of this chapter.

Direct Calorimetry

Animals lose heat from the body by radiation, convection and conduction from the body surface, by evaporation of water from the skin and lungs, and by excretion of urine and feces. In direct calorimetry the animal is enclosed in a well-insulated chamber which is equipped to measure these heat losses by the use of thermocouples or by circulation of water in pipes in the chamber. Newer types, called gradient-layer calorimeters, measure heat loss electrically as it passes through the wall of the chamber. The HI may be measured by feeding first at a low level and then at a higher level, or by fasting the animal and then feeding. These types of calorimeters are quite expensive to build and to operate and are seldom used where large animals are the main item of interest.

Indirect Calorimetry

In indirect calorimetry, heat production of the animal is estimated by determining O_2 consumption and, usually, CO_2 production. Some of the pioneer workers in animal nutrition (see Brody, 5) demonstrated that O_2 consumption and CO_2 production are closely correlated to heat production. It is well known that 1 mole of glucose requires 6 moles of O_2 for oxidation, produces 6 moles of CO_2, and yields 673 Kcal, and 5.007 Kcal/l. of O_2 used or/l. of CO_2 produced. Average values for carbohydrates are 5.047 Kcal/l. With respect to fats, relatively more O_2 is required and the caloric equivalent of a mixed fat is 4.69 Kcal/l. of O_2 consumed, or 6.6 Kcal/l. of CO_2 produced. Similar values for mixed proteins are 4.82 Kcal/l. of O_2 and 5.88 Kcal/l. of CO_2.

The ratio of the volume of CO_2 produced to the volume of O_2 consumed is known as the respiratory quotient (RQ). Thus, typical RQ's for carbohydrate are 1.0; for mixed fats, 0.7; and for mixed protein, 0.81. Note that each specific carbohydrate, fatty acid, or protein may have an RQ distinctive for that particular compound. For example, the RQ of fats with short-chain fatty acids is about 0.8, but that for long-chain fatty acids is about 0.7.

Because animals cannot completely oxidize the N in protein, the RQ is usually corrected by deducting an appropriate amount of energy to account for this. Methane production in the rumen is also a product of incomplete oxidation. A formula adopted by energy workers (7) for heat production is:

Heat (Kcal) = $3.866\ O_2 + 1.200\ CO_2 - 0.518\ CH_4 - 1.431\ N$, where O_2 = l. of oxygen consumed; CO_2 = l. of carbon dioxide produced;

CH_4 = l. of methane produced; and N = g of N excreted in the urine. Thus, with information on these various items, we can calculate how much protein was oxidized and estimate what percentage of the total heat was derived from carbohydrate or fat. For example, if we have an RQ of 0.78 that has been corrected for protein, we can look in an appropriate table (Brody, 5; Blaxter, 2) and see that this is equivalent to oxidation by body tissues of 26.3% carbohydrate and 73.7% fat.

As indicated, the RQ will usually be between 0.7 and 1.0. Two exceptions are worthy of note. RQ values in excess of 1.0 may result when carbohydrates are used for the synthesis of body fats, or where excess CO_2 is produced in animals with acidosis (ketosis).

Apparatus for Measuring Respiratory Gases

The most simple means of measuring O_2 consumption is with a face mask and a spirometer, a simple device into which the animal breathes for a short period of time. O_2 consumption is measured in the process. This apparatus has the disadvantage that it can only be used for short periods while an animal is confined. With modifications, spirometers can be used to measure O_2 consumption in active animals, or the same type of information may be obtained in animals with a cannulated trachea as shown in Fig. 9-3.

Most equipment for measuring respiratory gases involves the use of chambers in which the animal is maintained (Fig. 9-4). Provision is made for introducing feed and water and for collection of excreta. There are two types. The closed circuit type is designed so that air and respiratory gases are recirculated through the chamber. However, if the animal is kept in the chamber for more than a few hours, provision must be made to remove vaporized water, CO_2 and CH_4 and O_2 must be added to it. The expense of operation is relatively great. The other type, an open circuit, differs in that outside air is continuously passed through the chamber. This requires that careful measurements be made of the amount of air of known composition going through, and the outgoing air must also be carefully sampled for analyses for O_2, CO_2 and CH_4. This is the most common type in use with large animals at the present time. Automated sampling and analyzing instrumentation are available, which may be hooked up to a computer so that collection of data is relatively convenient, although relatively expensive.

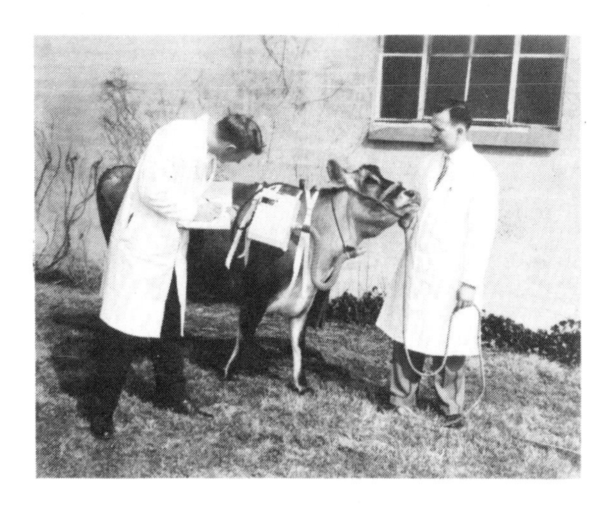

Figure 9-3. Photo illustrates the use of portable equipment to measure CO_2 production in a cow with a cannulated trachea. Similar equipment can be used to measure oxygen consumption when a face mask can be used. Courtesy of P.W. Moe, USDA, ARS, Beltsville, Md.

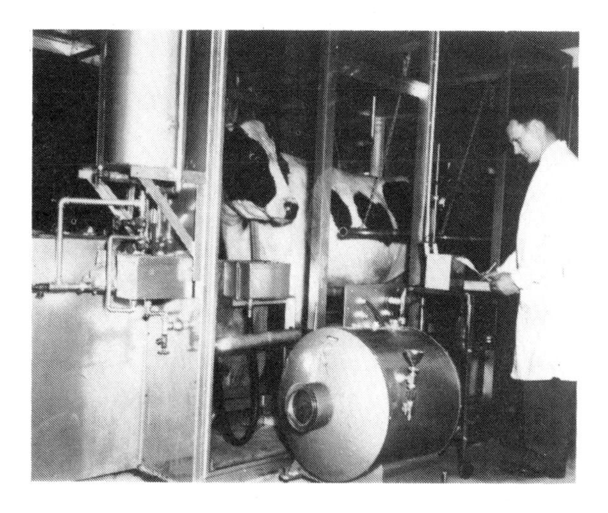

Figure 9-4. An example of an open circuit respiration chamber used at the USDA Experiment Station at Beltsville, Maryland. Courtesy of P.W. Moe.

Other Methods of Measuring Energy Retention

Comparative Slaughter Technique

As Flatt and Moe (7) point out, an accurate measurement of energy value of feeds requires the measurement of the actual amount of energy retained by the animal or produced as a useful product, or the measurement of all forms of energy loss. In growing and fattening animals the measurement of energy retention has proven to be useful for broilers, cattle, and swine. In work with broilers, the energy value of carcasses of day-old chicks is measured by bomb calorimetry. Similar chicks are then fed for a specified period of time and the energy values of their carcasses then determined. Thus, the amount of energy deposited can be calculated. In the work with cattle (10), body energy retention is measured as the difference in energy of the carcass of groups of animals slaughtered before and after a feeding period. In practice, one group is fed at maintenance and the other group at some high level. Energy retention is estimated based on carcass specific gravity, from which the amount of fat and protein in the tissues may be estimated. This type of procedure, although dependent upon several indirect measures, has merit in that it is measuring energy retention of the animal in a relatively normal environment as opposed to an animal enclosed in a respiration chamber.

Carbon-Nitrogen Balance

The principal forms in which energy is stored in the growing and fattening animal are as protein and fat; carbohydrate reserves are very low. If the C and N intake and excretion are known, then deposition in the tissues may be calculated. One must measure the amounts consumed and excreted and CO_2 and CH_4 in respiratory gases. Body protein is assumed to contain 16% N and 51.2% C; thus, the amount of protein storage can be computed when C and N retention are known. The remainder of the C is in fat which contains 74.6% C, so the amount of fat storage is then computed. The amount of protein stored is multiplied by 5.32 Kcal/g and fat by 9/37 Kcal/g (for cattle) to give an estimate of caloric storage.

Some Important Concepts in Energy Metabolism

Heat Production and Body Size

Most mammals and birds are called homeotherms; that is, they maintain a stable body temperature, although minor fluctuations of relatively short duration may occur as a result of extensive chilling, fevers, or vigorous exercise. With a constant body temperature, heat production over a period of time is equal to heat loss. In the nutritional physiology of animals, we are concerned with heat production (or loss) if we wish to relate information obtained with calorimetric studies (see later section) to more normal environmental conditions. Calorimetric data are obtained under very specific conditions, and it is not feasible to attempt to duplicate all of the situations — ration, temperature, humidity, activity, and disease — which are encountered by animals in their normal environment or to account for differences in age, size, species and breed.

The early nutrition research clearly showed that heat production was not well correlated to body weight of animals (see Table 9-3), and much research effort was expended to develop means of predicting heat production and establish some overall 'law' that applies to animals in general (Brody, 5). With inanimate objects, it is known that rate of cooling is proportional to surface area. Furthermore, surface area varies with the square of linear size or to the 2/3 power of weight if specific gravity is constant; thus, surface area varies with the square of linear size or the 2/3 power of volume, so heat production can be related to body surface or volume.

The body surface of a living animal is quite difficult to measure. A variety of different methods have been attempted, but repeatibility of such measurement is low, even that of such methods as measuring the surface of the skin after it is removed. This is partly due to the fact that surface area of a living animal is not constant. It may change with environmental temperature when the animal stretches out, rolls up in a ball, by fluffing up of feathers or otherwise changes its posture. Furthermore, the ability of most animals to constrict or dilate blood vessels in the skin effectively alters the normal skin temperature and heat loss. Insulation of the body by subcutaneous fat, thick skin, hair, wool, or feathers has a marked effect on heat loss, as well. In addition, heat loss has been shown to be related to the profile of the animal; thus a long-legged, long-necked animal, such as a giraffe, would have a markedly different exposed area as compared to an animal with a short, compact body, such as a pig.

Even though these various factors are involved in heat loss, it can be reasonably well related to surface area estimated by multiplying body weight by a fractional power. BW multiplied by a fractional power is referred to as **metabolic size.** Although there has been extensive controversy on this subject in the past, all current feeding standards in the USA and Great Britain now use body weight (BW) multiplied to the 0.73 or 0.75 power ($BW^{0.75}$) to estimate surface area. Note in Table 9-3 that heat production of different species ranges from 97 Kcal/kg of BW down to 15 for animals ranging from the rat to the cow. However, when heat production is expressed on the basis of surface area (estimated from BW), the heat production/kg of $BW^{0.73}$ is much more uniform, ranging from 61 to 80 Kcal/day. Obviously, these data are not identical, but expression on this basis gives a reasonably good means of comparing markedly different animals. Brody (5) shows data indicating similar estimates of heat production when expressed in this manner for animals ranging in size and diversity from the canary to the elephant.

The preceding comments apply to adult animals in a quiet state. Other factors that may affect heat production are discussed in a later section.

Heat Expenditure and the Environment

Animals lose heat by conduction, convection and radiation from the body surface, and by evaporation of water from the body surface and the lungs and oral surfaces. The rate at which heat is lost depends on the difference in temperature between the body surface and its environment. In addition, by constriction or dilation of blood vessels in the skin and extremities, the body surface may be cooled or warmed to some extent as compared to interior body temperature, thus increasing or decreasing surface temperature in relation to the environment. The effective surface temperature of the body is greatly influenced by insulation of subcutaneous fat, thickness of skin, and skin covering. On the other hand, insulation is greatly reduced by air movement or when the body surface is wet.

Important concepts in energy metabolism are those of a thermoneutral environment and critical temperature. A thermoneutral environment is defined as one in which the animal does not need to increase energy expenditure to either warm or cool the body. This area, or temperature range, is also called the comfort zone. The critical temperature (see Fig. 9-5) is usually defined as the point at which an animal must increase its heat production to prevent body temperature from falling, or (according to some writers) increase the rate of heat dissipation to prevent body temperature from rising. This upper point is sometimes called the point of hyperthermal rise.

The critical temperature for fasting animals in a cold environment is relatively high, being about 18-20°C for steers with a normal hair coat and about 30°C for a fasting sheep with a fleece of 5 mm in length. This is opposed to a critical temperature in steers of about 7°C when fed at maintenance and -1°C when fed to gain 500 g/day. The critical temperature of sheep fed at maintenance decreases from about 28°C with fleece of 1 mm in length to -3°C with fleece of 100 mm in length (2). Thus, the HI is being used to warm the body and greatly reduced the effective critical temperature, as does added skin insulation (wool).

Shivering is a means of increasing heat production by an involuntary contraction of muscles. Further heat production occurs by oxidation of fats or proteins from the tissues or diet. Heat loss may be reduced in cold climates as the animal adapts to it, probably by more efficiently reducing loss from the body surface by constriction of blood vessels near the surface. In addition, some animals, such as the sheep and eland, are more efficient in taking up O_2 from the air, with the result that less cold air must be inspired.

Table 9-3. Typical values for heat production of fasting adult animals of different species.

Animal	Body weight, kg	Heat production, Kcal/day			
		Per animal	Per kg of body weight	Per square meter of surface area	Per kg weight $^{0.73}$
Rat	0.29	28	97	840	69
Hen	2.1	115	55	701	67
Dog	14.0	485	35	745	71
Sheep	50	1060	21	890	61
Man	70	1700	24	950	77
Pig	122	2400	20	974	72
Cow	500	7470	15	1530	80

Most animals have much better means of protecting themselves from a cold than from a hot climate. In a hot climate the animal must cool itself by increasing evaporation from the body surface and/or by more rapid respiration and panting. The point at which heat stress occurs in animals varies with species and numerous environmental factors. The actual temperature that may cause heat stress is reduced by high humidity (which reduces evaporative cooling rate), by a high level of feeding, or by feeding any ration that produces a high HI (high protein or high fiber for ruminants), or by restriction of water intake.

HF do not enter into the picture. Only a few hours (overnight for man) of fasting are required in most monogastric species, for the GIT to approach this condition. In ruminants, however, food requires many hours to pass out of the GIT, and a true state of post-absorption is rarely obtained, although it may be approached, as indicated by a very low CH_4 production. A state of muscular repose is needed so that an unknown amount of activity does not increase heat production. This may be difficult to obtain in animals, particularly untrained ones.

Although perhaps not very precise, estimates of the needs for basal metabolism are that about 25% of the energy needs are required for circulation, respiration, secretion, and muscle tonus, and that the remaining 75% represents the cost of maintaining a stable body temperature (2).

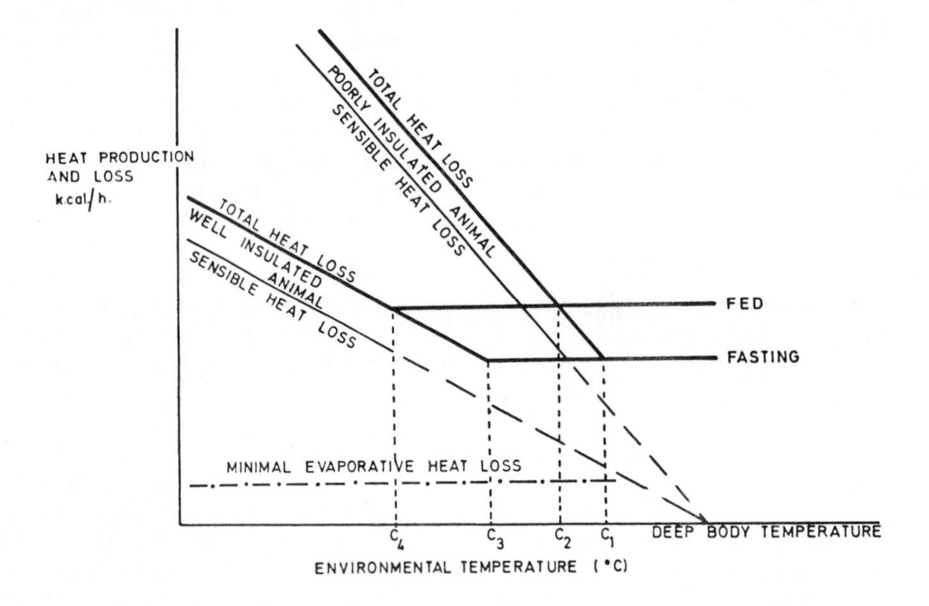

Figure 9-5. Schematic representation of the relation between environmental temperature and the energy exchanges of homeotherms in cold environments. The heavy lines show the total heat loss of animals well and poorly insulated, fasting, and fed. The horizontal part of this line defines the zone of thermal neutrality, in which range heat loss is unaffected by environmental temperature. The lower limit of the thermoneutral zone is the critical temperature, C_1, C_2, C_3, or C_4 below which more heat must be produced to maintain body temperature. Reproduced by permission from Hafez and Dyer [8].

Basal and Fasting Metabolism

Basal metabolism may be defined as the condition where a minimal amount of energy is expended to sustain the body. Determinations are carried out under standardized conditions, and such information provides comparative base values where energy requirements are not confounded by other factors. In order to meet the requirements for basal metabolism, the animal should be in (a) a post-absorptive state, (b) a state of muscular repose but not asleep, and (c) in a thermoneutral environment. The post absorptive state is used so that the HI and

Factors Affecting Basal Metabolism

Age. Age has a pronounced effect on basal metabolism in species that have been studied. For example, in man, heat production is about 31 Kcal/m at birth, this increases to about 50-55 Kcal at about 1 year of age and then gradually declines to 35-37 Kcal during the early to middle 20's. Further declines occur in older people. Effect of age on heat production in sheep is shown in Table 9-4. Note the marked decline in heat production, particularly between 9 and 15 wk of age, during which period the lambs were weaned. Part of the change with age may be related to differential development of tissues which have different O_2 requirements (see Mitchell, 11).

Neuro-endocrine Factors. It is well known that energy expenditure may be different in the two sexes. In man, basal metabolism of the male is typically 6-7% higher than the female, a difference which shows up at 2-3 years of age. In domestic animals, castration results in a 5-10% depression in basal metabolism. Thyroid activity has a pronounced effect as hypothyroid individuals may have a very low basal metabolism. Nervous, hyperactive animals have a high heat production, as might be expected.

Species and Breed Differences. A basal metabolism value of 70 Kcal $BW^{0.73}$ is considered to be an average value where BW is in kg. Note, however, in Table 9-3 that sheep tend to be about 15% below this and cattle about 15% above. Furthermore, data from the Rowett Research Institute (4) indicate that average values for Ayrshire steers were about 100 Kcal $BW^{0.73}$ as compared to 81 Kcal for Angus steers; similar differences have also been observed between breeds of sheep and dairy cows. Thus, breed differences may be almost as marked as species differences. With respect to the differences between sheep and cattle, Blaxter (2) argues that cattle originated and developed in cold norther climates where heat production was a critical factor for survival, while sheep probably originated in subtropical areas where low heat production has survival value.

Miscellaneous Factors. Other factors that have been shown to have some effect on basal metabolism include adaptation to fasting, where heat production/unit of surface area decreases with length of fast; muscular training (hypertrophy of muscles), which results in increased heat production; and mental effort, which causes a slight increase.

Maintenance

Maintenance may be defined as a condition where a nonproductive animal neither gains nor loses body energy reserves. In modern day agriculture, we are only rarely interested in just maintaining animals, as we are usually interested in keeping animals for some productive purpose. Nevertheless, information on maintenance energy needs serves as a more useful practical guideline, generally speaking, than does basal or fasting metabolism. If we are to establish the maintenance needs of an animal, several factors enter in. In addition to needs for basal metabolism, we must account for energy losses occurring during nutrient metabolism; we must, in some manner, account for increased physical activity by the animal associated with normal functions such as grazing, and for environmental factors which may alter energy needs.

In calorimetric experiments, the maintenance requirement of nonproductive animals may be measured with precision. In such situations, however, the animal is much less active than under more normal conditions, and it has proved to be quite difficult to put precise estimates on maintenance requirements that are reliable under different environmental conditions. Information on sheep (22) indicates that maintenance requirements of wethers are about 60-70% greater for grazing animals than for those housed in inside pens. The exact amount depended on condition of the sheep, the environment, and availability of herbage. If one assumes that heat production of maintenance for sheep in pens is about 70

Table 9-4. Fasting metabolism of lambs and sheep as affected by age. [a]

Age	Body wt, kg	$Kcal/kg^{0.73}$	Age	Body wt, kg	$Kcal/kg^{0.73}$
1 wk	5.8	132	6 mo.	28.2	63
3 wk	9.1	111	9 mo.	33.9	62
6 wk	13.0	119	1-2 yr.		63*
9 wk	18.0	116	2-4 yr.		58*
15 wk	24.3	68	4-6 yr.		55*
			over 6 yr.		52*

[a] From Blaxter (2)
*Wether sheep

Kcal/kg $BW^{0.73}$ (2), then this would indicate heat production when grazing of about 115 Kcal/kg $BW^{0.73}$. Work with Holstein dairy cows in calorimeters (13) indicates a maintenance requirement on the order of 114-122 Kcal of ME/kg $BW^{0.75}$, a value about equal to heat production of 100 Kcal/kg $BW^{0.75}$.

Frequently, energy requirements are estimated by feeding trials with animals under normal farm conditions. Although data of this type are less precise in that less information is available on actual tissue gain or loss of energy, such information may have more practical value than calorimetric data. Most feeding standards are based on the assumption that maintenance requirements under normal conditions are appreciably higher than basal or fasting metabolism rates, and 2 times the basal rate is frequently used, or 1.25-1.35 times fasting metabolism values when calculating maintenance requirements. Current NRC nutrient recommendations for cattle are based on maintenance estimates derived by Lofgreen and Garrett (10), which were estimated by the comparative slaughter technique. In their work, the maintenance requirement is estimated to be 77 Kcal/kg $BW^{0.75}$, a value which would seem to be too low as compared to calorimetric experiments on cattle. If this is the case, then their NE of maintenance values given for feedstuffs (see Ch. 16 on feeding standards) would over-estimate the relative value as compared to NE of gain.

Efficiency of Energy Utilization

Efficiency of energy utilization is of practical as well as academic concern to people involved in animal production as efficiency is often a vital factor in profitable production. In the USA, at least, the most common means of expressing efficiency is in terms of units of production/unit of feed (lb gain, eggs or milk/lb feed consumed). Where diets and type of production are similar, this gives a satisfactory comparison of gross efficiency. If we compare gain produced by a high-concentrate ration to that produced on a high-roughage ration, however, the comparison is poor is terms of utilization of available energy (DE, ME or NE), because the roughage is apt to be much lower in available energy than a concentrate.

Another factor must be considered when efficiency is measured is terms of units of production, particularly when dealing with body gain or loss in weight. Although the energy value of milk can easily be calculated if fat percentage is known, or that of eggs can be estimated with reasonable precision, the energy content of gain may vary widely. In studies with dairy cows, Reid and Robb (18) point out that caloric value of body gain ranged from 4.8 to 9.4 Mcal/kg and that for body loss from 6.3 to 7.9 Mcal/kg; they suggest that body tissue gain or loss could range from as much as 100% water to about 90% fat. In growing animals, most of the tissue gain in early post-natal growth is protein in nature, and lean meat contains about 75% water, but during fattening, added tissue may contain 90% fat. For these extremes, the combustion values of 1 g of added tissue would range from 1.4 Kcal to 8.5 Kcal, or a ratio of about 1:6, clearly indicating that weight gain may be of little value in estimating caloric efficiency. If gain resulted from water retention, the spread would be considerably greater.

Information from a variety of sources indicates that caloric efficiency is greatest for maintenance, followed by milk production and growth and fattening. The relative efficiency of maintenance may be due, largely, to the more efficient utilization of the HI and HF in maintaining body temperature. The less efficient use of energy for production as compared to maintenance is partly because of a decline in digestibility of a given diet as feed intake increases. Early post-natal growth is quite efficient, approaching that of maintenance, but the efficiency declines with age, partly because of gradually increased fat deposition. Biochemically, synthesis of proteins from amino acids can be shown to be a more efficient process, energetically, than synthesis of fat from glucose or glucose precursors, such as propionic acid or various amino acids. Examples of the range in efficiency that might be expected in ruminants and swine are shown in Table 9-5.

Gross efficiency $\left(\frac{\text{caloric value of product}}{\text{caloric intake}}\right)$ is greatly affected by age, as indicated, and level of production. Young animals generally eat considerably more/unit of metabolic size than older animals, so that maintenance requirements represent a smaller % of dietary intake. The same comment applies to thin animals as opposed to fat animals, or to any situation where animals are fed at less than maximum intake. Because of this, in most production situations it is economically feasible to feed for maximum intake and rate of production.

Net efficiency $\left(\frac{\text{caloric value of product}}{\text{intake above maintenance}}\right)$ however, is less affected by level of intake and, perhaps, more by genetic capability of the animal. In high-producing animals, size is not a factor, either.

Examples of the range in efficiency that might be expected in swine and ruminants as suggested by Reid (17) are shown in Table 9-5. Relative efficiencies for utilization of NE suggested by Kleiber (9) are: fat production in adult steers, 100; maintenance in cattle, 120; milk production by cow, 119; maintenance in swine, 145; growth and fattening in swine, 125; and fattening in swine, 130. Reid (17) has made interesting calculations on efficiency of production of food energy by different domestic species of animals. In these calculations, he has considered the cost of maintenance of the dam and has included credit for the carcass of the dam towards total food production. His calculated values in terms of percentage of DE recovered in food are: pork, 18; eggs, 12; broilers, 12; milk (3600 kg/year), 22; milk (5400 kg/year), 27; and beef, 6. One reason for the low overall efficiency of beef production is that a beef cow produces at a very low level compared to a dairy cow; thus, maintenance is a much greater proportion of total feed intake, and gross efficiency drops considerably.

Energy Terminology Used in Ration Formulation and Feeding Standards

As pointed out in previous discussions on the different categories of energy, GE has no value in itself for evaluation of feedstuffs or animal requirements. DE is often used in the USA, as is TDN, and these require no further explanation at this point. For poultry, the NRC recommends the use of ME for chickens and turkeys, and an appreciable amount of information is available on feedstuffs and animal requirements for these species. For swine, DE, TDN, or ME values are usually used in practice.

In Great Britain, the ARC (Agricultural Research Council) has gone along with suggestions of Blaxter and has adopted ME as the preferred base for energy. In the USA, three versions of NE are used in varying degrees for ruminants. These are discussed in succeeding paragraphs.

Table 9-5. Efficiency of energy utilization by pigs and ruminants ingesting diets of the usual range of quality fed in practice. [a]

Item	Pigs	Ruminants
Digestible energy, %	75-90	50-87
Utilization of ME above maintenance, %		
Body gain	75-80	30-62
Milk production	75-85	40-75
Fattening during lactation [b]		73

[a] From Reid (17). Data from a variety of sources. That for ruminants include diets ranging from fair quality forages to high-concentrate rations.

[b] Data from Moe et al (14).

Estimated Net Energy [ENE]

Based on some calorimetric work, Moore et al (15) came up with a formula which allows NE to be estimated from TDN; the resulting value is called ENE. The formula is:

$$\text{ENE (as Mcal/100 lb)} = 1.39 \times \%\text{TDN} - 34.63$$

or

$$\text{ENE (as Mcal/kg dry matter)} = 0.307 \times \%\text{TDN} - 0.764$$

Morrison (16) adopted this formula in the last edition of his book, and it has been used relatively widely as a result. Some dairy nutritionists still favor use of these values, partly because relatively few data are available on NE values of feedstuffs as compared to TDN values.

Net Energy for Lactation [NE_{milk}]

Calorimetric experiments have been carried out at the USDA Experiment Station at Beltsville in which the NE requirements for milk production have been studied in high-producing dairy cows, primarily Holsteins, and results have been summarized recently (13). NE_{milk} is, in effect, a measure of the total energy requirement of the nonpregnant cow for milk production. The average NE requirement for maintenance was calculated to be 73 Kcal/kg $BW^{0.75}$, which is roughly equivalent to 118 Kcal of ME/$kg^{0.75}$. The amount of NE required for milk production was calculated to be 0.74 Mcal NE_{milk}/kg of 4% fat-corrected milk. To calculate NE_{milk} values from DE or TDN, the authors suggest the equations:

$$NE_{milk} \text{ (as Mcal/kg dry matter)} = 0.68 \text{ DE (as Mcal/kg DM)} - 0.36$$

$$NE_{milk} \text{ (as Mcal/kg dry matter)} = 0.037 \times \%\text{ TDN} - 0.77$$

Adjustments can be made in requirements for tissue gain or loss of cows, for excess N intake, and for pregnancy. The authors agree that further refinement is needed, but the reader might note that the NRC has adopted this terminology for their most recent publication on dairy cows, although the NRC uses a different equation than the one given here.

Net Energy of Maintenance and Gain [NEm, NEg]

This system, also called the California system (see section on comparative slaughter technique), has come into use and has also been adopted by the NRC for growing beef and dairy cattle. In this scheme (10), the NE for maintenance, NEm, is calculated separately from the NEg, primarily because maintenance is more efficient process, energetically, than is gain. This causes some complications in formulation of rations, however (see Ch. 20). Relatively few feedstuffs have been experimentally evaluated with this procedure at this time, although NRC publications give calculated values which were derived with the following formulas. Where DE or TDN are known, ME is first calculated.

ME (as Mcal/kg of feed) = DE (as Mcal/kg) x 0.82

or ME = 3.615 TDN

For NE,

Log F = 2.2577 - 0.2213 ME (where F = g dry matter/kg $BW^{0.75}$)

NEm = 77/F

NEg = 2.54 - 0.0314 F

or NEm (as Mcal/kg dry matter) = 0.029 x % TDN - 0.29

NEg (as Mcal/kg dry matter) = 0.029 x % TDN - 1.01

Formulas Used in Calculating Different Energy Values

For the sake of convenience, the various formulas given throughout this chapter for calculating different energy values of feedstuffs are repeated in this section along with some added values for poultry. They are as follows:

DE (as Kcal) = TDN (in lb) x 2000

or

DE (as Kcal) = TDN (in g) x 4.4

ME for swine

$$\text{ME (in Kcal/kg)} = \frac{\text{DE (in Kcal/kg)} \times 0.96 - (0.202 \times \text{protein \%})}{100}$$

ME for poultry (Titus, 19)

ME (as Kcal/g) = digestibility (%) x energy equivalents (Kcal/g)

Energy equivalents suggested for poultry are (Titus, 19):

Protein	3.84	Carbohydrates (NFE)	
Ether extract		grains	4.2
meat and fish meals	9.49	legume seeds	4.0
grains and seeds	9.33	legume leaves & stems	3.8
milk products	9.21	milk products	3.7
		Crude fiber	2.1

For ruminants

ME (as Mcal/kg of feed) = DE (as Mcal/kg) x 0.82

ENE (as Mcal/kg dry matter) = 0.307 x % TDN - 0.764

NE_{milk} (as Mcal/kg dry matter) = 0.68 DE - 0.36

NEm (as Mcal/kg dry matter) = 0.029 x % TDN - 0.29

NEg (as Mcal/kg dry matter) = 0.029 x % TDN - 1.01

Note that several formulas are in use for calculating TDN, DE, and so on, based on analytical composition of different feedstuffs. These will be illustrated and discussed in Ch. 18 on feedstuffs.

References Cited

1. Armstrong, D.G. and K.L. Blaxter. 1957. Br. J. Nutr. 11:247.
2. Blaxter, K.L. 1962. The Energy Metabolism of Ruminants. Hutchinson & Co.
3. Blaxter, K.L. and J.L. Clapperton. 1965. Br. J. Nutr. 19:511.
4. Blaxter, K.L. and F.W. Wainman. 1966. Br. J. Nutr. 20:103.
5. Brody, S. 1945. Bioenergetics and Growth. Hafner Pub. Co.
6. Church, D.C. 1969. Digestive Physiology and Nutrition of Ruminants. Vol. 1 - Digestive Physiology. O & B Books, 1215 NW Kline Pl., Corvallis, Oregon.
7. Flatt, W.P. and P.W. Moe. 1969. In: Animal Nutrition and Growth. Lea & Febiger.
8. Hafez, E.S.E. and I.A. Dyer (ed.) 1969. Animal Nutrition and Growth. Lea & Febiger.
9. Kleiber, M. 1961. The Fire of Life. John Wiley & Sons.
10. Lofgreen, G.P. and W.N. Garrett. 1968. J. Animal Sci. 27:793.
11. Mitchell, H.H. 1962. Comparative Nutrition of Man and Domestic Animals. Vol. 1. Academic Press.
12. Mitchell, H.H. 1964. Comparative Nutrition of Man and Domestic Animals. Vol. 2. Academic Press.
13. Moe, P.W. and H.F. Tyrrell. 1972. Proc. Cornell Nutr. Conf., Cornell, University.
14. Moe. P.W., H.F. Tyrrell and W.P. Flatt. 1971. J. Dairy Sci. 54:548.
15. Moore, L.A., H.M. Irvin and J.C. Shaw. 1953. J. Dairy Sci. 36:93.
16. Morrison, F.B. 1956. Feeds and Feeding. 22nd ed. Morrison Pub. Co., Clinton, Ia.
17. Reid, J.T. 1970. In: Physiology of Digestion and Metabolism in the Ruminant. Oriel Press, Newcastle upon Tyne, England.
18. Reid, J.T. and J. Robb. 1971. J. Dairy Sci. 54:553.
19. Swift, R.W. et al. 1948. J. Animal Sci. 7:475.
20. Titus, H.W. 1955. The Scientific Feeding of Chickens. 3rd ed. Interstate Pub. Co.
21. Van Soest, P.J. 1971. Proc. Cornell Nutr. Conf., p 106.
22. Young, B.A. and J.L. Corbett. 1972. Austral. J. Agr. Res. 23:77.

Chapter 10 – Macrominerals

Seventeen mineral elements are known to be required by at least some animal species. They can be divided into two groups based on the relative amounts needed in the diet — macrominerals and micro or trace minerals. The macrominerals are: calcium (Ca); phosphorus (P); sodium (Na); chlorine (Cl); potassium (K); magnesium (Mg); and sulfur (S).

Some mineral elements — such as Ca and P — are required as structural components of the skeleton and others — such as Na, K and Cl — function in acid-base balance; still others — such as Zn and Cu — are contained in enzyme systems. Many minerals have more than one function.

Animal tissues contain many minerals in addition to those recognized as dietary essentials. Some of these may have a metabolic function not yet recognized, but others may be present as innocuous contaminants. Still others are toxic to the animal at relatively low concentrations. Even the required minerals are toxic if present in the diet in excess.

Each of the minerals known to be required is discussed with respect to distribution in body tissues, functions, metabolism, deficiency signs, and toxicity. The amounts of each of the macrominerals (except S) present in newborn and adult mammals of several species are shown in Table 10-1 (1). The proportions of each mineral, expressed as amount/kg of fat-free body tissue are strikingly similar among species in adults. Newborn animals also have similar composition, but differences in the degree of physiological maturity are reflected in the amounts of Ca, P, and Mg present among the species. Guinea pigs, humans, and pigs are physiologically more mature at birth than rabbits, rats, and mice and have higher concentrations of bone minerals. The Ca:P:Mg ratios of both newborn and adults are remarkably similar among species. One can infer from such data that macromineral metabolism and nutrition, with some exceptions, is similar among animal species and therefore that observations on one species may be extrapolated to others.

Calcium

Tissue Distribution

About 99% of the Ca stored in the animal body is in the skeleton as a constituent of bones and teeth. It occurs in about a 2:1 ratio, with P in

Table 10-1. Ca, P, Mg, Na, K, and Cl content of newborn and adult animals (amounts/kg of fat-free body tissue). [a]

	Man	Pig	Dog	Cat	Rabbit	Guinea pig	Rat	Mouse
					Newborn			
Body wt, g	3560	1250	328	118	54	80	5.9	1.6
Water, g	823	820	845	822	865	775	862	850
Ca, g	9.6	10.0	4.9	6.6	4.8	12.3	3.1	3.4
P, g	5.6	5.8	3.9	4.4	3.6	7.5	3.6	3.4
Mg, g	0.26	0.32	0.17	0.26	0.23	0.46	0.25	0.34
Na, meq	82	93	81	92	78	71	84	—
K, meq	53	50	58	60	53	69	65	70
Cl, meq	55	52	60	66	56	—	67	—
					Adult			
Body wt, kg	65	125	6.0	4.0	2.6	—	0.35	0.027
Water	720	750	740	740	730	—	720	780
Ca, g	22.4	12.0	—	13.0	13.0	—	12.4	11.4
P, g	12.0	7.0	—	8.0	7.0	—	7.5	7.4
Mg, g	0.47	0.45	—	0.45	0.50	—	0.40	0.43
Na, meq	80	65	69	65	58	—	59	63
K, meq	69	72	65	77	72	—	81	80
Cl, meq	50	—	43	—	32	—	40	—

[a] Source: Widdowson and Dickerson (1). To convert meq (milliequivalents) to mg, multiply by 23 for Na, 39 for K, and 35.5 for Cl; to covert g to meq, divide by 20,000 for Ca, 31,000 for P, and 12,000 for Mg.

bone primarily as hydroxyapatite crystals (2): $Ca^{++}_{10-x}(PO_4^{\equiv})_6\,(OH^-)_2\,(H_3O^+)_{2x}$ where x may vary from 0 to 2. When $x = 0$, the compound is called octacalcium phosphate; when $x = 2$ it is called hydroxyapatite. Ca is present in blood mostly in the plasma (extracellular) at a concentration of about 10 mg/100 ml and exists in three states (3): as the free ion (46-47%), bound to protein (45-46%), or complexed with organic acids such as citric or with inorganic acids such as phosphate (5-7%).

Functions

The most obvious function of Ca is as a structural component of the skeleton. A brief description of normal bone metabolism is needed to develop a full appreciation of the function of Ca as a major skeletal component. Bone is a metabolically active tissue with continuous turnover and remodeling both in growing and mature animals. The physiological control of bone metabolism is related to both endocrine and nutritional factors. Blood is the transport medium by which Ca is moved from the GIT to other tissues for utilization. Elaborate controls exist to maintain a relatively constant Ca concentration in the plasma. A decline in plasma Ca concentration triggers the parathyroid gland to release parathyroid hormone which raises plasma Ca by causing bone resorption. An increase in plasma Ca concentration triggers the thyroid gland to release calcitonin, a hormone from small glands in the thyroid, which depresses plasma Ca by inhibiting bone resorption (4). Thus, dietary factors that affect Ca absorption have an effect on the endocrine system in direct response to the amount of Ca reaching the blood from the GIT. The amount of Ca absorbed from the GIT depends on amount ingested and proportion absorbed. As the % of Ca in the diet increases, the proportion absorbed tends to decline. This is related to the fact that Ca absorption is an active process under the control of a Ca-binding protein (CaBP; see vitamin D, Ch. 13) which, at least in most species, is vitamin-D dependent. The absolute amount absorbed is generally greater in animals fed high-Ca diets, despite the lower % absorption. In vitamin-D deficiency, Ca absorption is reduced, because of impaired CaBP formation, so that skeletal abnormalities can occur even in the presence of adequate dietary Ca.

Bone apposition occurs by activity of osteoblasts (bone-forming cells) at growth plates and in other areas of rapid bone growth. Bone resorption occurs concomitantly by 2 processes, one by resorption of bone surfaces by activity of osteoclasts (bone-resorbing cells) and the other by resorption deep within formed bone (osteocytic resorption). The latter process is now generally believed to be the major process of bone resorption. The continuous apposition and resorption of bone is the basis for bone growth and for changes in its shape and density.

Ca controls the excitability of nerve and muscle. Reduced Ca^{++} concentration produces increased excitability of pre- and postganglionic nerve fibers. Higher than normal Ca^{++} concentrations have the opposite effect on nerves and muscles, causing them to be hypoexcitable. It has been postulated (5) that Ca imposes constraints on the ionic movements of Na and K by interacting with surface structures of the cell.

Ca is required for normal blood coagulation (see Table 13-3, page 161).

Metabolism

The overall metabolism of Ca is schematically diagrammed in Fig. 10-1. In the growing animal, net retention of Ca occurs in the body; that is, the amount stored in bone and other tissues exceeds that lost in feces, urine, and sweat. In nonpregnant, nonlactating adults, the amount of Ca ingested equals the amounts lost if the metabolic requirement is met.

Absorption. Dietary Ca is absorbed largely from the duodenum and jejunum of most animals. An exception appears to be the hamster in which most absorption is from the ileum. Absorption occurs both by active (energy-dependent) and passive (diffusion) transport. In rats and man, and presumably in other species, about 1/2 of the dietary Ca is absorbed by active transport. The importance of a vitamin D-dependent protein carrier of Ca in active transport has been elucidated by Wasserman et al (6) as described in Ch. 13. Other factors in addition to vitamin D affect efficiency of Ca absorption. Increased dietary Ca concentration decreases the % of Ca absorbed, although the absolute amount absorbed tends to remain relatively constant within the normal range of Ca concentration of the diet. Lactose as well as lysine and, to a lesser extent, several other amino acids have been shown to enhance Ca absorption in rats (7, 8), but phytic and oxalic acids decrease its absorption by forming insoluble complexes of Ca oxalate and Ca phytate. Such factors as high pH of the intestinal contents, high levels of dietary fat, which might be expected to form insoluble fatty acid-Ca soaps, and high fiber levels in the diet, are probably not of great significance (3).

Ossification of the skeleton to form hydroxyapatite crystals requires that the product of Ca ions and P ions in the fluid surrounding the bone matrix exceed a critical minimum level. Thus, if the product $(Ca^{++})(PO_4^{=})$ falls below the concentration required to precipitate Ca phosphate in the crystal lattice structure, ossification fails to occur and rickets or osteomalacia results, whether the deficiency is one of Ca or P, or both. Calcification is an active process, specifically requiring ATP.

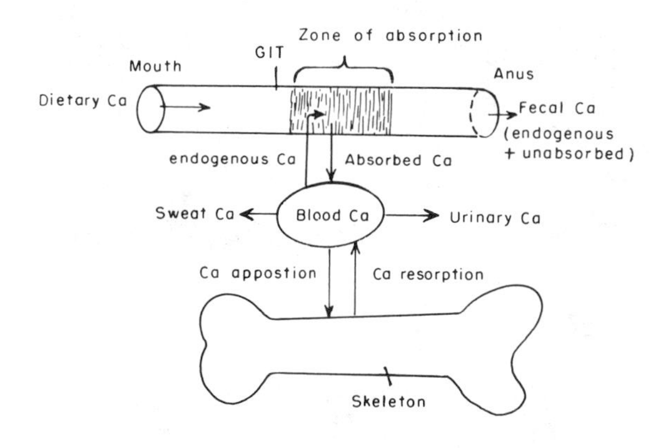

Figure 10-1. Schematic diagram of overall metabolism of calcium.

Excretion. The 3 major routes of Ca excretion are feces, urine and sweat. Fecal output includes both unabsorbed and endogenous Ca. The endogenous fraction, largely arising from secretions of the intestinal mucosa, is probably partially reabsorbed, as illustrated in Fig. 10-1. Therefore, that appearing in the feces is termed fecal endogenous Ca and represents about 20-30% of total fecal Ca. The apparent Ca absorbability (feed Ca minus fecal Ca) generally approximates 50%, although the % tends to decline as intake increases.

Urinary output of Ca is generally considerably less than that of fecal output in most species (Table 10-2). About 1/2 of the plasma Ca, mainly ionized Ca, is filtered in the kidney; more than 99% of this is reabsorbed under normal conditions. Diuretics do not generally affect Ca excretion, but Ca chelators, such as large intravenous doses of Na-citrate, Na-EDTA, or Ca-EDTA, greatly increase Ca excretion.

Loss of Ca in sweat is of only minor significance in most species, but in man, horses and other species in which sweating is prominent, large amounts of Ca can be lost by this route. Consolazio et al (9) estimated sweat loss of more than 1 g of Ca/day in adult men doing heavy physical work at high environmental temperatures. Sweat losses equalled or exceeded fecal and urinary losses in these men.

Table 10-2. Distribution of fecal and urinary Ca excretion in humans, cattle, and rats. [a]

Species	Age	Ca intake	Fecal Ca	Urine Ca
		mg/day		
Man				
Male	11-16 yr	1,866	1,018	127
	23	1,461	1,229	72
Female	14-16	874	655	194
	55-63	713	586	169
Cattle Young	adult	29,000	27,000	500
Rats	12 wk	44.1	20.8	0.9

[a] Source: Bronner (3)

Signs of Deficiency

The main effect of Ca deficiency is on the skeleton. In young, growing animals a simple Ca deficiency results in rickets and, in adults, the disease is called osteomalacia. In each case the bones become soft and often deformed due to failure in calcification of the cartilage matrix. The familiar skeletal changes associated with rickets are illustrated in Fig. 10-2. Simple Ca deficiency or vitamin D deficiency, which results in poor utilization of dietary Ca, even when the Ca level of the diet is adequate, may produce such abnormal bone development. The histological picture in rickets is one of the reduced calcification. The degree of change is related to the growth rate of the animal. In species such as the pig, dog, and chick, in which body weight may be doubled in a few days and the skeletal mineral turnover rate is therefore very rapid, a Ca deficiency may produce profound changes in bone after only a few days. In other species, such as sheep and cattle, a longer period is required to show deficiency signs.

A deficiency of Ca (or even normal amounts) in the presence of excess P also causes abnormal bone, but, in this instance, excess bone resorption by osteocytic osteolysis (resorption deep within bone) results in osteodystrophy fibrosa. This condition is characterized histologically by replacement of osseous tissue with fibrous connective tissue. The parathyroid gland is hyperactive in an attempt to maintain normal blood serum Ca

(10, 11). There is a generalized effect on the entire skeleton. "Big head" disease of horses (Fig. 10-3), simian bone disease of monkeys and twisted snouts of pigs (Fig. 10-4) may result from feeding excess P.

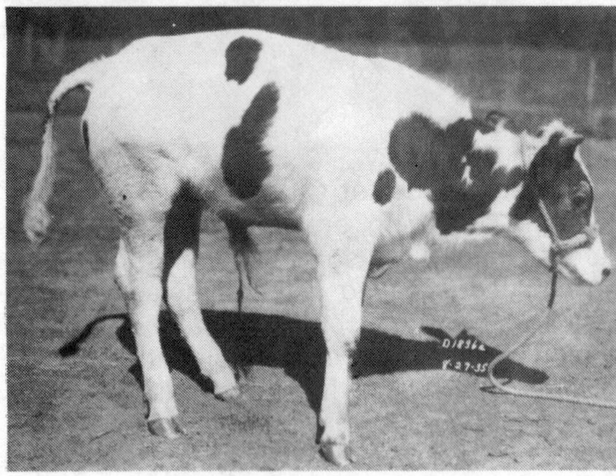

Figure 10-2. A. Bowed legs in a pig with Ca-deficiency rickets. Courtesy of L. Krook. From Brown et al [1966. Cornell Vet. 56, supplement 1]. B. An advanced stage of rickets in the calf. Courtesy of J.W. Thomas, Michigan State University.

The histological picture of the nasal turbinate is similar to that often observed in growing pigs with atrophic rhinitis (AR). Although AR is considered by many to be strictly of infectious origin, apparently dietary Ca-P imbalance is partially or completely responsible in many instances (12, 13).

Lameness and spontaneous bone fractures often, but not necessarily, accompany both osteomalacia and nutritional secondary hyperparathyroidism is adult animals. Reproduction and lactation are adversely affected in rats fed Ca deficient diets (14). The demands of the fetus for Ca are tremendous during late gestation. In rats, fetal uptake of Ca/hr in late gestation is equal to the total maternal Ca blood content (15). Thus, inadequate dietary intake necessitates resorption of Ca from the maternal skeleton to meet fetal needs.

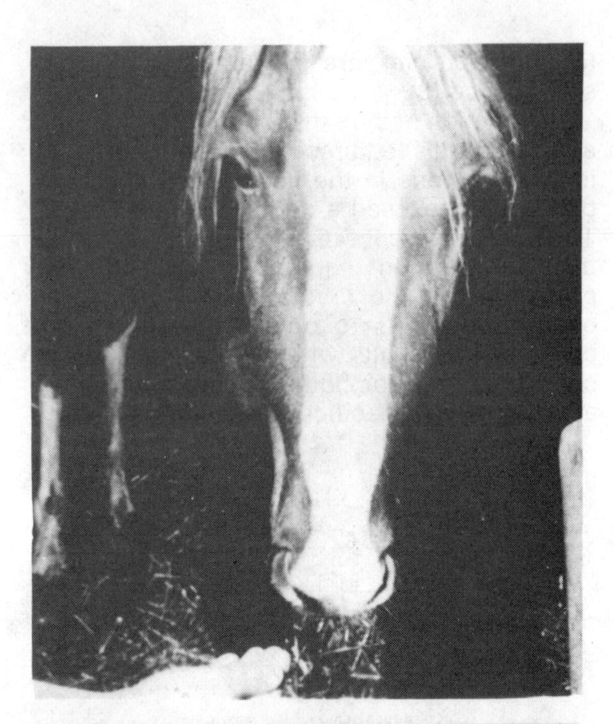

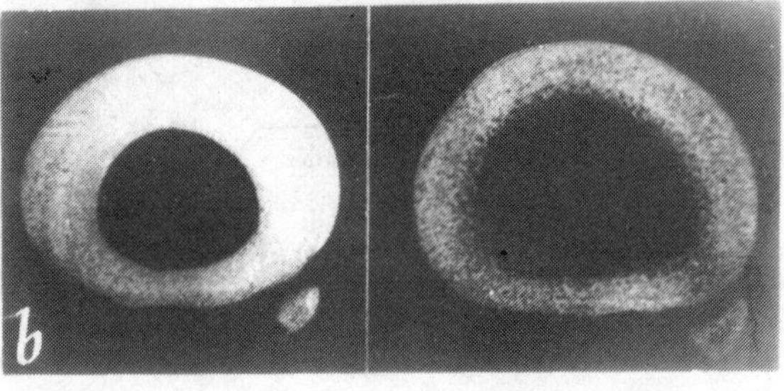

Figure 10-3. A. Nutritional secondary hyperparathyroidism [NSHP] in a horse; "big head," a spontaneous case. The facial bones are enlarged and the facial crest is no longer visible. B. Radiographs of section of the metacarpus of horse [left] fed normal Ca:P diet; right, horse fed excess P. In NSHP the cortex is thinnger and the radiographic density is markedly decreased. From Krook and Lowe [1964, Vet. Path. 1, suppl. 1].

Although blood serum Ca concentration may decline slightly during the early weeks of dietary Ca deficiency, control by parathyroid hormone (increases bone resorption) and calcitonin (inhibits bone resorption) renders this a relatively useless index of Ca nutrition. Only by frequent serial blood sampling over an extended period of weeks or months is monitoring of serum Ca meaningful.

A reduction in bone ash content occurs in all forms of dietary Ca deficiency or Ca-P imbalance. All bones of the skeleton are affected, although the magnitude of change may vary. The proportion of Ca and P remains constant (about a 2:1 Ca-P ratio). The bone ash may be determined directly by ashing the fat-free bone or indirectly by determining specific gravity or by expressing density as g ash/unit volume. This latter method eliminates the need to ether-extract the bone before ashing. Changes in bone density of large magnitude can be identified by radiography, but this method is too insensitive to perceive differences not exceeding about 30%.

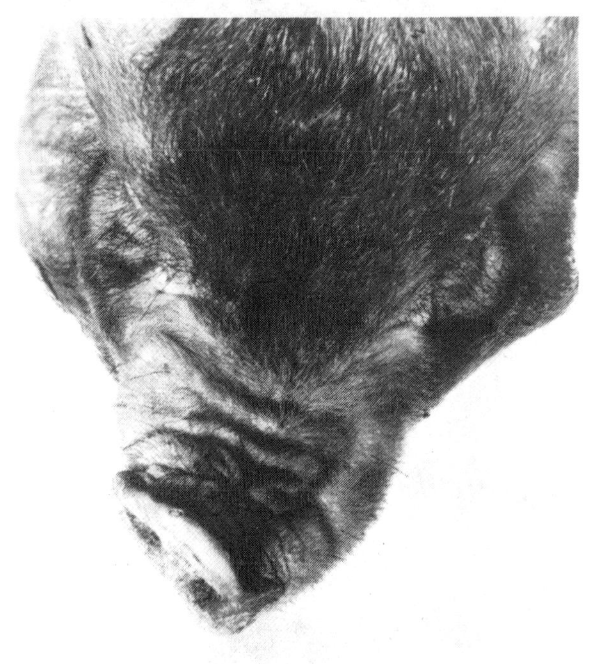

Figure 10-4. Pig with distorted snout due to NSHP caused by a dietary Ca:P imbalance.

Recently, interest has been shown in the effects of physical inactivity on Ca excretion. Striking increases in Ca losses in urine and feces were noted in astronauts confined to small space capsules in the early space flights. Salton et al (16) observed a 2-fold increase in Ca excretion by healthy human subjects after 3 wk of bed rest. The possible implications for animal production in close confinement systems are under study (17). Available evidence indicates that degree of exercise may have important effects on Ca utilization.

Severe Ca deficiency may produce hypocalcemia which results in tetany and convulsions. Follis (18) outlined the pathogenesis of Ca tetany in animals as follows. Calcium deficiency: deficient intake; disturbance in absorption, with vitamin D deficiency or diarrhea; excessive kidney excretion; formation of complexes; and parathyroid effects, either hypofunction (maldevelopment) or removal. Ca tetany is related to the requirement of Ca in normal transmission of nerve impulses and in muscle contraction. Ca tetany may be transient or may culminate in sudden death. Death presumably results from failure in normal heart muscle contractions. Ringer (19) first demonstrated the need for Ca in association with Na and K in normal heart muscle contraction.

The classical example of Ca tetany is the "milk fever" or parturient paresis syndrome of dairy cattle. The condition usually occurs early in lactation during the period of large drains on body Ca reserves for milk production. The etiology is probably related to endocrine function. Both the parathyroid gland and calcitonin-secreting cells of the thyroid may be involved. Dietary Ca intake prior to and during early lactation is important in influencing the capability of these glands to respond appropriately to the sudden change in metabolic demands for Ca imposed by lactation. Ca tetany in dairy cattle and other species responds dramatically to intravenous injection of $CaCl_2$, Ca lactate or other Ca salts to elevate serum Ca above the concentration of 5 or 6 mg/ml that is associated with onset of tetany.

Blood-clotting time is influenced by Ca. The clotting process consists of a complicated series of reactions as outlined in Ch. 13. Such substances as oxalate, citrate, and EDTA are commonly used to prevent blood coagulation in vitro and these same compounds, if administered in large quantities, may also form salts with Ca in vivo and interfere with blood clotting. A reduction in ionized Ca in blood of sufficient magnitude to become a limiting factor in blood clotting is not always seen in simple dietary Ca deficiency. Widespread hemorrhage has been observed in tissues of rats (20) and dogs (21) fed diets severely deficient in Ca.

Toxicity

Acute Ca toxicity has not been reported but chronic ingestion of Ca in excess of metabolic requirements results in abnormalities in bone which can be regarded as manifestations of toxicity. The tendency toward hypercalcemia resulting from continued absorption of excess Ca stimulates calcitonin production by the thyroid. The inhibitory effect of calcitonin on bone resorption is a homeostatic mechanism aimed at minimizing the sources of serum Ca of other than dietary origin. Sustained calcitonin secretion leads to excess bone mass in response to inadequate bone resorption relative to bone apposition. This abnormal thickening of bone cortex is termed osteopetrosis. It has been produced in growing dogs (22) (Fig. 10-5) by feeding 2.0% Ca-1.4% P diets from weaning to young adulthood, and in mature dairy bulls (23) by prolonged feeding of high Ca diets designed for lactating dairy cows. Tumors of the ultimobranchial (calcitonin-producing cells) tissue of the thyroid of these bulls were observed, presumably in response to the sustained hyperactivity resulting from continued hypercalcemia.

Calcification of soft tissues may occur in high Ca feeding, but such calcification only occurs in sites of cellular damage such as in atherosclerosis or inflammation. Tissue damage from Mg deficiency is associated with soft tissue calcification, but high Ca in itself usually does not induce it. A diet containing 2.5% Ca and 0.4% available P has been shown to produce nephrosis, visceral gout, and Ca urate deposits in the ureters of growing pullets (24). Increasing the P level to achieve a more nearly optimum Ca-P ration in such high Ca diets prevented the kidney lesions. Because Ca is present in egg shells, nearly all as $CaCO_3$, that Ca requirement of laying hens are much higher than in most species is not surprising. The hen requires about 3.3 g Ca/day for maximal egg production. The amount in relation to body size would be considered toxic to mammals, but is required by the caged hen to prevent cage fatigue, a condition resulting in excess removal of Ca phosphate from medullary bone and proneness to fracture. Thus, clearly the level of a mineral required to produce toxicity is dependent on species as well as on physiological (productive) state of the animal. This principle applies generally to most minerals.

Urinary calculi (kidney stones) which block the kidney tubules or ureters may occur in animals. The calculi are not believed to be formed by high Ca alone, but rather to require imbalances of other minerals or formation of abnormal complexes with cholesterol or other steroids. In ruminants, excess P in relation to Ca is more likely to cause calculi.

Excess Ca in the diet reduces the absorption and utilization of other minerals. The classical example of a nutritional disease precipitated by high dietary Ca in pigs is the Zn deficiency disease, parakeratosis. The Zn deficiency may not appear in animals fed a normal level of Ca but can become a serious problem when dietary Ca is increased without changing Zn intake. Similar reductions in mineral utilization induced by excess Ca have been reported for Mg, Fe, I, Mn, and Cu (25, 26).

Phosphorus

Tissue Distribution

The P content of adult humans approximates

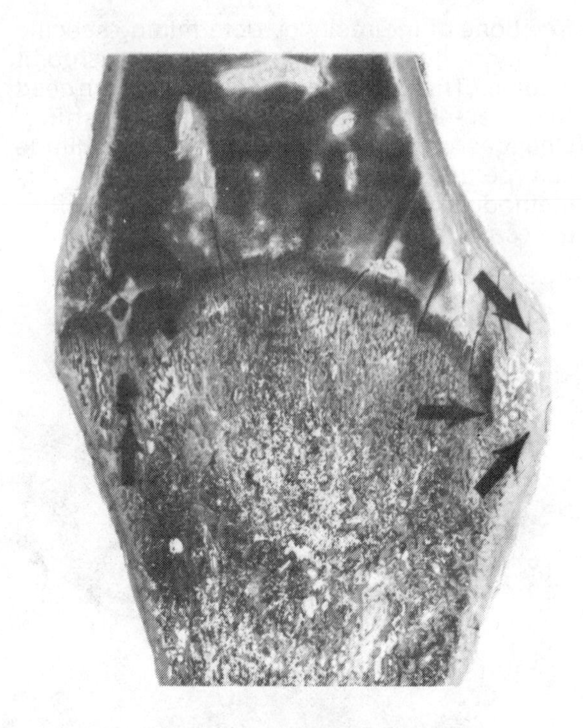

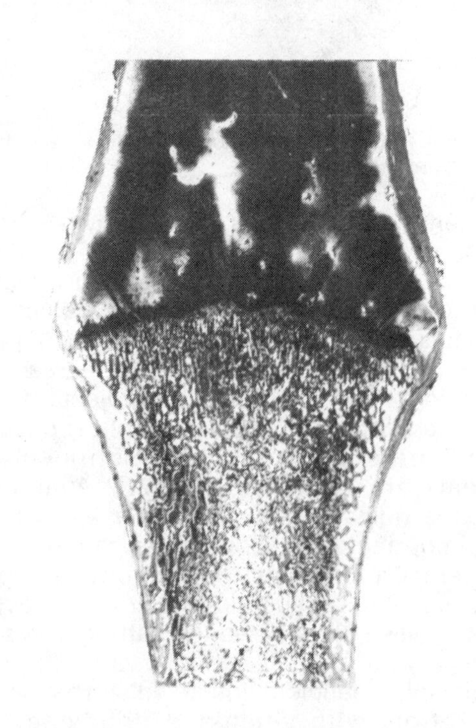

Figure 10-5. Excess bone mass [osteopetrosis] in dogs fed excess Ca [upper] or normal Ca [lower] level showing costochondral junction with irregular growth of cartilage at horizontal arrow, cartilage cells island at vertical arrow and thick perichondral ring at oblique arrows. Courtesy of L. Krook and Cornell Vet.

1.1% of the fat-free body, of which about 80% is in the skeleton (27, 28, 29). Bone ash contains about 18% P. The % of P in the body and the proportion of total P in the skeleton increases throughout prenatal and postnatal life as ossification of the skeleton progresses to maturity. The P in the skeleton is present as part of the hydroxyapatite crystal (see Ca section) while that in the soft tissues is present mostly in organic forms. Table 10-3 gives the P content of an array of organs and tissues of adult man (29). In blood serum, P exists in both inorganic and organic form, the latter as a constituent of lipids. Of the inorganic P, about 10% is bound to serum proteins and 50-60% is ionized (34). P in red blood cells is present as inorganic P, organic acid-soluble P, lipid P, and RNA P, the proportions varying with age and species (35). Total serum P concentration under normal conditions in most species is 6-9 mg/100 ml. The transfer rates of P among spatially distinct compartments of the body have been calculated by Lax et al (30), using the assumption that each compartment exchanges P with others through a central compartment (blood). The exchange rate of P among several of the important organs was highest for the skeleton, followed in descending order by muscle and heart. The average time spent by a single P atom in each compartment was also calculated (2.8 hr for blood cells vs. 393 hr for brain); the length of time in a compartment was found to be related to the metabolic activity of the tissue. Thus, P movement within the body is clearly in a dynamic state.

Functions

As with Ca, the most obvious function of P is as a component of the skeleton. In this role, it provides structural support for the body. Because Ca and P occur together in bone, the discussion of the function of Ca (see Ca section) in bone also applies to P. P is a component of phospholipids which are important in lipid transport and metabolism and cell-membrane structure. As such, P is present in virtually all cells (see Ch. 8). P functions in energy metabolism as a component of AMP, ADP, and ATP and of creatine phosphate. The importance of high-energy phosphate bonds in normal life processes was discussed in Ch. 8. P is a component as phosphate of RNA and DNA, the vital cellular constituents required for protein synthesis (see Ch. 6). It is a constituent of several enzyme systems (cocarboxylase, flavoproteins, NAD).

Metabolism

The metabolism of P can be discussed in terms of bone metabolism, the metabolism of phospholipids, and of high-energy phosphate

Table 10-3. Calcium, phosphorus, and magnesium content of organs and tissues of adult humans.[a]

Tissue or organ	Ca	P	Mg
		% of fresh tissue	
Skin	0.015	0.083	0.0102
Skeleton	11.51	5.19	0.191
Teeth	25.46	13.24	0.618
Striated muscle	0.014	0.116	0.0198
Nerve tissue	0.015	0.224	0.0107
Liver	0.012	0.127	0.0081
Heart	0.018	0.123	0.0168
Lungs	0.017	0.228	0.0069
Spleen	0.010	0.169	0.0124
Kidney	0.019	0.124	0.0169
Alimentary tract	0.009	0.111	0.131
Adipose tissue	0.009	0.055	0.0060
Remaining tissue (solid)	0.047	0.163	0.0170
Remaining tissue (liquid)	0.008	0.088	0.0084
Total composition			
Whole body	1.98	1.06	0.045
Fat-free	2.07	1.11	0.047
Bone ash	39.44	17.80	0.654

[a] Source: Forbes et al (29)

compounds such as ATP, ADP, and creatine phosphate. Bone metalbolism was discussed earlier and phospholipid metabolism and metabolism of high-energy phosphate compounds are discussed in Ch. 8 (Lipids) and in other sections relating to energy utilization.

Absorption of P from the GIT occurs by passive diffusion. Vitamin D apparently has no effect on P absorption (27). No active transport system is known for P as it is for Ca, although in vitro work has shown that P may traverse the intestinal cell membrane against a concentration gradient in the presence of Ca. Thus, P absorption is related directly to dietary P concentration. An excess of dietary P in relation to Ca depresses Ca absorption (31). This may be because of formation of insoluble Ca phosphate salts and/or to Ca being bound by phytic acid (the hexaphosphoric acid ester of inositol). Many plant seeds are high in P, much of which is present as phytic acid.

P absorption from the GIT is rapid as demonstrated by radioisotope studies with ^{32}P (32). Much of the labeled P is incorporated into phospholipids in the intestinal mucosal cells (33).

Secretion of P into the intestinal lumen (endogenous fecal P) occurs, but this loss does not represent as high a proportion of the daily loss as for Ca. Most of the P excretion occurs through the kidneys and renal excretion appears to be the main regulator of blood P concentration. When intestinal absorption of P is low, urinary P falls to a low level with reabsorption by the kidney tubules approaching 99% (27). Kidney excretion of P is under control of parathyroid hormone and 1,25-dihydroxy vitamin D as a part of the overall blood homeostatic mechanism for Ca and P.

Signs of Deficiency

The most common sign of P deficiency in growing animals is rickets (Fig. 10-2). The gross and histological changes in rickets have been described earlier (see Ca section and Ch. 13). Fecal output of P tends to remain relatively unchanged while urinary excretion is reduced, but the total excretion may still exceed intake when Ca intake is relatively high. Calcium excretion in both urine and feces is increased in P deficiency as a manifestation of reduced calcification of bone. As the deficiency progresses, appetite fails and growth is retarded. Deficient animals (Fig. 10-6) often have a depraved appetite and may chew on wood. This abnormal behavior of eating or chewing is termed **pica** (Fig. 10-7). Adults fed low P diets may exhibit pica, and bone density is decreased as in rickets. Impaired fertility has been reported in P-deficient cattle (36). Blood-serum Ca is increased and serum P is decreased by P deficiency.

Blood P homeostasis is more complicated than that for blood Ca because blood P is in equilibrium not only with bone P but also with several organic P compounds. Nevertheless, kidney excretion of P is sufficiently controlled by parathyroid hormone secretion and 1,25-dihydroxy vitamin D to result in relatively stable serum P concentration even with severe dietary P deficiency.

Toxicity

An excess of dietary P results in nutritional secondary hyperparathyroidism manifested in excessive bone resorption (fibrous osteodystrophy) which may result in lameness and spontaneous fractures of long bones. Growing pigs with severe bone resorption produced by low Ca-high P diets may suffocate because of softening of the ribs to the extent that normal respiratory movements are inhibited (12).

Ca-P ratios greater that 1:2 may produce fibrous osteodystrophy is growing or adult animals. Lean meat and many cereal grain by-products (notably wheat bran) contain several times as much P as Ca (see section on Ca deficiency).

High P has a laxative effect so that dietary excesses result in diarrhea and high fecal loss of P as well as other nutrients.

Figure 10-6. A phosphorus deficient cow. Note stiffness and thin condition. Bones were depleted to such an extent that one rib was broken in a casual examination. Courtesy of R.B. Becker, Florida Agr. Expt. Sta.

Figure 10-7. Pica. A P deficiency, as well as deficiencies of other nutrients, may result in a depraved appetite [pica]. This example shows some of the material recovered from the stomach of a deficient cow. It includes oyster shells, porcelain, teeth, a section of cannon bone, inner tube, tire casing, pieces of metal and pebbles. Courtesy of R.B. Becker, Florida Agr. Expt. Sta.

Magnesium

Tissue Distribution

Magnesium is widely distributed in the body and, except for Ca and P, is present in larger amounts in the body (see Table 10-1) than any other mineral. About ½ of body Mg is in bone at a concentration of 0.5-0.7% of the bone ash (38). Mg in soft tissues is concentrated within cells; highest concentration is in liver and skeletal muscle (39, 40). Blood Mg is distributed about 75% in red blood cells (6 meq/l.) and 25% in serum 1.1/2.0 meq/l.). The concentration in serum seems to vary among species as shown in Table 10-4. Of the serum Mg, about 35% is protein-bound in mammals and birds, even though total Mg is variable among species.

Functions

Mg is required for normal skeletal development as a constituent of bone; it is required for oxidative phosphorylation by mitochondria of heart muscle (41) and probably by mitochondria of other tissues. Mg is required for activation of enzymes which split and transfer phosphatases and the many enzymes concerned in the reactions involving ATP. Because ATP is required in such diverse functions as muscle contraction, protein, nucleic acid, fat, and coenzyme synthesis; glucose utilization; methyl-group transfer; sulfate, acetate, and formate activation; and oxidative phorphorylation, to name but a few. By inference, the activating action of Mg extends to all of these functions (38). Mg is a cofactor in decarboxylation and is required to activate certain peptidases. Specific examples of enzyme systems requiring Mg for activity have been described by Wacker and Vallee (38).

Metabolism

Examination of the functions of Mg show clearly that its metabolism is complex and varied (38, 42, 43). Absorption from the GIT occurs mostly from the ileum. No carrier is known for Mg absorption nor has vitamin D been shown to enhance its absorption. Although some have suggested a common pathway for Ca and Mg absorption, the fact that Ca is absorbed mainly from the upper small intestine and is associated with a vitamin D-dependent binding protein, suggests no common pathway.

Homeostatic control of blood and tissue Mg is not clearly understood. Hyperparathyroidism is associated with increased urinary excretion and reduced serum Mg, but a specific effect on Mg aside from the concomitant release of Mg from bone when Ca is released in response to parathyroid hormone has not been shown.

Mg excretion occurs via the feces and urine. About 55 to 60% of ingested Mg is absorbed, and the absolute amount absorbed is proportional to dietary intake (44). Urinary excretion accounts for about 95% of losses of absorbed Mg, and fecal excretion accounts for most of the remainder. Endogenous fecal excretion is largely into the proximal small intestine, so that, as with Ca, probably some reabsorption occurs as it traverses the GIT.

Signs of Deficiency

Mg deficiency in growing rats results in anorexia, reduced weight gain, reduced serum Mg, hypomagnesemic tetany and, within 3-5 days, characteristic hyperemia of the ears and extremities (45, 46). Repka (47) found a close association between occurrence and severity of cutaneous hyperemia and plasma Mg concentration. Continued and severe hypomagnesemia was accompanied by slight hypercalcemia after 3 wk and depression of some of the liver enzyme systems requiring Mg. Bunce and Bloomer (48) recently showed that kidney Ca elevation in Mg deficiency is accompanied by a decrease in total serum Mg, but an increase in the ratio of free to total Mg. Concentration of Mg in liver was not depressed. Severe leukocytosis develops concurrently with hyperemia of extremities. Mg deficiency in pigs results in weak and crooked legs, hyperirritability, muscular twitching, reluctance to stand, tetany, and death. Plasma and urinary histamine are elevated in Mg deficiency, and serum and liver glutamic oxalacetic transaminase activity are increased (49, 50, 51, 57).

Tufts and Greenberg (52) demonstrated the calcification and necrosis of the kidney that occur in Mg deficiency. High dietary Ca and P appear to aggravate the Mg deficiency and accentuate the calcification of soft tissues associated with inadequate Mg.

Red blood-cell Mg concentration is reduced as well as plasma Mg concentration, but the decline is more gradual and reaches 50% of control values by the 10th day as compared to the 4th or 5th day for plasma Mg concentration. Although liver Mg concentration is unaffected, skeletal muscle concentration is reduced by 25% or more (53, 54). Bone Mg content is reduced in Mg deficiency (54, 55), when expressed either as % of Mg in whole bone or in the bone ash.

A decline in tissue K and a rise in tissue Ca and Na occur in Mg-deficient animals (56). Activities of several Mg-dependent enzymes, including plasma alkaline phosphatase, muscle enolase and pyruvate phosphokinase, are depressed in Mg deficiency, but many others are not (53, 57). Mitochondria of kidney tubule cells are swollen in Mg deficiency as shown by electron microscopy (58). Chow and Pond (59) showed mitochondrial swelling in liver from ammonia-intoxicated rats and observed loss of intramitochondrial Mg and PO_4 ions. Head and Rook (60) proposed that high ammonia interferes with Mg absorption by forming the insoluble Mg ammonium phosphate (struvite) at alkaline pH.

A common problem of grazing cattle is a syndrome called grass tetany, Mg tetany, or wheat-pasture poisoning. It occurs most frequently in cattle grazing cereal forages or native pastures in periods of lush growth (usually in spring months), but it is also a problem at times in cattle fed conventional wintering rations (61, 62, 63, 64). Sjollema (65) first described the symptoms of tetany which have since been ascribed to hypomagnesemia (low blood Mg) (66). The etiology of Mg tetany is not completely understood, although it is generally agreed that hypomagnesemia, whatever its underlying cause, is the triggering factor. Phytic acid P decreases Mg absorption in nonruminants (67), but no difference between organic and inorganic P was observed in affecting Mg absorption in sheep (68). The high level of K and protein usually present in lush pastures has suggested the possibility of an antagonism with Mg (69, 70). Newton et al (71) found that sheep fed high K tended to excrete more ^{28}Mg in urine and feces, and that high K interfered with Mg absorption, but not its re-excretion into the GIT. Moore et al (72) reported elevated urinary Mg excretion in sheep fed high N (urea) diets, and Colby and Frye (73) reported an increased severity of Mg deficiency in rats fed high Ca, K, or protein diets.

High concentrations of *trans*-aconitate, an intermediate in the citric acid cycle, have been observed in early season forages and have been suggested as having a role in Mg tetany (74, 75, 76). Intravenous administration of either *trans*-aconitic or citric acid into cattle produces tetany resembling clinical Mg tetany (77). The full implications of these findings in relation to grass tetany are not fully realized, but the syndrome apparently is more complicated than a simple dietary Mg deficiency. In practice, supplemental Mg from a variety of inorganic sources is effective in preventing Mg tetany. The problem is one of the practical difficulties in administering Mg to free-ranging cattle and sheep on pasture.

Evidence from rats and monkeys (78) suggests that submarginal Mg intake increases susceptibility to atherosclerosis in the presence of high cholesterol intake. The significance of this relationship in the incidence of atherosclerosis in man and animals is unknown.

Toxicity

Reports of Mg toxicity by feeding have not been found, probably because of the ability of the kidney to excrete excess Mg in response to

Table 10-4. Serum Mg concentration of man and animals. [a]

Species	Mg concentration, meq/l.
Cow	1.6
Dog	1.6
Goat	1.9
Horse	1.5
Hen	1.6
Human	2.0
Mouse	1.1
Pig	1.3
Rabbit	2.1
Rat	1.6
Sheep	1.7

[a] Source: Wacker and Vallee (38)

elevated serum levels. Intravenous injection of $MgSO_4$ was shown in 1869 (79) to result in peripheral muscle paralysis of dogs. Subsequent early work confirmed this and led to the use of Mg as an anesthetic for surgery. Mg decreases the release of acetylcholine at the neuromuscular junction and sympathetic ganglia (80, 81).

Mg induces a drop in blood pressure and high serum concentrations (greater than 5 meq/l.) affect the electrocardiogram and may cause the heart to stop in diastole (82, 83).

Potassium, Sodium and Chlorine

These 3 minerals are considered together because they are all electrolytes that play a vital role in maintaining osmotic pressure in the extracellular and intracellular fluids, and in maintaining acid-base balance. Each has its own special functions, in addition, which will be discussed separately.

Distribution in Body Tissues

The total body content of K, Na, and Cl in newborn and adults of several mammals including man are shown in Table 10-1. Normal ratios among electrolytes are remarkably constant among species. Houpt (84) has described the tissue distribution and interrelationship of K, Na, and Cl in maintaining acid-base balance. K is present mainly within cells (about 90% of body K is intracellular) and is readily exchangeable with the extracellular fluid. On the other hand, Na is present largely in extracellular fluid with less than 10% within cells. Of the other 90%, about ½ is absorbed to the hydroxyapatite crystal of bone and ½ is present in plasma and interstitial fluids. Cl acts with bicarbonate (HCO_3) to electrically balance the Na of the extracellular fluid. Excess excretion of Na by the kidney is accompanied by Cl excretion. Cl is present almost exclusively in the extracellular fluid. Gamble (85) diagrammed the concentrations of K^+, Na^+, Cl^-, HCO_3 and organic constituents of blood plasma, interstitial fluid, and intracellular fluid (Fig. 10-8). The meq of cations/l. within each compartment exactly equals the meq of anions/l. to ensure electrical neutrality. The composition depicted for cell fluids is only an approximation, because some individual tissues have cell electrolyte composition considerably different than man.

Functions

K is located mostly within cells and, by means of an energy-requiring system related to Na movement, it influences osmotic equilibrium; K functions in maintenance of acid-base balance in the body; it is required in enzyme reactions involving phosphorylation of creatine and is required for activity of pyruvate kinase (18). K facilitates uptake of neutral amino acids by cells (86), and influences carbohydrate metabolism by affecting uptake of glucose into cells (87). It is required for normal tissue protein synthesis in protein depleted animals (88, 89) and is required for normal integrity of the heart and kidney muscle (18) and for a normal electrocardiogram (90).

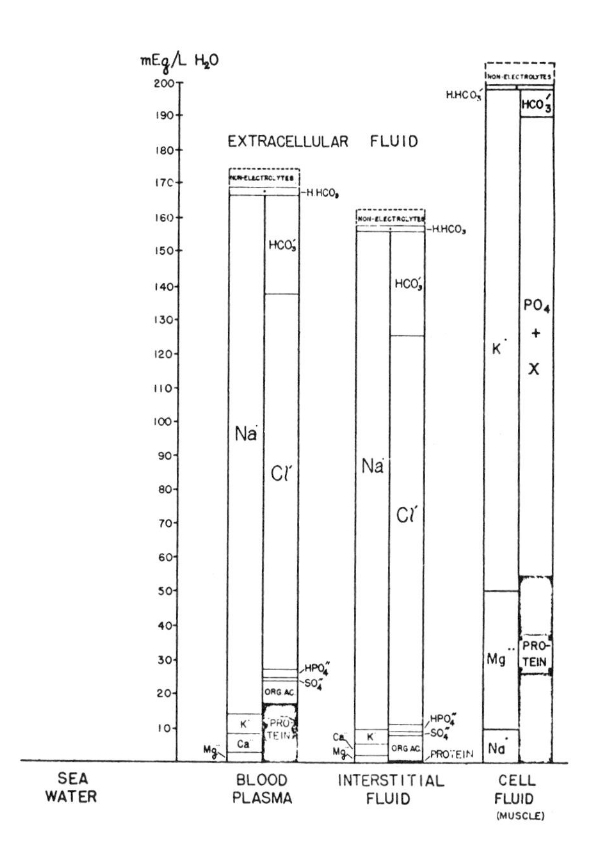

Figure 10-8. Electrolytes in blood plasma, interstitial fluid, and intracellular fluid. From Gamble [85].

Na functions as the extracellular component through an energy dependent Na "pump," along with K and Mg and Intracellular components, in maintaining osmotic pressure. It functions in maintaining acid-base balance in the body and in transfer of nerve impulse transmission by virtue of the potential energy associated with its separation from K by the cell membrane.

Cl functions in regulation of extracellular osmotic pressure and in maintaining acid-base balance in the body. It is the chief anion of gastric juice where it unites with H ions to form HCl.

Metabolism

K, Na, and Cl ions are not absorbed in appreciable amounts from the stomach, but considerable absorption takes place from the upper small intestine and lesser amounts from the lower small intestine and large intestine. The daily secretion of fluids into the GIT from saliva, gastric juice, bile, and pancreatic juice is 4-5 times the daily oral intake (91). Thus, large variation in intake has a relatively small effect on the total load of fluids and electrolytes entering the GIT. About 80% of the NaCl load and 50% of the K load on the GIT is from secretions. Cl is secreted in large quantities in the stomach, Na is secreted into the upper small intestine, and K mainly into the ileum and large intestine (92). All 3 ions are absorbed both by passive and active processes; Na and K cross the mucosa by active transport in the intestine but largely by diffusion in the stomach, whereas Cl is transferred by active processes in the stomach and upper intestine, but by passive diffusion in the large intestine (92).

The regulation of K, Na, and Cl ion concentrations within rather narrow limits in extra- and intracellular compartments of the body is not completely understood (84, 93). Ingestion of more of each mineral than needed results in ready excretion by the kidneys. Thus, toxicity is not likely unless water intake is restricted, which impedes urine output. Excretion of Na by the kidney and partial, but not complete, reabsorption from the kidney tubules. That not reabsorbed is lost in the urine. The plasma level of the Na tends to be controlled by the action of the hormone, aldosterone, secreted by the adrenal cortex. Its action is to increase Na reabsorption from the kidney tubule. A fall in plasma Na, as a result of reduced intake, results in increased aldosterone release. Aldosterone production is, in turn, under the control of the adrenocorticotrophic hormone of the anterior pituitary gland, and its secretion is impaired in hypophysectomized animals.

The antidiuretic hormone of the posterior pituitary also plays a part in Na excretion through its response to changes in osmotic pressure of the extracellular fluid induced by water deprivation. Osmoconcentration results in increased antidiuretic-hormone release as a means of conserving water. Thus, the complex relationships between body Na content, aldosterone secretion, and antidiuretic hormone secretion, although incompletely understood, are of utmost importance in maintaining fluid electrolyte homeostasis. In animals that perspire extensively, such as man and the horse, large amounts of Na are lost from the body by this route. K intake usually exceeds by several times the metabolic requirements for it, yet K toxicity does not ordinarily occur because of the ability of the kidney to regulate its excretion. Aldosterone also affects K excretion; high K in extracellular fluid stimulates aldosterone secretion in the same way that low Na does. A rather constant ratio of Na to K in the extracellular fluid is maintained. In K deficiency, some Na is transferred inside the cell to replace K, and in that way preserve osmotic and acid-base equilibrium. When intake of either Na or K is inadequate, the deficiency is aggravated by increasing the intake of the other (95). Cl concentration in the extracellular fluid tends to be controlled in relation to Na. Excess kidney excretion is also affected by bicarbonate ion (HCO_3) concentration. If plasma bicarbonate rises, an equal amount of Cl is excreted in the urine to maintain an equal concentration of cations and anions in the plasma.

K and Cl homeostasis are closely related A deficiency of one leads to a metabolic deficiency of the other (94). K reabsorption by the kidney tubule requires the presence of Cl. Thus, KCl is more effective than other K salts in repletion from K deficiency.

Deficiency Signs

K. Deficiency of K results in abnormal electrocardiograms in calves (90), chicks (97), and pigs (98) as well as other species (96). Postmortem examination may not always reveal pathological changes in heart muscle (98), but in rats tiny gray opacities on the ventricles of the heart and loss in striations have been observed. Kidney lesions have been also observed (99, 100, 101). Growth retardation, unsteady gait, general overall muscle weakness, pica (Fig. 10-9), and emaciation followed by death are the symptoms of K deficiency in animals and birds. Mg deficiency results in failure to retain K and, in this way, may lead to K deficiency in animals and man (94). Diarrhea is associated with loss of electrolytes, notably K, in abnormally high quantities in the feces, thereby upsetting osmotic pressure relationships and acid-base balance.

Figure 10-9. K deficient lamb. This picture illustrates the wool biting and pulling that may occur in deficient animals. University of Manitoba photo, courtesy of W.K. Roberts.

Na. The main signs of Na deficiency are reduced growth rate and efficiency of feed utilization in growing animals and reduced milk production and weight loss in adults (Fig. 10-10). A cloudy appearance of the cornea of the eye has been reported in rats (18, 99). Smith and Aines (102) studied salt (NaCl) deficiency in lactating dairy cows and found that Na, K, and Cl concentration of milk remained unchanged as did the plasma Na concentration. Urinary excretion of Na declined to near zero within a month and appetite and milk production dropped sharply. Animals deprived of Na display a great craving for it and have been observed to drink urine in an effort to satisfy their craving. The Na ion appears to be the critical mineral in NaCl deficiency (102, 103).

Figure 10-10. A salt deficient cow. Note the dehydrated appearance of the animal. From Smith and Aines [102].

Cl. Depressed growth rate seems to be the major sign of Cl deficiency. Low dietary Cl results in a decline in urinary Cl concentration to near zero, and Cl concentration of skin, muscle, liver, kidney, brain, viscera, and total carcass is reduced (104). Kidney lesions develop within one month of feeding on a Cl-deficient diet (18). Leach and Nesheim (105) described a syndrome in chicks fed Cl-deficient diets. They show a characteristic nervous reaction induced by sudden noise; they fall forward with their legs extended backward.

In practice, Na and Cl are supplied in the diet together as common salt and it appears that the metabolic requirement for each is proportional to the amounts contributed by NaCl.

Toxicity

Because the kidney normally regulates its excretion of K, Na, and Cl in accord with variations in dietary intake, a toxicity of any of the 3 electrolytes is unlikely except when water intake is restricted, drinking water is saline, or as a result of renal malfunction. Chronic K excess induced by one of the above means results in hypertrophy of the adrenal cortex with accompanying changes in aldosterone output (106).

Chronic, excess Na ingestion results in hypertension associated with degenerative vascular disease. The glomeruli of the kidney are affected (107). Excess salt ingestion has been suggested to be in the etiology of hypertension in man (108). Continued high intakes of salt increase total extracellular fluid volume; this is the basis for the use of low salt diets in hypertension and congestive heart failure.

Acute salt toxicity has been produced experimentally in growing pigs (109) by severely restricting water intake while simultaneously providing feed containing 2% NaCl for 1 or 2 days. Symptoms included staggering, marked weakness, paralysis of hind limbs or general paralysis, violent convulsions, and death. Postmortem examination revealed no lesions except a slight increase in volume of fluid in the pericardium and small hemorrhages in the liver.

Excess Cl is not likely, but the use of purified diets containing amino acids and other salts added in the hydrochloride form may increase the total acidity of the diet. Scott et al (110) have recommended that extra K and Na be added to such diets to assure maintenance of acid-base balance.

Sulfur

Tissue Distribution

S is required by animals mainly as a component of organic compounds. These include the amino acids methionine, cystine, cysteine; the vitamins biotin and thiamin; certain mucopolysaccharides, including the chondroitin sulfate and mucoitin sulfates; heparin; glutathione; coenzyme A. Because proteins are present in every cell of the body, and S-containing amino acids are components of virtually all proteins (usually 0.6 to 0.8% of the protein), S is distributed widely throughout the body and in every cell (111, 94) and makes up about 0.15% of body weight (25).

Functions

S functions mainly through its presence in organic metabolites, but some evidence has shown that inorganic sulfate ($SO_4^=$) is important in the diet of the rat (112). Its functions in organic compounds will be listed in terms of the metabolic function of the compounds containing it.

Inorganic SO_4 functions in acid-base balance as a constituent of intracellular, and to a lesser extent, extracellular fluid; as a component of S-containing amino acids, it is required for protein synthesis (see Ch. 6); as a component of biotin, it is important in lipid metabolism (see Ch. 14); and as a component of thiamin, it is important in carbohydrate metabolism (see Ch. 14); as a component of coenzyme A, it is important in energy metabolism (see Ch. 7, 8, 14); and as a component of mucopolysaccharides, it is important in collagen and connective tissue metabolism; and as a component of heparin, it is required for blood clotting; S is a component of ergothioneine of red blood cells and of glutathione, a universal cell constituent; and S is a component of certain hormones including estrogen and protein-containing hormones as a constituent of amino acids.

Metabolism

Absorption of inorganic sulfate from the GIT is inefficient. Active transport of $SO_4^=$ takes place from the upper small intestine. Kun (113) placed little importance on inorganic sulfur in nonruminant animal nutrition, but Michels and Smith (112) provided evidence in rats that inorganic S is needed in the diet to prevent an increase in the S-amino acid requirement. Both inorganic and organic forms of S are used for sulfation of cartilage mucopolysaccharides.

Organic forms of S are readily absorbed, as discussed elsewhere (Ch. 6, 7, 8, 14) in relation to the compounds that contain S. Inorganic S is excreted via the feces and urine. Unabsorbed sulfate is probably reduced in the lower GIT and excreted as sulfate (114). Endogenous fecal S enters the GIT largely through the bile as a component of taurocholic acid. Urinary S is present mainly as inorganic $SO_4^=$, but also as a component of thiosulfate, taurine, cystine, and other organic compounds. Because the bulk of body S is present in amino acids, that urinary S excretion tends to parallel urinary N excretion is not surprising. High protein diets are associated with large amounts of urinary S and N.

In ruminants fed nonprotein N, a growth response may be obtained by inorganic S supplementation. Rumen microflora incorporate N into cellular protein, but synthesis is limited if S is not available in sufficient quantities for formation of S-amino acids. Thus, inorganic S may be important in ruminant nutrition, but only in terms of its need by microflora of the rumen for protein synthesis. Sheep, which produce wool high in S-containing amino acids, need a higher S:N ratio than nonwool-producing ruminants (115). Birds, whose feathers are high in S-amino acids, likewise have a higher S requirement than mammals.

Deficiency Signs

Inorganic S has not been shown to be essential for animals for normal maintenance or productive functions. Deficiency of methionine, thiamin, or biotin, however, each of which contains S, certainly produces functional and morphological lesions as discussed elsewhere (Ch. 6, 14). Also, the absence of inorganic S from the diet may increase the requirement for S-containing amino acids, implying that S from these sources is used for synthesis of other organic forms of S in the absence of inorganic S (112).

Sheep fed nonprotein N to replace protein without concomitant supplementation with S show reduced wool growth, and weight gain of sheep and cattle is depressed in S deficiency. These effects on animal performance are manifestations of inadequate microbial nutrition on which the host depends for synthesis of organic metabolites and therefore cannot be considered as direct S deficiency.

Toxicity

Because the intestinal absorption of inorganic S compounds is low, S toxicity is not a practical problem. Excesses of S-amino acids cause anorexia and growth depression, but such effects are observed with excesses of amino acids in general, not specifically the S-amino acids.

References Cited

Calcium

1. Widdowson, E.M. and J.W.T. Diskerson. 1964, In: Mineral Metabolism. Vol. 2, Part A. Academic Press, N.Y.
2. Neuman, W.F. and M.W. Neuman. 1958. The Chemical Dynamics of Bone Mineral. Univ. Chicago Press, Chicago.
3. Bronner, F. 1964. In: Mineral Metabolism. Vol. 2, Part A. Academic Press, N.Y.
4. Copp, D.H. et al. 1962. Endocrin. 70:638.
5. Brink, F. 1954. Pharmacol. Rev. 6:243.
6. Wasserman, R.H., R.A. Corradino and A.N. Taylor. 1968. J. Biol. Chem. 243:3978.
7. Lengemann, F.W. 1959. J. Nutr. 69:23.
8. Wasserman, R.H., C.L. Comar and M.M. Nold. 1956. J. Nutr. 59:371.
9. Consolazio, C.F. et al. 1962. J. Nutr. 78:78.
10. Minkin, C. and J.M. Jennings. 1972. Science 176:1031.
11. Waite, L.C. 1972. Endocrinology 91:1160.
12. Brown, W.R., L. Krook and W.G. Pond. 1966. Cornell Vet. 56: supplement 1, p 108.
13. Logomarsino, J.V. 1973. Ph.D. Thesis, Cornell University, Ithaca, N.Y.
14. Boelter, M.D.D. and D.M. Greenberg. 1943. J. Nutr. 26:105.
15. Comar, CM. 1956. Ann. N.Y. Acad. Sci. 64:281.
16. Salten, B. 1968. Circulation 38, Suppl VII.
17. Anon. 1961. Nutr. Rev. 19:42, 27:103.
18. Follis, R.H., Jr. 1958. Deficiency Disease. C.C. Thomas, Springfield, Ill.
19. Ringer, S. 1882. J. Physiol. 3:380.
20. Boelter, M.D.D. and D.M. Greenberg. 1941. J. Nutr. 21:61; 21:75.
21. Martin, G.J. 1937. Growth 1:175.
22. Wu Fu-Ming. 1973. Ph.D. Thesis, Cornell University, Ithaca, N.Y.
23. Krook, L. et al. 1971. Cornell Vet. 61:625.
24. Scott, M.L., M.C. Nesheim and R.J. Young. 1969. Nutrition of the Chicken. M.L. Scott and Assoc., Pub., Ithaca, N.Y.
25. Maynard, L.A. and J.K. Loosli. 1969. Animal Nutrition. McGraw-Hill Book Co., N.Y.
26. Davis, G.K. 1959. Fed. Proc. 18:1119.

Phosphorus

27. Irving, J.T. 1964. In: Mineral Metabolism. Vol. 2, Part A. Academic Press, N.Y.
28. Forbes, R.M., A.R. Cooper and H.H. Mitchell. 1953. J. Biol. Chem. 203:359.
29. Forbes, R.M., A.R. Cooper and H.H. Mitchell. 1956. J. Biol. Chem. 223:969.
30. Lax, L.C., S. Sidlofsky and G.A. Wrenshall. 1956. J. Physiol. 132:1.
31. Nicolaysen, R., N. Eeg-Larsen and O.J. Malm. 1953. Physiol. Rev. 33:424.
32. Weissberger, L.H. and E.S. Nasset. 1942. Amer. J. Physiol. 138:149.
33. Artom, C., G. Sarzana and E. Segre. 1938. Arch. Intern. Physiol. 47:245.
34. Walser, M. 1961. J. Clin. Invest. 40:723.
35. Widdowson, E.M. and J.W.T. Dickerson. 1964. In: Mineral Metabolism. Vol. 2, Part 2. Academic Press, N.Y.
36. Theiler, A. and H.H. Green. 1931-32. Nutr. Abs. Rev. 1: 359.
37. Day, H.G. and E.V. McCollum. 1939. J. Biol. Chem. 130:269.

Magnesium

38. Wacker, W.E.C. and B.L. Vallee. 1964. In: Mineral Metabolism. Academic Press, N.Y.
39. Eichelberger, L. and F.C. McLean. 1942. J. Biol. Chem. 142:467.
40. Baldwin, D., P.K. Robinson, K.L. Zierler and J.L. Lilienthal. 1952. J. Clin. Invest. 31:350.
41. Vitale, J.J., M. Nakamura and D.M. Hegsted. 1957. J. Biol. Chem. 228:573.
42. Arkawa, J.K. 1965. In: Electrolytes and Cardiovascular Disease. S. Karger, N.Y.
43. Walser, M. 1967. In: Rev. Physiol. Biochem. and Exptl. Pharmacol. Springer-Verlag, N.Y.
44. Chutkow, J.G. 1964. J. Lab. Clin. Med. 63:80.
45. Kruse, H.D., E.R. Orent and E.V. McCollum. 1932. J. Biol. Chem. 96:519.
46. Follis, R.H. 1958. Deficiency Disease. C.C. Thomas Pub., Springfield, Ill.
47. Repka, F.J. 1972. Ph.D. Thesis, Cornell Univ., Ithaca, N.Y.
48. Bunce, G.E. and J.E. Bloomer. 1972. J. Nutr. 102:863.
49. Battifora, A.A., et al. 1968. Arch. Path. 86:610.
50. Mayo, R.H., M.P. Plumlee and W.M. Beeson. 1959. J. Animal Sci. 18:264.
51. Bois, P. 1963. Brit. J. Exptl. Path. 44:151.
52. Tufts, E.V. and D.M. Greenberg. 1938. J. Biol. Chem. 122:693; 122:715.
53. Elin, R.J., W.D. Armstrong and L. Singer. 1971, Amer. J. Physiol. 220:543.
54. Forbes, R.M. 1966. J. Nutr. 88:403.
55. MacIntyre, I. and D. Davidsohn. 1958. Biochem. J. 70:456.
56. Martin, H.E. and M.L. Wilson. 1960. Metabolism 9:484.
57. Gunther, T.H. 1970. Zeitschrift fur Klinische Chemie und Klinische Biochemie 8:69. Cited by Repka, 1972.

58. Hess, R., I MacIntyre, N, Alcock and A.G.E. Pearse. 1959. Brit. J. Exptl. Path. 40:80.
59. Chow, Kye-Wing and W.G. Pond. 1972, Proc. Soc. Exptl. Biol. Med. 139:150.
60. Head, M.J. and J.F. Rook. 1955. Nature 176:262.
61. Blaxter, K.L. and F.F. McGill. 1956. Vet. Rev. Annot. 2:35.
62. Sims, F.H. and H.R. Crookshank. 2956. Texas Agric. Expt. Sta. Bul. 842.
63. Leffel, E.C. and K.R. Mason. 1959. In: Magnesium and Agric. Symposium, Univ. W. Virginia, Morgantown.
64. Fontenot, J.P., et al. 1965. Feedstuffs 37(12):66.
65. Sjollema, B. 1932. Nutr. Abst. and Rev. 1:621.
66. Rook, J.A.F. and J.E. Storey. 1972. Nutr. Abst. and Rev. 32:1055.
67. Meinstzer, R.B. and H. Steenbock. 1955. J. Nutr. 56:285.
68. Dutton, J.E. and J.P. Fontenot. 1967. J. Animal Sci. 26:1409.
69. Ward, G.M. 1966. J. Dairy Sci. 49:268.
70. Metson, A.J., W.M.H. Saunders, T.W. Collie and V.W. Graham. 1966. New Zeal. J. Agric. Res. 9:410.
71. Newton, G.L., J.P. Fontenot, R.E. Tucker and C.E. Polan. 1972. J. Animal Sci. 35:440.
72. Moore, W.F., J.P. Fontenot and K.E. Webb. 1972. J. Animal Sci. 35:1046.
73. Colby, R.W. and C.M. Frye. 1951. Amer. J. Physiol. 166:209; 166:408.
74. Burt, A.W.A. and D.C. Thomas. 1961. Nature 192:1193.
75. Burau, R. and P.R. Stout. 1965. Science 150:766.
76. Stout, P.R., J. Brownell and R.C. Burau. 1967. Agron. J. 59:21.
77. Bohman, V.R., A.L. Lesperance, G.D. Harding and D.L. Gruner. 1969. J. Animal Sci. 29:29.
78. Vitale, J.J. et al. 1963. Circ. Res. 12:642.
79. Jolyet, F. and M. Cahouns. 1869. Arch. physiol. norm et pathol. 2:113 (cited by Wacker and Vallee, 1964).
80. Del Costillo, J. and L. Engbaek. 1954. J. Physiol. 124:370.
81. Hutter, C.F. and K. Kostral. 1954. J. Physiol. 124:234.
82. Haury, V.G. 1939. J. Pharmacol. Exptl. Therap. 65:453.
83. Stanbury, J.B. and A. Farah. 1950. J. Pharmacol. Exptl. Therap. 100:445.

K, Na, and Cl

84. Houpt, T.R. 1970. In: Duke's Physiology of Domestic Animals. 8th ed. Cornell Univ. Press, Ithaca, N.Y.
85. Gamble, J.L. 1954. Chemical Anatomy, Physiology and Pathology of Extracellular Fluid. Harvard Univ. Press, Cambridge, Mass.
86. Riggs, T.R., L.M. Walker and H.N. Christensen. 1958. J. Biol. Chem. 233:1479.
87. Gardner, L.I., et al. 1950. J. Lab. Clin. Med. 35:592.
88. Cannon, P.R., L.E. Frazier and R.H. Hughes. 1952. Metabolism 1:49.
89. Howard, J.E. and R.A. Carey. 1949. J. Clin. Endocrinol. 9:691.
90. Sykes, J.F. and B.V. Alfredson. 1940. Proc. Soc. Exp. Biol. Med. 43:575.
91. Carter, C.W., R.V. Coxon, D.S. Parsons and R.H.S. Thompson. 1959. Biochemistry in Relation to Medicine. Longmans, Green & Co., London.
92. Wilson, T.H. 1962. Intestinal Absorption. W.B. Saunders, Philadelphia, Pa.
93. Walser, M. and G.H. Mudge. 1961. In: Mineral Metabolism. Vol. 1, Part A. Academic Press, N.Y.
94. Sandstead, H.H. 1967. In: Present Knowledge in Nutrition. 3rd ed. The Nutr. Found., New York.
95. Burns, C.H., W.W. Cravens and P.H. Phillips. 1953. J. Nutr. 50:317.
96. Ringer, S. 1882. J. Physiol. 3:380.
97. Sturkie, P.D. 1950. Amer. J. Physiol. 162:538.
98. Jensen, A.H., S.W. Terrill and D.E. Becker. 1961. J. Animal Sci. 20:464.
99. Follis, R.H., E. Orent-Kieles and E.V. McCollum. 1942. Amer. J. Path. 18:29.
100. MacPherson, C.R. and A.G.E. Pearse. 1957. Brit. Med. Bul. 13:19.
101. Milne, M.D. and R.C. Muehrcke. 1957. Brit. Med. Bul. 13:15.
102. Smith, S.E. and P.D. Aines. 1959. Cornell Agr. Expt. Sta. Bul. 938.
103. Mayer, J.H., R.H. Grummer, R.H. Phillips and G. Bohstedt. 1950. J. Animal Sci. 9:300.
104. Greenberg, D.M. and E.M. Cuthertson. 1942. J. Biol. Chem. 145:179.
105. Leach, R.M. and M.C. Nesheim. 1963. J. Nutr. 81:193.
106. Hartraft, P.M. and E. Sowa. 1964. J. Nutr. 82:439.
107. Meneely, G.R., J. Lemley-Stone and W.J. Darby. 1961. Amer. J. Cardiol. 8:527.
108. Dahl, L.K., M.G. Smilay, L. Silver and S. Spraragen. 1962. Circ. Res. 10:313.
109. Bohstedt, G. and R.H. Grummer. 1954. J. Animal Sci. 13:933.
110. Scott, M.L., M.C. Nesheim and R.J. Young. 1969. Nutrition of the Chicken. M.L. Scott and Assoc. Pub., Ithaca, N.Y.

Sulfur

111. Hays, V.W. and M.J. Seveson. 1970. In: Duke's Physiology of Domestic Animals. 8th ed. Cornell Univ. Press, Ithaca, N.Y.
112. Michels, F.G. and J.T. Smith. 1965. J. Nutr. 87:217.
113. Kun, E. 1961. In: Metabolic Pathways. Academic Press, N.Y.
114. Hays, V.W. 1972. In: The Searching Seventies. Natl. Feed Ingred. Assoc. Proc., Moorman Mfg. Co., Quincy, Ill.
115. Thomas, W.E., J.K. Loosli, H.H. Williams and L.A. Maynard. 1951. J. Nutr. 43:515.

Chapter 11 – Micro (Trace) Minerals

The distinction between macrominerals and micro or trace minerals is based on the relative amounts required in the diet for normal body function. The trace minerals function as activators of enzyme systems or as components of organic compounds and, as such, are required in small amounts. A vast array of enzyme systems require trace elements for activation. Examples of enzymes requiring specific trace minerals are given for each mineral. Note that Mg, a macromineral, functions in many respects like the trace elements.

The list of trace minerals required by animals continues to grow. The latest one added to the list is Si which was shown in 1972 (4) to be required for growth of chicks. The following are known now to be required by one or more animal species for normal life processes: Co (cobalt); I (iodine); Fe (iron); Cu (copper); Zn (zinc); Mn (manganese); Se (selenium); Cr (chromium); F (fluorine); Mo (molybdenum); and Si (silicon).

Each is discussed in terms of tissue distribution, functions, metabolism, deficiency signs, and toxicity. Trace minerals in animal nutrition and metabolism have been extensively reviewed by Bowen (1) and Underwood (2) and in part of a series on mineral metabolism edited by Comar and Bronner (3).

Cobalt

Animals have no known requirement for Co other than as a constituent of vitamin B_{12} (see Ch. 14), even though it was first recommended as an important diet constituent as early as 1929 when large amounts were shown to be capable of stimulating red blood cell synthesis in rats. Later Underwood and Filmer (5) showed that "wasting disease" of sheep and cattle in Australia could be cured or prevented by Co, and Smith et al (6) showed that vitamin B_{12} injections brought about dramatic improvement in sheep showing Co deficiency.

Tissue Distribution

Liver, kidney, adrenal, and bone tissue contain Co in the highest concentration in all species studied (2, 8). The forms in which it exists in tissues other than as a constituent of vitamin B are not completely known, although other bound forms do exist. Askew and Watson (9) reported 150 ppb of Co in dry liver from normal sheep compared with 20 ppb in liver of Co-deficient sheep; dried spleen, kidney, and heart of normal sheep also contained more Co.

Co supplementation of gestation diets of cows results in a higher concentration of Co in the tissues of calves; similarly, Co content of milk is increased by supplementation of the lactation diet. This extra Co is apparently not present as a constituent of vitamin B_{12}, as it is not utilized until the young animal has developed a functioning rumen.

Functions

All evidence indicates that Co functions only as a constituent of vitamin B_{12}. Co deficiency, in the presence of adequate vitamin B_{12} intake, has never been produced in nonruminants. Likewise, ruminants with normal rumen function apparently do not require dietary Co greater than that needed for microbial vitamin B_{12} synthesis. Although Co can activate certain enzymes, there is no evidence that it is required for this purpose because the enzymes will function in the presence of other divalent metals.

Metabolism

Inorganic Co is poorly absorbed from the GIT in animals, although adult man may be an exception. Comar et al (10, 11) found 80% of an orally administered dose of radioactive Co in the feces of rats and cattle. Injected Co is excreted mainly through the kidney. Although some loss occurs via the GIT, the amount is small and probably through the bile and intestinal wall, but not from the pancreas. Tissue Co appears to be eliminated slowly. No synthesis of vitamin B_{12} in animal tissues from inorganic Co is known, even though tissue levels are high.

Large oral doses of Co induce polycythemia (increased RBC concentration) in several species. The mechanism is apparently by production by anoxia, possibly by binding -SH compounds, with increased RBC synthesis as a compensatory response.

Deficiency Signs

Because Co functions as a constituent of vitamin B_{12}, the deficiency signs described for vitamin B_{12} (see Ch. 14) apply for Co. Ruminants grazing in Co-deficient areas show loss of appetite, reduced growth, or loss in body weight followed by emaciation (Fig. 11-1), normocytic normochromic anemia, and eventually death. Fatty degeneration of the liver occurs and hemosiderosis of the spleen (2, 12).

Microbial synthesis of B_{12} in the rumen of deficient ruminants is greatly depressed. The first notable response to Co feeding is an improved appetite followed by increased blood hemoglobin concentration. Critical levels of Co in diets for ruminants below which Co deficiency signs appear are 0.07 to 0.11 ppm of dry matter (2, 13). Soil maps (2, 7) show large areas in the USA, Australia and other parts of the world where soil is deficient in Co, resulting in low Co content of plant tissues and deficiencies in animals consuming the plants.

Figure 11-1. Co-deficient sheep. Note the severe emaciation and wool chewing that has occurred. The lamb on the left received an adequate diet. By permission of S.E. Smith, Cornell University.

A method of preventing Co deficiency in grazing ruminants is the use of small dense pellets composed of cobalt oxide and iron which are administered orally with a balling gun (14). The pellets lodge in the rumen and are gradually dissolved over a period of months, yielding a constant supply of Co for vitamin B_{12} synthesis. Such a technique eliminates the need for a mineral supplement which may not be voluntarily consumed in appropriate amounts by free-ranging animals. Australian work indicates that the use of 2 pellets offers enough abrasion to provide a steady supply of Co for vitamin B_{12} synthesis for more than 5 years in sheep (15).

Toxicity

Because of its low absorption rate, Co toxicity is not likely. A daily intake of 3 mg Co/kg body weight for 8 wk is tolerated by sheep without harmful effects (12). Higher doses result in appetite depression and anemia. Co given as a soluble salt to provide 300 mg Co/kg body weight is lethal to sheep (16). Cattle may be less tolerant than sheep (2). A concentration of about 2 mg/kg in the diet of growing pigs produced no toxicity over a 100-day feeding period (17).

Iodine

It was suggested nearly 150 years ago that I deficiency caused goiter in man, but the theory was rejected and lay dormant until about 1900 when it was observed that I was concentrated in the thyroid gland and its concentration reduced in persons with goiter (2). Kendall (18) isolated thyroxin in 1919 and Harington and Barger (19) synthesized it in 1927. The structure of thyroxin (3,3',5,5'-tetraiodothyronine or T_4) is as follows:

Thyroxine

Of the several compounds showing thyroid activity, thyroxin is present in blood in the highest concentration (20). The thyronine nucleus (structure shown above minus I's) is essential for activity. Other compounds with thyroid activity include: 3,5-3'-triodothyronine or T_3 (3-5 times as active as thyroxin); 3,3'-diiodothyronine (similar to thyroxin in activity); 3,5-diiodo-3'-5'-dibromothyronine (nearly as active); and 3,5,3'-triiodothyropropionic acid (300 times as active as thyroxin) (2).

Tissue Distribution

The thyroid gland contains the highest concentration of I (0.2-5% of dry weight) and in the largest amount (70-80% of total body I). I is also preferentially concentrated in the stomach (or abomasum), small intestine, salivary glands, skin, mammary gland, ovary, and placenta (19, 20). Species differences exist in this capacity. The total I content of the adult human is 10-20 mg. The concentration of inorganic I in most tissues is 1-2 mcg/100 g, and that of organically bound I is about 5 mcg/100 g for muscle with higher concentration in tissues that concentrate I.

The only known function of I is as a constituent of thyroxin and other thryoid-active compounds. Thus, it is intimately associated with basal metabolic rate.

Metabolism

Inorganic I is absorbed from the GIT by 2 processes, one in common with other halides (Cl, Br) and one specific for I. The specific I transport system is present in the stomach as well as the small intestine. The stomach and a midsection of the small intestine secrete I from the serosal to the mucosal surface. In fact, the gastric juice reaches an I concentration up to 40 fold that of the plasma in humans (21). The I-specific transport mechanism is saturated by high I concentrations. This is in common with the similar system in the thyroid gland, but differs in that GIT absorption is not affected by thyroid stimulating hormone, although thyroid tissue is profoundly affected.

Salivary secretion of I is an active process in most species, as evidenced by the 40-fold concentration of radioactive I in the saliva as compared to plasma of animals given radioiodine. The rat is apparently an exception (22).

The I supply to the developing embryo is apparently enhanced by 2 mechanisms. The ovary and placenta both concentrate I by an active process. The role of the ovary is best illustrated in birds. Radioiodine injected into the hen is taken up by the yolk of eggs laid afterward, indicating that the yolk is an I store for the developing chick embryo (20). Similarly, placental tissue of mammals concentrates I in late pregnancy. A second mechanism favoring fetal I uptake is the presence in fetal serum of a specific thyroxin-binding protein which increases in concentration in late fetal development and has a higher affinity for thyroxin than the thyroxin-binding protein of the maternal plasma (23). This favors the fetus receiving an adequate I supply if the combined thyroxin content of the maternal and fetal plasma is low (20). The mammary gland also concentrates I, and the transfer of inorganic I is by an active process resulting in a 40-fold concentration of I in milk as compared to plasma. Although some T_4 and T_3 are found in milk, the amounts are so small as to be of doubtful value (24).

The key organ for I metabolism is the thyroid gland. It concentrates I by an active process, which is enhanced by thyroid stimulating hormone (TSH) secreted by the anterior pituitary gland. The I-concentrating ability of the thyroid is expressed as a ratio of the I concentration of the thyroid to the I concentration of the serum (T/S value). Normal animals have a T/S value of 20. The T/S value is decreased by hypophysectomy and increased by stimulating TSH release.

I reaching the thyroid from the plasma is concentrated in the lumen of follicles of the thyroid, each of which is composed of a single layer of cells arranged as a sphere. The I so stored is contained in a colloid protein, thyroglobulin. Inorganic radioiodine given intravenously appears largely as protein-bound I (thyroproteins) in the thyroid gland shortly after injection (20). The iodinated proteins of the thyroid include mainly thyroglobulins (the thyroactive fraction), but also small amounts of others.

I is oxidized by peroxidase to a reactive form for thyroglobulin formation. Gross (20) outlined the steps in biosynthesis of the thyroid active compounds, thyroxin (T_4), 3,5,3'-triiodothyronine (T_3), 3-monoiodotyrosine (MIT), and 3,5-diiodotyrosine (DIT). Tryosine present in the thyroglobulin molecule can be iodinated to form MIT, which is in turn used to form 3,3'-diiodothyronine or T_3, or MIT can be further iodinated to form DIT, which is in turn used to form T_3 or T_4 (20). The amount of I in thyroglobulin is, therefore, dependent on the proportions of these tyrosine derivatives present.

Thyroglobulin does not appear in the plasma but is hydrolyzed by thyroid proteases in the thyroid follicle. On hydrolysis, only the iodothyronines are detected in the plasma. Free iodotyrosines are de-iodinated by enzymes in the thyroid, and the free I is available for recycling through thyroglobulin.

Iodothyronines are transported in the plasma bound to a globulin (thyroxin-binding globulin or TBG) or to a thyroxin-binding prealbumin (20). Some binding of thyroxin by plasma albumin, also occurs. TBG, however, binds most of the iodothyronines at normal plasma concentrations. No evidence has shown that thyroxin and its derivatives leave the plasma, except for entrance into lymph (25). Thus, the concentration of free iodothyronines in the blood plasma and extracellular fluids probably controls the rate of transfer to sites of action (20, 26). Only about 0.05% of total plasma thyroxin is present in the free state (27).

The protein bound I (PBI) level of the serum varies with level of thyroid activity, as well as with species and age. Plasma PBI levels are increased with hyperthyroid activity and are generally higher in young animals and in pregnancy. Although plasma PBI levels provide a general assessment of thyroid activity, consistent results have not been obtained in attempts to relate them to growth rate or milk production.

About 80% of thyroid hormones entering the tissues are broken down by de-iodinization of liver, kidney, and other tissues with the liberated I recycled for further use and the tyrosine residues catabolized or used for tissue protein synthesis (2). The remaining 20% is lost to the body through excretion via the bile, by conjugation to form glucuronides or sulfate esters, or by oxidative deamination (2, 20). Inorganic I is excreted mainly via the kidneys. Smaller amounts are lost to sweat and in feces. The salivary gland secretes large amounts, but most of this is reabsorbed from the GIT.

Deficiency Signs

Because I functions as a constituent of thyroid-active compounds which, in turn, play a major role in controlling the rate of cellular oxidation, it is not surprising that a dietary deficiency of I has profound effects on the animal. Dietary I deficiency reduces basal metabolic rate (BMR). I deficiency in young animals is called **cretinism**, and, in adults, **myxedema**. Tissues of thyroid-deficient animals consume less oxygen and the reduced BMR is associated with reduced growth rate and reduced gonadal activity. Skin becomes dry and hair brittle. Reproductive problems associated with I deficiency include resorbed fetuses, abortions, stillbirths, irregular or suppressed estrus in females, and a decreased libido and deterioration of semen quality in males (2). Pigs and calves produced by I-deficient dams may be hairless with dry, thick skin (Fig. 11-2) or, if hair is present, the coat is harsh and sparse, or in sheep the wool is scanty (28). The fleece of adult sheep recovered from I deficiency in early life may be of poor quality because of interference with normal development of the wool producing cells (29).

Thyroid enlargement (goiter) is induced by an attempt of the thyroid gland to secrete more thyroxin in response to TSH stimulation. TSH is released in response to a reduced thyroxin production. Thyroid hormones, in a negative feedback arrangement, inhibit release of thyrotropic hormone releasing factor (TRF) by the hypothalamus and TSH by the anterior pituitary (30). In the absence of adequate thyroxin for inhibiting TSH release, the thyroid gland becomes hyperactive and enlarged. Goiter is a common problem in human and animal populations living in inland areas in many parts of the world. The use of iodized salt has reduced the problem, but endemic goiter still remains as a major nutritional disease in many areas.

In addition to simple I deficiency, several goitrogenic substances are present in common foods. These antithyroid compounds act by interfering with the iodinization of tyrosine, thus blocking iodothyronine synthesis. Kingsbury (31) has described goitrogens important in animal nutrition and listed plants containing them.

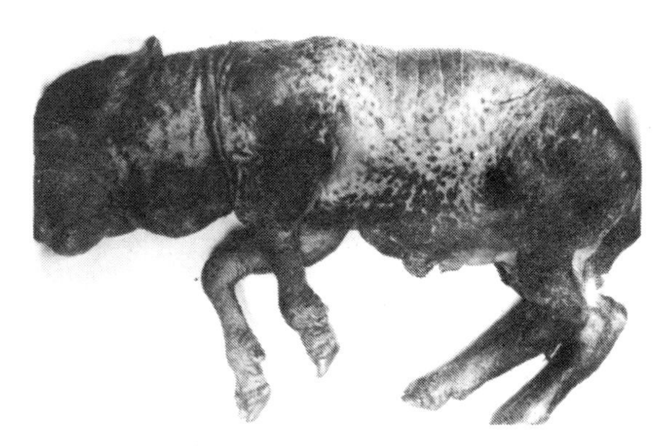

Figure 11-2. An iodine-deficient lamb showing hairlessness, thick wrinkled skin, and a marked enlargement of the thyroid gland. Photo courtesy of W.M. Hawkins, Montana State University.

Toxicity

Long term chronic intake of large amounts of I reduces thyroid uptake of I. Marked species differences exist in tolerance to high intake of I. Fertility of male rats fed 2400 ppm I for 200 days was unaffected (32); reproduction of female swine was unaffected by the same intake during gestation (13), but rabbits showed increased prenatal mortality. Egg production of hens was dramatically reduced by feeding 312 ppm and stopped with 5000 ppm I, and hatchability was reduced (34). An effect on thyroxin production seems unlikely, as egg production was resumed within 1 week after withdrawal of the I from the diet. The absence of an effect on thyroid function has since been confirmed. Apparently, the levels of I normally encountered in nutrition are far less than levels necessary to cause toxic symptoms. Single oral massive doses of I are, of course, toxic and may be lethal, but such toxicity must be categorized as poisoning in the general sense.

Zinc

Zn was first shown to be an essential nutrient for animals in 1934 when Todd et al (35) produced deficiency in the rat. In farm animal nutrition, the disease of swine termed parakeratosis by Kernkamp and Ferrin (36) was later shown to result from Zn deficiency (37). Subsequently, Zn deficiency has been produced experimentally in other farm animals, and Zn deficiency in humans has been reported as a practical problem (38, 39). The nutrition, physiology, and metabolism of Zn has been reviewed in detail in several recent publications (2, 38, 39, 40).

Tissue Distribution

Zn is widely distributed in body tissues but is in highest concentration in liver, bone, kidney, muscle, pancreas, eye, prostate gland, skin, hair, and wool. Radioactive Zn given orally or intravenously to normal or Zn-deficient cattle reached a peak concentration in the liver within a few days, but concentration in red blood cells, muscle, bone, and hair did not reach a peak until several weeks later (41). Deficient animals retained more ^{65}Zn in skin, testes, scrotum, kidney, muscle, heart, lung, and spleen than did normal animals, suggesting tissue specificity in meeting needs when Zn supply is short. Zn is a constituent of a large number of enzymes, including alcohol, lactate, malate, and glutamate dehydrogenases, alkaline phosphatase, carboxypeptidases A and B, and carbonic anhydrase, and is effective in activation of a large number of other enzymes (39, 40). The latter role is shared by several other metal ions and is not specific for Zn. The tissue distribution of Zn is roughly associated with the tissue distribution of enzyme systems to which it is related. For example, bone Zn is high and alkaline phosphatase of bone is high. The high concentration of Zn in the pancreas is probably related both to its presence in digestive enzymes and to its association with the hormone, insulin, secreted by the pancreas.

The Zn concentration in blood is divided between the cells and plasma in a 9:1 ratio (39). Plasma Zn is bound loosely to albumins (1/3) and more firmly to globulins (2/3) and is responsive to dietary levels. Most of the Zn in RBC's is present as a component of carbonic anhydrase.

Functions

Zn is a component of an array of enzyme systems in several body tissues; the biochemical functions of Zn are therefore related to the functions of these enzymes. Zn is an activator of several metalloenzyme systems and probably shares with other metal ions, which it can replace, the function of binding reactants to the active site of the enzyme.

Zn is required for normal protein synthesis and metabolism (2, 42, 43), and is a component of insulin and, in this way, functions in carbohydrate metabolism.

Metabolism

Absorption of Zn from the GIT occurs throughout the small intestine and amounts to 5-40% of the intake (2). There is disagreement as to whether active transport of Zn occurs. It was (40) suggested that Zn was actively taken up by intestinal mucosal cells, but the release from the mucosa was determined by the concentration of Zn on the serosal side of everted gut sacs in rats; others were unable to show active transport across intestinal mucosa in rats using similar techniques (40, 44). The absorption of Zn is adversely affected by high dietary Ca concentration, and the presence of phytate further aggravates it (45). The complexing of Zn by phytate forms an insoluble and unabsorbable compound and is believed to be the mechanism whereby Zn availability to animals is reduced. A chelating agent, ethylene diamine tetra acetate, or EDTA, improves Zn availability by competing with phytate to form an EDTA-Zn complex that allows absorption of Zn either as free Zn ion or as the EDTA-Zn complex. The presence of phytate or of other chelating agents in natural feedstuffs such as soybean meal and corn is of practical importance (38, 46); diets containing similar amounts of Zn may produce a variable incidence of parakeratosis in swine and variable weight gain in chicks and rats (47, 48). Phosphate also binds Zn and may be a factor in the observed difference among protein sources in Zn availability (49, 50).

Zn metabolism and movement in tissues has been studied by Pekas (51) who used ^{65}Zn infusion to follow Zn movement in the body. He reported that Zn is removed rapidly from the blood by the tissues and that tissues saturated with Zn (muscle) translocate Zn to unsaturated tissues (liver, pancreas, and kidney). Saturated tissues were characterized by no net change in Zn content, and unsaturated tissues showed a marked increase in Zn content. Injected Zn is excreted largely in the feces. Thus, fecal Zn in animals on an adequate Zn intake includes both unabsorbed and endogenously secreted Zn. The pancreatic juice appears to be the main route of excretion of endogenous Zn. Endogenous losses are reduced during dietary Zn deficiency, serving to conserve body stores (52, 53). Except in abnormal conditions such as nephrosis or hypertension, urinary losses of Zn are very low. Excretion can be increased 10-fold by administration of EDTA. Placental transfer of Zn is directly related to maternal dietary intake (2).

In animals that sweat freely, Zn loss by this route can be extensive in hot environments. Prasad et al (54) showed that man may lose 5 mg Zn/day on a diet adequate in Zn; Zn-deficient individuals lose less than half that under the same environmental conditions, again illustrating the homeostatic control of body Zn stores.

Deficiency Signs

The most conspicuous sign of Zn deficiency is growth retardation and anorexia in all species studied and a reduction in plasma alkaline phosphatase activity. Thickening or hyperkeratinization of the epithelial cells is common (parakeratosis in swine, see Fig. 11-3). Rats show scaling and cracking of the paws, rough hair coat, and alopecia (loss of hair); sheep show abnormal changes in wool and horns; poor feathering and dermatitis occur in poultry (2).

Zn deficiency retards bone formation and is associated with reduced division and proliferation of cartilage cells in the epiphyseal growth plate (2, 55). Bone alkaline phosphatase is reduced, bone density is depressed, and Zn content of bone and of liver are reduced (55).

Figure 11-3. A Zn-deficient pig showing marked parakeratosis. Photo courtesy of W.M. Beeson, Purdue University.

Hens fed Zn-deficient diets for several months show no abnormalities, but chicks produced show poor livability and a high incidence of malformations. Chicks from adequately fed hens develop a perosis-like leg abnormality when fed Zn-deficient diets. This defect has been prevented by feeding histidine or histamine (56). The mechanism of this protective effect is unknown. Feeding female rats Zn-deficient diets during the first 18 days of gestation results in high early mortality as well as difficulty in parturition and abnormal changes in maternal behavior of the dams (57). Pups from Zn-deficient dams show lower liver and total body Zn concentrations than controls and have a smaller birth weight.

Zn deficiency has drastic effects on the male reproductive organs. Hypogonadism is observed in Zn-deficient males of all species studied. In man, hypogonadism, suppressed development of secondary sex characteristics and dwarfism have been observed in Iranian and Egyptian young men (40, 58). Recovery of testes size and sperm production are achieved in young animals repleted with Zn (2).

Wound healing of Zn-deficient animals is severely delayed (59). The exact mode of action of Zn in tissue repair is unknown, but the role of Zn in normal protein synthesis probably provides a clue to the relationship.

Zn deficiency causes impairment of glucose tolerance in rats, supporting the known relationship between Zn and insulin (2).

A striking feature of Zn deficiency is the dramatic remission of clinical signs when Zn is administered. This is well illustrated in parakeratosis in Zn-deficient pigs. The skin of animals with severe lesions shows dramatic improvement with a few days of Zn refeeding and complete disappearance of the lesion within 2-3 wk; feed intake improves immediately after Zn is added to the diet.

Toxicity

A wide margin of safety exists between the required intake of Zn and the amount that will produce toxic effects. Although the requirement for most species is less than 50 mg/kg of diet, levels of 1-2.5 g/kg have been fed rats with no deleterious effects. Smith and Larson (60) observed anemia, growth failure, and death in rats fed 1% Zn, although lower levels (0.7%), allowed life but induced anemia and reduced growth. Zn fed at 1 g/kg showed no ill-effects in pigs, but levels of 4 and 8 g/kg produced depressed growth, stiffness, hemorrhages around bone joints, and excessive bone resorption (61, 62). Birds are similar to swine in their tolerance to Zn, but sheep and cattle are less tolerant. Levels of 0.9-1.7 g/kg of Zn depressed appetite and induced depraved appetite manifested by wood-chewing and excessive consumption of mineral supplements (63). The lower tolerance of ruminants to high dietary Zn may be related to changes in rumen metabolism brought about by a toxic effect of Zn on the rumen microflora (64).

Because the anemia produced by excess Zn can be prevented by extra Cu and Fe, it has been suggested that the anemia is an induced Cu and Fe deficiency as a result of interference with their absorption from the GIT in the presence of high Zn (2).

Iron

Iron has been recognized as a required nutrient for man and animals for more than 100 years. Despite this fact, Fe deficiency remains a common disease affecting nearly 1/2 of the human population in some areas of the world and persisting as a major problem in livestock production. Reviews of Fe metabolism and nutrition are numerous (2, 65, 66, 67).

Tissue Distribution

Sixty to 70% of body Fe is present in hemoglobin in red blood cells and myoglobin in muscle; 20% is stored in labile forms in liver, spleen, and other tissues and is available for hemoglobin formation, and the remaining 10-20% is firmly fixed in unavailable forms in tissues as a component of muscle myosin and actomyosin and as a constituent of enzymes and associated with metalloenzymes (65). The absolute amounts of Fe present in various forms in several species are summarized in Table 11-1 (68). Fe is present in hemoglobin, myoglobin, and heme enzymes, cytochromes, catalases, and peroxides as heme, an organic compound consisting of Fe in the center of a porphyrin ring. In hemoglobin, which contains 0.34% Fe, an atom of ferrous Fe in the center of the porphyrin ring connects heme, the prosthetic group, with globin, the protein. Hemoglobin contains 4 porphyrin rings and combines reversibly with atmospheric oxygen brought into the blood via the lungs. Hemoglobin is solely within the red blood cell. Myoglobin, which has 1/4 the molecular weight of hemoglobin (16,500), is present in muscle and has an affinity for oxygen greater than that of hemoglobin. The heme enzymes — catalases and peroxidases — contain Fe in the ferric state (Fe^{+++}). Each of these classes of enzyme liberates oxygen from peroxides. Cytochrome enzymes are heme enzymes of the cell mitochondria and act in electron transfer by virtue of the reversible oxidation of Fe ($Fe^{++} \rightleftharpoons Fe^{+++}$). Cytochromes a, b, and c, cytochrome oxidase, and others have been identified, but only cytochrome c is readily extractable from tissues. Other enzymes containing Fe include xanthine oxidase, succinic dehyrodgenase, and NADH-cytochrome reductase.

Fe in blood plasma is bound in the ferric state (Fe^{+++}) to a specific protein, transferrin, a β_1-globulin. Transferrin is the carrier of Fe in the blood and is normally saturated only to the extent of 30-60% of its total iron-binding capacity.

Fe is stored in liver, spleen, and bone marrow as an Fe-protein complex, ferritin, and as a component of hemosiderin. Fe makes up 20% of the ferritin-Fe complex and is present in the ferric state. Hemosiderin contains 35% Fe as ferric hydroxide. It is present in tissues as a brown, granular pigment. Ferritin can be considered the soluble and hemosiderin the insoluble form of storage Fe. Under normal conditions, and in Fe deficiency, Fe is stored in about equal amounts in each form, but in Fe excess, hemosiderin Fe predominates.

Functions

Fe is required for hemoglobin and myoglobin formation and therefore for normal oxygen transport to tissues. Fe is important in electron transport mechanisms in cells as a component of several heme enzyme systems, and Fe is a component of several nonheme metalloenzymes including xanthine oxidase which is involved in release of Fe from ferritin.

Table 11-1. Amounts and distribution of Fe in the rat, dog, calf, horse, and man. [a]

Species	Body wt	Total body Fe	Hemoglobin Fe	Myoglobin Fe	Cytochromic Fe	Storage Fe
	kg	g	g	g	g	g
Calf	182.0	11.13	7.55	1.060	0.0053	2.517
Dog	6.35	0.44	0.300	0.040	0.00059	0.100
Horse	500.0	33.0	19.85	6.45	0.0715	6.617
Man	70.0	4.26	3.10	0.120	0.00336	1.03
Rat	0.25	0.015	0.011	0.0003	0.00006	0.0036

[a] Adapted from Moore (68)

Metabolism

Absorption. Fe is absorbed mainly from the duodenum in the ferrous state (Fe^{++}), and usually only to the extent of 5-10%. The body tenaciously holds absorbed Fe for reuse. Thus, the Fe released from hemoglobin breakdown associated with RBC destruction (RBC life is 60 to 120 days in most species) is recycled for resynthesis of hemoglobin. This conservation of Fe is illustrated in the diagrammatic outline of Fe metabolism in Fig. 11-4. Absorption is more efficient under acid conditions; thus, the amount of Fe absorbed from the stomach and duodenum, where HCl from stomach secretion results in a low pH, is greater than from the ileum. Using radioiron, Brown and Justus (69) showed in rats a gradient from upper to lower end of the small intestine in the absorption of Fe, with Fe uptake more than 10 times greater in the proximal duodenum than in the distal ileum. Maximum absorption occurred at pH 2 to 3.5. Fe absorption is greater in Fe-depleted animals than in animals receiving adequate Fe. The absolute amount of Fe absorbed is increased as the size of an oral dose increases, but the % absorption decreases (70). Ascorbic acid increases Fe absorption by reducing Fe^{+++} to the more soluble Fe^{++} form; histidine and lysine also increase Fe absorption, probably by forming a soluble Fe-amino acid chelate. Fe present in hemoglobin and myoglobin is readily absorbed as heme Fe. High levels of inorganic phosphates reduce Fe absorption by forming insoluble salts; phytate has also been reported to reduce Fe absorption, but the practical importance in normal diets is unknown. High levels of other trace elements, including Zn, Mn, Cu, and Cd, also reduce Fe absorption, presumably by competing at protein binding sites in the intestinal mucosa (71).

The ultimate regulation of Fe absorption is apparently dependent on the Fe concentration of the intestinal mucosal cells; that is, of the Fe taken up by the mucosal cells, only a small part is transferred from the cell to the blood; the bulk is retained in the cell and lost into the intestinal lumen when the cell is sloughed off in the normal process of regeneration of the intestinal epthelium (72). In deficient animals Fe is more readily transferred from the mucosal cell to blood until saturation of the tissues results in a return to normal retention of Fe in the mucosal cell. Parenterally administered Fe decreases Fe absorption in some species by transfer into the intestinal epithelial cell from the serosal side (from the blood). Thus, the "mucosal block" theory of Fe absorption, which was advanced by Hahn et al in 1943, remains as an important general concept, even though the mechanisms

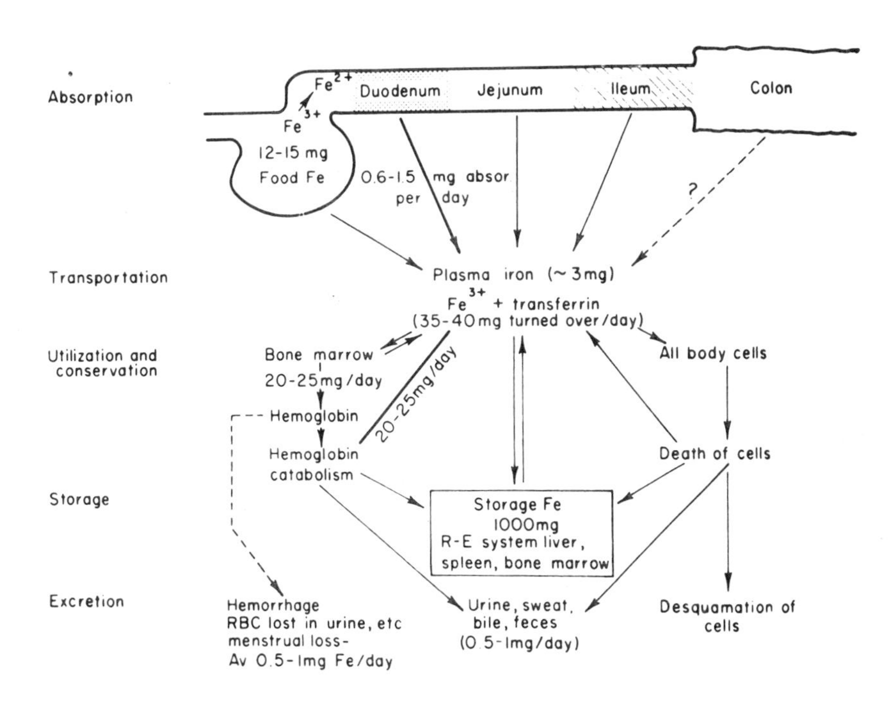

Figure 11-4. Schematic representation of iron metabolism in man. RBC = red blood cells; R-E system = reticuloendothelial system. From Moore and Dubach [65].

may not be exactly as originally envisioned (73). Recent evidence suggests that a factor in gastric juice is needed for optimal Fe absorption in rats (74); the presence of such a control mechanism in other species remains to be determined.

Excretion. Body Fe is tenaciously retained (see Fig. 11-4). Fecal Fe is mainly unabsorbed dietary Fe, but a small amount (0.3-0.5 mg/kg in man) is lost through bile and sloughed intestinal mucosal cells. Even when Fe is injected, very little of it is excreted in either feces or urine. Urinary Fe loss does occur when parenteral Fe is given in excess of the plasma Fe-binding capacity or when chelating agents are given (75). Small amounts of Fe are lost in sweat in man (88).

Placental and Mammary Transfer. Considerable species variability exists in the efficiency of transfer of Fe to the fetus across the placenta. Fe is transported to the fetus by an active process and concentration in the fetal circulation exceeds that in maternal plasma. Transferrin does not cross the placenta; rather, Fe is dissociated from transferrin on the maternal side of the placenta and reassociated with transferrin on the fetal side. As gestation progresses, more and more Fe is transferred to the fetus. Although the newborn of some species have a relatively high liver Fe concentration, the newborn pig is not well supplied with Fe and is especially prone to Fe deficiency. The poor placental transfer of Fe is not appreciably increased even by high levels of supplementation of the maternal diet or by parenteral administration to the sow (76).

Milk of all species is notably low in Fe. Attempts to increase the Fe concentration of sow milk by parenteral Fe or by feeding extra Fe to the sow during lactation have not resulted in appreciable increases in milk Fe content (77).

Storage and Mobilization. Fe is stored within the cells of the liver, spleen, bone marrow, and other tissues as ferritin and hemosiderin in roughly equal proportions. Incorporation of plasma Fe (transferrin) into ferritin in liver cells is energy-dependent (ATP) and is related to the reduction of Fe^{+++} of transferrin to Fe^{++}, making it available for ferritin formation (2, 78). Release of Fe^{++} in liver ferritin to plasma is catalyzed by xanthine oxidase. Similar reactions are presumed to occur in other Fe-storage tissues. The corresponding mechansims for release and uptake of hemosiderin between transferrin and storage tissues are not known.

The turnover rate of Fe in the plasma is very rapid; about 10X the amount of Fe in the plasma at any one time is transported each day. Most of this is used for hemoglobin synthesis. Inorganic radioiron appears almost entirely as a component of hemoglobin in 7-14 days in humans. The continuous redistribution of body Fe was summarized by Underwood (2) as follows:

Quantitatively most important:
plasma → erythroid marrow → red cell → senescent red cell → plasma

Less important:
plasma → ferritin and hemosiderin → plasma

plasma → myoglobin and Fe-containing enzymes → plasma

Enzymes contain Fe bound to the protein, to porphyrin, or to flavins (metalloenzymes)* or as a loosely bound activator (metal activated enzymes).* A partial list of Fe-containing enzymes in mammals and birds, along with some important Fe-containing proteins is given in Table 11-2.

Deficiency Signs

The most common sign of Fe deficiency is a microcytic hypochromic anemia which is characterized by smaller than normal red cells and less than normal hemoglobin. Fe-deficiency anemia is a common problem in newborn animals because of inefficient placental and mammary transfer. In pigs unsupplemented with Fe, blood hemoglobin declines from about 10 g/100 ml at birth to 3 or 4 g/100 ml at 3 wk. The extremely rapid growth rate of young pigs (5X birth wt at 3 wk) results in a dilution effect of total body Fe stores unless iron is fed or injected. Intramuscular injection of 150-200 mg of Fe-dextrin at 2 or 3 days of age maintains normal hemoglobin level at 3 wk of age at which time consumption of dry feed provides ample Fe. Anemic pigs show pallor, labored breathing, rough hair coats, poor appetite, reduced growth rate, and increased susceptibility to stresses and infectious agents. Heavy infestation with internal parasites of the blood-sucking type leads to an induced Fe-deficiency anemia in many animal species, including sheep, pigs, cattle, and man.

*Metalloenzymes have the active metal firmly bound in a constant stoichometric ratio to the protein of the enzyme; in metal-activated enzymes, the activating metal is loosely bound and readily lost on processing. Often metals loosely bound in this way can be replaced by other metal ions without loss of the activity of the enzyme.

Table 11-2. Fe-containing enzyme and proteins in animals.[a]

Metalloporphyrin enzymes	Metalloproteins other than enzymes
Cytochrome oxidase	Hemoglobin
Cytochrome C	Myoglobin
Other cytochromes	Transferrin
Peroxidase	Ovotransferrin
Catalase	Lactotransferrin
Aldehyde oxidase	Ferritin
Metalloflavin enzymes	
NADH cytochrome c reductase	
Succinic dehydrogenase	
Lactic dehydrogenase	
α-glycerophosphate dehydrogenase	
Choline dehydrogenase	
Aldehyde dehydrogenase	
Xanthine oxidase	

[a] Adapted from Bowen (1)

Suckling lambs and calves also become anemic if fed exclusively on milk. Veal calves have light-colored muscle because of low myoglobin content and low blood hemoglobin. This is a desired characteristic of veal on most markets and has led to the practice of feeding low-Fe milk replacers.

Fe deficiency anemia in humans is most common in children and among women of child-bearing age because of large losses of iron during menstruation. Menstrual loss accounts for 16-32 mg of Fe during the cycle or 0.5-1.0 mg/day in addition to the 0.5-1.0 mg excretion by other routes. Thus, the Fe loss in adult women may be twice that of adult men. Most surveys indicate that Fe deficiency anemia in humans is a major disease problem worldwide; in some countries, 30-50% of the population may be affected. Fe deficiency in humans is associated with pallor, chronic fatigue, and general lack of the sense of well-being. The fact that the response to Fe supplementation includes a rapid return to a sense of well-being, which occurs before increased hemoglobin synthesis can occur, suggests that Fe-containing enzymes are affected (65). Although changes in activities of Fe-containing enzymes in Fe deficiency in man are not well documented, Lahey et al (79) have found reduced liver catalase in Fe deficient swine.

Toxicity

Fe overload has been produced in animals by injection or by long periods of excessive intake and in humans by repeated blood transfusions or after prolonged oral administration of Fe. Excess Fe is found in the tissues as hemosiderin. Transferrin concentration is normal and plasma Fe is increased only until transferrin is saturated. The reticuloendothelial cells rather than the parenchyma accumulate the excess Fe. Liver fibrosis is common in some cases of human Fe toxicity because of genetic defect in control of Fe absorption and excretion (idiopathic hemochromatosis), but it is normally not associated with Fe toxicity in animals (65). In the genetic defect, Fe accumulates in the parenchyma, but in excess Fe intake in man and animals it accumulates in reticuloendothelial cells.

Baby pigs have been administered 10X the amount of Fe normally injected for prevention of anemia with no ill effects (81). Hemoglobin tended to be elevated, but growth rate was not affected. Thus, when metabolic Fe demand is high as in rapidly growing suckling animals, a wide margin of safety between adequate and toxic dosages apparently exists.

Copper

Cu was first shown to be a dietary essential in 1928 by Hart et al (82). The role of Cu in nutrition was reviewed in 1950 (83) and more recently by Adelstein and Vallee (84) and Underwood (2).

Tissue Distribution

The liver, brain, kidneys, heart, pigmented part of the eye, and hair or wool are highest in Cu concentration in most species; pancreas, spleen, muscles, skin, and bone are intermediate; and thyroid, pituitary, prostate, and thymus are lowest (2, 84). The concentration of Cu in tissues is highly variable within and among species. Young animals have higher concentrations of Cu in their tissues than adults, and dietary intake has an important effect on the Cu content of liver and blood. Cu in blood is 90% associated with the α_2-globulin, ceruloplasmin, and 10% in the red blood cells as erythrocuprein. Pregnancy is associated with increased plasma Cu in the form of ceruloplasmin, apparently in response to elevated blood estogens. These tissue relationships are illustrated in Fig. 11-5.

Functions

Cu is required for normal red blood cell formation (hematopoiesis) apparently by allowing normal Fe absorption from the GIT and release of Fe from the reticuloendothelial system and the liver parenchymal cells to the blood plasma. This function appears to be related to the required oxidation of Fe from the ferrous to ferric state for transfer from tissue to plasma. Ceruloplasmin is the Cu-containing enzyme required for this oxidation. Cu is required for normal bone formation by promoting structural integrity of bone collagen (86) and for normal elastin formation in aorta and the remainder of the cardiovascular system. This appears to be related to the presence of Cu in amine oxidases, the enzymes required for removal of the epsilon amino group of lysine in the normal formation of desmosine and isodesmosine, key cross-linkage groups in elastine (87). Cu is required for normal myelination of the brain cells and spinal cord as a component of the enzyme cytochrome oxidase which is essential for myelin formation (88).

Cu is required for normal hair and wool pigmentation, presumably as a component of polyphenyl oxidases which catalyze the conversion of tyrosine to melanin and for incorporation of disulfide groups into keratin in wool and hair growth.

Metabolism

Absorption. Site of Cu absorption from the GIT varies among animal species; in the dog, it is absorbed mainly from the jejunum, in man

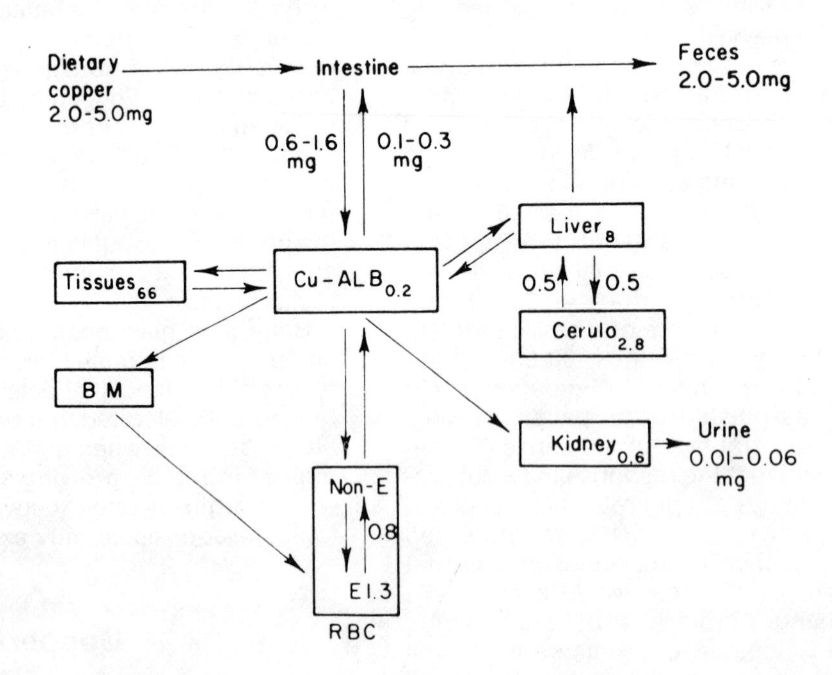

Figure 11-5. Schematic representation of some matabolic pathways of copper in man. The numbers in the boxes refer to mg of Cu in the pool. The numbers next to the arrows refer to mg of Cu transversing the pathway each day. Cu-ALB - direct-reading fraction; cerulo = ceruloplasmin; NON-E = nonerythrocuprein; BM = bone marrow; RBC = red blood cell. From Cartwright and Wintrobe [85].

from the duodenum, in the rat from the stomach and small intestine and in the pig from the small intestine and colon. Degree of absorption is also variable, exceeding 30% in man, but less in other species studied. A Cu-binding protein has been found in the duodenal epithelial cells of the chick (89) which presumably plays a part in Cu absorption. The pH of the intestinal contents affects absorption; Ca salts reduce Cu absorption by raising pH. Absorption is also affected by other elements; ferrous sulfide reduces Cu absorption by forming insoluble CuS; Hg, Mo, Cd, and Zn all reduce Cu absorption. The last two have been shown to displace Cu from a Cu-binding protein in the intestinal mucosa of chicks (89); the mode of action of Hg and Mo is not clear, although the formation of insoluble $CuMoO_4$ has been suggested as the explanation for the effect of Mo.

Some forms of Cu are absorbed more readily than others; cupric sulfate is better absorbed than cupric sulfide; cupric nitrate, chloride, and carbonate are better absorbed than cuprous oxide (2).

Transport and Tissue Utilization. Absorbed Cu is loosely bound to plasma albumin and is distributed to the tissues and taken up by the bone marrow in red blood cell formation where it is present partly as erythrocuprein (see Fig. 11-5). Cu reaching the liver is taken up by parenchymal cells and stored or released to the plasma as Cu-albumin and in larger quantities as a component of ceruloplasmin or is used for synthesis of a large array of Cu-containing enzymes (see Table 11-3) and other Cu-containing proteins.

Excretion. Bile is the major pathway of Cu excretion (see Fig. 11-5). Smaller amounts are lost in intestinal cell and pancreatic secretions, in urine, and negligible amounts in sweat. Radiocopper studies indicate that the main source of urinary Cu is that loosely bound to albumin in the plasma.

Deficiency Signs

Dietary Cu deficiency is associated with a gradual decline in tissue and blood Cu concentration. Blood Cu levels below 0.2μg/ml result in interference with normal hematopoiesis and anemia. The anemia is hypochromic microcytic in some species (rat, rabbit, pig) but in some it is hypochromic macrocytic (cattle, sheep) or normochromic (chicks, dogs) (2). Cu deficiency shortens RBC life span and reduces Fe absorption and utilization. Thus, the anemia appears to be related partly to a direct effect on RBC formation arising from the need for Cu as a component of RBC's and partly to an indirect effect related to the reduced ceruloplasmin concentration of plasma which in this way reduced Fe absorption and utilization. Apparently no interference with heme biosynthesis occurs in Cu deficiency (90).

A widespread problem in lambs that has been traced to a Cu deficiency results in incoordination and ataxia (91). Low levels of Cu in pastures used for grazing, coupled with high intakes of Mo and S, precipitate the condition which is known as swayback or enzootic neonatal ataxia. Newborn lambs are most often affected, but a similar condition can also be produced in the young of goats, guinea pigs, pigs, and rats (2). Degeneration and failure in myelination of the nerve cells of the brain and spinal cord are the basis for the observed nervous disorder. The Cu content of the brain is reduced, leading to a reduction in cytochrome oxidase activity which is necessary for phospholipid synthesis. Intramuscular injection of Cu-glycine, Cu-EDTA or Cu-methionine complexes into pregnant ewes has

Table 11-3. Metalloprotein enzymes, metalloporphyrin enzymes and nonenzyme metalloproteins containing Cu in animals.[a]

Metalloprotein enzymes	Metalloproteins other than enzymes
Tyrosinase	Erythrocuprein
Monoamine oxidase	Hepatocuprein
Ascorbic acid oxidase	Cerebrocuprein
Ceruloplasmin	Milk copper protein
Galactose oxidase	
Metalloporphyrin enzymes	
Cytochrome oxidase	

[a] From Bowen (1)

been used successfully to prevent swayback in lambs (92).

Cu deficiency results in bone abnormalities in many species — pigs, chicks, dogs, horses, rabbits (2). A marked failure of mineralization of the cartilage matrix occurs. The cortex of long bones is thin, although the Cu, P, and Mg concentrations of the ash remain normal. The defect appears to be related to a change in the cross-link structure of collagen, rendering it more soluble than collagen from normal bones (86). The amine oxidase activity of bones from Cu-deficient chicks is reduced 30-40% (86).

The hair and wool of Cu-deficient animals fail to develop normally, resulting in alopecia and slow growth of fibers (Fig. 11-6). Wool growth in sheep is sparse and normal crimping is impeded, leading to straight, hair-like fibers termed steely wool. The change in wool texture is related to a decrease in disulfide groups and increase in sulfhydryl groups and an interference with arrangement of the polypeptide chains. The pigmentation process is extremely sensitive to Cu. Levels of dietary Cu that fail to produce anemia, nerve damage, or bone changes can produce pigmentation failure in the wool of black sheep or in the hair of pigmented cattle. Achromotrichia (loss of hair pigmentation) can be produced in wool in alternating bands by feeding a Cu-deficient or Cu-adequate diet to sheep in alternating intervals. Blockage of the pigmentation process was achieved by Dick (93) within 2 days by feeding Mo and inorganic sulfate to sheep in the presence of marginal Cu. Presumably, the effect of Cu deprivation on pigmentation is related to its role in conversion of tyrosine to melanin.

Cu deficient chicks, pigs, and cattle show cardiovascular lesions and hemorrhages. Falling disease in cattle in Australia was found more than 30 years ago to be related to a progressive degeneration of the myocardium of animals grazing on Cu deficient plants. Later, the histological changes in blood vessels of chicks and pigs were found to be related to a derangement of the elastic tissue of the aorta and other blood vessels (94, 95). The role of Cu in cardiovascular lesions associated with coronary heart disease in man is not know, but deserves study.

Fetal death and resorption in rats and reduced egg production and hemorrhage and death of the embryos in poultry have been shown to be produced by Cu deficiency (96, 97). Reproductive failure of cattle in Cu-deficient locations has also been reported, but a specific effect of Cu is contributing to the syndrome has not been established. In poultry, the primary lesion appears to be a defect in RBC and connective tissue formation in the embryo, possibly induced by a reduction in monoamine oxidase activity (97).

Toxicity

Sheep and calves appear to be more susceptible to Cu toxicity than other species (2). Hemoglobinuria, jaundice, and tissue necrosis have been observed in calves fed a milk substitute containing 115 ppm Cu. Death of sheep, with accompanying hemoglobinuria from excess Cu, has been reported on pasture because of continued free-choice consumption of a trace-mineral salt mixture containing the recommended Cu level (98). Underwood (2) has summarized the conditions under which Cu toxicity occurs in sheep: when the Cu content of soil and pasture is high, when Mo of plant is low, or when liver damage from consumption of certain poisonous plants predisposes to Cu poisoning by decreasing the ability of the liver to dispose of Cu.

Much research has been done on supplementation of growing swine diets with high levels of Cu since the report by Braude of improved growth rate with 250 ppm Cu added to practical diets (99). The mode of action of Cu in producing this growth response is still not known, and the occasional toxicity associated with its use at this level has resulted in no FDA approval of its use above 15 ppm in the USA. Signs of toxicity vary from a slight growth depression and mild anemia to sudden death

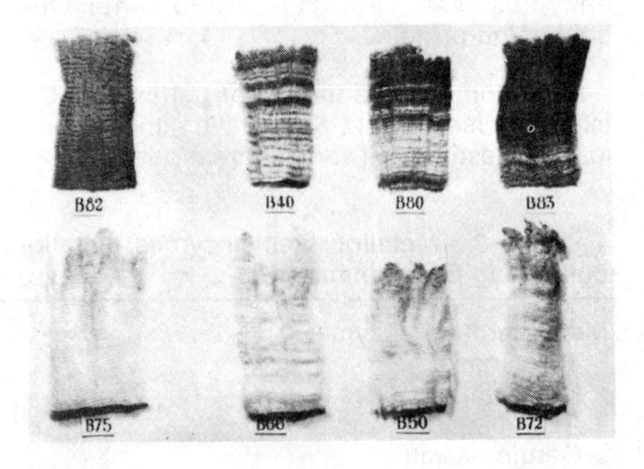

Figure 11-6. Wool from Cu-deficient [or excess Mo] sheep. Normal wool on the left. The other samples show banding of black wool, loss of definition of crimp, and, in some, secondary waves. By permission of G.L. McClymont.

accompanied by liver damage and hemorrhage. Death losses with 250 ppm Cu are now believed to be the result of inadequate diet mixing, as only isolated cases of such toxicity have been reported. Levels of 425+ ppm Cu produce marked anemia, jaundice, and liver damage. When a level of 250 ppm of Cu has resulted in depressed weight gain, there is a microcytic hypochromic anemia associated with it, but usually no liver damage. Gipp et al (100) reported reduced liver Fe and mean corpuscular volume and hemoglobin concentration in growing pigs fed a semipurified diet containing dried skimmilk. These changes were prevented by feeding extra Fe, indicating an induced Fe deficiency. Source of protein is an important factor in the response to added Cu; milk protein is associated with more severe anemia and growth depression than soybean protein (101). Presumably, the interaction between Cu and Fe and the effect of protein source on Cu toxicity are both related to effects on absorption of Cu and Fe from the GIT.

Cu toxicity in man seems highly unlikely, except by accidental contamination of foodstuffs, as no cases of chronic disorders in man traceable to excessive Cu intakes have been reported (2). However, metabolic Cu intoxication in man is a well-known pathological condition termed Wilson's disease. The primary defect is thought to be the failure of the liver to remove albumin-bound Cu from the plasma for incorporation into ceruloplasmin because of a genetically controlled absence of the necessary liver enzyme system (102). The result is massive accumulation of liver Cu, liver cirrhosis, renal damage, reduced plasma ceruloplasmin, and elevated levels of Cu in brain and kidney. Administration of chelating agents to induce urinary excretion of Cu have been employed to minimize tissue damage.

Manganese

Mn was first recognized as a dietary essential for animals in 1931 (103-105). Although the widespread occurrence of Mn in foodstuffs makes a deficiency less likely than for many other mineral elements, sufficient examples exist of practical problems with Mn deficiency in animals and birds to justify detailed consideration. The metabolism, functions and deficiency and toxicity signs of Mn have been reviewed in detail (2, 106).

Tissue Distribution

The total body supply of Mn is less than that of most other required minerals; for example, the total Mn content of adult man is only 1% of the amount of Zn and 20% of Cu. It is widespread throughout the body and does not tend to accumulate in liver and other tissues in high concentrations when ingested in large amounts, in contrast to most other trace elements. Bone, kidney, liver, pancreas, and pituitary gland have the highest concentration. Bone Mn content is more responsive to dietary intake than that of other tissues; it represented 25% of the total body content in one study with pullets and was less mobilizable from the skeleton than from liver, skin, feathers, and muscle (107). A large proportion of soft tissue Mn is present in labile intracellular forms, but in bone it is associated mainly with the inorganic fraction (108).

Functions

Mn is essential for formation of chondroitin sulfate, a component of mucopoly saccharides of bone organic matrix (109, 110). Thus, it is essential for normal bone formation. Mn is necessary for prevention of ataxia and poor equilibrium in newborn animals and birds. Biochemical and histological evidence relating to the role of Mn in this role is lacking except for the presence of a structural defect in the inner ear of affected animals (111). This abnormality appears to be related to defective cartilage mucopolysaccharide formation and is, therefore, probably a deficiency sign associated with the previous function listed. It is required for normal estrus and ovulation in females and for libido and spermatogenesis in males, but no biochemical evidence as yet explains this effect.

Mn is a component of the metalloenzyme, pyruvate carboxylase, and, as such, plays a role in carbohydrate metabolism; it is necessary as a cofactor in the enzyme that catalyzes the conversion of mevalonic acid to squalene and stimulates synthesis of cholesterol and fatty acids in rat liver (2). Thus, it plays a vital role in lipid metabolism. Mn stimulates arginase activity in mammals and in vitro it activates a variety of other enzymes. The biological significance of these in vitro studies remains to be determined.

Metabolism

Absorption. The mode and control of Mn absorption from the GIT are poorly understood. The amount absorbed appears to be proportional to the amount ingested, and absorption is usually less than 10% (106). Excessive dietary Ca and P reduces Mn absorption. Lassiter et al

(112), however, have recently shown that both urinary and fecal excretion of parenterally administered radioactive Mn is greater in rats previously fed a low Ca diet than those fed a high Ca diet. Although this may mainly reflect the greater bone resorption of rats fed low Ca, thereby resulting in less Mn retention by bone, the results do illustrate that dietary Ca and P content affects not only Mn absorption but also Mn tissue losses. Divalent Mn is absorbed about as well as the oxide, carbonate, chloride, and sulfate salts. Mn absorption is increased in Fe deficiency.

Transport and Storage. Mn is absorbed from the GIT as Mn^{++} oxidized to Mn^{+++}, transported rapidly to all tissues, and is concentrated in mitochondria-rich tissues. Liver, pancreas, and brain mitochondria have been shown in several species to accumulate radioactive Mn (106). Mn is carried in blood as Mn^{+++}, loosely bound to a plasma β_1-globulin (other than transferrin), and is quickly removed from blood, first appearing in mitochondria and later in the nucleus of cells. Mn content of milk and fetuses and of eggs closely reflects dietary intake, illustrating the dissimilarity between Mn and such minerals as Fe and Cu in transport across mammary and reproductive tissues.

Excretion. The main route of Mn excretion is via the bile, with smaller amounts lost through pancreatic secretions and sloughed intestinal mucosal cells and still smaller amounts in urine and sweat. Much of the bile Mn is reabsorbed. Bile concentration parallels that of blood; this may explain the failure of body tissues to accumulate large concentrations of Mn. Thus, excretion rather than absorption is the primary means by which body Mn homeostasis is normally maintained.

Deficiency Signs

A wide range of skeletal abnormalities are associated with Mn deficiency in several animal species. These abnormalities are all undoubtedly related to the role of Mn in mucopolysaccharide synthesis in bone organic matrix. Lameness, shortening and bowing of the legs, and enlarged joints are common in Mn-deficient rodents, pigs, cattle, goats, and sheep, and perosis or slipped tendon is common in poultry (Fig. 11-7). Choline deficiency also produces perosis (see Ch. 14). This condition includes malformation of the tibiometatarsal joint, bending of the long bones of the leg, and slipping of the gastrocnemius tendon from its condyle (Fig. 11-7). Parrot beak, shortened and thickened legs and wings — manifestations of the same defect in bone formation and often termed chondiodystrophy — are common signs of Mn deficiency in birds. Impairment in calcification of bones in Mn-deficient animals is not a factor in the bone malformations. Although plasma alkaline phosphatase is sometimes reduced in Mn-deficient animals, no change occurs in the composition of bone ash, except an occasionally observed decline in Mn concentration (113). Eggshell thickness and strength are also reduced in Mn deficiency. The ataxia and poor coordination and balance commonly observed in newborn mammals and newly hatched birds whose dams have been fed Mn-deficient diets is irreversible and is related to abnormal bone formation of the inner ear bones during prenatal life (111). In chicks the characteristic head retraction is sometimes referred to as the "star-gazing" posture. A genetically related congenital ataxia of mice caused by defective development of inner ear bones has been shown to be preventable by Mn supplementation of the maternal diet at a specific time in gestation (114). This represents a unique example of an interaction between a nutrient and a mutant gene affecting bone development.

Delayed estrus, poor conception, decrease in litter size, and livability have been reported in farm and laboratory animals, and decreased egg production and hatchability occur in birds deprived of Mn. In males, Mn deficiency produces absence of libido and failure in spermatogenesis (2, 113).

Newborn guinea pigs from Mn-deprived dams show marked hypoplasia of the pancreas and adult Mn-deficient guinea pigs show changes in pancreas histology and impaired glucose tolerance (115). Whether this effect of Mn on the pancreas and glucose tolerance is related to some specific effect on insulin production or to an indirect action through the presence of Mn in pyruvate carboxylase, which is involved in gluconeogenesis, is not known. The blood of diabetics is low in Mn and Mn administration to these subjects has a hypoglycemic effect (116, 117). Thus, Mn appears to play a role in glucose metabolism, but its practical importance in this role in animal nutrition is not known. Mn deficiency has been reported to be associated with reduced liver and bone lipids in rats and reduced back fat in pigs, but the exact role of Mn in lipid metabolism remains unknown.

Although Mn deficiency is not considered

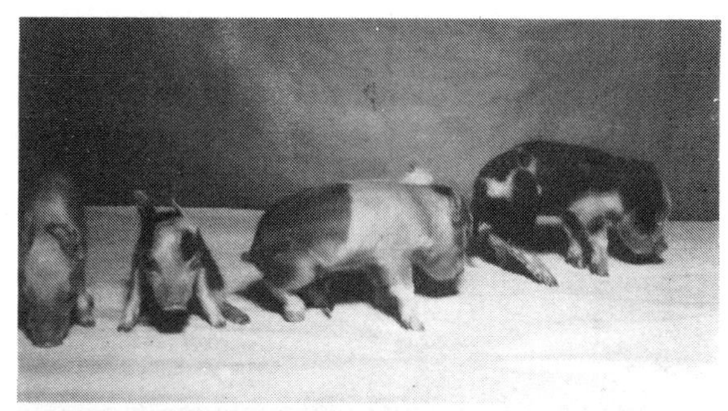

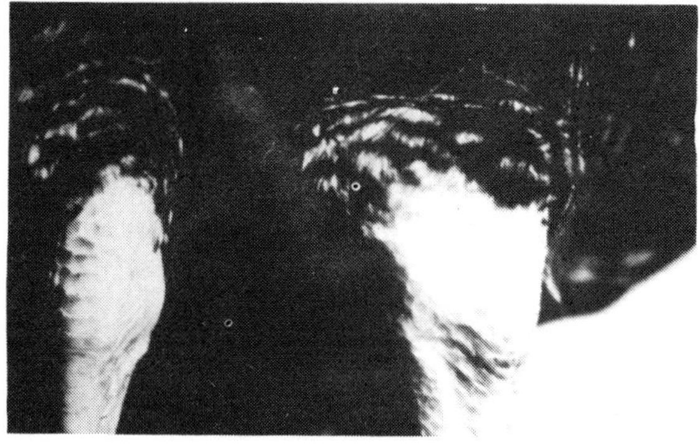

Figure 11-7. Manganese deficiency. Top. Litter from sow fed 0.5 ppm of Mn. The pigs showed weakness and poor sense of balance at birth. Courtesy of W. M. Beeson. Bottom. Perosis in the chick, which results in enlargement of the hock.

likely in man, Schroeder et al (118) have suggested the possibility of a relationship between several human disorders and Mn nutriture. The relatively high levels of Mn in many foods commonly available to man in relation to metabolic requirements would suggest that such relationships are unlikely but worthy of examination.

Toxicity

High dietary levels of Mn are tolerated well by most species; the toxic effects of Mn appear to be related more to interference with utilization of other minerals than to a specific effect of Mn itself. Rats, poultry, and calves show no ill effects of diets containing 820 to 1000 ppm Mn, but pigs show reduced growth at 500 ppm. The reduced growth rate associated with excess Mn is mainly a reflection of reduced appetite.

Ca and P utilization are adversely affected by excess Mn. Rickets has been produced in rats fed 1.73% Mn in the diet and hypoplasia of tooth enamel has been observed in rats, cows, and pigs fed excess Mn. Fe deficiency anemia is produced in cattle, pigs, rabbits, and sheep fed 1000 to 5000 ppm Mn (2). In humans, Cotzias (106) reviewed the signs of chronic excess Mn intake by lung inhalation in workers mining Mn ores. Neurological disorders and behavior changes are observed.

Selenium

Selenium was recognized as a toxic mineral many years before its essentiality for animals and birds was discovered. Schwarz and Foltz (119) first reported in 1957 that Se prevents liver necrosis in rats and in the same year Schwartz et al (120) and Patterson et al (121) showed that it also prevents exudative diathesis in chicks. The following year Muth et al (122) cured muscular dystrophy in lambs with supplemental Se. All of these lesions had been known previously to be produced by vitamin E deficiency (see Ch. 13). Thus, a new era of nutrient interrelationships was developed. Knowledge of Se and vitamin E in nutrition and metabolism has been reviewed extensively (2, 123, 124, 125, 126).

Tissue Distribution

Se is present in all cells of the body, although the concentration is generally less than 1 ppm; thus, the total body supply is relatively low. Liver, kidney, and muscle generally contain the highest concentrations of Se and the values for these and other tissues are affected by dietary intake (Tables 11-4, 11-5). Concentration of Se in liver and kidney of animals fed toxic levels (5-10 ppm) may be as high as 5-7 ppm (2).

Functions

Although tremendous research efforts in several laboratories during the past 15 years have clearly established a role of Se in preventing some of the deficiency diseases known to be induced by vitamin E deficiency (muscular dystrophy, exudative diathesis), the work of Thompson and Scott was the first to firmly establish Se as a required nutrient in its own right (128). Since then Noguchi et al (129) and Rotruck et al (130) have provided further data elucidating the specific biochemical functions of Se.

Se is a component of the enzyme glutathione peroxidase and in this role is involved in breakdown of peroxides arising from tissue lipid oxidation, and thus it plays a central role in maintaining integrity of cellular membranes. Se is also required for normal pancreatic morphology and through this effect on pancreatic lipase production is responsible for normal absorption of lipids and tocopherols from the GIT (131). Whether the effect of Se on the pancreas is really only a manifestation of its role as a component of glutathione peroxidase remains to be determined. Neither is it known if additional functions exist for Se which will explain the diverse signs of deficiency in animals.

Table 11-4. Selenium content of the longissimus muscle of pigs fed diets from different sources in the USA.[a]

Source[b]	Diet Se, ppm	Muscle Se, ppm, dry basis
Arkansas	0.152	0.817
Idaho	0.086	0.392
Illinois	0.036	0.223
Indiana	0.052	0.232
Iowa	0.235	0.977
Michigan	0.040	0.206
Nebraska	0.330	1.178
New York	0.036	0.163
North Dakota	0.412	1.430
South Dakota	0.493	1.893
Virginia	0.027	0.118
Wisconsin	0.178	0.501
Wyoming	0.158	1.098

[a] Ku et al (127)
[b] Corn-soybean meal diets of similar composition.

Table 11-5. Selenium content of tissues of pigs fed Se at different levels from sodium selenite.[a]

Tissue	Dietary Se, ppm				
	0.05[b]	0.10	0.15	0.25	1.05
	Tissue Se, ppm fresh basis				
Longissimus muscle	0.05	0.08	0.09	0.11	0.13
Myocardium	0.11	0.21	0.21	0.23	0.33
Liver	0.14	0.51	0.57	0.59	0.80
Kidney	1.37	2.07	2.06	1.95	2.10

[a] Groce et al (127)
[b] Corn-soybean meal basal diet

Metabolism

Absorption. The main site of Se absorption is from the duodenum; it is not absorbed from the rumen or abomasum of sheep or the stomach of pigs (132). It is absorbed relatively efficiently either from natural Se-containing feedstuffs or as inorganic selenite (35-85%). Absorption of Se in ruminants is probably largely as selenomethionine and selenocystine as a result of the incorporation of inorganic dietary Se into amino acids by rumen microflora. Because Se-depleted animals retain more Se than adequately fed animals, it has been suggested that increased absorption occurs in response to tissue needs. However, excretion is also responsive to tissue needs (2). The absorption of Se is not affected by dietary tocopherol (133).

Transport and Storage. After absorption, Se is transported in the plasma in association with a plasma protein and enters all tissues where it is stored mainly as selenomethionine and selenocystine (134, 135). Se is incorporated into RBC's, leucocytes, myoglobin, nucleoproteins, myosin, and several enzymes, including cytochrome c and aldolase (2).

Excretion. Tissue Se is relatively labile as illustrated by its rapid loss from tissues of animals fed low-Se diets following consumption of high-Se diets (136). Loss occurs via the lungs, feces, and urine. The proportion excreted by each route depends on the route of administration, tissue levels, and animals species (2). In sheep injected Se is excreted mainly in the urine in proportion to the amount administered; fecal loss is small and constant with dosage level; expired Se is similar to fecal loss at low levels of administration, but rises with dosage level and exceeds fecal loss by several fold at high dosage levels (137). Orally administered Se is excreted in feces in largest quantities at low intake levels; as intake rises, fecal loss remains relatively stable and expired Se increases steadily; urinary loss rises at moderate levels of supplementation, then declines.

Fecal losses represent mostly unabsorbed Se, but also that excreted via the bile, pancreatic duct, and intestinal mucosal cells. Arsenic (As), which is known to reduce absorption of Se from the GIT (138), also increases bile and urinary excretion of Se when As is injected (139, 140).

The form in which Se is provided in the diet has an important effect on its absorption. Scott (141) recently summarized the utilization of plant vs. animal sources of Se and concluded that the generally superior utilization of plant Se results from the presence of Se in the inorganic form in plants; animal tissues contain Se mainly in the organic form as selenomethionine and selenocystine.

The sulfate ion has important effects on Se metabolism under some conditions. Urinary excretion of Na selenate is increased by parenterally or orally administered sulfate. The effect of sulfate is less marked when Se is provided as selenite than as selenate; sulfate probably has no effect on utilization of Se provided in the organic form.

Placental transfer of Se is extensive in all species studied as evidenced by prevention of muscular dystrophy in lambs and calves by Se administration to the pregnant dam and by radioactive tracer studies. Selenomethionine is more readily transferred to the fetus than inorganic Se (2).

Deficiency Signs

Several deficiency diseases of animals and birds that were earlier considered to be from vitamin E deficiency have more recently been shown to be preventable by Se as well. These include nutritional muscular dystrophy (NMD) in lambs, poultry, pigs, and calves (see Fig. 11-8); exudative diathesis (ED) in poultry, and liver necrosis in rats and pigs. Encephalomalacia in chicks is cured or prevented by vitamin E or antioxidants, but not by Se.

Se-responsive NMD is common in calves and lambs in many parts of the world. Allaway et al (142) have mapped geographic areas with soil Se deficiency and Se excess in the USA. Undoubtedly, widely differing levels of soil Se are present in other parts of the world as well. Affected lambs may show a stiff gait and arched

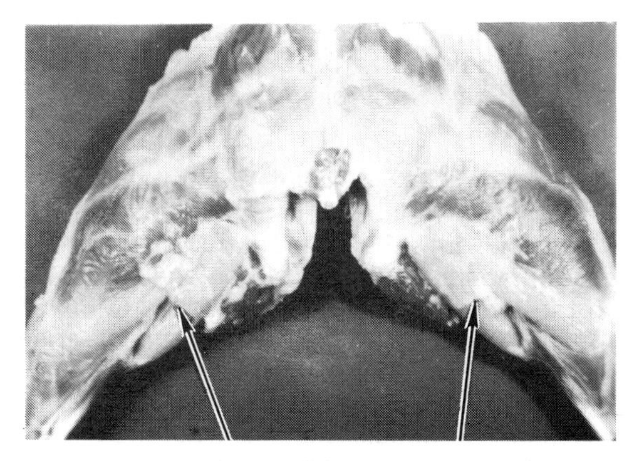

Figure 11-8. Selenium deficiency resulting in nutritional muscular dystrophy in a lamb carcass. Note severly affected muscles pointed out by the arrows. Photo by H. Muth, Oregon State University.

back giving rise to the name stiff lamb disease. Severe cases result in death. Injection or oral administration of vitamin E or Se to the pregnant or lactating dam or the newborn protects against NMD. Postmortem examination reveals whitish streaks in the striated muscle, which are caused by degeneration of the muscle fiber (Zenker's degeneration) and account for the common name, white muscle disease. If the heart muscle is affected, sudden death may occur. This is an increasingly common problem in growing pigs and is often referred to as mulberry heart disease. Animals with NMD show drastic rises in blood plasma concentrations of several enzymes that are normally intracellular but are released into the plasma when tissue damage occurs. These enzymes include glutamic oxaloacetic transaminase (GOT) and lactic dehydrogenase isoenzymes (LDH) (2, 143). In pigs, the chief target organ, in addition to the heart muscle, is the liver. Massive liver necrosis and sudden death is a common sign of Se deficiency in growing pigs (Fig. 13-7). Serum ornithine carbamyltransferase is elevated because of liver cell damage which releases the intracellular enzyme into the blood. Vitamin E supplementation of the diet of pigs fed marginal Se prevents this liver damage. The syndrome in pigs is called hepatosis dietetica and has become a practical field problem in Scandinavia, New Zealand (2), and the USA (127). This is due, presumably, to soil depletion of Se or changes in methods of fertilization or harvesting and processing of cereal grains which may affect their tocopherol or Se content or biological availability. Corn grown on low-Se soil fertilized with Se contains a higher level of Se than that grown on adjacent unfertilized plots and the Se is biologically available for chicks (144).

Chicks fed diets deficient in vitamin E and Se develop ED which is prevented by supplementation with either nutrient. ED is characterized by accumulation of subcutaneous fluids on the breast, resulting from escape of blood constituents to the extracellular spaces after capillary damage. Noguchi et al (129) recently showed that glutathione peroxidase activity in plasma of Se-deficient chicks declined to less than 10% of normal within ca. 5 days. The acitivity of this enzyme correlates well with the protection by Se of chicks from ED. Vitamin E, although it protects against ED in the absence of Se, has no effect on glutathione peroxidase activity. Thus, the metabolic role of Se in protecting against ED is distinct from that of vitamin E. Scott (145) has hypothesized that plasma glutathione peroxidase acts as the first line of defense against peroxidation of lipids in the capillary membrane by destroying peroxides formed in the absence of vitamin E. Peroxidation of membrane lipids is prevented with adequate vitamin E, thus blocking the destruction of the capillary membranes. In its absence, Se functions to prevent peroxidation by its presence in glutathione peroxidase. Thus, the disruption of capillary membrane integrity which results in ED is prevented by either vitamin E or Se, but by slightly different mechanisms.

Se deficiency in chicks fed adequate vitamin E results in pancreatic degeneration and fibrosis which impedes vitamin E absorption from the GIT (128). Gries and Scott (131) have shown that the pancreas is the target organ for Se deficiency in chicks, as skeletal muscle, heart muscle, liver, and other tissues are unaffected if vitamin E is adequate.

Infertility in cattle, sheep, and birds has been reported to respond favorably to Se feeding or injection (2). Knowledge is incomplete as to whether such effects are related distinctly to Se or whether a sparing effect on the vitamin E requirement is involved. Experiments with rats over successive generations have demonstrated reduced fertility in females and males even in the presence of adequate vitamin E (146). Thus, inconsistent results in farm animals may reflect differences in tissue stores of Se and vitamin E as well as species differences.

Toxicity

Animals grazing on seleniferous soils or fed crops grown on these soils develop a fatal disease known as blind staggers or alkali disease. This syndrome was described in horses as early as 1856 and has since been studied in other species. Soils containing more than 0.5 ppm Se are potentially dangerous to livestock (147) as the plants harvested from them may contain 4 ppm of more. Blind staggers is characterized by emaciation, loss of hair, soreness and sloughing of hooves, and erosion of the joints of long bones leading to lameness, heart atrophy, liver cirrhosis, anemia, excess salivation, grating of the teeth, blindness, paralysis, and death (2).

In poultry, egg production and hatchability are reduced and deformities are common, including lack of eyes and deformed wings and feet. Abnormal embryonic development is also prominent in rats, pigs, sheep, and cattle fed excess Se (2).

Diets containing 5 ppm of Se produce toxic signs in most species. High-protein diets tend to protect against Se toxicity, and inorganic sulfate has been reported to relieve the growth depression in rats fed 10 ppm Se (148). Arsenic (As) added to the feed or drinking water alleviates Se toxicity in dogs, pigs, cattle, and chicks (2). Liver succinic dehydrogenase activity in rats is reduced by high Se and restored by As (2). This is the only enzyme known to be affected by excess Se. The mode of action by which toxicity signs are brought about is unknown.

Other Trace Elements

In addition to those already discussed in detail, four other elements may be considered essential, based on the broad definition that their absence from the diet results in metabolic or functional abnormalities in one or more tissues of an animal species. These are: chromium (Cr), fluorine (F), molybdenum (Mo), and silicon (Si).

A detailed description of the tissue distribution, functions, metabolism, and deficiency signs of these minerals in animals is beyond the scope of this chapter. Excellent detailed coverage has been provided by Underwood (2) and in the books edited by Comar and Bronner (149). Only selected references are included in the present discussion to explain the reasoning on which each of the 4 minerals has been interpreted as being a dietary essential. Much more information is available on toxic effects, especially of F and Mo. Toxic aspects of these and other minerals are considered in Ch. 12.

Chromium

Evidence that Cr might be required by animals was first reported in 1954 when synthesis of cholesterol and fatty acids by rat liver was shown to be increased by Cr (150). In 1959, Schwarz and Metz (151) demonstrated the importance of trivalent Cr in glucose utilization. This has since been shown to be related to the presence of Cr as a cofactor in insulin. Cr utilization not only depends on its valence (trivalent is utilized, hexavalent is not) but on its chemical form. Mertz et al (152) showed that radioactive Cr transfer to the rat fetus was accomplished only when Cr was incorporated into yeast and not when it was administered to the mother as inorganic salts. The importance of Cr in glucose metabolism of other animals and man needs to be established; no clear evidence has been found for a practical need for Cr supplementation of diets for animals fed typical natural ingredients (153).

Fluorine

F can qualify as an essential nutrient for animals and man mainly on the basis of its protective effect on dental caries (154). No reports of a positive growth response to F addition to F-low diets appeared until 1972 when Schwarz and Milne (155) obtained improved growth in rats by applying strict control of dietary and environmental contamination. Although low levels in the diet or in drinking water (1 ppm) protect against dental caries (tooth decay), higher levels produce mottled tooth enamel and enlarged bones (156, 157). Low dietary F has been reported to protect against osteoporosis in humans, but this point is still controversial as animal experiments have failed to support the observation (158, 159). Growth and remodeling of bones is certainly affected by level of dietary F, so the conditions under which F is beneficial or harmful to bone and teeth may be related not only to level of F intake but to other unexplained factors. High dietary Ca depresses F uptake of bone (160, 161), and probably other dietary variables are also important (2).

Molybdenum

Evidence that Mo is essential for animals was first reported in 1953 when it was shown to be a component of the metalloenzyme, xanthine oxidase (162). Later, chicks, turkey poults, and lambs were shown to perform more favorably when Mo was added to semipurified diets (163-165). The growth response in lambs was obtained using a basal diet containing 0.36 ppm Mo which resulted in a 2.5 fold improvement in daily gain. Because most feedstuffs contain considerably more Mo, it is not surprising that Mo has not been generally recognized as an essential nutrient. Its presence in feeds at toxic levels is of greater practical concern.

Silicon

Si has recently been recognized as an essential nutrient for chicks. It is one of the most abundant elements on earth and is present in large amount in soil and plants and, as a result, is ingested in large quantities by animals. Absorption occurs as monosilicilic acid, which represents only a small fraction of the total Si ingested (2). Carlisle (166) first reported a growth response in chicks fed Si by using purified diets carefully prepared to minimize Si contamination and by raising the animals in a rigidly controlled environment low in Si. Subsequently, Schwarz and Milne (167), using similarly controlled conditions, obtained a 25-34% increase in growth of rats by Si supplementation to a low-Si purified diet. Si appears to be involved in some way in the initiation of the mineralization process in bones (168, 169). From a practical viewpoint, adverse effects of high Si intake rather than Si deficiency appear to be of greater importance.

Figure 11-9. Controlled environment plastic isolator for studying trace element requirements. Courtesy of K. Schwarz.

Possibility of Additional Required Minerals

Many minerals are ubiquitous in nature and are commonly present in animal tissues, but no metabolic role has been designated for them. Some of these appear to be harmless, some toxic. Among them, vanadium and tin have been suggested as being required for growth of animals (170, 171). Further conformation is needed to establish these and other minerals as dietary essentials, but clearly newer, more sensitive analytical methods and closer control of environmental contamination (see Fig. 11-9), along with a greater awareness of inter-relations among minerals, favor the discovery of additional essential minerals for animals and man. These possibilities are explored in several recent publications (2, 171, 172, 173).

References Cited

General

1. Bowen, H.J.M. 1966. Trace Elements in Biochemistry. Academic Press, N.Y.
2. Underwood, E.J. 1971. Trace Elements in Human and Animal Nutrition. 3rd ed. Academic Press, N.Y.
3. Comar, C.L. and F. Bronner. 1962. Mineral Metabolism. Vol. II, Part B. Academic Press, N.Y.
4. Carlisle, E. 1972. Fed. Proc. 31:700.

Cobalt

5. Underwood, E.J. and J.F. Filmer. 1935. Aust. Vet. J. 11:84.
6. Smith, S.E., B.A. Koch and K.L. Turk. 1951. J. Nutr. 44:455.
7. Maynard, L.A. and J.K. Loosli. 1969. Animal Nutrition. McGraw-Hill Book Co., N.Y.
8. Braude, R., A.A. Free, J.E. Page and L.E. Smith. 1949. Br. J. Nutr. 3:289.
9. Askew, H.D. and J. Watson. 1943. New Zeal. J. Sci. Tech. A25:81.
10. Comar, C.L., G.K. Davis and R.F. Taylor. 1946. Arch. Biochem. 9:149.
11. Comar, C.L. and G.K. Davis. 1947. Arch. Biochem. 12:257.
12. Smith, S.E., D.E. Becker, J.K. Loosli and K.C. Beeson. 1950. J. Animal Sci. 9:221.
13. Andrews, E.D., B.J. Stephenson, J.P. Anderson and W.C. Faithful. 1958. N.Z.J. Agr. Res. 1:125.
14. Dewey, D.W., H.J. Lee and H.R. Marston. 1958. Nature 181:1367.
15. Dewey, D.W., H.J. Lee and H.R. Marston. 1969. Aust. J. Agr. Res. 20:1109.
16. Andrews, E.D. 1965. N.Z. Vet. J. 13:101.
17. Dinussen, W.E., E.W. Klosterman, E.L. Lasley and M.L. Buchanan. 1953. J. Animal Sci. 12:623.

Iodine

18. Kendall, E.C. 1919. J. Biol. Chem. 39:125.
19. Harington, C.R. and G. Barger. 1927. Biochem. J. 21:169.
20. Gross, J. 1962. In: Mineral Metabolsim. Vol. II, Part B. Academic Press, N.Y.
21. Myant, N.B., B.D. Corbett, A.G. Honour and E.E. Pochin. 1950. Clin. Sci. 9:405.
22. Cohen, B. and N.B. Myant. 1959. J. Physiol. 145:595.
23. Osorio, C. and N.B. Myant. 1960. Br. Med. Bul. 16:159.
24. Potter, G.D., W. Tong and J.L. Chaikoff. 1959. J. Biol. Chem. 234:330.
25. Roche, J. and R. Michel. 1960. Ann. N.Y. Acad. Sci. 86:454.
26. Robbins, J. and J.E. Rall. 1960. Physiol. Rev. 40:415.
27. Sterling, K. and M.A. Bremmer. 1966. J. Clin. Invest. 45:153.
28. Andrews, F.N., et al. 1948. J. Animal Sci. 7:298.
29. Ferguson, K.A., P.G. Schinckel, H.B. Carter and W.H. Clarke. 1956. Aust. J. Biol. Sci. 9:575.
30. Guillemin, R., et al. 1963. Endocrinol. 73:564.
31. Kingsbury, J.M. 1964. Poisonous plants in the United States and Canada. Prentice-Hall, Inc., Englewood Cliffs, N.Y.
32. Ammerman, C.B., et al. 1964. J. Nutr. 84:107.
33. Arrington, L.R., L.N. Taylor, C.B. Ammerman and R.L. Shirley. 1965. J. Nutr. 87:394.
34. Perdomo, J.T., R.H. Harms and L.R. Arrington. 1966. Proc. Soc. Exp. Biol. Med. 122:758.

Zinc

35. Todd, W.R., C.A. Elvehjem and E.B. Hart. 1934. Amer. J. Physiol. 107:146.
36. Kernkamp, H.C.H. and E.F. Ferrin. 1953. J. Amer. Vet. Med. Assoc. 123:217.
37. Tucker, H.F. and W.D. Salmon. 1955. Proc. Soc. Exp. Biol. Med. 88:613; Raper, J.T. and L.V. Curtin. 1953. Cited by Hoekstra, ref. 46.
38. Hoeskta, W.G. 1967. In: Present Knowledge of Nutrition. 3rd ed. The Nutr. Fdn., Inc., N.Y.
39. Vallee, B.L. 1962. In: Mineral Metabolsim. Vol. II, Part B. Academic Press, N.Y.
40. Prasad, A.S. 1966. Zinc Metabolism. C.C. Thomas Pub. Co., Springfield, Ill.
41. Miller, W.J., D.M. Blackmon, R.P. Gentry and F.M. Pate. 1970. J. Animal Sci. 31:149.
42. Hsu, J.M. and R.L. Woosley. 1972. J. Nutr. 102:1181.
43. Theuer, R.C. and W.G. Hoekstra. 1966. J. Nutr. 89:448.

44. Methfessel, A.H. and H. Spencer. 1968. Fed. Proc. 27:422.
45. Oberleas, D., M.E. Muhrer and B.L. O'Dell. 1966. J. Nutr. 90:56.
46. O'Dell, B.L. and J.E. Savage. 1960. Proc. Soc. Exp. Biol. Med. 103:304.
47. Pond, W.G. and J.R. Jones. 1964. J. Animal Sci. 23:1057.
48. O'Dell, B.L., C.E. Burpo and J.E. Savage. 1972. J. Nutr. 102:653.
49. Lewis, P.K., R.H. Grummer and W.G. Hoekstra. 1957. J. Animal Sci. 16:927.
50. Vohra, P. and F.H. Kratzer. 1966. J. Nutr. 89:106.
51. Pekas, J.C. 1968. J. Animal Sci. 27:1559.
52. Miller, W.J., et al. 1966. J. Dairy Sci. 49:1446.
53. Miller, W.J., et al. 1967. J. Nutr. 92:71.
54. Prasad, A.S., et al. 1963. J. Lab. Clin. Med. 62:84.
55. Norrdin, R.W., L. Krook, W.G. Pond and E.F. Walker. 1973. Cornell Vet. 63:264.
56. Nielsen, F.H., M.L. Sunde and W.G. Hoekstra. 1967. Proc. Exp. Biol. Med. 48:521.
57. Apgar, J. 1972. J. Nutr. 102:343.
58. Prasad, A.S., J.A. Halsted and M. Nadine. 1961. Amer. J. Med. 31:532.
59. Miller, W.J., J.D. Morton, W.J. Pitts and C.M. Clifton. 1965. Proc. Soc. Exp. Biol. Med. 118:427.
60. Smith, S.E. and E.J. Larson. 1946. J. Biol. Chem. 163:29.
61. Brink, M.F., D.E. Becker, S.W. Terrill and A.H. Jensen. 1959. J. Animal Sci. 18:836.
62. Hsu, F. 1974. Ph.D. Thesis, Cornell University, Ithaca, N.Y.
63. Ott, E.A., W.H. Smith, R.B. Harrington and W.M. Beeson. 1966. J. Animal Sci. 25:414.
64. Ott, E.A., et al. 1966. J. Animal Sci. 25:432.

Iron

65. Moore, C.V. and R. Dubach. 1962. In: Mineral Metabolism. Vol. II, Part B. Academic Press, N.Y.
66. Peden, J.C. 1967. In: Present Knowledge in Nutrition. 3rd ed. The Nutr. Foundation, Inc., N.Y.
67. Gross, F. 1964. Iron Metabolism. Springer Verlag, Berlin, Germany.
68. Moore, C.V. 1951. Harvey Lect. 55:67.
69. Brown, E.B. and B. Justus. 1958. Amer. J. Physiol. 194:319.
70. Smith, M.D. and I.M. Pannaciulli. 1958. Br. J. Haematol. 4:428.
71. Forth, W. 1970. In: Trace Element Metabolsim in Animals. E. & S. Livingstone, Edinburgh, Scotland.
72. Conrad, M.E., L.R. Weintraub and W.H. Crosby. 1964. J. Clin. Invest. 43:963.
73. Hahn, P.F., et al. 1943. J. Exp. Med. 78:169.
74. Murry, J. and N. Stein. 1970. In: Trace Element Metabolism in Animals. E. & S. Livingstone, Edinburgh, Scotland.
75. Figueroa, W.G. 1960. In: Metal Binding in Medicine. Lippincott, Philadelphia, Pa.
76. Pond, W.G., R.S. Lowrey, J.H. Maner and W.G. Pond. 1960. J. Animal Sci. 19:1286.
77. Pond, W.G., T.L. Veum and V.A. Lazar. 1965. J. Animal Sci. 24:668.
78. Mazar, A. and A. Carlton. 1965. Blood 26:317.
79. Lahey, M.E., et al. 1952. Blood 7:1053.
80. Dowdle, E.B., D. Schachter and H. Schenker. 1960. Amer. J. Physiol. 198:609.
81. Maner, J.H., W.G. Pond and R.S. Lowrey. 1959. J. Animal Sci. 18:1373.

Copper

82. Hart, E.B., et al. 1928. J. Biol. Chem. 77:797.
83. McElroy, W.D. and B. Glass. 1950. Symposium on Copper Metabolism. Johns Hopkins Press, Baltimore, Md.
84. Adelstein, S.J. and B.L. Vallee. 1962. In: Mineral Metabolism. Vol. II, Part B. Academic Press, N.Y.
85. Cartwright, G.E. and M.M. Wintrobe. 1964. Amer. J. Clin. Nutr. 14:224; 15:94.
86. Rucker, R.B., H.E. Parker and J.C. Rogler. 1969. J. Nutr. 98:57.
87. Partridge, S., et al. 1964. Biochem. J. 93:30c.
88. Fell, B.F., C.F. Mills and R. Boyne. 1965. Res. Vet. Sci. 6:10.
89. Starcher, B. 1969. J. Nutr. 97:321.
90. Lee, G.R., G.E. Cartwright and M.M. Wintrobe. 1968. Proc. Soc. Esp. Biol. Med. 127:977.
91. Roberts, H.E., B.M. Williams and A. Harvard. 1966. J. Comp. Path. 76:279, 285.
92. Hemingway, R.G., A. MacPherson and N.S. Ritchie. 1970. In: Trace Element Metabolism in Animals. E. & S. Livingstone, Edinburgh, Scotland.
93. Dick, A.T. 1953. Aust. Vet. J. 29:18, 30:196.
94. O'Dell, B.L., B.C. Hardwick, G. REynolds and J.E. Savage. 1961. Proc. Soc. Exp. Biol. Med. 108:402.
95. Coulson, W.F. and W.H. Carnes. 1963. Amer. J. Path. 43:945.
96. Howell, J. McC. and G.A. Hall. 1969. Br. J. Nutr. 23:47.
97. Simpson, C.F., J.E. Jones and R.H. Harms. 1967. J. Nutr. 91:283.
98. Kowalczyk, T., A.L. Pope, K.C. Berger and B.A. Muggenberg. 1964. J. Amer. Vet. Med. Assoc. 145:352.
99. Braude, R. 1967. World Rev. Animal Prod. 3:69.
100. Gipp, W.F., et al. 1973. J. Nutr. 103:713.
101. Gipp, W.F., W.G. Pond and E.F. Walker. 1973. J. Animal Sci. 36:91.
102. Peden, J.C. 1967. In: Present Knowledge in Nutrition. The Nutr. Foundation, Inc., N.Y.

Manganese

103. Kemmerer, A.R., C.A. Elvehjem and E.B. Hart. 1931. J. Biol. Chem. 92:623.
104. Waddell, J., H. Steenbock and E.B. Hart. 1931. J. Nutr. 4:53.
105. Orent, E.R. and E.V. McCollum. 1931. J. Biol. Chem. 92:651.
106. Cotzias, G.C. 1962. In: Mineral Metabolsim. Vol. II, Part B. Academic Press, N.Y.
107. Mathers, J.W. and R. Hill. 1968. Br. J. Nutr. 22:635.
108. Borg, D.C. and ,G.C. Cotzias. 1958. J. Clin. Invest. 37:1269.
109. Leach, R.M. 1967. Fed. Proc. 26:118.
110. Leach, R.M., A.M. Muenster and E.M. Wien. 1969. Arch. Biochem. Biophys. 133:22.
110. Hurley, L.S., E. Wooten, G.J. Everson and C.W. Asling. 1960. J. Nutr. 71:15.
112. Lassiter, J.W., W.J. Miller, F.M. Pate and R.P. Gentry. 1972. Proc. Soc. Esp. Biol. Med. 139:345.
113. Smith, S.E., M. Medlicott and G.H. Ellis. 1944. Arch. Biochem. 4:281.
114. Erway, L., L.S. Hurley and A. Fraser. 1966. Sci. 152:1766.
115. R.E. Schrader and G.J. Everson. 1968. J. Nutr. 94:269.
116. Rubenstein, A.H., N.W. Levin and G.A. Elliot. 1962. Nature 194:188.
117. Kosenko, L.G. 1965. Fed. Proc. 24, Transl. Supp. T237.
118. Schroeder, H.A., J.J. Balassa and I.H. Tipton. 1966. J. Chronic Dis. 19:545.

Selenium

119. Schwarz, K. and C.M. Foltz. 1957. J. Amer. Chem. Soc. 79:3293.
120. Schwarz, K., J.G. Bieri, G.M. Briggs and M.L. Scott. 1957. Proc. Exp. Biol. Med. 95:621.
121. Patterson, E.L., R. Milstrey and E.L.R. Stokstad. 1957. Proc. Soc. Exp. Biol Med. 95:617.
122. Muth, O.H., J.E. Oldfield, L.F. Remmert and J.R. Schubert. 1958. Sci. 28:1090.
123. Schwarz, K. 1965. Fed. Proc. 24:58.
124. Scott, M.L. 1962. In: Mineral Metabolism. Vol. II, Part B. Academic Press, N.Y.
125. Muth, O.H. 1964. Symposium: Selenium in Biomedicine. AVI Publishing Co., Westport, Conn.
126. Rosenfeld, I. and O.A. Beath. 1964. Selenium. Academic Press, N.Y.
127. Ku, P.K., W.T. Ely, A.W. Groce and D.E. Ullrey. 1972. J. Animal Sci. 34:208; Groce, A.W., et al. 1973. J. Animal Sci. 37:948.
128. Thompson, J.N. and M.L. Scott. 1970. J. Nutr. 100:797.
129. Noguchi, T., A.H. Cantor and M.L. Scott. 1973. J. Nutr. 103:1502.
130. Rotruck, J.T., et al. 1973. Sci. 179:588.
131. Gries, C.L. and M.L. Scott. 1972. J. Nutr. 102:1287.
132. Wright, P.L. and M.C. Bell. 1966. Amer. J. Physiol. 211:6.
133. Ehlig, C.F., D.E. Hogue, W.H. Allaway and D.J. Hamm. 1967. J. Nutr. 92:121.
134. McConnell, K.P. and R.S. Levy. 1962. Nature 195:775.
135. McConnell, KP.P 1963. J. Agr. Food Chem. 11:385.
136. Anderson, H.D. and A.L. Moxon. 1941. J. Nutr. 22:219.
137. Lopez, P.C., R.L. Preston and W.H. Pfander. 1968. J. Nutr. 94:219.
138. Moxon, A.L. 1938. Sci. 88:81.
139. Levander, O.A. and C.A. Baumann. 1966. Toxicol. Applied Pharmacol. 9:98.
140. Palmer, I.S., D.D. Fischer, A.W. Halverson and O.F. Olson. 1969. Biochem. Biophys. Acta 77:336.
141. Scott, M.L. and A.H. Cantor. 1972. Proc. Cornell Nutr. Conf. p 66. Cornell University, Ithaca, N.Y.
142. Allaway, W.H., D.P. Moore, J.E. Oldfield and O.H. Muth. 1966. J. Nutr. 88:401.
143. Whanger, P.D., P.H. Weswig, O.H. Muth and J.E. Oldfield. 1969. J. Nutr. 99:331.
144. Pond, W.G., W.H. Allaway, E.F. Walker and L. Krook. 1971. J. Animal Sci. 33:996.
145. Scott, M.L. 1973. Proc. Cornell Nutr. Conf., p 123. Cornell University, Ithaca, N.Y.
146. McCay, K.E.M. and P.H. Weswig. 1967. J. Nutr. 98:383.
147. Moxon, A.L. 1937. So. Dak. Agr. Exp. Sta. Bul. 311.
148. Halverson, A.W. and K.J. Monty. 1960. J. Nutr. 70:100.
149. Comar, C.L. and F. Bronner. 1962-72. Mineral Metabolism. Vol. I, II, III. Academic Press, N.Y.

Chromium

150. Curran, G.L. 1954. J. Biol. Chem. 210:765.
151. Schwarz, K. and W. Mertz. 1959. Arch. Biochem. Biophys. 85:292.
152. Mertz, W., E.E. Roginski, F.J. Feldman and D.E. Thurman. 1969. J. Nutr. 99:363.
153. Mertz, W. 1969. Physiol. Rev. 49:163.

Fluoride

154. Dean, H.T. 1942. In: Fluorine and Dental Health. A.A.A.S. Symposium Proc., Washington, D.C.
155. Schwarz, K. and D.B. Milne. 1972. Bioinorganic Chem. 1:331.
156. Hodge, H.C. and F.A. Smith. 1965. In: Fluorine Chemistry. Vol. 4. Academic Press, N.Y.
157. Forsyth, D.M., W.G. Pond, R.H. Wasserman and L. Krook. 1972. J. Nutr. 102:1623.
158. Bernstein, D.S., et al. 1966. J. Amer. Med. Assoc. 198.85.
159. Henrikson, P.A., et al. 1970. J. Nutr. 100:631.

160. Forsyth, D.M., W.G. Pond and L. Krook. 1972. J. Nutr. 102:1639.
161. de Renzo, E.C., et al. 1953. Arch. Biochem. Biophys. 45:247.

Molybdenum

162. Rickert, D.A. and W.W. Westerfeld. 1953. J. Biol. Chem. 203:915.
163. Higgins, E.S., D.A. Rickert and W.W. Westferfeld. 1956. J. Nutr. 59:539.
164. Reid, B.L., A.A. Kurnich, R.L. Svacha and J.R. Couch. 1956. Proc. Soc. Exp. Biol. Med. 93:245.
165. Ellis, W.C., W.H. Pflander, M.E. Muhrer and E.E. Pickett. 1958. J. Animal Sci. 17:180.

Silicon

166. Carlisle, E.M. 1972. Science 178:619.
167. Schwarz, K. and D.B. Milne. 1972. Nature 239:333.
168. Carlisle, E.M. 1970. Science 167:279.
169. Schwarz, K. 1971. In: Newer Trace Elements in Nutrition. Dekker, N.Y.

Additional minerals

170. Schwarz, K. and D.B. Milne. 1971. Science 174:426.
171. Schwarz, K. 1970. In: Trace Element Metabolism in Animals. E.& S. Livingstone, Edinburgh, Scotland.
172. Mertz, W. and W.E. Cornatzer. 1971. In: Newer Trace Elements in Nutrition. Dekker, N.Y.
173. Millers, C.F. 1970. In: Trace Element Metabolism in Animals. E.& S. Livingstone, Edinburgh, Scotland.

Chapter 12 – Toxic Minerals

All minerals, indeed all nutrients, may be toxic to animals when ingested in excess amounts. The margin of safety between the minimum amount required in the diet and the amount that produces adverse effects varies among minerals and according to conditions. For example, NaCl may produce convulsions and death in pigs if fed at only 4-5X the required concentration in the diet when access to water is restricted, but the tolerance is much greater with adequate water intake. On the other hand, although the amount of Zn required by pigs is about 25-50 ppm in the diet, 20-40X this concentration is required to produce toxic signs.

Although many minerals may be shown to be toxic under experimental conditions, only a few are of considerable practical importance because of their wide distribution in the environment or their pronounced toxic properties. Schroeder et al (1-4) have discussed the concepts of mineral toxicity and suggested the possibility of toxicity to animals on a wide array of trace minerals, some of which are ubiquitous in the environment. Our discussion here is limited to some of the minerals that represent more pressing practical problems for animals and man and about which considerable knowledge is available. These minerals include Pb, Cd, Hg, F, and Mo. Other minerals that may be hazardous to health under some conditions are also discussed.

Lead

Lead (Pb) poisoning in man was recognized several hundred years ago and became a clinical problem in Western Europe in the 17th and 18th centuries where cooking utensils and other household articles contained high concentrations of Pb (5). In more recent times, Pb poisoning in children has become more common as a result of ingestion of high-Pb paint which flaked and peeled from walls and fixtures, and of using eating and drinking utensils of high-Pb clay (6, 7, 8). Pb toxicity is presently considered the most common cause of accidental death by poisoning in both man and animals worldwide (6, 9, 10); Pb toxicity may also occur in areas where contamination results from agricultural spray residues or near smelting plants, mines, and roadways, the latter as a result of tetraethyl lead from motor fuels.

Signs of Pb toxicity include pallor, lassitude, anorexia, and irritability in children and adults (11), but the most definitive diagnosis is elevated blood Pb levels and urinary excretion of amino-levulinic acid in man and animals (7). Pb toxicity affects several organs and tissues. Anemia is produced in Pb toxicity as a result of a decrease in survival time of red blood cells (excessive hemolysis) and a decrease in red cell formation from a block in heme synthesis (12). RBC from Pb-intoxicated animals show strippling and nucleation. In the kidney, pathological changes occur resulting in amino aciduria, glycosuria, and hyperphosphaturia (13, 14). Necrosis, hemorrhage, and ulceration of the stomach and small intestine occur in animals, and, in the nervous system, petechial hemorrhages and loss of myelin from nerve sheaths in the brain accompanied by cerebro-cortical softening after prolonged exposure have been reported (13, 15). Pb toxicity also affects the skeleton, causing osteoporosis in sheep (16) and reduced bone matrix formation and excess resorption of mineralized bone in children and rabbits (17). Enlarged joints of long bones are common in swine (18; Fig. 12-1) and horses. In pigs (18), a low-Ca diet (0.7%), in the presence of 1000 ppm of Pb, caused greater uptake of Pb by the tissues and more severe symptoms of toxicity than with a high-Ca diet (1.2%). In the liver, Pb toxicity causes petechial hemorrhages and necrosis in horses, dogs, and pigs (18, 19).

Figure 12-1. Lameness and flexed carpal joints in a pig fed 1000 ppm of lead. Courtesy of L. Krook, Cornell, University.

Pb may be absorbed into the body via the GIT, lungs and skin. Absorption from the GIT is low (20%) and does not seem to be affected by level of Pb in the diet (20,21). Retention of inhaled Pb in the body is 37-47% (22).

Continuous oral administration of Pb results in highest concentration in the skeleton (up to 90% of body burden), with smaller amounts in liver and kidney and still smaller concentrations in other tissues (5, 14, 15, 18). Tissue uptake of Pb is affected by dietary Ca and P level; high Ca-P depresses tissue Pb in pigs and rats (18, 22). Transport of Pb in blood is as an aggregate of Pb phosphate absorbed to the surface of red blood cells. Major routes of Pb excretion are the GIT and kidneys. That lost through the GIT is largely via the bile (2). Urinary Pb is in 2 forms, inorganic and organic, the latter being of greater importance in Pb intoxication (23). Urinary Pb output is proportional to blood Pb concentration.

Cadmium

Cadmium (Cd) is toxic to a wide range of animal life and has specific adverse effects on the testes and kidney (24-30). The suggestion by Schroeder (31) of a relationship between Cd and human hypertension has raised concern about the consequences of Cd contamination of the environment. The atmosphere around some cities of the USA contains a significant amount of Cd (as high as 6.2 mg Cd/m^3), and water may be an important dietary source (32, 38). Vegetables, fruits, and nuts are poor sources of Cd (31), but some seafoods and cereal grain byproducts exceed 1 ppm Cd (33).

Total Cd in tissues of adults in the USA has been estimated at 30 mg, with 10 mg in kidney and 4 mg in liver. Kubota et al (34) detected Cd in less than 1/2 of the human blood samples tested; the concentration of Cd in samples with detectable amounts was less than 1 mcg/100 ml in 83% of the cases. Cd concentration of testes, liver, spleen, kidneys, teeth, and hair of calves is increased by high levels of dietary Cd (35), but mammary transfer is low (36).

Specific adverse effects of Cd, when fed to experimental animals, include kidney damage and hypertension (37) and microcytic hypochromic anemia (25, 30, 35, 38, 39). The anemia and growth depression can be partially alleviated by concomitant oral or parenteral administration of addition Fe. Parisek (29) has shown that Cd injection into the testes of male rats causes tissue necrosis and sterility; in pregnant females, it causes destruction of the placenta and reproductive failure. The specific sensitivity of the pregnant females to Cd is apparently related to the presence of the placenta, for the removal of the fetuses or placenta or both prevents some of the symptoms observed in pregnant animals.

Cd toxicity is partially alleviated by high dietary Zn (35, 40, 41) and Fe (30), and a complex interaction is apparent between levels of dietary Cu, Fe, and Zn in protecting against Cd toxicity (39). A series of unrelated compounds protect the testes against injected Cd. These include Zn (42, 43), Se (43), cysteine, glutathione, and estrogen (44). The mode of protection is unknown, although it has been suggested (24) that Se may protect by forming relatively unstable Se salts of Cd which are stored harmlessly in the body.

At present, only limited information is available on the importance of environmental Cd contamination on human and animal health, but more complex knowledge of Cd metabolism and intoxication clearly should be a high priority goal for nutritionists and pathologists.

Mercury

The toxicity of mercury (Hg) has been recognized for many years; it is a hazardous environmental contaminant through its uses in industry and agriculture (24, 45). Its occurrence in nature and its toxic properties have been reviewed thoroughly (45, 46).

Hg combines preferentially with -SH groups and in this way inhibits enzyme systems containing such groups (47). Hg accumulates in the lysosomes within cells and has been associated with their rupture; this has been suggested (48) as the basis for the cell destruction by Hg, as destruction of lysosomes releases hydrolytic enzymes. Hg poisoning produces kidney necrosis and death; concomitant administration of Se protects against the necrosis and improves survival (49). The reason for this protection by Se is not clear, although it may be related to formation of relatively insoluble compounds of low toxicity, such as $HgSeO_3$, which are stored in the body. This would fit with the observation that Hg decreases the respiratory excretion of volatile Se compounds by rats (24).

Organic Hg compounds such as methyl Hg and phenyl Hg are more toxic than inorganic Hg compounds such as $HgCl_2$. All of these compounds cause severe growth depression of rats and liver and kidney accumulation of Hg (50).

Hg content of hair, fingernails, and teeth (51) gives a useful index of the degree of exposure of man and animals. Head hair and fingernails of control subjects vs. those of dental assistants were 5.5 and 7.3 ppm vs. 32.3 and 68.8 ppm, respectively, in two studies (52, 53). Kidneys consistently have a higher Hg concentration than all other tissues in humans (54). Urinary excretion is less than fecal excretion, although genetic differences have been observed among chickens (55). Some of the fecal Hg is a result of excretion through the bile.

Knowledge of the forms in which Hg is present in body tissues is meager. Living organisms can methylate Hg compounds (50); methyl Hg is retained in tissues longer than inorganic forms and appears to be more toxic (24, 50). The conversion of Hg compounds to methyl Hg by microorganisms has been suggested as a factor in the high concentration of Hg present in fish (56, 57). The high concentration of Hg in fish is apparently not a result of recent environmental contamination, as samples of fish stored for many years have also been shown to be high in Hg.

The practical dangers of Hg poisoning in man and animals appear to be the result of a few specific industrial practices and processes, and the awareness of these problems should minimize such problems in the future.

Fluorine

Aside from its role in preventing dental caries, fluorine (F) can be considered a toxic mineral for man and animals. Underwood (24) has described in detail the tissue distribution, metabolism, and toxic effects of F and has presented a concise historical summary of the early reports of F toxicosis. Chronic fluorosis was reported from several parts of the world in the 1930's, mostly as a result of F-containing dusts from steel mills and other industries processing F-rich substances (such as reduction of aluminum ore), and secondarily in association with the use of high-F rock phosphate as mineral supplements for animals. Stained teeth of fluorotic cattle are shown in Fig. 12-2. Toxic effects in humans exposed to F in air, water, or food are as follows (58): 1 ppm F in water reduces dental caries, but 2 ppm or more F induces mottled enamel, 8 ppm in water induces osteosclerosis in some subjects, and 110 ppm in food or water produces growth retardation and kidney changes; 20-80mg/day or more ingested from water or air produces crippling fluorosis, and 2.5 g or more in an acute oral dose is fatal. Detailed discussion of the effects of F on teeth is inappropriate here, but may be found in several reviews (58, 59, 60).

The effects of high F on bone are somewhat contradictory; reported effects of fluorosis in animals range from osteosclerosis to osteoporosis (60, 61, 62). Jowsey et al (63) reported increased bone formation, and Ramberg et al (64) reported increased bone resorption in response to F. Forsyth et al (65, 66) fed levels of up to 450 ppm of F in the presence of 1.2% or 0.5% Ca (Ca:P ratio kept constant at 1.2:1) to pigs through 2 generations; they found that high F did not increase bone density but did cause mottled bone (Fig. 12-3) and interfered with collagen metabolism (Fig. 12-4). The lower level of Ca-P was associated with greater F retention in bone, but there was no evidence of soft tissue lesions or of hyperostosis. Growth rate was depressed by high F, but high dietary Ca-P did not reduce growth. Newborn pigs from sows fed high F had decreased length, volume, and weight and increased F concentration of the humerus, demonstrating the effective placental transfer of F. Further increase occurred between birth and weaning as a result of increased F concentration in the milk of F-fed dams.

F is a potent enzyme inhibitor (60). Fatty acid oxidase activity in kidney is decreased in fluorosis (67), suggesting impaired lipid metabolism. Carbohydrate metabolism is also disturbed (68, 69). Marked species difference in the tolerance to F has been shown (24). The NRC (Phillips et al, 70) suggested that poultry and turkeys tolerate 150-400 ppm soluble F in the diet, but sheep and swine are less tolerant (70-100 ppm), and cattle still less tolerant (30-50 ppm). In general, all animals can tolerate slightly higher intakes of F from rock phosphate or phosphatic limestones (70). High dietary Ca (65, 66) or aluminum (71) or a high concentration of Ca in the water, when F is administered in the water (72), inhibit the toxic effects of F.

Almost complete absorption of soluble F from the GIT occurs. F tends to follow the distribution of Cl in the body; it freely crosses cell membranes and thus leaves the blood quickly after absorption. Some F enters red blood cells so that about 25% of total blood F is within the red cells (73), but the skeleton readily takes up F. Absorbed F that escapes retention in the skeleton is excreted mainly in the urine, but also to a smaller extent in sweat

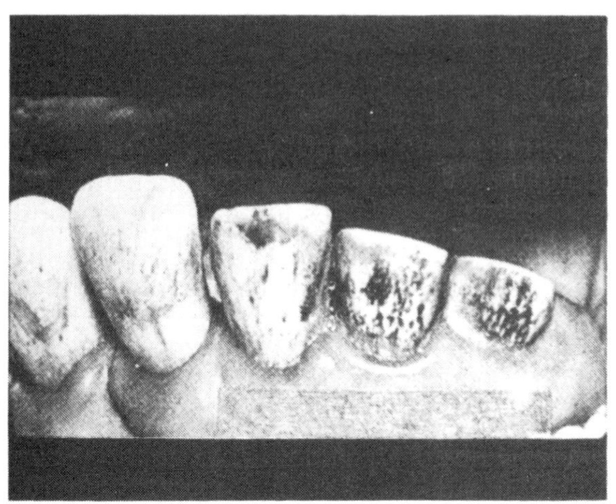

Figure 12-2. Stained teeth of fluorotic cattle. Vegetative staining is typical of excessive flourine intake at moderate levels. University of Tennessee photograph.

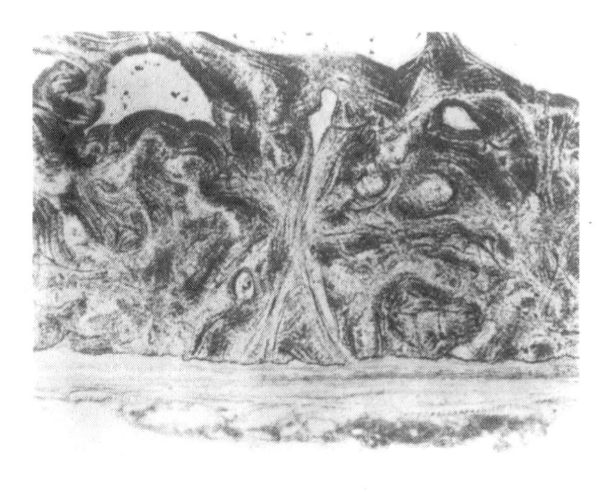

Figure 12-3. Normal bone of control sow [top] vs. mottled bone of sow fed diet containing 450 ppm fluoride [bottom]. From Forsyth et al [59].

and by re-excretion into the GIT (24). Urinary excretion in cattle and sheep normally represents 50-90% of the dietary intake, depending on the solubility and the proportion of F absorbed (74).

Molybdenum

Molybdenum (Mo) toxicity was first recognized in England when Ferguson et al (75) reported that cattle grazing certain pastures high in Mo developed severe diarrhea. The syndrome became known as peat scours or teart and was later found to be prevented by Cu supplementation of the diet (76). Dick (77) subsequently reported that high Mo and sulfate intake produces Cu deficiency in ruminants. Thus, the primary toxic effect of high Mo intake is the precipitation of Cu deficiency.

Striking species differences exist in susceptibility to high Mo intakes. Cattle appear to be the most susceptible, sheep less so, and horses and pigs still less. Mo has been fed at 1000 ppm in the diet of pigs for 3 mo. with no ill effects (78); levels of 50-100 ppm produce severe diarrhea in cattle. Poultry, guinea pigs, rabbits, and rats are between pigs and cattle in their tolerance to high Mo. Signs of toxicity, in addition to diarrhea in cattle and rats, include anemia, dermatosis, anorexia, deformed front legs in rabbits, and reduced lactation and testicular degeneration in several species (24).

Liver alkaline phosphatase activity is decreased in kidney and intestine of Mo-intoxicated rats, and liver sulfide oxidase activity is decreased (79, 80). This interference with S metabolism has been suggested (24) as being a factor in the mechanism of Mo toxicity. The observed relationship of dietary sulfate level to Mo tolerance fits this suggestion. Mo-Cu-sulfate interrelationships have been reviewed in detail by Underwood (24). The reduction in liver sulfide oxidase activity in Mo toxicity may lead to precipitation of insoluble cupric sulfide, rendering Cu physiologically unavailable (81).

Species differences in both absorption and excretion of Mo probably account for differences in tolerance to high dietary Mo. Hexavalent forms of Mo and that present in forages are water soluble and well absorbed by cattle; Mo is more rapidly absorbed by pigs than by cattle, but is quickly lost in the urine of pigs (24). Mo content of milk of ruminants is increased by high-Mo forage; milk Mo of ewes grazing forage containing 1 ppm Mo was 10 mcg/l. compared to 980 mcg/l. for milk of ewes grazing forage containing 3 ppm Mo (82). The Mo of milk is apparently bound entirely to xanthine oxidase, so that Mo concentration is proportional to enzyme activity (83).

The chemical form in which Mo exists in blood is unknown, although it is present both in red cells and in plasma. The proportion in red cells varies from 70% at low Mo intake to much lower values as dietary Mo increases. The absorption of Mo from the GIT is greatly reduced by the presence of high dietary sulfate; this in turn affects the concentration of Mo in blood and tissues (24).

The practical importance of Mo as a toxic mineral appears to relate mainly to cattle and sheep grazing pastures in areas of the world

Table 12-1. Toxic properties of various trace minerals having no known dietary requirement.[a]

Trace element	Concentration in drinking water mg/liter	Test species	Weight gain reduction	Life span reduction	Carcinogenic	Reference
Cadmium (Cd)	5	mice	no	yes	no	8,10
Cadmium	5	rats	no	yes	no	1,9,11
Lead (Pb)	5	mice	no	yes	no	8,10
Lead	5	rats	no	yes	no	1,9,11
Lead	25	rats	no	no	no	6
Selenate (Se^{+6})	.02	rats	no	no	yes	2,12
Selenite (Se^{+4})	.02	rats	yes	yes	no	2,12
Selenite	.02	mice	no	no	—	2
Tellurium (Te)	.02	rats	no	no	no	2,12
Tellurium	.02	mice	yes	yes	no	2
Titanium (Ti)	5	mice	no	yes	no	8,10
Arsenic (As)	5	mice	no	yes	no	3
Arsenic	5	rats	no	no	no	5
Germanium (Ge)	5	mice	no	yes	no	3
Germanium	5	rats	no	yes	no	5
Tin (Sn)	5	mice	no	no	no	3
Tin	5	rats	no	yes	no	3
Vanadium (V)	5	mice	no	no	no	3
Vanadium	5	rats	no	no	no	6
Zirconium (Zn)	5	mice	no	no	no	4
Zirconium	5	rats	no	no	no	6
Niobrium (Nb)	5	mice	yes	yes	no	4
Niobrium	5	rats	no	no	no	6
Antimony (Sb)	5	mice	yes	yes	no	4
Antimony	5	rats	no	yes	no	6
Fluorine (F)	10	mice	no	no	no	4
Nickel (Ni)	5	mice	no	no	no	10
Scandium (Sc)	5	mice	yes	no	no	7
Chromium (III) (Cr^{+6})	5	mice	no	no	no	10
Chromium (VI) (Cr^{+3})	5	mice	yes	no	no	7
Gallium (Ga)	5	mice	yes	yes	no	7
Ythrium (Y)	5	mice	yes	no	no	7
Rhodium (Rh)	5	mice	yes	no	yes	7
Palladium (Pd)	5	mice	yes	no	yes	7
Indium (Im)	5	mice	yes	no	no	7

[a] From Schroeder et al (1-4, 84). Se, V, F, Ni and perhaps others may be required under some conditions.

[b] Survival of males was decreased slightly.

[c] Tumorogenic (13).

with high soil Mo contents. Such areas, however, are widespread, and of significant worldwide concern; molybdenosis has been reported in cattle grazing in such widely separated geographic locations as England, New Zealand, Ireland, and the western USA.

Other Minerals

Schroeder et al (1-4, 84) and Browning (85) have characterized several trace minerals as to their innate toxic effects as judged by growth, life span, and carcinogenic activity in mice. The minerals were added to the drinking water continuously throughout life. Table 12-1 summarizes the results obtained in this extensive series of experiments. Based on records of weight gain, longevity, and incidence of malignant tumors in rats and mice, Schroeder has concluded that the following minerals appear to have innate toxicity: Cd, Pb, F, Se, Te, Ti, As, Ge, Sn, Nb, Sb, Ni, Sc, Cr^{+6}, Ga, Y, Rh, Pd, and In. In addition, Hg, V and Sr are known from other work to be toxic, and each of the required minerals is toxic when fed in excess of needs (see Ch. 10 and 11). Schroeder (1) has suggested that suppression of growth rate induced in mice by some of the toxic elements is in the following order: Ga>Y>Sc>In>Cr^{+6}>Pd>Rh. The following have been shown to be tumerogenic: Se, Ni, Pd, and Rh.

In addition to these, vanadium (V) has been shown to depress growth of chicks when fed at 10+ ppm (86) and Franke and Moxon (87) produced growth depression and diarrhea in rats fed 25 ppm. The latter authors estimated V to be more toxic than As, Mo, or Te and less toxic than Se when each mineral was fed at 25 ppm. Aluminum (Al), boron (B), and rubidium (Rb) have produced toxicity when consumed in excessive amounts (88, 89, 90).

Some of these toxic minerals as well as others have received attention as possible essential trace minerals for man and animals because of some favorable response to their addition to the diet (24, 91). V has been implicated in feathering in birds and cholesterol metabolism in animals. Ni activates several enzyme systems; Al may be involved in one or more enzyme systems; tin (Sn) and barium (Ba) may be required for growth of some species; bromine (Br) may replace part of the Cl requirement of chicks; and Rb and cesium (Cs) may replace part of the K requirement. None of these effects qualify any of these minerals, on the basis of present knowledge, as dietary essentials according to the criteria outlined by Cotzias (92).

Future research with the aid of more sensitive analytical methods and more carefully controlled nutritional and environmental variables may enable the reclassification of one or more of the ubiquitous and, in some instances, toxic minerals as essential nutrients.

Radionuclides in Nutrition

Considerable public and medical concern has been voiced about the effects of exposure to radioactive isotopes of minerals on human and animal health. This general subject is outside the realm of our interests, except for a consideration of the consequences of contamination of the food chain with dangerous radionuclides. For example, the presence of ^{131}I and ^{90}Sr in food products, especially milk, has led to an appreciable body of literature on the physiological effects (93). No attempt is made here to enumerate these effects in detail.

Chronic effects of low levels of radiation of mammals are increased incidence of leukemia, cancer, and genetic mutations (94). With respect to the rare earth minerals (95), although their distribution in nature is widespread, their significance in animal and human nutrition is in doubt.

The protective ability of animal organisms against uptake of ^{90}Sr and other radio nuclides, as well as potentially toxic rare earths, is important. Of course, radioactive isotopes as tracers in studying metabolism of minerals is an established practice in nutrition research in which a valuable tool is utilized. The hazards must be recognized, however. The preferential utilization of Ca over ^{90}Sr has been determined in many species (96). Hardy et al (97) compared the skeletal and soft-tissue burdens of Mg, Si, ^{90}Sr, Ba, and ^{226}Ra in pigs and sheep at various stages of growth in relation to Ca burden. Ca utilization by pigs was 47% and 17% by sheep, but utilization of all other minerals was considerably less.

A measure of the discrimination by the body against these minerals is the observed ratio (OR) which is an expression of the overall discrimination that is observed in the movement of 2 elements from a source into a biological system (98). Table 12-2 shows the OR of each of these nuclides as the nuclide: Ca ratio in the whole body divided by the nuclide: Ca ratio in the diet. The pig and sheep both clearly discriminated against stable Sr, ^{90}Sr, Ba and ^{226}Ra, as evidenced by the low OR values.

Table 12-2. Ratio of nuclide to Ca in whole body and in diet of pigs and sheep (observed ratio, OR)[a]

Nuclide/Ca	Nuclide/Ca in whole body/Nuclide:Ca in diet (OR)	
	Pig	Sheep
Sr/Ca	0.14	0.45
^{90}Sr/Ca	0.17	0.20
^{226}Ba/Ca	0.02	0.62
Ra/Ca	0.01	0.52

[a] Hardy et al (97)

The pig discriminated against Ba 7X more effectively than against Sr, and against ^{226}Ra 14X more than against Sr. In contrast, the sheep showed little difference in OR for Sr, Ba, and Ra. More ^{226}Ra and Ba occurred in soft tissue than in bone in the pig, but the sheep only 3-4% was found in soft tissue. Thus, species differences apparently occur not only in relative utilization of rare earths but in tissue distribution. Thus, with our present knowledge, we can not generalize about the response to be expected to contamination of diets of man and animals with potentially hazardous rare earths, either stable or radioactive isotopes.

Potential Pollution of the Environment by Minerals

Based on amounts of minerals present in the earth's surface and their use patterns in industry, Bowen (94) has divided mineral elements into 4 categories as follows with respect to pollution potential:

This classification is of only limited use in relation to nutrition and metabolism of man and animals because of the ever-changing nature of technology which results in variable exposure to specific minerals. Nevertheless, the concept of patterns of movement of toxic minerals in the environment in response to changing technology must be recognized by nutritionist and biologists concerned with controlling pollution while providing an adequate feed and food supply for animals and man.

Very high potential pollution		High potential pollution		Moderate potential pollution		Low potential pollution	
Ag	Pb	Ba	Mo	Al	Ge	Ga	Si
Au	Sb	Bi	P	As	K	I	Sr
Cd	Sn	Ca	Ti	B	L	La	Ta
Cr	Tl	Fe	U	Be	Na	Mg	Zr
Cu	Zn	Mn		Br	Ni	Nb	
Hg				Cl	Rb		
				Co	V		
				F	W		

References Cited

1. Schroeder, H.A. and M. Mitchener. 1971. J. Nutr. 101:1431.
2. Schroeder, H.A., J.J. Balassa and W.H. Vinton, Jr. 1964. J. Nutr. 83:239.
3. Schroeder, H.A. and J.J. Balassa. 1967. J. Nutr. 92:245.
4. Schroeder, H.A. et al. 1968. J. Nutr. 95:95.

Lead

5. Aub, J.C., L. Faihall, A.S. Minot and P. Reznikoff. 1962. Lead poisoning. Medicine Monography, Vol. 7, William and Wilkins, Baltimore, Md.
6. Reddick, L.P. 1971. Southern Med. J. 64:446.
7. Alpert, J.J. 1969. Pediatrics 44:291.
8. Lin-Fu, J.S. 1970. U.S. Public Health Serv. Pub. No. 2108:1.
9. Lin-Fu, J.S. 1971. U.S. Public Health Serv. Pub. No. 72-5105:1.
10. U.S. Dept. Health, Education and Welfare. 1973. DHEW Pub. No. (HSM) 73:10005. 101 pp.
11. Barltrop, D. 1969. J. Hosp. Med. 2(No. 9):1567.
12. Granick, S. and D. Mauzerall. 1960. In: Metabolic Pathways. Vol. II. Academic Press, Inc., N.Y.
13. Goyer, R.A. 1971. Current Top. in Pathol. 55:147.
14. Zook, B.C., J.L. Carpenter and E.P. Leeds. 1969. J. Amer. Vet. Med. Assoc. 155:1329.
15. Christian, R.G. and L. Tryphonas. 1971. Amer. J. Vet. Res. 32:203.
16. Clegg, F.G. and J.M. Rylands. 1966. J. Comp. Path. 76:22.
17. Hass, G.M., D.V.L. Browan, R. Eisentein and A. Hemmons. 1964. Amer. J. Path. 45:691.
18. Hsu, F. S-Y. 1974. Ph.D. Thesis, Cornell University, Ithaca, N.Y.
19. Link, R.P. and R.R. Pensinger. 1966. Amer. J. Vet. Res. 27:118.
20. Blaxter, K.L. 1950. J. Comp. Path. 60:140.
21. Kehoe, R.A. 1961. Publ. Health Hygiene. 24:101.
22. Sobel, A.E., H. Yaska, D.D. Peters and B. Kramer. 1940. J. Biol. Chem. 132:239.
23. Dinischiotu, G.T. et al. 1960. Br. J. Indust. Med. 17:141.

Cadmium

24. Underwood, E.J. 1971. Trace Elements in Human and Animal Nutrition. 3rd ed. Academic Press, N.Y.
25. Axelsson, B. and M. Piscator. 1966. Arch. Environ. Health. 12:374.
26. Jacobs, R.M., M.R.S. Fox and M.H. Aldridge. 1969. J. Nutr. 99:119.
27. Schroeder, H.A., J.J. Balassa and W.H. Vinton, Jr. 1964. J. Nutr. 83:239.
28. Schwartze, E.W. and C.L. Alsberg. 1923. J. Pharmacol. Exp. Therap. 21:1.
29. Parizek, J. 1960. J. Reprod. Fert. 1:294; 7:263; 9:111.
30. Pond, W.G., E.F. Walker, Jr. and D. Kirtland. 1973. J. Animal Sci. 36:1122.
31. Schroeder, H.A. 1965. J. Chronic Dis. 18:647.
32. Carrol, R.E. 1966. J. Amer. Med. Assoc. 198:177.
33. Schroeder, H.A., A.P. Nason, I.H. Tipton and J.J. Balassa. 1967. J. Chronic Dis. 20:179.
34. Kubota, J., V.A. Lazar and F.L. Losee. 1968. Arch. Environ. Health. 16:788.
35. Powell, G.W., W.J. Miller, J.D. Morton and C.M. Clifton. 1964. J. Nutr. 84:204.
36. Miller, W.J. et al. 1967. J. Dairy Sci. 50:1404.
37. Schroeder, H.A. et al. 1966. Arch. Environ. Health. 13:788.
38. Hill, C.H., G. Matrone, W.L. Payne and C.W. Barber. 1963. J. Nutr. 80:227.
39. Bunn, C.R. and G. Matrone. 1966. J. Nutr. 90:395.
40. Jacobs, R.M., M.R. Spiney-Fox and M.H. Aldridge. 1969. J. Nutr. 99:119.
41. Miller, G.W., W.J. Miller and D.M. Blackmon. 1967. J. Nutr. 93:203.
42. Gunn, S.A., T.C. Gould and W.A.D. Anderson. 1961. Arch. Path. 71:274.
43. Mason, K.E., J.O. Young and J.E. Brown. 1964. Anat. Res. 148:309.
44. Gunn, S.A., T.C. Gould and W.A.D. Anderson. 1965. Proc. Soc. Esp. Biol. Med. 119:901; 122:1036.

Mercury

45. Peakall, D.B. and R.J. Lovett. 1972. Biosci. 22:20.
46. Montague, P. and K. Montague. 1971. Mercury. The Gruin Co. Inc., N.Y.
47. Clarkson, T.W. 1968. J. Occupat. Med. 10:351.
48. Norseth, T. 1968. Biochem. Pharmacol. 17:581.
49. Parizek, J. and I. Ostadalova. 1967. Experientia. 23:142.
50. Johnson, S.L. and W.G. Pond. 1974. Nutr. Reports Internatl. 9:135.
51. Nixon, G.S., H. Smith and H.D. Livingston. 1967. In: Symposium on Nuclear Activation Techniques in the Life Sciences, I.A.E.A., Vienna, Austria.
52. Rodger, W.J. and H. Smith. 1967. Forensic Sci. Soc. J. 7:86.
53. Howie, R.A. and H. Smith. 1967. Forensic Sci. Soc. J. 7:90.
54. Joselow, M.M., L.J. Goldwater and S.B. Weinberg. 1967. Arch. Environ. Health. 15:64.
55. Miller, V.L. et al. 1967. Poult. Sci. 46:142.

56. Jensen, S. and A. Jernelov. 1969. Nature. 223:753.
57. Westoo, G. 1966. Acta Chem. Scand. 20:2131.

Fluorine

58. Bhussry, B.R. et al. 1970. In: Fluorine and Human Health. World Health Organization, Geneva, Switzerland.
59. Adler, P. 1970. In: Fluorine and Human Health. World Health Organization, Geneva, Switzerland.
60. Hodge, H.C. and F.A. Smith. 1965. Fluorine Chemistry, Vol. IV. Academic Press, Inc., N.Y.
61. Kick, C.H. et al. 1935. Ohio Agric. Exp. Sta. Bul. 558.
62. Roholm, K. 1937. Fluorine Intoxication. H.K. Lewis, London.
63. Jowsey, J.R., R.F. Schenk and F.W. Reutter. 1968. J. Clin. Endocrin. Metab. 28:869.
64. Ramber, C.F., Jr., J.M. Phang, G.P. Mayer. 1970. J. Nutr. 100:981.
65. Forsyth, D.M., W.G. Pond, R.H. Wasserman and L. Krook. 1972. J. Nutr. 102:1623.
66. Forsyth, D.M., W.G. Pond and L. Krook. 1972. J. Nutr. 102:1639.
67. Sievert, A.H. and P.H. Phillips. 1960. J. Nutr. 72:429.
68. Carlson, J.R. and J.W. Suttie. 1966. Amer. J. Physiol. 210:79.
69. Zebrowski, E.J. and J.W. Suttie. 1966. J. Nutr. 88:267.
70. Phillips, P.H. et al. 1960. Natl. Acad. Sci.-Natl. Res. Council Pub. 824.
71. Becker, D.E., J.M. Griffith, C.S. Hobbs and W.M. McIntyre. 1950. J. Animal Sci. 9:647.
72. Weddle, D.A. and J.C. Muhler. 1954. J. Nutr. 54:437.
73. Bell, M.C., G.M. Merriman and D.A. Greenwood. 1970. J. Nutr. 73:379.
74. Hobbs, C.S. et al. 1954. Tenn. Agr. Exp. Sta. Bul. 235.

Molybdenum

75. Ferguson, W.S., A.H. Lewis and S.J. Watson. 1938. Nature 141:553.
76. Dick, A.T. and L.B. Bull. 1945. Austral. Vet. J. 21:70.
77. Dick, A.T. 1953. Austral. Vet. J. 29:233; 1954. Austral. Vet. J. 30:196.
78. Davis, G.K. 1950. In: Symposium on Copper Metabolism. Johns Hopkins Press, Baltimore.
79. Van Reen, R. 1954. Arch. Biochem. Biophys. 53:77.
80. Williams, M.A. and R. Van Reen. 1956. Proc. Soc. Exp. Biol. Med. 91:638.
81. Siegel, L.M. and K.J. Monty. 1961. J. Nutr. 74:167.
82. Hogan, K.G. and A.J. Hutchinson. 1965. New Zealand J. Agr. Res. 8:625.
83. Hart, L.I., E.C. Owen and R. Proudfoot. 1967. Br. J. Nutr. 21:617.

Other minerals

84. Schroeder, H.A. et al. 1963. J. Nutr. 80:39; 1965. J. Nutr. 86:51; 86:31; 1967. J. Nutr. 92:334; 1968. J. Nutr. 96:37; 1971. J. Nutr. 101:1531.
85. Browning, E. 1961. Toxicity of Industrial Metals. Butterworth and Co., Ltd., London.
86. Berg, L.R. 1963. Poult. Sci. 42:766.
87. Frankie, K.W. and A.L. Moxon. 1937. J. Pharmacol. Exp. Therap. 61:89.
88. Deobold, H.J. and C.A. Elvehjem. 1958. Amer. J. Physiol. 111:118.
89. Pfeiffer, C.C., L.F. Hallman and I. Gersh. 1945. J. Amer. Med. Assoc. 128:266.
90. Glendenning, B.L., W.G. Schrenk and D.B. Parrish. 1956. J. Nutr. 60:563.
91. Schwarz, K. 1972. In: Nuclear activation techniques in the life sciences. Internatl. Atomic Energy Agency, Vienna.
92. Cotzias, G.C. 1967. Proc. First Annual. Conf. Trace Substances in Environ. Health. Univ. Missouri, Columbia, Mo.

Radionuclides in Nutrition

93. Comar, C.L. 1966. In: Radioactivity and Human Diet. Pergamon Press, Oxford, England.
94. Bowen, H.J.M. 1966. Trace Elements in Biochemistry. Academic Press, Inc., N.Y.
95. Kyker, G.C. 1962. In: Mineral Metabolism. Vol. II, Part B. Academic Press, Inc., N.Y.
96. Comar, C.L. and R.H. Wasserman. 1960. In: Radioisotopes in the Biosphere Center for Continuation Study, University of Minnesota, Minneapolis.
97. Hardy, E. J. Rivera, I. Fisenore, W. Pond and D.E. Hogue. 1970. Health and Safety lab., U.S. Atomic Energy Commission Symp., pp 183-190, N.Y.
98. Comar, C.L., R.H. Wasserman and M.M. Nold. 1956. Proc. Soc. Exp. Biol. Med. 92:859.

Chapter 13 – Fat-Soluble Vitamins

The term vitamin was coined in 1912 by Funk (1) who suggested that food contained special organic constituents that prevented certain of the classical human diseases of that time — beriberi, pellagra, rickets, scurvy. Since that time a long list of vitamins has been discovered and isolated.

Vitamins are required in minute quantities for normal body function, yet each has its own specific function and the omission of a single vitamin from the diet of a species that requires it produces specific deficiency symptoms and ultimately results in death of the animal. Although many of the vitamins function as coenzymes (metabolic catalysts), others have no such role but perform other essential functions. The known vitamins can be divided on the basis of solubility properties into fat-soluble and water-soluble. The fat-soluble vitamins — A, D, E, and K — are considered in this chapter and the water-soluble vitamins are considered in Ch. 14.

Vitamin A

Structure

Vitamin A is required in the diet of all animals thus far studied. It can be provided as the vitamin or as its precursor, carotene. The nomenclature and formulas for vitamin A and many different carotenes are described in detail by Harris (2).

The structures of vitamin A alcohol and β-carotene are shown. Vitamin A is composed of a β-ionone ring and an unsaturated side chain. β-carotene is composed of 2 vitamin A molecules joined as shown.

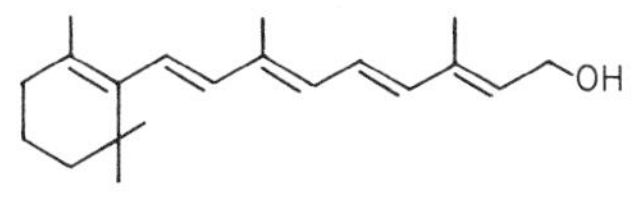

OH

Vitamin A Alcohol (all *trans*)

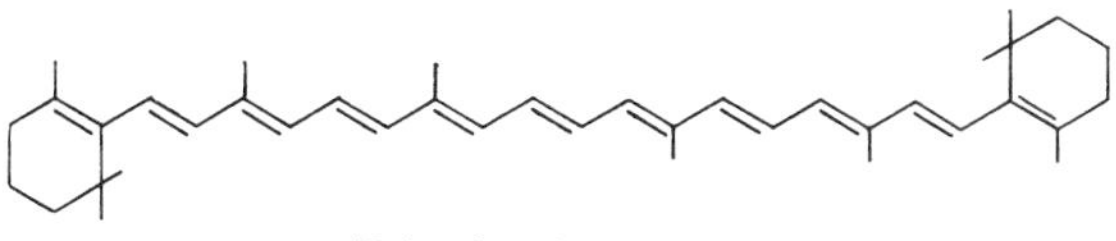

Beta - Carotene

Vitamin A can occur as vitamin A alcohol (retinol), as shown, as the aldehyde (retinal), or as the acid (retinoic acid) in the free form or esterfied with a fatty acid (for example, as vitamin A palmitate). It can occur as the all *trans* form (as shown), all *cis* form or as a mixture of *cis* and *trans* forms. The *trans* form of retinol is considered to have 100% biological activity. Biological activities of 2 isomers of retinal, expressed as a % of potency of all *trans* retinol are (8); all *trans*, 91; and 2-mono *cis* (neo), 93.

A great number of carotenoid pigments exist in nature in addition to β-carotene, but plants contain essentially no vitamin A. These carotenoids differ from each other in the configuration of the ring portion of the molecule. They include α- and γ-carotene, cryptoxanthin, zeaxanthin, and xanthophyll. The vitamin A precursors must be split to release biologically available vitamin A. Zeaxanthin and xanthophyll possess no vitamin A activity but others have some activity, the amount depending on the animal species. For the rat (34) the relative biopotency of retinol is 100; β-carotene, 50; α-carotene, 25; γ-carotene, 14; and cryptoxanthin, 29.

The conjugated double bonds in vitamin A and carotene cause a characteristic yellow color. Exposure to ultraviolet light destroys the double bonds and the biological activity of vitamin A and its precursors. Some enzymes present in natural feedstuffs destroy carotenoids. Esterified vitamin A is more stable than retinol or retinal. The stability of vitamin A added to feeds can be increased by covering minute droplets of vitamin A with gelatin or wax or by adding an antioxidant such as ethoxyquin to the feed; the antioxidant is oxidized in preference to vitamin A. In current nutrition practice in the USA, most vitamin A is supplied by synthetic sources which can be produced very cheaply.

Functions

At least 3 distinct functions of vitamin A have been identified (28, 29). Vitamin A is required (as retinol) for normal night vision (formation of rhodopsin or visual purple in the eye); this is the only specific chemical reaction that has been identified. Vitamin A combines as a prosthetic group of rhodopsin which breaks down on exposure to light. This reaction is part of the physiological process of sight as described by Wald (37). Vitamin A is required for normal epithelial cells which line or cover body surfaces or cavities — respiratory, urogenital and digestive tracts, and skin; and it is required for normal bone growth and remodeling (normal osteoblastic activity).

Deficiency Signs

The practical significance of hypovitaminosis A (vitamin A deficiency) is perhaps greater than for any other vitamin (12). Because one function of vitamin A is to allow formation of rhodopsin in the eye, night blindness is a symptom of vitamin A deficiency in all animals studied. The degree of failure in dark adaptation has been used as a measure of the quantitative needs in man and some animals.

The essentiality of vitamin A for normal epithelium creates a wide variety of deficiency symptoms in animals deprived of the vitamin (Fig. 13-1). Some of the common deficiency signs in various animals are: xerophthalmia in children and in growing animals (this condition is characterized by dryness and irritation of the cornea and conjunctiva of the eye and results in cloudiness and infection); keratinization of respiratory epithelium, resulting in greater severity of respiratory infections; reproduction difficulties, including abortions and birth of weak offspring, and associated thickening of vaginal epithelium; reproductive failure in males because of effects on spermatogenic epithelium; embryonic death in chicks and mammalian embryos; poor growth in surviving young (Fig. 13-2) and uric acid deposits in kidneys, heart, liver, and spleen and keratinixation of epithelium of respiratory tract of chicks.

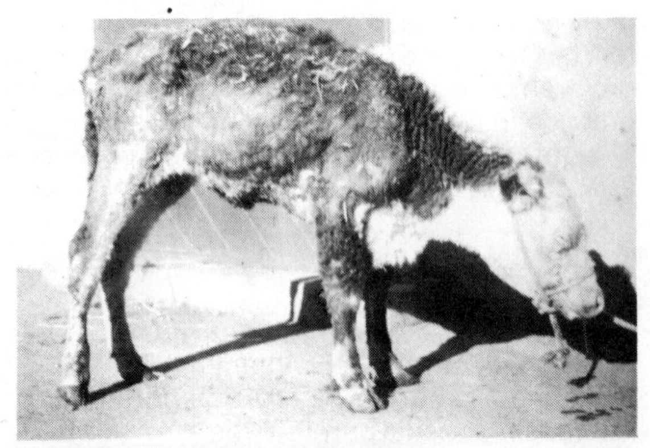

Figure 13-1. Severe vitamin A deficiency in a calf. Animal shows excessive lacrimation, a lethargic appearance, and appears to have diarrhea and some respiratory involvement. Courtesy of Chas. Pfizer Co.

Figure 13-2. Vitamin A-deficient chick showing xerophthalmia. Courtesy of Poultry Sci. Dept., Cornell U.

The importance of vitamin A in normal bone formation relates to a variety of deficiency signs which, although seemingly unrelated, have a common basis involving abnormal skeletal development. Gallina et al (15) provided evidence in calves that vitamin A deficiency produces increased osteoblastic activity. This agrees with earlier reports in other species. Nerve disorders such as unsteady gait, ataxia, and convulsions occur as a result of partial occlusion of the spinal cord by the surrounding vertebral column in growing animals. Exophthalmia (Fig. 13-3) and elevated cerebrospinal fluid pressure exists in vitamin A deficiency, apparently as a result of excess pressure on the brain stem associated with a constricted spinal column and optic foramen. Blindness and skeletal abnormalities occur in deficient newborn pigs. A wide range of additional manifestations (Table 13-1) of vitamin A deficiency have been reported, but there is no definitive knowledge as to the exact mode of action on which the metabolic changes are based. Bone changes occurring in vitamin A deficiency are associated with changes in chondroitin sulfate, mucopolysaccharide synthesis, and increased urinary excretion of inorganic sulfate in rats. Vitamin A metabolism has been linked with (5, 6) vitamin E (as an antioxidant in the stability of biological membranes); vitamin D (in bone metabolism); sterols (deficiency reduces cholesterol synthesis); squalene (deficiency increases squalene synthesis); and coenzyme Q or ubiquinone (deficiency increases ubiquinone synthesis in liver). Vitamin A deficiency in rats causes adrenal gland atrophy and reduced gluconeogenesis. Vitamin A is somehow involved with biosynthesis of adrenal steroids and of glycogen. A recent report suggested that vitamin A deficiency may be related to kidney stone formation in rats (40), based on the observed reduction in urinary Ca excretion.

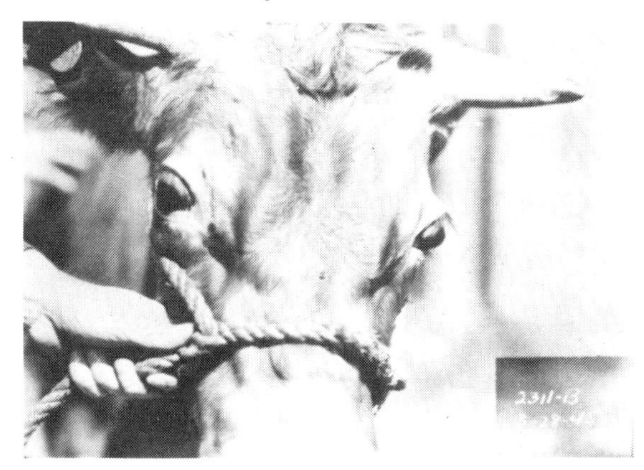

Figure 13-3. Heifer showing exopthalmia or bulging eye condition due to high cerebrospinal fluid pressure and partial closure of the optic foramen. Courtesy of L.A. Moore, U.S.D.A.

Metabolism

Vitamin A from synthetic sources or animal tissues in feedstuffs is present primarily as the palmitate ester which is hydrolyzed by pancreatic enzymes which are activated by bile salts. Free vitamin A is incorporated into the lipid micelle (see Ch. 8) and reaches the microvilli of the upper jejunum where it is transferred into the mucosal cell by active transport as retinol. Within the mucosal cell it is re-esterified to palmitate and other esters, incorporated into chylomicrons, and transported to the lymph (5, 16, 22, 27, 34, 36, 54, 62, 79, 89).

Carotenoids (19, 20) are split within the rat intestinal mucosal cell by a specific enzyme (19, 20) to form retinal which is reduced to retinol. Some retinol is converted to retinal and retinoic acid and absorbed into the portal blood as a glucuronide (11, 26). Tissues other than intestinal mucosal cells are capable of splitting carotenoids to vitamin A. Liver contains an enzyme with the same properties as the β-carotene-splitting enzyme of the intestine (31) and lung and kidney may also be involved (36). The details of absorption in other species are less well known, and species differences seem certain, as some animals deposit very little carotene in depot fat even though their diets are high in carotenoids, but others have appreciable amounts of carotene in their depot fat and milk. Rats, cats, dogs, sheep, goats, and guinea pigs apparently convert most or all carotene to vitamin A, but cattle, (especially Guernseys), horses, some rabbits, chickens, and man have blood and depot fats high in carotenoids if dietary carotenoids are high. In fact, chickens absorb only hydroxycarotenoids unchanged and deposit them in tissues. A wide variation in efficiency exists even between species considered to be efficient converters of carotenoids to vitamin A. For example, the pig is only 1/3 as efficient as the rat in converting β-carotene to vitamin A (10). Some recycling of vitamin A by enterohepatic circulation occurs, but this is probably not a major conservation mechanism.

The degree of absorption of vitamin A and its precursors varies, depending on the animal species and type of diet. Apparent absorbability of vitamin A in dairy cattle fed a variety of forages averaged 78% in one report (39). Evidence in the literature, however, shows considerable degradation in the rumen. In man, vitamin A absorption and storage may be improved by adding fat to a low-fat diet.

Transport of vitamin A and its precursors from the GIT to the liver is in the lipoprotein form, as chylomicrons. The distribution of vitamin A, its esters and carotenoids in human blood serum during absorption from the GIT is summarized in Table 13-2. The liver hydrolyzes the retinyl ester and free retinol is carried by the blood as a protein complex to tissues requiring it for normal metabolism. In vitamin A deprivation, liver stores are depleted in an effort to maintain blood concentration of vitamin A within normal limits. For this reason deficiency signs may not occur for several months after a vitamin A-low diet is instituted. Liver concentrations, therefore, are of more diagnostic value in assessing vitamin A status than are blood concentrations.

Liver biopsy techniques have been developed for use in cattle as a means of studying vitamin A requirements (Fig. 13-4). The turnover rate of vitamin A in liver of beef cattle has been estimated at 132 days (21). Protein level of the diet did not significantly affect turnover rate. Kohlmeier and Burroughs (24) suggested that cattle entering finishing lots with 20-40 μg/g of vitamin A in the liver have

Table 13-1. Vitamin A deficiency signs in animals.[a]

Abnormality	Animals Studied	Abnormality	Animals Studied
General			
Anorexia	Rat, fowl, farm animals	Nervous system incoordination	Rat, bovine, pig
Growth failure and weight loss	Rat, fowl, farm animals	Paresis	Rat, pig
Xerosis of membranes	Rat, fowl	Nerve degeneration or twisting	Rat, dog, rabbit, bovine, bird, pig
Roughened hair or feathers	Rat, birds, farm animals	Constriction of optic foramina	Bovine, dog
Infections	Rat, birds, farm animals	Bone formation	
Death	Rat, birds, farm animals	Defective modeling	Dog, bovine
Eyes		Restriction of brain cavity	Dog
Night blindness	Rat, farm animals	Reproduction	
Xerophthalmia	Rat, bovine	Degeneration of testes	Rat
Keratomalacia	Rat		
Opacity of cornea	Rat, bovine	Abnormal estrus cycle	Rat, bovine
Loss of lens	Rat, bovine		
Papilloidema	Bovine	Resorption of fetuses	Rat
Constriction of optic nerve	Bovine, dog		
Respiratory system		Congenital abnormalities	
Metaplasia of nasal passages	Fowl	Anophthalmia	Pig, rat
Pneumonia	Rat, bovine	Microophthalmia	Pig, rat
Lung abscesses	Rat	Cleft palate	Pig, rat
GIT		Aortic arch deformation	Rat
Metaplasia of forestomach	Rat		
Enteritis	Rat, farm animals	Kidney deformities	Rat
Urinary system		Hydrocephalus	Rabbit, bovine
Thickened bladder wall	Rat	Miscellaneous	
Cystitis	Rat	Increased cerebrospinal fluid pressure	Bovine, pig
Urolithiasis	Rat		
Nephrosis	Rat		
Liver		Cystic pituitary	Bovine
Metaplasia of bile ducts	Rat		
Degeneration of Kupffer cells	Rat		

[a] Adapted from Roels (5).

sufficient reserves for 90-120 days under normal feeding conditions and that no dietary vitamin A is required for good feedlot performance if plasma vitamin A is maintained above 25 μg/ml and liver vitamin A exceeds 2 μg/g. Such factors as vitamin A destruction in feedstuffs and initial stores of liver vitamin A will, of course, affect the level of dietary vitamin A needed to maintain these minimum tissue levels. The data apply here to cattle, but the same type of consideration must be applied to other species when determining appropriate levels of supplementation.

Dietary protein intake affects vitamin A utilization. Protein deficiency causes reduced vitamin A concentrations in plasma and reduced liver storage. Signs of vitamin A deficiency may appear in protein-deficient animals even in the presence of adequate liver vitamin A storage. This has been suggested to

Table 13-2. Distribution of retinol, vitamin A esters, and carotenoids in human blood serum during absorption. [a]

Serum fraction	Retinol %	Vitamin A ester %	Carotenoids %
Chylomicrons	5.3	7.5	0
Lipoprotein Sf 10-100	3.9	79.4	0
Lipoprotein Sf 3-9	20.2	8.6	78.3
Other proteins	70.6	4.4	21.7

[a] From Krinsky et al (25)

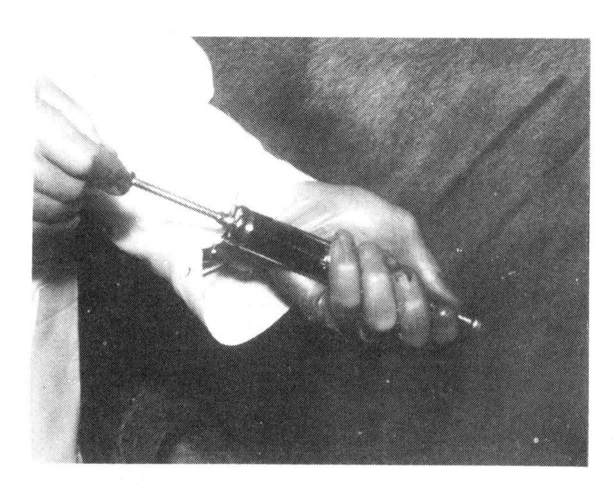

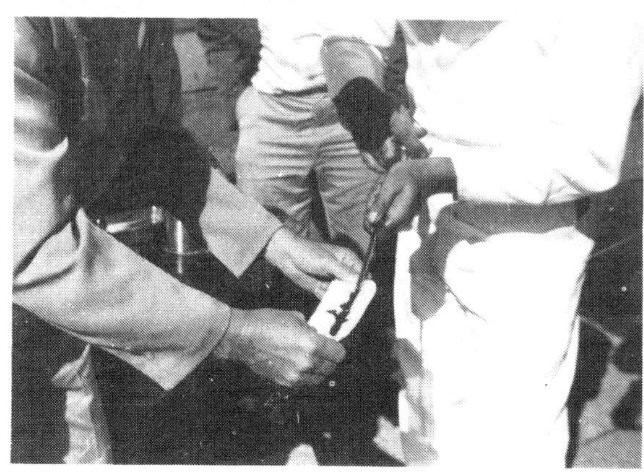

Figure 13-4. Liver biopsy. Top: Liver sample being aspirated with a syringe. Bottom: Sample of liver obtained. Courtesy of J.F. Bone, Oregon State U.

result from reduced transport of vitamin A from liver because of reduced serum albumin, the carrier protein for vitamin A in blood (14). Impaired conversion of carotene to vitamin A in protein deficiency may also occur, but the dominant factor appears to be a defect in transport (5, 6).

Thyroid function is affected by vitamin A intake (13). Vitamin A deficiency reduces thyroxin secretion and causes thyroid hyperplasia. Conversely, thyroxin stimulates conversion of carotenoids to vitamin A and increases storage of vitamin A, but also increases depletion of vitamin A reserves when a vitamin A-deficient diet is fed (5, 6).

Vitamin A is readily transported by the mammary gland to the milk of swine (30, 35) and goats (35). Cattle transfer both vitamin A and carotene to the milk in response to dietary intake, the proportions depending partly on breed. Vitamin A concentration of human milk tends to be related to maternal intake of the vitamin (33).

Toxicity

Vitamin A is not easily excreted, so long-term ingestion of larger amounts than needed or acute dosage with a large excess may result in toxic symptoms. The toxic range is reached when daily intake reaches 50-500 fold the metabolic requirement. Death has been reported in humans after a single dose of 500,000-1,000,000 IU of vitamin A (34). Chronic toxicity manifests itself as anorexia, weight loss, thickening skin, scaly dermatitis, swelling and crusting of the eyelids, patchy hair loss, hemorrhaging, decreased bone strength, spontaneous fractures, thinning of the bone cortex, and death. Excess mucous forms and normal keratinization is inhibited in hypervitaminosis A (17). The bone changes described in young pigs fed excess vitamin A have been attributed to destruction of epiphyseal cartilage and decreased matrix formation in the presence of normal remodeling (38). Excess vitamin A apparently causes hemolysis of red blood cells and may cause disruption of lipoprotein membranes (7).

The dietary levels causing toxic symptoms will, of course, vary according to species, age, body store, degree of absorbability, and degree of conversion of carotene to vitamin A where the free vitamin is not fed. In pigs, toxic symptoms include rough hair coat; hyperirritability and sensitivity to touch; petechial hemorrhages over the legs and abdomen; cracked bleeding skin above the hooves; blood in urine and feces; loss of control of legs; periodic tremors; and death (9).

Excess vitamin A during pregnancy results in malformed young in rats, mice, and pigs, but less so for guinea pigs and rabbits (18, 32). During early gestation an excess induces embryonic death, but, if begun later in gestation, abnormalities may occur, the severity and type differing according to species. In pigs, a single excess dose of vitamin A injected on day 18 or 19 of pregnancy causes cleft palate, abnormal skulls and skeleton, and, sometimes eyelessness in the newborn, but injection of the same dose before or after this stage of development has no such effect. Clearly, the effect is related to the relative rate of growth and differentiation of a particular organ at the time the excess vitamin A is given.

Vitamin D

Structure

Several sterols have biological vitamin D activity, but only 2, vitamin D_2 (irradiated ergostrol or calciferol) and vitamin D_3 (irradiated 7-dehydrocholesterol) are of major importance. Most mammals can use either vitamin D_2 or D_3 efficiently, but birds require vitamin D_3. The structure of vitamin D_2 and its precursor, ergosterol, are shown:

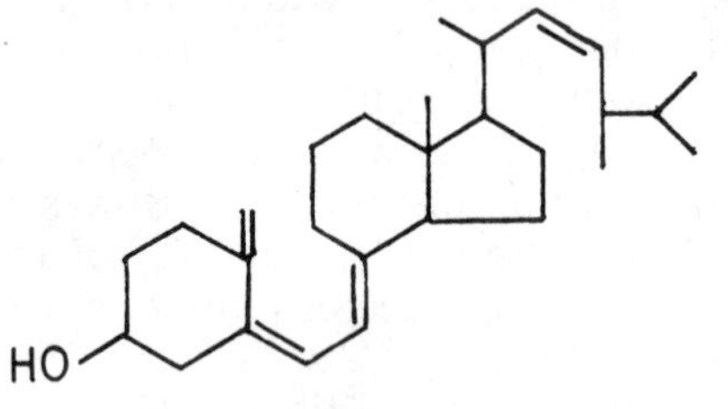

HO

Vitamin D_2
(Calciferol or Ergocalciferol)

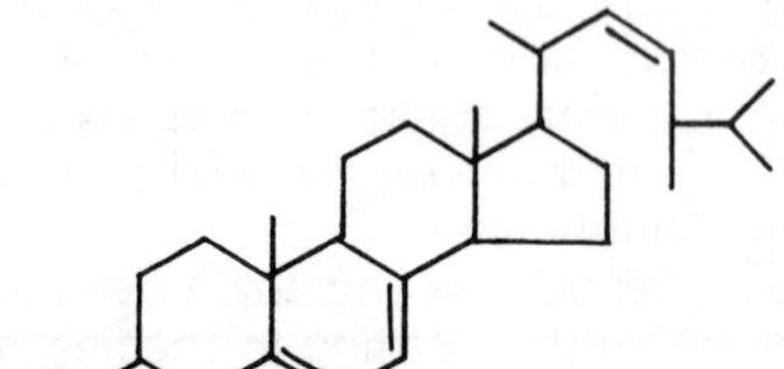

HO

Ergosterol

Ergosterol is the chief plant source and dehydrocholesterol is found in animal tissues. Ultraviolet light converts each provitamin to its respective biologically active form. Exposure of harvested green forage to sunlight for several hours converts enough plant sterols to vitamin D_2 to provide a good source of the vitamin for animals. Similarly, exposure of animals to sunlight for a few minutes/day is sufficient to convert skin sterols to vitamin D_3 which eliminates the need for a dietary source.

Functions

Vitamin D was named by McCollum who had shown earlier (53) that it differed from vitamin A. Its ability to prevent rickets in growing animals resulted in its first being called the antirichitic factor. At least 2 mechanisms exist by which it prevents rickets: (a) it increases absorption of Ca from the intestinal tract; this has been shown to result from a vitamin D-dependent Ca-binding protein in the intestinal tract (57, 59); (b) a metabolite, 1,25-dihydroxycholecalciferol, is responsible for mobilization of Ca from bone and stimulates Ca absorption from the GIT (44).

Metabolism

Although the importance of vitamin D in normal Ca and P metabolism has been recognized for many years, during the past decade outstanding work by 2 groups, Wasserman and associates at Cornell University and DeLuca and associates at University of Wisconsin, has provided a much more complete understanding of the mechanisms. Wasserman et al (57, 59) have isolated a Ca-binding protein from intestinal mucosa of several species of animals (birds, rats, dogs, cattle, pigs, monkeys, guinea pigs) which requires vitamin D. Many of the properties of this protein have been determined. A deficiency of dietary Ca increases the Ca-binding protein activity, and a high level of dietary Ca supresses it.

Vitamin D is necessary for formation of this Ca-binding protein, so that Ca absorption is controlled by dietary vitamin D in relation to the amount of Ca-binding protein activity produced in response to vitamin D. Such a vitamin D-dependent protein may be involved in transport of Ca across other tissue membranes in the body, but this is not known at present.

The elegant work of DeLuca and associates was recently summarized (41). Lund and DeLuca (51) demonstrated that more biologically active metabolites of vitamin D exist than vitamin D itself. One of these (25-hydroxycholecalciferol or 25-OH-D_3) was isolated in 1968 (42). This metabolite is formed in the liver from vitamin D_3 by a hydroxylation reaction. The amount of 25-OH-D_3 formed is apparently controlled by a feedback mechanism in the liver itself so than, in normal animals, only the amount of 25-OH-D_3 required is formed and released into the blood. In this way extra vitamin D_3 ingested or formed by solar irradiation of the skin can be conserved for use during times of inadequate supply.

The biological activity of 25-OH-D_3 is 1½-5X greater than that of vitamin D in prevention or treatment of rickets. Moreover, it acts more quickly than vitamin D in stimulating intestinal Ca absorption and bone mobilization. Oral administration of 25-OH-D_3 daily at 1 mg/animal appears to reduce the incidence of milk fever, a disease of Ca metabolism, in dairy cattle. Vitamin D has also been used in this way, but is less effective.

The 25-OH-D_3 does not act directly on the target tissues, but must be converted to another metabolite, 1,25-dihydroxycholecalciferol or 1,25-$(OH)_2D_3$, by the kidney. This compound was isolated from intestinal tissue by Holick et al (48, 49), but has been shown (44-46) to be synthesized in the kidney from 25-OH-D_3; 1,25-$(OH)_2D_3$ is about 3.6X more active than 25-OH-D_3 (about 5.5X more active than D_3). The structures of 25-OH-D_3 and 1,25-$(OH)_2D_3$ are shown below.

High dietary Ca decreases the production of 1,25-$(OH)_2D_3$ by the kidney, and low dietary Ca stimulates it. Regulation of 1,25-$(OH)_2D_3$ synthesis is also related to serum Ca concentration, through the action of parathyroid hormone (PTH) which catalyzes the conversion of 25-OH-D_3 to 1,25-$(OH)_2D_3$. Thus, 1,25-$(OH)_2D_3$ appears to be the ideal compound for treatment of diseases related to parathyroid insufficiency. Various analogs of this very expensively isolated and purified compound are presently being tested in several laboratories for their biological activity for use in parathyroid therapy.

Nephrectomized animals cannot make 1,25-$(OH)_2D_3$ and, therefore, do not respond to physiological doses of vitamin D in terms of increased Ca absorption or serum Ca concentration. Administration of 1,25-$(OH)_2D_3$ to nephrectomized animals, however, induces the same response as in normal animals, suggesting that this compound does not have to be further metabolized for metabolic activity (47, 55). The specific biochemical events that occur after localization of 1,25-$(OH)_2D_3$ in the target tissue to produce the physiological response still remain to be determined. The overall mechanism of vitamin D movement in the body and the functions of its metabolites are schematically diagrammed in Fig. 13-5.

Vitamin D is stored mainly in the liver, but also in kidney and lung and perhaps other tissues. Placental and mammary transfer are limited, when compared with vitamin A, but sufficiently high levels are present in the newborn and milk of most species to prevent early rickets. Only limited data are available on quantitative transfer of vitamin D across placental and mammary tissue, probably because the only reliable assay of the vitamin to date is the biological rat "line test." This is an expensive, slow procedure involving measurement of the Ca and P deposition in long bones or rachitic rats fed test diets. A relationship between maternal diet vitamin D content and the amount of vitamin D of the newborn has

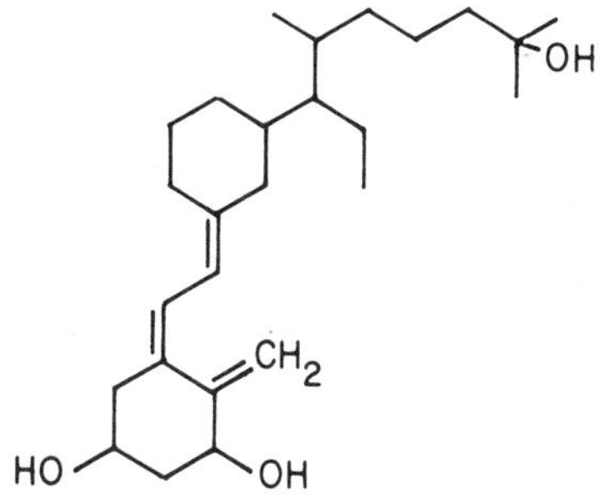

OH, CH_2, HO, OH

1,25-Dihydroxycholecalciferol
(1,25-$(OH)_2$-D_3)

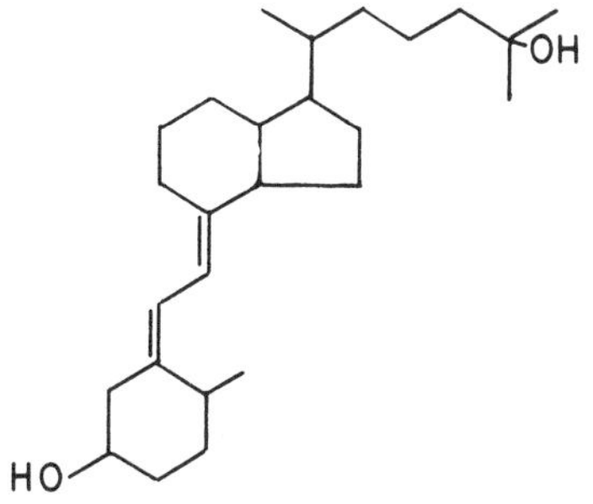

OH, HO

25-Hydroxycholecalciferol
(25-OH-D_3)

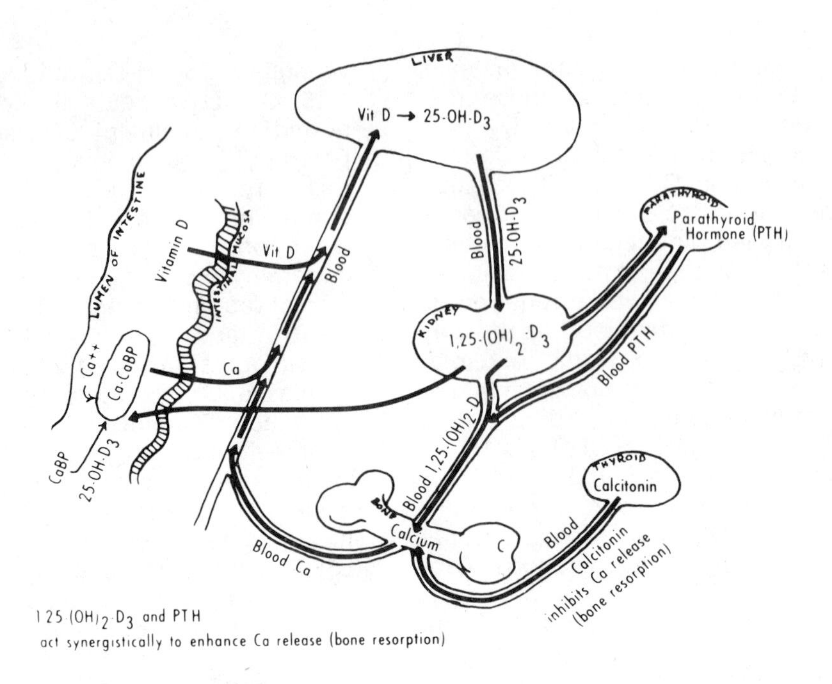

Figure 13-5. Schematic diagram of vitamin D metabolism in the body in relation to Ca metabolism.

been reported in humans (56).

Although the vast majority of research has been directed at calculating the effect of vitamin D on Ca and P metabolism, available evidence indicates that vitamin D also promotes absorption from the GIT of beryllium, cobalt, iron, magnesium, strontium, zinc, and, perhaps, still other elements (52, 60). Whether the effect is caused by the same protein as the vitamin D-dependent CaBP, to another single protein or series of protein carriers, or to other mechanisms, is not known.

Deficiency Signs

The main effect of a deficiency of vitamin D is abnormal skeletal development. Normal caicification cannot occur in the absence of adequate Ca and P. Therefore, either a deficiency of vitamin D, which results in impaired utilization of Ca, or a deficiency of Ca or of P, will produce similar abnormalities in the skeleton (the roles of Ca and P are discussed in Ch. 10).

The term applied to vitamin D deficiency in young, growing animals is rickets; the comparable condition in adults is osteomalacia. In each instance inadequate calcification of organic matrix occurs, which results in lameness, bowed and crooked legs (Fig. 13-6), spontaneous fractures of long bones and ribs and beading of the ribs. In growing animals the bone ash % is reduced (Ca:P ratio tends to remain constant), and weight gain may be depressed. In adults, negative mineral balance occurs, and bone ash concentration declines.

Species differences exist in the response to deprivation of vitamin D by dietary means or by protection from ultraviolet light (sunlight). Calves and pigs can grow normally on a level of vitamin D in the diet which quickly produces rickets in chicks. Pheasants and turkeys have a higher vitamin D requirement than chicks. On the other hand growing guinea pigs kept in the dark and fed vitamin D-free diets do not

Figure 13-6. Lameness and sore joints in pig fed a vitamin D-deficient diet and kept indoors without access to sunlight.

develop rickets over a 100-day period if adequate Ca and P are provided (43). Part of the difference in species requirements may be related to growth rate, as animals and birds with very rapid growth tend to be more susceptible to rickets. Protein source affects the vitamin D requirement of pigs. Soybean protein, high in phytate, is associated with a higher vitamin D requirement than milk protein.

In vitamin D deficiency, serum Ca concentration tends to be reduced, although hormonal mechanisms (parathyroid hormone and calcitonin) are quite efficient in maintaining a relatively constant range. Although parathyroid hormone is under the influence of vitamin D, calcitonin is not (54). Serum alkaline phosphatase may be increased in vitamin D-deficiency rickets. This enzyme is present in bone and is associated with bone resorption.

Vitamin D deficiency can be prevented by only a few minutes of exposure to direct sunlight, although skin pigmentation affects the amount of sunlight required to prevent rickets and white-skinned animals require less sunlight than dark-skinned ones. Loomis (50) has suggested that skin color in humans is an adaptation which provides for protection against rickets in inhabitants of northern regions as well as protection against excess vitamin D synthesis in skin of inhabitants of equatorial regions.

Toxicity

Excess vitamin D causes abnormal deposition of Ca in soft tissues. This Ca is resorbed from bone, resulting in brittle bones subject to deformation and fractures. Ca deposits are frequent in kidneys, aorta, and lungs. Such lesions can be produced in rats with dosages of 300,000 to 600,000 IU, in chicks with 4,000,000 IU/kg of diet and with pigs at 250,000 IU/animal/day for 30 days. Human infants show toxic signs with levels as low as 3000-4000 IU/day (only 10X the requirement). Hypervitaminosis D can lead to death, usually from uremic poisoning resulting from severe calcification of kidney tubules. Excess vitamin D during pregnancy apparently does not cause severe abnormalities in the fetus, but is not harmless, as premature closing and shortening of the skull bone of rabbits and abnormal teeth have been observed in newborn rabbits whose dams were given excess vitamin D.

Vitamin E

Vitamin E was first recognized as a fat-soluble factor necessary for reproduction in rats (70). Alpha tocopherol, the most active biological form of vitamin E, was isolated by Evans et al (71). Other forms of tocopherol have been designated with prefixes such as β and γ. The structure of α-tocopherol is shown (4):

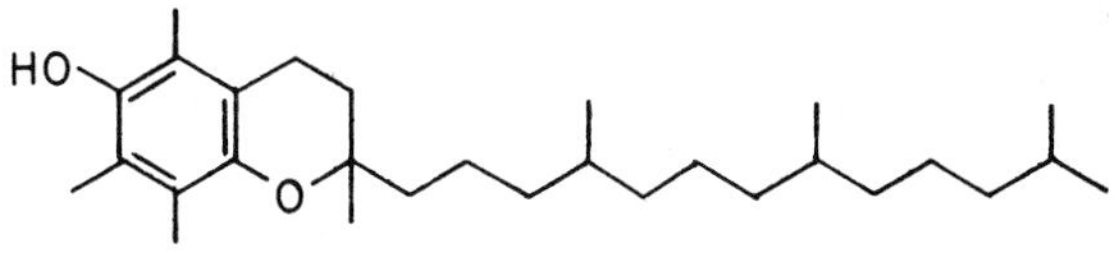

HO O

Alpha-Tocopherol

Other compounds with chemical structures similar to tocopherols are found in animal tissues but have limited biological activity. Vitamin E is very unstable; its oxidation is increased by the presence of minerals and of polyunsaturated fatty acids (PUFA), and decreased by esterification to form tocopheryl acetate. The d-isomer is more active than the l-form. Most commercially available vitamin E is available as dl-α tocopheryl acetate.

Knowledge of vitamin E nutrition and metabolism was reviewed in a symposium in honor of H.M. Evans (86), and more recent information has been summarized in several other reviews (62, 78, 87, 88).

Functions

The activity of α-tocopherol has been classified under two main headings (62): Effects attributable to the hydroxy group of the molecule (antioxidant effect) and effects brought about by metabolites of α-tocopherol.

Apparently, the central function of the vitamin in preventing a wide variety of deficiency signs is as an antioxidant. It plays an important role in maintaining the integrity of biological cellular membranes (5, 80). No active metabolites of α-tocopherol have been found and no proof exists of a tocopherol-dependent enzyme (69). The recognition that the mineral element, Se, can, under some conditions, protect against most vitamin E deficiency symptoms has led to an enormous body of literature attempting to elucidate the relationship between the 2 nutrients. These interrelationships are described in greater detail in the section on Se in Ch. 11.

Deficiency Signs

The manifestations of vitamin E deficiency are varied and some are species specific, but can be divided into 3 broad categories as follows (7, 81): reproductive failure; derangement of cell permeability; and muscular lesions (myopathies).

Reproductive failure associated with vitamin E deficiency can be related to embryonic degeneration, as in the rat and bird, or to sterility from testicular atrophy as in the chicken, dog, guinea pig, hamster, and rat, or to ovarian failure in female rats (84).

Derangement of cell permeability may affect liver, brain, kidney, or blood capillaries. Liver necrosis occurs in vitamin E deficiency of rats and pigs (Fig. 13-7). Se can prevent it, even in the absence of supplemental vitamin E. Red blood cell hemolysis occurs in vitamin E-deficient rats, chicks, and human infants; kidney degeneration in deficient mink, monkeys, and rats; and steatitis (inflammation of adipose tissue) in chicks, mink, and pigs. Kidney degeneration and steatitis were prevented by Se in the species studied.

Figure 13-7. Necrotic liver from pig fed a vitamin E-Se deficient diet.

Two common manifestations of vitamin E deficiency in chicks are encephalomalacia (abnormal Purkinje cells in cerebullum) and exudative diathesis (ED). Se prevents ED but not encephalomalacia. In encephalomalacia, incoordination and ataxia result from hemorrhages and edema in the cerebellum. Synthetic antioxidants, such as ethoxyquin and DPPD, prevent encephalomalacia. Polyunsaturated FA enhance the incidence of encephalomalacia in chickens fed diets marginal in vitamin E. In ED severe edema results from increased capillary permeability. A bluish-green subcutaneous exudate resulting from blood protein loss is evident on the breast of the affected birds preceding death.

Nutritional muscular dystrophy (NMD, stiff-lamb disease, white muscle disease, muscular dystrophy) is common in growing lambs, calves, pigs, chicks, turkeys, and rabbits fed vitamin E-deficient diets. The lesions involve degeneration of skeletal muscle fiber (Zenker's degeneration) which are seen grossly as whitish streaks on the surface of muscles. Degeneration of heart muscle (Mulberry heart disease) and of liver (liver necrosis) are common in vitamin E-deficient pigs. Sudden death is a common occurrence in calves and in affected pigs which show the heart and liver lesions on necropsy. In NMD peroxidation of muscle tissues (68) and activities of lysosomal enzymes are increased. Glutamic oxalacetic transaminase (SGOT) in blood serum is increased drastically in animals with NMD. This is an index of damage to muscle or heart tissue resulting in release of this enzyme into the blood, although it is not specific for NMD. A common sign of vitamin E deficiency in the turkey is myopathy of the gizzard as well as of heart and skeletal muscles. Se can prevent these lesions as well as those of liver, heart, and skeletal muscle in most types of NMD in mammals. NMD of birds is apparently only partly responsive to Se. Work by Hathcock et al (74) indicates that the amino acid cysteine has a primary role in prevention of NMD in chicks, as NMD is increased by factors that deplete body cysteine and decreased by factors that spare it. Scott et al (7) have suggested that cysteine and vitamin E act 2 different ways to prevent NMD in chicks. The complete story on vitamin E, cysteine, Se, and other factors as agents in the prevention of NMD in birds and animals is gradually being unraveled.

Metabolism

Absorption, Transport and Storage. Tocopherols are absorbed mainly by micellar formation in the upper intestinal tract in the presence of bile salts. Both the d- and l-isomer are absorbed, but the d-isomer has a relative potency 1.2X greater than that of the dl-mixture. Tocopherols are absorbed into the lymphatics and are transported as a part of lipoproteins. Storage occurs in liver, skeletal muscle, heart, lung, kidney, spleen, and pancreas in similar amounts and in pituitary, testes, and adrenals in even higher concentrations (5, 6). Because tocopherols are fat-soluble, concentrations are best expressed in units/g of fat rather than/g of tissue. The wide tissue distribution of tocopherol would be expected in view of its role in preventing peroxidation and in maintaining cell membrane integrity.

Vitamin E-PUFA Interrelationships. Diets high in PUFA (polyunsaturated fatty acids) clearly increase the vitamin E requirement. Young dogs fed vitamin E-deficient diets with safflower oil (high in PUFA) developed browning of the fat around the intestine, hemolysis of RBC, and depressed plasma tocopherol in direct relation to dietary fat level (76); however other work (63, 83) has failed to demonstrate that high levels of vitamin E can prevent this pigment formation. Testes, heart, adipose tissue, adrenal, brain, skeletal muscle, and bone marrow accumulate large amounts of flourescent pigment in vitamin E-deficient animals fed PUFA. Nevertheless, recent work presents strong evidence that vitamin E is effective (85, 88). For example, Reddy et al (85) found that rats fed 10% lard-1% cod liver oil diets without vitamin E for 4 mo. accumulated twice the fluorescent pigments in adipose tissue as those fed the same diet but supplemented with vitamin E. Furthermore, adipose tissue of rats fed cod liver oil, which is higher in PUFA than corn oil, had 3X the flourescence of adipose tissue from those fed corn oil. Fluorescent pigments in adipose tissue extracts may be the products of in vivo lipid peroxidation. A proposed pathway of formation of fluorescent (ceroid) pigment is as follows (85):

The apparent increase in consumption of fats high in PUFA in the USA increases the possibility that vitamin E may be becoming a limiting nutrient in human nutrition (61, 64). Harman suggested that peroxidation (free radical reactions) of adipose tissues play a significant role in aging (72). He hypothesized that human productive lifespan could be extended by dietary manipulation to reduce peroxidation in body tissues.

Lifespan of mice is shortened as much as 10% by increasing the amount and/or degree of unsaturation of dietary fat (72) and it has been shown that several antioxidants reduce the incidence of congenital malformation (mostly skeletal deformities) in vitamin E-deficient rats (65, 67, 79).

Very limited information is available on the role of PUFA in affecting the vitamin E requirement for reproduction in animals.

Placental and Mammary Transfer of Vitamin E. The placental transfer of vitamin E is inefficient (65, 82), whereas FA cross the placenta more efficiently. Therefore, a reduction of vitamin E-PUFA ratio in fetal tissues may predispose the newborn to vitamin E deficiency, but no information is available on the question. Some placental transfer of vitamin E does occur, but malformations have been observed in young rats fed low-vitamin E even though vitamin E was found in liver, blood, and carcasses of the young (66).

The tocopherol content of the milk of both ruminant and nonruminant animals is affected by concentration in the diet. Thus, susceptibility of newly weaned animals to vitamin E deficiency is dependent on the body stores they have accumulated in response to gestation and lactation diet of the dam.

Toxicity

Few reports have been made of vitamin E toxicity in animals or man (77, 81). Scattered descriptions of hemorrhagic syndromes, nervous disorders, edema, changes in endocrine glands, and antagonism of vitamin K have appeared in experimental animal research, but detailed pathology of hypervitaminosis E is not available. Mason and Horwitt (81) report that human adults have tolerated oral doses of 1 g/day for months without undesirable effects Apparently the range of safe level of intake is wider than for other fat-soluble vitamins, but the wholesale consumption of vitamin E as well as of all nutrients should be avoided.

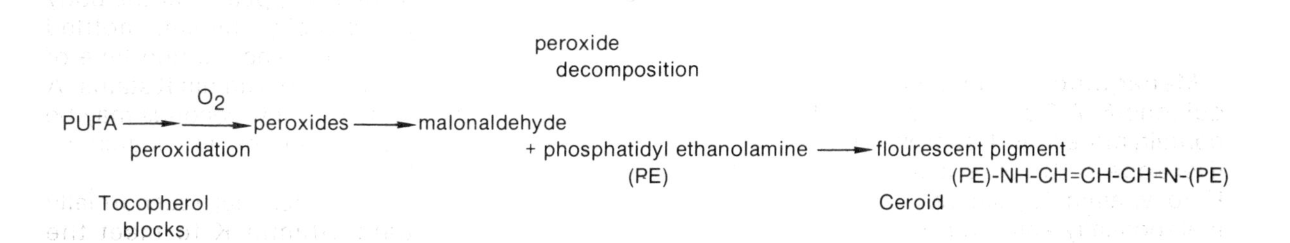

Vitamin K

Structure

A hemorrhagic syndrome was described by Dam (93) in chicks fed a diet low in sterols, and in 1935 Dam (94) showed that the missing factor was a fat-soluble vitamin which he named vitamin K. Vitamin K is really a group of compounds (89). The structure of the two most important natural sources of vitamin K, phylloquinone (vitamin K_1) and menaquinone (vitamin K_2) are shown (3). Vitamin K_1 is common in green vegetables and vitamin K_2 is a product of bacterial flora in the GIT of animals and man. A synthetic, menadione (vitamin K_3), is widely used commercially; its structure is also shown.

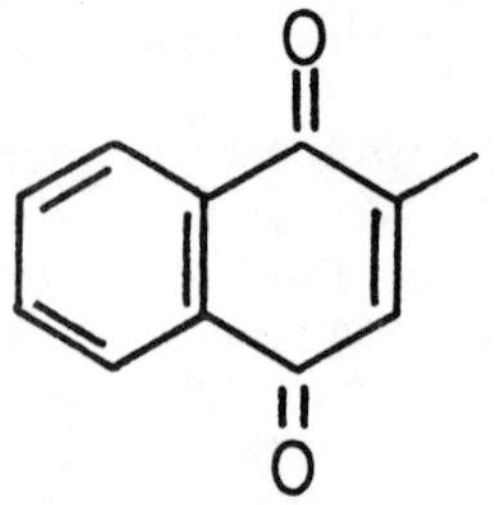

$CH_2CH{=}C(CH_3)(CH_2)_3CH(CH_3)(CH_2)_3CH(CH_3)(CH_2)_3CHCH_3(CH_3)$

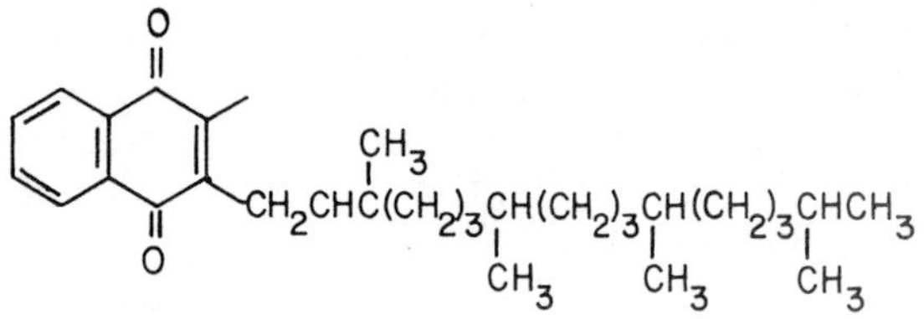

Phylloquinone (K_1)

$CH_2CH{=}C(CH_3)CH_2(CH_2CH{=}C(CH_3)CH_2)_5CH_2CH{=}C(CH_3)CH_3$

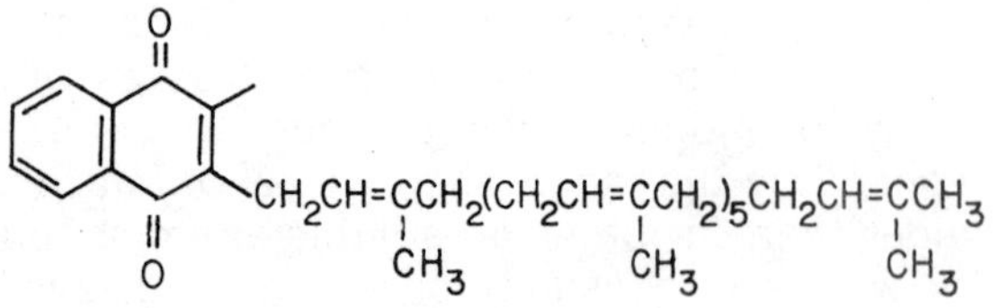

Menaquinone (K_2)

Menaquinone can also occur as menaquinone-6, 7, 8 or 9; that is, the side chain may contain more than 4 isoprene units shown. The liver apparently converts these forms of vitamin K to vitamin K_2, suggesting that this is the metabolically active form.

Menadione (K_3)

Functions

Vitamin K is required for normal blood clotting. Specifically, it is required for synthesis of prothrombin in the liver. It is not a component of prothrombin, but acts in some way on enzyme systems involved in prothrombin synthesis and in the synthesis of other factors involved in the total blood clotting mechanism. Olsen (98) suggested that vitamin K acts by influencing messenger RNA formation needed for prothrombin synthesis. Such a role could help to explain the rapid synthesis of prothrombin that occurs in vitamin K-deficient animals treated with vitamin K (2-6 hr). Vitamin K is involved in at least 4 steps in clot formation: a plasma thromboplastin component (Factor IX), a tissue thromboplastin component (Factor VII), "Stuart" factor (Factor X), and prothrombin (Factor II). The exact mode of action in each reaction is unknown (99).

The clotting mechanism (Table 13-3) is stopped by oxalate and citrate, which precipitate Ca^{++}, and by heparin which apparently blocks formation of the Stuart factor.

Deficiency Signs

A deficiency of vitamin K results in prolonged blood clotting time, generalized hemorrhages, and death in severe cases. Often, subcutaneous hemorrhages appear over the body surface, giving a blotchy, bluish, mottled appearance to the skin. The clotting time of the blood is a good index of vitamin K status. A normal clotting time of a few seconds may be extended to several minutes in vitamin K-deficient animals.

The microflora of the intestinal tract normally produce adequate vitamin K to meet the

Table 13-3. Vitamin K involvement in blood clotting.

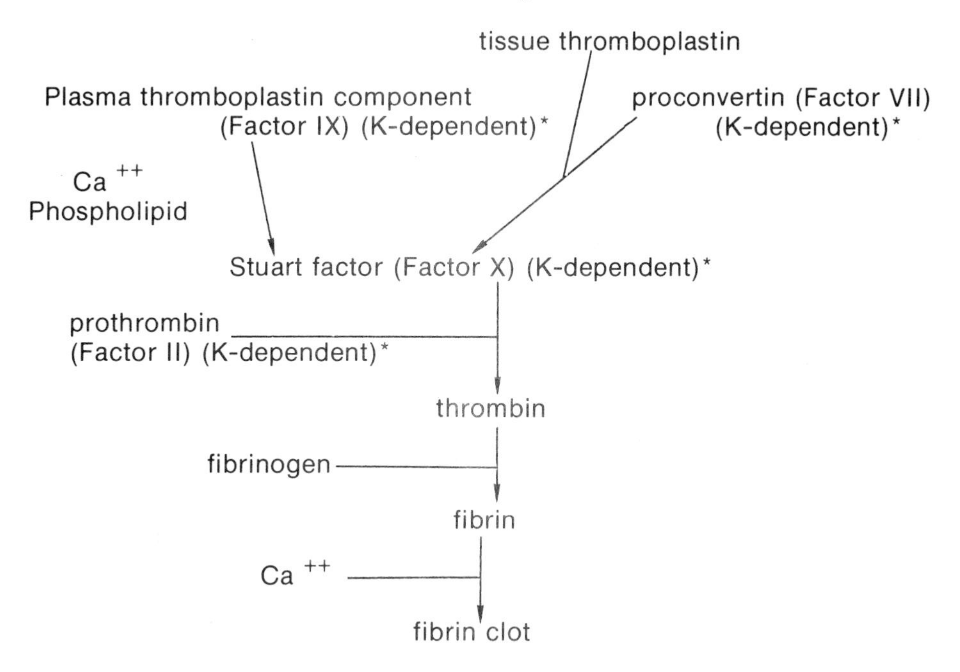

*Synthesis of each is inhibited by vitamin K antagonists.

metabolic needs of the host. The vitamin is obtained either by absorption from the lower GIT after synthesis or by coprophagy (feces eating), which serves as an important means of supplying nutrients to many animal species. Barnes and Fiala (90) showed that prevention of coprophagy in rats produced vitamin K deficiency. Nutritionists have not, until recently, greatly concerned themselves with supplying adequate vitamin K in the diet, because the GIT microflora play such an important role in synthesizing vitamin K. Newer technology has, however, created problems with deficiency of vitamin K under practical conditions. Chicks fed sulfaquinoxaline to control coccidiosis develop vitamin K deficiency (7). This sulfa drug is an antagonist of vitamin K. Pigs fed in wire floor cages, in which access to feces is prevented, develop increased prothrombin clotting times (101) and growing-finishing pigs fed 60-65% cane sugar developed deficiency symptoms (91). Field reports of a hemorrhagic syndrome preventable by supplemental vitamin K in growing-finishing swine suggest that vitamin K deficiency is an increasingly serious practical problem. Cunha (92) suggested several factors that may be contributing to this increase. Mycotoxins (from moldy grain) may antagonize vitamin K; increased confinement feeding of swine has been associated with less use of pasture and forages that are high in vitamin K; and the use of slotted floors lessens the opportunity for coprophagy. Clearly, much remains to be discovered concerning factors that affect vitamin K requirement and which produce deficiencies.

Metabolism

Absorption and Utilization. The naturally occurring and synthetic vitamin K's are absorbed most efficiently in the presence of dietary fat and bile salts, and menadione requires bile salts for absorption. The efficiency of absorption of vitamin K depends on the form in which the vitamin is combined as well as on the specific isomer involved. The mechanisms by which vitamin K is transported in blood are not clearly defined, probably because of the variable solubility in water vs. fats of the several compounds having vitamin K activity.

The biological activity of several compounds showing vitamin K potency is listed in Table 13-4. The activity of menadione and its derivatives depends on the relative stabilities of the preparations used and on whether or not sulfaquinoxaline is present in the test diet. Menadione and its derivatives are less effective than phylloquinone in counteracting the antagonistic effects of sulfaquinoxaline. Menadione is very reactive and may be toxic; for this reason, and because of its lower biological activity, it is far less commonly used in feeds than are its derivatives.

Vitamin K compounds are very unstable under alkaline conditions. Menadione dimethylpyrimidinol bisulfate is more stable as well as higher in biological activity than menadione.

Vitamin K Antagonists and Inhibitors

A well-known natural antagonist of vitamin K is dicoumarol, often present in weather-damaged sweet clover hay (97). The presence of dicoumarol in hay or silage causes massive internal hemorrhages and death in calves. Another important antagonist of vitamin K is Warfarin®, which is a competitive inhibitor of menaquinone (K_2). The effect of Warfarin in increasing clotting time can be reversed by simultaneously increasing the amount of K_2 given to the animal. The inhibition of vitamin K function by Warfarin has resulted in the large scale marketing of Warfarin (a product of the Wisconsin Alumni Research Foundation) as a rat poison.

The structures of dicoumarol and of Warfarin are shown below.

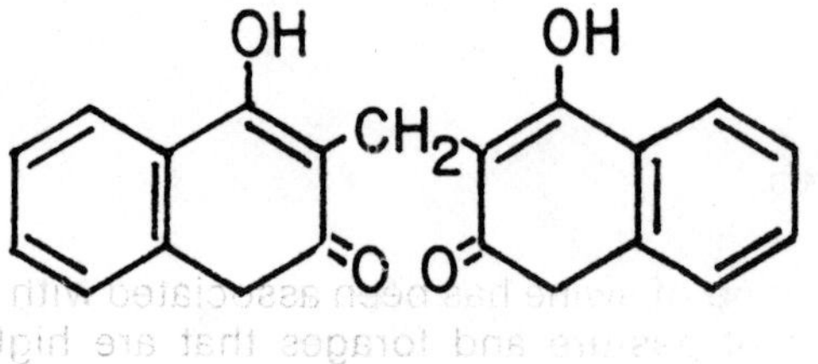

Dicoumarol

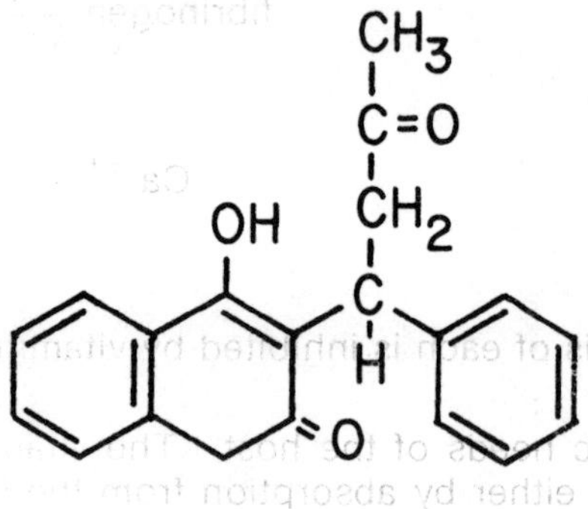

Warfarin

Table 13-4. Relative biological activity of several forms of vitamin K.[a]

	Activity relative to that of natural vitamin K_1 (phylloquinone-4)
Phylloquinone-1	5
Phylloquinone-2	10
Phylloquinone-3	30
Phylloquinone-4	100
Phylloquinone-5	80
Phylloquinone-6	50
Menaquinone-2	15
Menaquinone-3	40
Menaquinone-4	100
Menaquinone-5	120
Menaquinone-6	100
Menaquinone-7	70
Menadione	40-150[b]
Menadione sodium bilsulfate	50-150[b]
Menadione dimethylpyrimidinol bisulfate	100-160[b]

[a] From Griminger (96)

[b] Activity depends on relative stabilities of preparations used and on presence or absence of sulfaquinoxaline in the test diet.

Other vitamin K antagonists, including α-tocopheryl quinone, sulfaquinoxaline (previously discussed), and some napthoquinone derivatives may be of importance in affecting the dietary requirement for vitamin K.

Toxicity

Phylloquinone and menaquinone derivatives are nontoxic even in high dosage levels, but menadione is toxic to skin and respiratory tract of several animal species; its bisulfate derivatives are not (7, 95). Menadione given in prolonged high doses produces anemia, porphyrinuria, and othe abnormalities in animals and chest pains and shortness of breath in man (100).

References Cited

1. Funk, C. 1912. J. State Med., London 20:341.
2. Harris, R.S. 1967. In: The Vitamins, Vol. I. 2nd ed. Academic Press, N.Y.
3. Harris, R.S. 1971. In: The Vitamins, Vol. III. 2nd ed. Academic Press, N.Y.
4. Harris, R.S. 1972. In: The Vitamins, Vol. V. 2nd ed. Academic Press, N.Y.
5. Roels, O.A. 1967a. In: The Vitamins, Vol. I. 2nd ed. Academic Press, N.Y.
6. Roels, O.A. 1967b. In: Present Knowledge of Nutrition. 3rd ed. The Nutrition Foundation, N.Y.
7. Scott, M.L., M.C. Nesheim and R.J. Young. 1969. Nutrition of the chicken. M.L. Scott and Associates, Ithaca, N.Y.
8. Ames, S.R., W.J. Swanson and P.L. Harris. 1955. J. Amer. Chem. Soc. 77:4136.
9. Anderson, M.D., V.C. Speer, J.T. McCall and V.W. Hays. 1966. J. Animal Sci. 25:1123.
10. Braude, R., et al. 1941. Biochem. J. 35:693.
11. Fidge, N.H., T. Shiratori, J. Ganguly and D.S. Goodman. 1968. J. Lipid Res. 9:103.
12. Follis, R.H. 1958. Deficiency Disease. C.C. Thomas Pub., Springfield, Ill.
13. Frape, D.L., V.C. Speer, V.W. Hays and D.V. Catron. 1959. J. Nutr. 68:333.
14. Friend, C.J., et al. 1960. Proc. Nutr. Soc. 19:xxxiv.
15. Gallina, A.M., et al. 1972. J. Nutr. 100:129.
16. Ganguly, J. 1960. Vitamins and Hormones 18:387.
17. Gaylor, J.L. 1964. N.Y. State J. Med. 64:905.
18. Giroud, A. 1970. The Nutrition of the Embryo. C.C. Thomas Pub., Springfield, Ill.
19. Goodman, D.S. and H.S. Huang. 1965. Science 149:879.
20. Goodman, D.S., H.S. Huang, M. Kansi and T. Shiratori. 1967. J. Biol. Chem. 242:3543.
21. Hayes, B.W., G.E. Mitchell, Jr. and C.O. Little. 1967. J. Animal Sci. 27:516.
22. Huang, H.S. and D.S. Goodman. 1965. J. Biol. Chem. 240:2839.
23. Johnson, B.C. and G. Wolf. 1960. Vitamins and Hormones. 18:457.
24. Kohlmeir, R.H. and W. Burroughs. 1970. J. Animal Sci. 30:1012.
25. Krinsky, N.I., D.G. Cornwell and J.L. Oncley. 1958. Arch. Biochem. Biophys. 73:233.
26. Lippel, K. and J.A. Olson. 1968. J. Lipid Res. 9:168.
27. Mahadevan, S., P.S. Sastry and J. Ganguly. 1963. Biochem. J. 88:531; 534.
28. Moore, T. 1957. Vitamin A. Elsevier, Amsterdam, The Netherlands.
29. Moore, T. 1960. Vitamins and Hormones 18:499.
30. Neilsen, H.E., N.J. Hajgaard-Olsen, W. Hjarde and E. Leerbeck. 1965. Acta. Agric. Scand. 15:235.
31. Olson, J.A. and O. Hayaishi. 1965. Proc. Natl. Acad. Sci. U.S. 54:1364.
32. Palludin, B. 1966. In: Swine in Biomedical Research. Frayn Printing Co., Seattle, Washington.
33. Rodriguez, M.S. and M.I. Irwin. 1972. J. Nutr. 102:909.
34. Smith, S.E. 1970. In: Duke's Physiology of Domestic Animals. 8th ed. Comstock Pub. Co., Ithaca, N.Y.
35. Thomas, J.W., J.K. Loosli and J.P. Willman. 1947. J. Animal Sci. 6:141.
36. Ullrey, D.E. 1972. J. Animal Sci. 35:648.
37. Wald, G. 1954. Amer. Scientist 42:73.
38. Walke, R.E., S.W. Nielsen and J.E. Rousseau. 1968. Amer. J. Vet. Res. 29:1009.
39. Wing, J.M. 1969. J. Dairy Sci. 52:479.
40. Zile, M., H.F. DeLuca and H. Ahrens. 1973. J. Nutr. 102:1255.
41. Anonymous. 1973. Nutr. Rev. 31(2):58.
42. Blunt, J.W., H.F. DeLuca and H.K. Schnoes. 1968. Biochem. 7:3317.
43. Chapman, M.W., et al. 1973. J. Nutr. 103(7):xxv.
44. DeLuca, H.F. 1972. Proc. Cornell Nutr. Conf., p 20.
45. Fraser, D.R. and E. Kodicek. 1970. Nature 228:764.
46. Gray, R., J. Boyle and H.F. DeLuca. 1971. Science 172:1232.
47. Holick, M.F., H. Garabedian and H.F. DeLuca. 1972. Science 176:1146.
48. Holick, M.F., H.K. Schnoes and H.F. DeLuca. 1971. Proc. Natl. Acad. Sci. U.S. 68:803.
49. Holick, M.F., et al. 1971. Biochem. 10:2799.
50. Loomis, W.F. 1967. Science 157:501.

51. Lund, J. and H.F. DeLuca. 1966. J. Lipid Res. 7:739.
52. Maynard, L.A. and J.K. Loosli. 1969. Animal Nutrition. 6th ed. McGraw-Hill Book Co., N.Y.
53. McCollum, E.V., et al. 1922. J. Biol. Chem. 53:293.
54. Morii, H. and H.F. DeLuca. 1967. Amer. J. Physiol. 213:358.
55. Tanaka, Y., H. Frank and H.F. DeLuca. 1972. J. Nutr. 102:1569.
56. Toverced, K.U. and F. Ender. 1935. Acta Paediat 18:174.
57. Wasserman, R.H. and A.N. Taylor. 1966. Science 152:791.
58. Wasserman, R.H. and A.N. Taylor. 1968. J. Biol. Chem. 243:3987.
59. Wasserman, R.H., A.N. Taylor and F.A. Kallfelz. 1966. Amer. J. Physiol. 211:419.
60. Worker, N.A. and B.B. Migicorsky. 1961. J. Nutr. 75:222.
61. Antar, M.A., M.A. Ohlson and R.E. Hodges. 1964. Amer. J. Clin. Nutr. 14:169.
62. Boguth, W. 1969. Vitamins and Hormones 27:1.
63. Bunyan, J., A.T. Diplock, M.A. Cawthorne and J. Green. 1968. Br. J. Nutr. 22:165.
64. Call, D.L. and A.M. Sanchez. 1967. J. Nutr. 93 (Suppl. 1, Part III): 1-28.
65. Cheng, D.W., T.A. Bacinson, A.M. Rao and S. Serbbammal. 1960. J. Nutr. 71:54.
66. Cheng, D.W., K.G. Braun, B.J. Braun and K.H. Udine. 1961. J. Nutr. 74:111.
67. Cheng, D.W., L.F. Change and T.A. Bacinson. 1957. Anat. Rec. 129:167; Cheng, D.W. and B.H. Thomas. 1953. Proc. Iowa Acad. Sci. 60:290.
68. Dasai, J.D., C.C. Calvert, M.L. Scott and A.L. Tappel. 1964. Proc. Soc. Exptl. Biol. Med. 115:462.
69. Draper, H.H. and A.S. Csallany. 1969. Fed. Proc. 28:1690.
70. Evans, H.M. and K.S. Bishop. 1922. J. Metabolic Res. 1:319, 335.
71. Evans, H.M., J.H. Emerson and G.A. Emerson. 1936. J. Biol. Chem. 113:319.
72. Harman, P. 1969. J. Amer. Geriatrics Soc. 17:721.
73. Harris, P.L. and N.D. Embree. 1963. Amer. J. Clin. Nutr. 13:385.
74. Hathcock, J.N., S.J. Hull and M.L. Scott. 1968. J. Nutr. 94:147.
75. Hathcock, J.N., M.L. Scott and J.N. Thompson. 1968. Proc. Soc. Exptl. Biol. Med. 127:935.
76. Hayes, K.C., S.W. Nielson and J.E. Rousseau. 1969. J. Nutr. 99:195.
77. Hillman, R.W. 1957. Amer. J. Clin. Nutr. 5:597.
78. Horwitt, M.K., C.C. Harvey and E.M. Harmon. 1968. Vitamins and Hormones 26:487.
79. King, O.W. 1964. J. Nutr. 83:123.
80. Malenaar, I., J. Vos and F.A. Hommes. 1972.
81. Mason, K.E. and M.K. Horwitt. 1972. In: The Vitamins V. 2nd ed. Academic Press, N.Y.
82. Nitowski, H.M., K.S. Hsu and H.H. Fordon. 1962. Vitamins and Hormones. 20:559.
83. Potek, A.J., F.E. Kendall, N.M. de Fritch and R.L. Husch. 1967. Arch. Path. 84:295.
84. Raychauduti, C. and I.D. Desai. 1971. Science 173:1028.
85. Reddy, K.B. Fletcher, A. Tappel and A. Tappel. 1973. J. Nutr. 103:908.
86. Symposium in Honor of H.M. Evans. 1962. Vitamins and Hormones 20:375.
87. Tapel, A.L. 1970. In: The Fat Soluble Vitamins. U. Wisconsin Press, Madison.
88. Witting, L.A. 1970. In: Progress in Chemistry of Fats and Other Lipids 9:517.
89. Ansbacher, S. and E. Fernholz. 1939. J. Amer. Chem. Soc. 61:1924.
90. Barnes, R.H. and G. Fiala. 1958. J. Nutr. 68:603.
91. Brooks, C.C., R.M. Nakamura and A.Y. Miyahara. 1973. J. Animal Sci. (in press).
92. Cunha, T.J. 1971. Feedstuffs 43(9):20.
93. Dam, H. 1929. Biochem. Ztschr. 215:475.
94. Dam, H. 1935. Biochem. J. 29:1273.
95. Day, E.J. 1967. Proc. Texas Nutr. Conf., p 32.
96. Griminger, P. 1966. Vitamins and Hormones 24:605.
97. Link, K.P. 1959. Circulation 19:97.
98. Olson, R.E. 1964. Science 145:926.
99. Olson, R.E. 1970. Nutr. Rev. 28:171.
100. Owen, C.A., Jr. 1971. In: The Vitamins. 2nd ed. Academic Press, N.Y.
101. Shendel, H.E. and B.C. Johnson. 1962. J. Nutr. 76:124.

Chapter 14 – Water-Soluble Vitamins

Unlike the fat-soluble vitamins, water-soluble vitamins (except B_{12}) are not stored in appreciable quantities in body tissues. They must be supplied in the diet on a day-to-day basis for those animals having a GIT in which microbial synthesis is not a prominent feature. In ruminants the water-soluble vitamin requirement is met almost entirely from microbial synthesis in the rumen and lower GIT; in herbivores such as the horse and rabbit, microbial synthesis occurs in the colon or cecum. Thus, the nutritionist need not be deeply concerned about providing a dietary source for these species. For pigs, poultry, and other simple-stomached animals, including man, however, a dietary source is essential.

Most of the water-soluble vitamins are required in minute amounts. They are all organic compounds, but are unrelated to each other in structure. Most function as metabolic catalysts, usually as coenzymes. All produce profound aberration in metabolism if unavailable to the tissues in sufficient amounts. Because of their ready excretion by the kidney, acute toxicity of the water-soluble vitamins is unlikely.

Each water-soluble vitamin will be discussed as to structure, function, and deficiency signs. The IUNS Committee on Nomenclature (1) provided rules for naming vitamins. These rules are needed to minimize confusion and misinterpretation in discussing vitamin nutrition and are followed here.

Thiamin [aneurin, B_1]

Structure

Thiamin consists of one molecule of pyrimidine joined with one of thiazole. Williams (2) established its structure, as shown:

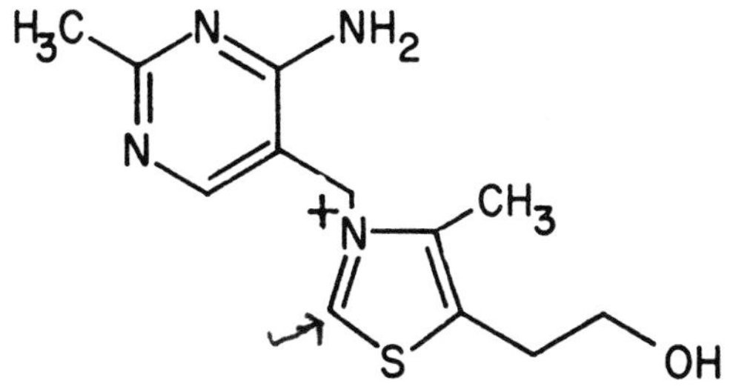

H3C N NH2 N +N CH3 S OH

Functions

Thiamin is phosphorylated in the liver to form coenzymes, cocarboxylase or thiamin pyrophosphate (TPP) and lipothiamide (LTPP). TPP is involved in the decarboxylation of α-keto acids. Among the reactions in which TPP participates, the decarboxylation of pyruvic acid to acetaldehyde is important and can be illustrated as:

$$\text{Pyruvic acid} \xrightarrow{\text{TPP}} \text{acetaldehyde} + CO_2$$

LTPP is involved in oxidative decarboxylation of α-ketoglutarate and other metabolites of the citric acid cycle.

The pig is somewhat unique in that it stores appreciable thiamin in its tissues. This accounts for the fact that pork is an excellent source of dietary thiamin.

Deficiency Signs

In thiamin deficiency, blood pyruvic acid and lactic acid concentration increase; this has been used in the assessment of thiamin status. Since pyruvic acid is a key metabolite in energy utilization in the citric acid cycle (see Fig. 7-1), it is evident that thiamin deficiency seriously upsets carbohydrate and lipid metabolism.

The immediate effect of a thiamin deficiency is a reduction in appetite (anorexia). In humans, the thiamin deficiency syndrome, beriberi, has been recognized for centuries. The syndrome includes numbness of extremities, weakness and stiffness in the thighs, edema of the feet and legs, unsteady gait, rigidity and paralytic symptoms, and pain along the spine (3). In animals, the classical disease is polyneuritis in chicks (retraction of the head; Fig. 14-1) and rats (walking in circles). Bradycardia (slow heart rate) is a common manifestation in all species studied and myocardial damage results in dilation of the heart (Fig. 14-2) and heart failure in swine (4). Body temperature is reduced and the adrenal gland is enlarged. Whether or not the symptoms of thiamin deficiency, including the heart lesion, result from pyruvic or lactic acid accumulation or to other factors is not certain (5).

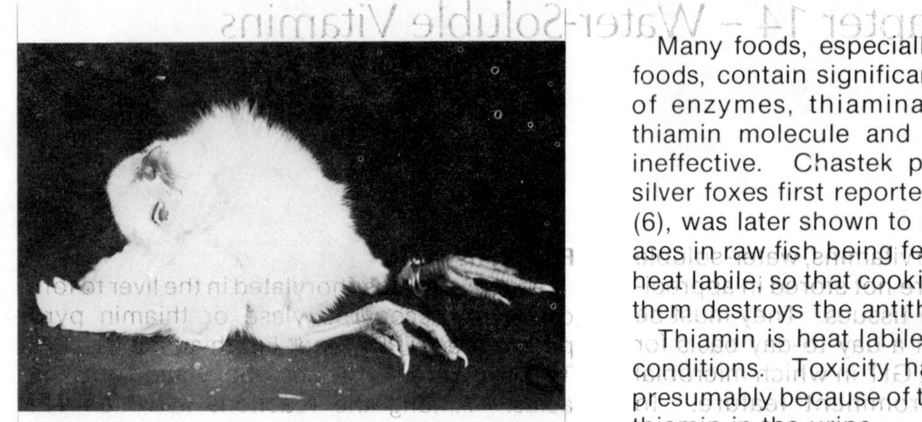

Figure 14-1. Polyneuritis in the chick showing the typical retraction of the head and rigidity of the legs as well as the marked effect on growth. Courtesy of Poultry Sci. Dept., Cornell U.

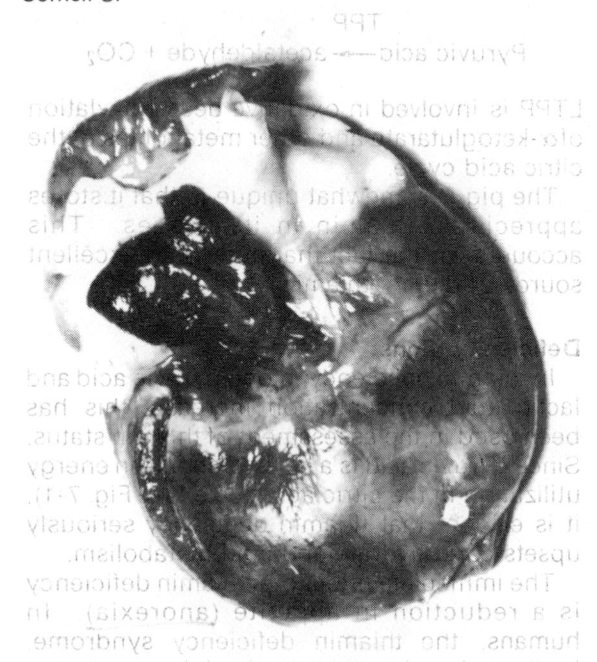

Figure 14-2. Enlarged flabby heart from a thiamin-deficient pig. Courtesy of D.E. Ullrey, Michigan State U.

Antagonists

There are several compounds that resemble thiamin in chemical structure but have no thiamin activity. Addition to the diet or injection of these antimetabolites* results in thiamin deficiency, which can be overcome by increasing the thiamin intake. One such thiamin antagonist is pyrithiamin.

Many foods, especially certain fish and sea foods, contain significant amounts of a group of enzymes, thiaminases, which split the thiamin molecule and render it biologically ineffective. Chastek paralysis, a disease in silver foxes first reported by Green and Evans (6), was later shown to be caused by thiaminases in raw fish being fed. These enzymes are heat labile, so that cooking of foods containing them destroys the antithiamin activity (7).

Thiamin is heat labile, especially in alkaline conditions. Toxicity has not been reported, presumably because of the rapid loss of excess thiamin in the urine.

*The concept of antimetabolites in nutrition has been reviewed by Woolley [1959], Sci. 129:615. Antimetabolites are used in clinical medicine [for example, in control of certain types of cancer] and in nutritional research.

Riboflavin [lactoflavin, B_2]

Structure

Riboflavin is a yellow fluorescent pigment consisting of ribose and isoalloxazine.

$CH_2(CHOH)_3CH_2OH$

N N O

NH

N

O

Riboflavin was synthesized by Karrer et al (8) and Huhn et al (9). It is heat stable but readily destroyed by light (10). Riboflavin, riboflavin 5-phosphate (also called flavin mononucleotide or FMN) and flavin adenine dinucleotide (FAD) are the only naturally occurring substances with riboflavin activity, although several synthetic derivatives show some activity (11).

Functions

Riboflavin functions in the coenzymes, FAD and FMN, which occur in a large number of enzyme systems. The interconversions of these biologically active forms can be illustrated as follows (12):

$$\text{Riboflavin} + \text{ATP} \xrightarrow{\text{flavokinase}} \text{FMN} + \text{ADP}$$

$$\text{FMN} + \text{ATP} \xrightarrow{\text{FAD pyrophosphyorylase}} \text{FAD} + \text{pyrophosphate}$$

FMN and FAD are present in most if not all animal cells and occur as coenzymes in the flavoprotein enzyme systems (12) such as oxidases (aerobic dehydrogenases) and anaerobic dehydrogenases. Both FAD and FMN are closely related in several reactions with niacin coenzymes, coenzymes I (NAD) and coenzyme II (NADP).

Deficiency Signs

The primary general effect, as with most of the water-soluble vitamin deficiencies, is reduced growth rate in young animals. Various pathological lesions accompany a deficiency (11, 13). In rats, conjunctivitis, corneal opacity, and vascularization of the cornea are common signs of deficiency and hair loss — related to a failure in regeneration of hair follicles — may occur. Mature female rats fed riboflavin-deficient diets have abnormal estrous cycles and give birth to a high proportion of pups with congenital malformations of the skeleton. In deficient dogs, fatty infiltration of the liver and corneal opacity have been reported. Changes in the cornea have also been reported as typical of riboflavin deficiency in pigs; in addition, hemorrhages of the adrenals, kidney damage, anorexia, vomiting, and birth of weak and stillborn piglets have been reported in deficient swine. A specific sign of deficiency in poultry is curled-toe paralysis (Fig. 14-3), which results from degenerative changes in the myelin sheaths of the nerve fibers.

In mature ruminants, the microflora of the rumen synthesize sufficient riboflavin and other water-soluble vitamins (as well as vitamin K) to eliminate a dietary need. Young preruminant animals (lambs and calves) have a dietary requirement early in life, however, and deficiency signs can be induced with synthetic milk diets. Diets low in riboflavin produce lesions in the corners of the mouth and on the lips, anorexia, loss of hair, and diarrhea in calves and lambs. The type of diet fed to immature ruminants appears to be a controlling factor in the development of rumen microbial populations that synthesize riboflavin.

Riboflavin deficiency has also been reported in monkeys, mice, cats, horses, and other species; the symptoms resemble in one or more ways those described for other animals. Riboflavin deficiency has been reported to cause moderate suppression of antibody production in rats (14, 15) and in pigs (16).

In man, riboflavin deficiency causes lesions of the lips and mouth (angular stomatitis), scrotal dermatitis, conjuctivitis, and burning of the eyes. Corneal vascularization has not been noted in deficient humans, even though it is a widespread phenomenon in other animals.

Metabolism

Riboflavin is rapidly absorbed by phosphorylation in the intestinal mucosal cell, and is used directly by cells throughout the body (17, 18). The maintenance requirement is dependent mainly on excretion rather than decomposition. A half life of 16 days has been reported for the rat under normal conditions; turnover is accelerated with high riboflavin diets (19).

Dietary composition and environmental temperature may affect the dietary requirement. High-fat diets or low-protein diets tend to increase the requirement, and the requirement of growing pigs has been reported to be increased by high environmental temperature (20).

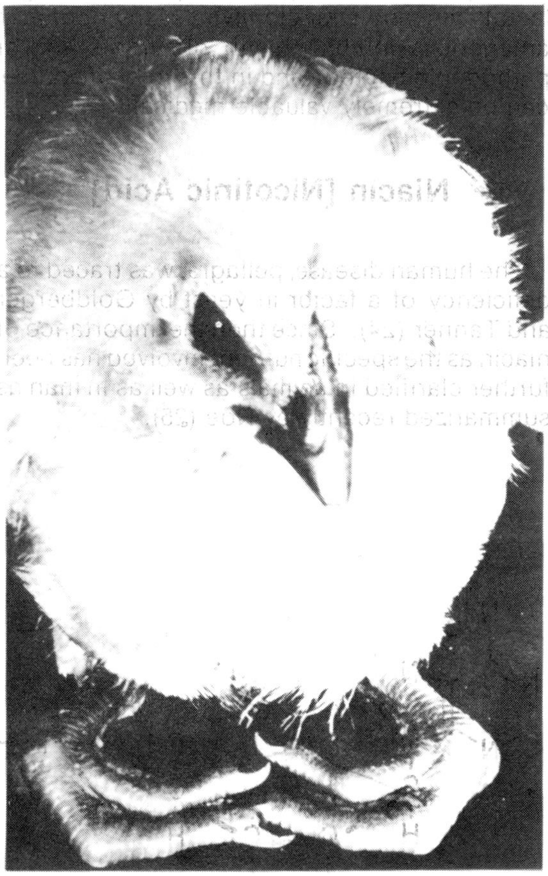

Figure 14-3. Curled-toe paralysis in a ribofalvin-deficient chick, a typical sign of a severe deficiency. Courtesy of Chas. Pfizer Co.

Toxicity

Riboflavin toxicity is extremely unlikely because of its rapid loss in urine. Rats given 560 mg/kg of body weight in an intraperitoneal injection show some mortality from kidney damage (21), but such a massive dose is certainly not physiological. High-protein diets may be toxic in riboflavin deficiency (22). This may be related to the reduction in D-amino acid oxidase activity of liver observed in riboflavin deficiency (23).

Analogs and Antagonist

Some analogs show riboflavin activity (5). At least 4 derivatives of isoalloxazine have about half the biological activity of riboflavin for rats. A long list of other isoalloxazine derivatives have no activity. Several analogs of riboflavin have been shown to be inhibitors of riboflavin. These include isoriboflavin (5, 6-dimethyl-9-D-1'-ribitylisoalloxazine), dinitrophenazine, galactoflavin, and D-araboflavin. Many riboflavin antagonists inhibit growth of tumors and/or pathogenic bacteria and in this role, of course, can be extremely valuable medically.

Niacin [Nicotinic Acid]

The human disease, pellagra, was traced to a deficiency of a factor in yeast by Goldberger and Tanner (24). Since then the importance of niacin as the specific nutrient involved has been further clarified in animals as well as in man as summarized recently by Roe (25).

Structure

Niacin (nicotinic acid) and its amide derivative have the following structures:

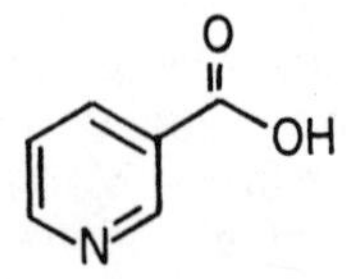

Niacin (Pyridine-3-carboxylic acid)

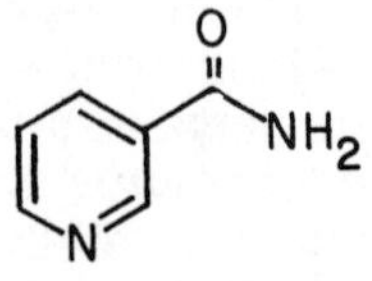

Nicotinamide

Niacin and niacinamide are equivalent to each other in biological activity for animals and man. Both are very stable in heat, light and alkali and are, therefore, stable in feeds.

Functions

Niacin functions as a constituent of 2 important coenzymes which act as codehydrogenases. These enzymes are NAD or nicotinamide adenine dinucleotide and NADP or nicotinamide adenine dinucleotide phosphate. These niacin-containing enzymes are important links in the transfer of hydrogen from substrates to molecular oxygen, resulting in water formation. As such, they are vital in several chemical reactions that occur in animal tissues. Their formulas are shown:

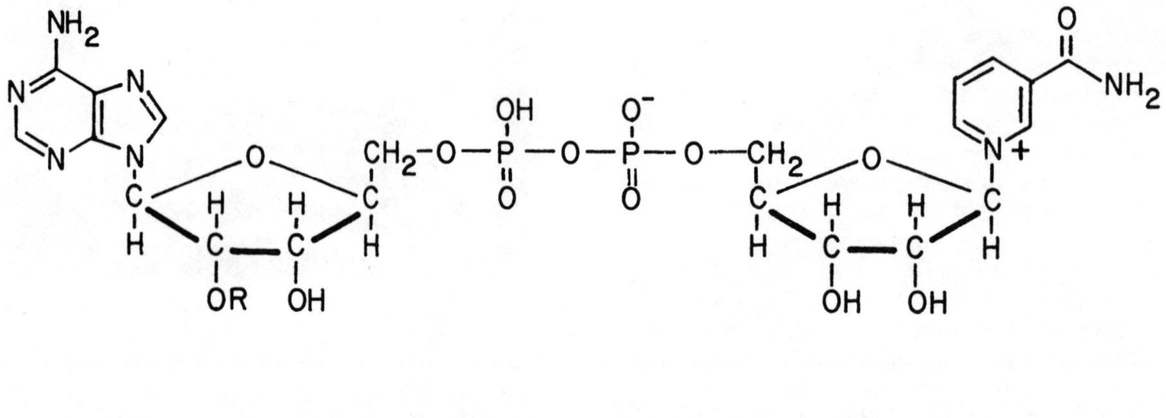

NAD (R=H) or NADP (R=PO_3H_2)

NAD and NADP probably combine with a variety of protein carriers (apoenzymes) with specificity for a particular reaction. Thus, a large number of substrates can be dehydrogenated. All animal cells contain NAD and NADP; the ratio of the 2 varies among tissues. NAD is convertible to NADP and vice versa. Similarly, many of the reactions above that involve NAD and NADP are reversible. The enzymes (NAD and NADP) are reduced to the dihydro forms which are, in turn, dehydrogenated by flavin (riboflavin-containing) enzymes.

Deficiency Signs

The general sign of niacin deficiency is reduced growth and appetite. Specific deficiency signs in animals include diarrhea, vomiting, dermatitis, unthriftiness, and an ulcerated intestine (necrotic enteritis) in swine; poor feathering, a scaly dermatitis and, sometimes, "spectacled eye" (Fig. 14-4) in chicks; and a preculiar darkening of the tongue (black tongue) and mouth lesions in dogs. In man, a bright red tongue, mouth lesions, anorexia, and nausea are the common syndrome of pellagra (3, 26).

Niacin present in most cereal grains is not biologically available for the pig and other simple-stomached animals (27-30). This probably helps to account for the high incidence of pellagra among human populations subsisting on diets composed mainly of cereal grains.

Adult ruminants do not need dietary niacin under normal conditions because of synthesis by rumen microflora, but niacin deficiency has been produced in calves (31). Robinson (5) has summarized the evidence that supports the view that intestinal synthesis of niacin in man may provide some of the metabolic requirement for the vitamin under some conditions.

Figure 14-4. "Spectacled eye" in a niacin-deficient chick illustrating the loss of feathers around the eye. Courtesy of M.L. Sunde, Univ. of Wisconsin.

Niacin-Tryptophan Interrelationships

If animals are deprived of dietary niacin, the metabolic requirement for the vitamin can be met by conversion of tryptophan, an amino acid, to niacin (see Ch. 6). If tryptophan is provided in the diet in slight excess of the amount needed for tissue protein synthesis, the excess can be used to satisfy the niacin requirement. The synthesis of 1 mg of niacin in this way requires about 60 mg of tryptophan for the pig (32). The efficiency of conversion probably varies with species and with dietary and metabolic states. For example, Harmon et al (33) showed that the dietary niacin requirement of growing pigs was higher when the diet contained corn protein instead of milk protein, presumably because of the lower tryptophan availability from corn in addition to the known lack of availability of niacin from corn. N_1-methylnicotinamide and N_1-methyl-6-pyridone-nicotinamide are the main excretory products of niacin in the pig, dog, rat, and man, but in the chick it is dinicotinylornithine (26). Ruminants appear to excrete the vitamin largely unchanged. Levels of niacin metabolites in urine are often used in studies of niacin requirements and metabolism.

Analogs and Antagonists

Several compounds structurally related to niacin possess some vitamin activity. These include several fatty acid esters of niacin (34) and 3-hydroxyxanthranilic acid, an intermediate in the conversion of tryptophan to niacin (35).

Mice fed 3-acetyl-pyridine show signs of niacin deficiency (36). Although it appears to have some vitamin activity for normal dogs, it is toxic for niacin-deficient dogs (37). Considerable variation exists among species in utilization of or antagonism by various analogs of niacin (5).

Pantothenic Acid [Pantoyl-β-alanine]

A growth factor for yeast was first identified by Williams (38) as pantothenic acid. Later, Elvehjem and Koehn (39) and Lepkovsky and Jukes (40) prepared it from liver. It has since been shown to be required by a large number of animal species.

Structure

The structure of pantothenic acid, the peptide of a butyric acid derivative and β-alanine, is shown:

$$HO-CH_2-C(CH_3)_2-CH(OH)-C(=O)-NH-CH_2-CH_2-C(=O)OH$$

It is usually present as the Ca or Na salt of the d-isomer (active form) of pantothenic acid. The Ca salt is the most common form in which the vitamin is added to diets because it is less hygroscopic than the Na salt.

Functions

Pantothenic acid functions as a component of coenzyme A (CoA) — the coenzyme required for acetylation of numerous compounds in energy metabolism. The role of CoA in metabolism of fatty acids was discussed in Ch. 8. CoA is required in the formation of two-C fragments from fats, amino acids, and carbohydrates for entry into the citric acid cycle and for synthesis of steroids. A deficiency of pantothenic acid precludes the synthesis of CoA because the CoA molecule contains pantothenic acid at 10% of its weight. The structure of CoA is shown:

adenosine phosphate

pantothenic acid

β-mercaptoethylamine

Coenzyme A

The transformations of lipids and carbohydrates that involve S-acylated CoA include (15): hydrolysis of the thiol ester bond; racemization; dehydrogenation; reduction; hydration; carboxylation; condensation reactions; and transferases.

Deficiency Signs

In addition to causing reduced growth rate as a generalized effect in all animals studied, pantothenic-acid deficiency results in dermatitis in chicks (39, 40), graying of the hair (achromotrichia) in rats and foxes (41, 42), hemorrhaging and degeneration of the adrenal cortex of rats and mice (43-45), fetal death and resorption in rats (46), and embryonic death in chicks (47). In guinea pigs, diarrhea, rough hair coat, anorexia, and enlarged and hemorrhagic adrenals have been reported (48). A striking feature of pantothenic-acid deficiency is fatty infiltration of the liver. Other signs in dogs, are vomiting, gastritis, enteritis, and hemorrhage of the thymus, and kidney (49, 50). A prominent feature a deficiency in pigs is the effect on the nervous system. In addition to the impaired growth, loss of hair, and enteritis noted in other species, the deficient pig develops a peculiar gait call "goose-stepping" (Fig. 14-5). Follis and Wintrobe (51) showed that the abnormal gait was a result of degenerative damage to various nerves. Teague et at (52) observed abnormalities in newborn pigs of sows fed deficient diets through 2 reproductive cycles. Although no effects appeared on litter size, incidence of stillbirths, or vigor of the newborn pigs, one or more pigs in all litters from pantothenic acid-deprived sows showed difficulty and incoordination in walking.

Ruminants do not normally require dietary pantothenic acid because of rumen synthesis, but deficiency has been produced in calves by feeding a synthetic milk diet low in the vitamin. Signs of deficiency are rough hair coat, dermatitis, anorexia, loss of hair around the eyes (spectacle eye), and demyelinization of the sciatic nerve and spinal cord.

Figure 14-5. "Goose-stepping" in a pig fed a pantothenic acid-deficient diet.

Pantothenic acid appears to be closely related to antibody production. Ludovici et al (53) showed a marked impairment in antibody formation to several types of antigens in deficient rats. Reduced formation of antibodies to diptheria toxoid in rats (54), to salmonella pullorum in pigs (16), and to tetanus toxoid in man (55) has been observed in pantothenic-acid deficiency.

Although clearcut deficiency of pantothenic acid has not been described in man, deficiency signs have been induced by feeding a pantothenic acid antagonist (omega-methyl-pantothenic acid). Signs included fatigue, apathy, gastrointestinal disturbances, cardiovascular instability (tachycardia and variable blood pressure), and increased susceptibility to infections. Caution must always be used in interpreting results of experiments, in which vitamin antagonists are used to induce deficiency of the vitamin from those caused by metabolism of the antagonist is impossible.

Vitamin B_6

Birch and Gyorgy (56) first established the properties of a component of a crude concentrate that was found to cure dermatitis in rats fed diets supplemented with thiamin and riboflavin. This compound was named pyridoxine. Two other compounds with pyridoxine activity were later identified, pyridoxal and pyridoxamine. The IUNS Committee on Nomenclature (1) now recommends that the term B be used for all 2-methyl pyridine derivatives with the biological activity of pyridoxine. The structures of the major compounds with vitamin B activity (pyridoxine, pyridoxal, and pyridoxamine) are shown:

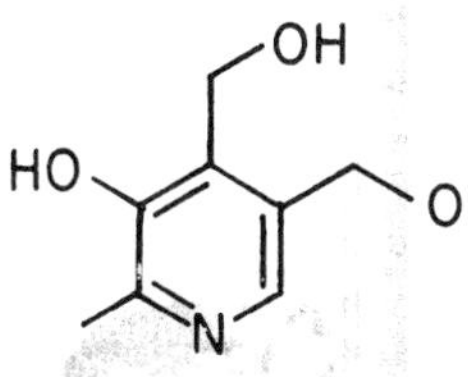

Pyridoxine

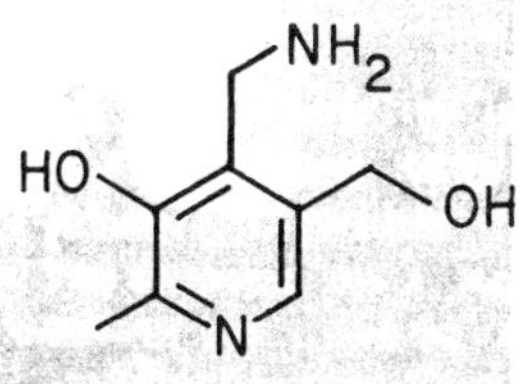

Pyridoxamine

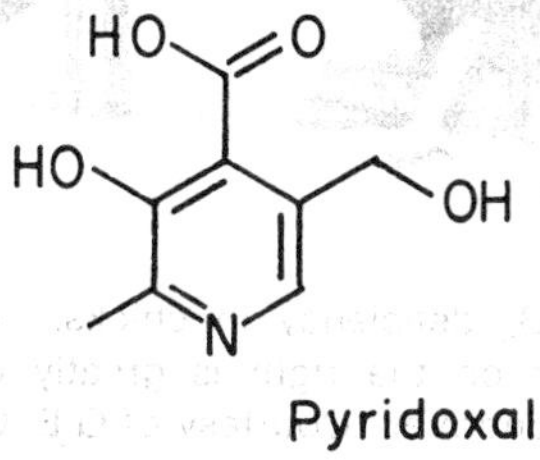

Pyridoxal

The chemistry of each of these forms of vitamin B_6 has been described by Harris (57), and Sauberlich (58) has described the biosynthesis and metabolic interconversions of them. A specific enzyme is involved in each conversion and niacin and riboflavin are both associated with one or more reactions as components of coenzymes (NADP and FMN, respectively). Vitamin B_6 is stable to heat, light, and alkali solutions and maintains it activity well in mixed diets.

Functions

Vitamin B_6 functions as a coenzyme of a vast array of enzyme systems associated with protein and nitrogen metabolism. As early as 1945, the B_6 requirement was recognized to increase with high dietary protein. This is explained by the fact that pyridoxal phosphate is the coenzyme of transaminases, which must be increased in activity to metabolize increased quantities of dietary protein. Sauberlich (58) tabulated a list of more than 50 pyridoxal-5'-phosphate-dependent enzymes and the reactions catalyzed and has listed the major types of enzymatic reactions involving pyridoxal phosphate-dependent enzymes as follows: transamination; decarboxylation; racemization (not in animal tissues); amine oxidation; aldol reaction; cleavage; dehydration (deamination); and desulfhydration.

Aspartic transaminase (glutamic oxaloacetic transaminase) is the most abundant transaminase in mammalian tissues. Amino acid decarboxylases are widespread in microorganisms, but the following are important in animal tissues: glutamic acid decarboxylase, cysteine-sulfonic acid decarboxylase, and aromatic L-amino acid decarboxylases. Tryptophan metabolism is heavily dependent on pyridoxal phosphate dependent enzymes, and a vitamin B_6 deficiency results in urinary excretion of abnormal metabolites of tryptophan — xanthurenic acid, kynurenine, and 3-hydroxykynurenine. Tyrosine and phenylalanine metabolism also involve a number of pyridoxal phosphate-dependent enzymes and deaminases (dehydrases) requiring pyridoxal phosphate are important in catabolism especially of serine, threonine, and homoserine (58).

Many phosphorylases in animal tissues contain pyridoxal phosphate, and total phosphorylase activity in skeletal muscle of rats is decreased in vitamin B_6 deficiency (59).

Vitamin B_6 is somehow involved in red blood cell formation, probably because pyridoxal phosphate is required for porphyrin synthesis (60). Vitamin B_6 is also important in the endocrine system; deficiency affects activities of growth hormone, insulin, and gonadotrophic, adrenal, thyroid, and sex hormones, but the mechanisms are not well understood (61).

Absorption of amino acids from the GIT somehow involves vitamin B_6, but the mechanisms have not been established; no specific enzyme systems requiring pyridoxal phosphate are known.

Deficiency Signs

The most common sign of B_6 deficiency involves the nervous system. Convulsions have been observed in all species studied. Demyelinization of the peripheral nerves, swelling and fragmentation of the myelin sheath, and later other degenerative changes occur (3). The convulsive seizures observed in B_6-deficient animals have resulted in many studies of brain metabolites and enzymes. The administration of γ-amino-butyric acid temporarily controls the seizures in B_6-deficient animals (62).

Vitamin B_6 deficiency results in reduced antibody response to various antigens in rats (63), guinea pigs (64), and pigs (16, 70-72), and in skin lesions on the feet, around the face and ears (acrodynia), atrophy of the hair follicles, and abscesses and ulcers around the sebaceous glands from secondary infection in rats, mice and monkeys (65). Skin lesions have not been reported in B_6-deficient swine, calves, or poultry.

Vitamin B_6 is required for normal reproduction in the rat and for normal egg laying and hatchability in hens (26). It is synthesized by rumen bacteria and good evidence has been shown for some intestinal synthesis in simple-stomached animals (15). Although deficiency of B_6 is unlikely in farm animals because of its wide distribution in feedstuffs, the pig fed a B_6-deficient experimental diet exhibits the classical symptoms of deficiency, including convulsive seizures (Fig. 14-6, 14-7), poor growth, a brown exudate around the eyes, and low urinary excretion of the vitamin (3, 66).

Vitamin B_6 deficiency produces a wide variety of changes related to lipid metabolism including reduced carcass fatness, fatty liver, and elevated plasma lipids and cholesterol (58). The metabolic mechanism by which these effects occur in B_6 deficiency is unclear.

Oxalic acid is excreted in large quantities in the urine of B_6-deficient animals and humans; it is derived from metabolism of glycerine, serine, alanine, and other compounds which, under normal conditions, are metabolized to compounds other than oxalic acid.

Toxicity and Analogs

Vitamin B_6 toxicity is very unlikely; rabbits, rats, and dogs tolerate doses up to 1 g/kg body weight (76). High doses given orally or parenterally produce convulsions, impaired coordination, paralysis, and death. All 3 forms of the vitamin are readily absorbed from the GIT and are widely distributed in tissues. Most of an administered dose appears in the urine with negligible amounts found in feces or sweat. The only known excretory product of B_6 in animals is 4-pyridoxic acid which is derived directly from pyridoxal (68).

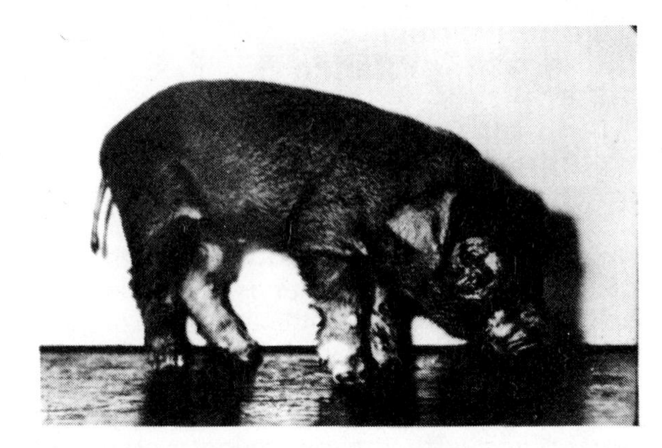

Figure 14-6. A B_6 deficient piglet. Deficient pigs typically show anorexia, poor growth and convulsions. Courtesy of D.E. Ullrey, Michigan State University.

Figure 14-7. B_6 deficiency in chicks. Growth of the deficient chick on the right is greatly depressed and feathering is abnormal. Courtesy of G.E. Combs.

Structural analogs, including 4-deoxypyridoxine and 2-methylpyridoxine, function as B_6 antagonists in animals, and agents such as hydrazine derivatives (including isonicotinylhydrazine) form inactive complexes with pyridoxal or pyridoxal phosphate and produce B_6 deficiency symptoms, including convulsions in man and animals (67).

Vitamin B_{12} [Cyanocobalamin]

Vitamin B_{12} is the most recently discovered vitamin. It was first known as the animal protein factor (APF), because animals fed diets not containing animal protein were prone to the hematological and neurological signs that were later shown to be prevented or cured by vitamin B_{12}. It has since been shown to be present in animal tissues and excreta, although it is probably not formed by animal or plant tissues. The only known primary source of vitamin B_{12} is from microorganisms; it is synthesized by a wide range of bacteria but not in appreciable amounts by yeast and fungi (69). The chemistry and properties of B_{12} have been reviewed by Moore and Folkers (70).

Structure

The compound, α-(5, 6-dimethylbenzimidazolyl) cobamid cyanide, is designated vitamin B_{12} or cyanocobalamin. Several derivatives, including hydroxycobalamin and nitritocobalamin, also have B_{12} activity (1). The structure for cyanocobalamin is shown.

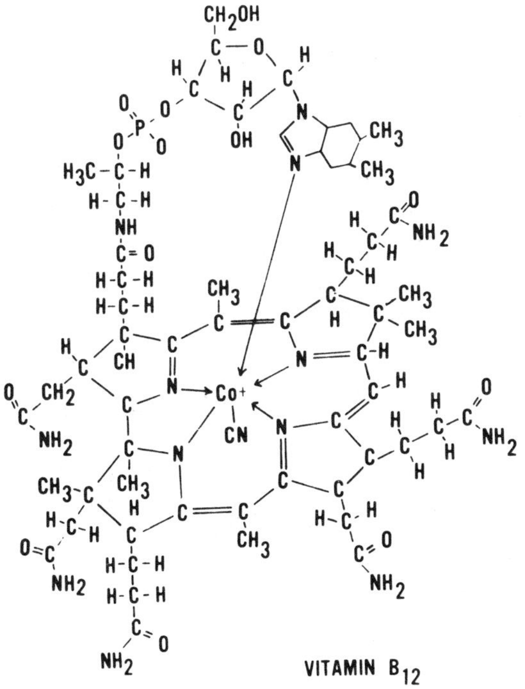

VITAMIN B_{12}

Vitamin B_{12} is a dark red crystalline compound, unstable at temperatures above 115°C for 15 minutes, unstable to sunlight, and stable within the pH range of 4-7 at normal temperatures. It contains cobalt and phorphorus in a 1:1 molar ratio.

The vitamin was first isolated in 1948 by Rickes et al (71) and by Smith and Parker (72). Since then a variety of isolation sources have been used with high yields coming from fermentation liquors of streptomycin, aureomycin, or by special fermentations carried out with selected microorganisms (70).

Functions

Vitamin B_{12} functions as a coenzyme in several important enzyme systems. These include isomerases (mutases), dehydrases, and enzymes involved in methionine biosynthesis. The oxidation of propionate in animal tissues involves a series of reactions requiring B_{12} as well as pantothenic acid as components of coenzymes. These reactions are illustrated (73). Vitamin B_{12} is a component of a coenzyme for the enzyme, methylmalonyl-CoA isomerase, which catalyzes the conversion of methylmalonyl-CoA to succinyl-CoA, which in turn is converted to succinate for entrance into the TCA cycle.

Propionyl-CoA
↑↓ (CO_2 + ATP)
Methylmalonyl-CoA
↑↓ (B_{12} coenzyme) (methylmalonyl-CoA isomerase)
Succinyl-CoA
↑↓ (-CoA)
Succinate

Vitamin B_{12} is closely linked with another water-soluble vitamin, folic acid,* in the synthesis of methionine in both bacterial and animal cells. Larrabee et al (74) proposed a scheme for describing the overlapping metabolic effects of vitamin B_{12} and folic acid (THF) in methionine synthesis from homocysteine. Niacin (as NAD) is involved in the series of reactions as well.

A more complete understanding of the role of vitamin B_{12} in animal metabolism and its relationship to folic acid awaits further study (73), but the general statement can be made that it is a component of coenzymes required for methyl-group synthesis and metabolism and that it is required for nucleoprotein synthesis in tandem with folic acid.

Absorption, Transport, Storage and Excretion of Vitamin B_{12}

Vitamin B_{12} absorption requires the presence of an enzyme secreted by the mucosal cells of the stomach and upper small intestine (cardiac region of the stomach in man; pyloric region of stomach and upper duodenum of pigs and probably other animals) (75). This enzyme has been designated the intrinsic factor (IF), and in its absence B_{12} deficiency occurs. IF from one species may inhibit absorption of B_{12} in another species. For example, IF from pig stomach renders orally administered B_{12} less available to chicks and rats (76).

The B_{12}-IF complex passes down the GIT to the ileum where it is further complexed with Ca and Mg ions and is adsorbed to the surface of the mucosa (77). It is then disassociated from the Ca or Mg by an apparently specific releasing enzyme contained in the intestinal secretions, and the B is absorbed. The exact mechanism by which it passes through the mucosal cell membrane is not known, but it has been suggested that it passes through the cell membrane by pinocytosis before the complex is broken by the releasing enzyme. After reaching the blood, B_{12} is bound to an α-globulin designated as transcobalamin I or, if injected, to a β-globulin designated transcobalamin II, from which it is subsequently shifted to transcobalamin I. The exact role of each form in the transport and storage of B_{12} is not known.

The amount of B_{12} not needed for immediate use is stored in liver and other tissues. In humans, 30-60% is stored in the liver, 30% in muscle, skin and bone, and smaller amounts in other tissues (75). When the protein-binding capacity of the blood serum for B_{12} is surpassed, excretion of the free vitamin occurs through the kidney and bile. More B_{12} is excreted daily in the bile in humans than is contained in the blood. The daily fecal excretion is less than the daily bile excretion, showing that reabsorption of a high proportion of B_{12} from the GIT occurs.

Deficiency Signs

Although B_{12} is widely distributed in animal tissues and products, a deficiency is inevitable when monogastric animals are maintained for long periods of time on diets entirely of plant origin. In ruminants, whose rumen microflora are capable of synthesizing the vitamin for use by the host, a B_{12} deficiency can be induced by feeding a Co-deficient diet. Because Co is a constituent of the B_{12} molecule, the vitamin cannot be synthesized when Co is not available. The relationships between B_{12} and Co in ruminants have been detailed in reviews by Underwood (78) and Church et al (79) and are discussed in Ch. 12.

Growth failure is a general symptom of deficiency in all animal species studied. In B_{12}-deficient chicks, poor feathering, kidney damage, impaired thyroid function, lower level of sulfhydryl groups in the thyroid, perosis, depressed plasma proteins, and elevated blood nonprotein N and glucose occur (76). Eggs from B_{12}-deprived hens fail to hatch; embryos die at about day 17 and show multiple hemorrhages, fatty livers, enlarged hearts and thyroids, and lack of myelination of the sciatic nerves and spinal cord (80, 81).

Baby pigs deprived of B_{12} show rough hair coats, incoordinated hind leg movements, normocytic anemia, and enlarged liver and thyroids (82). Sows fed B_{12}-deficient diets during gestation have a high incidence of abortion, small litters, abnormal fetuses, and inability to rear the young (83, 84). In deficient rats, retarded heart and kidney development, fatty liver, reduced blood sulfhydryl compounds, a decrease in activities of liver cytochrome oxidase and dehydrogenases, and an increase in liver CoA activity occur (76). The increased pantothenic acid content of liver in B_{12} deficiency is a general observation among the animal species studied. Vitamin B_{12} deficiency in man is manifested by a megaloblastic anemia (pernicious anemia), but is observed mainly when absorption is impaired by the absence of IF.

Cobalt-deficient cattle and sheep develop deficiency signs reversible by B_{12} Signs include reduced appetite, emaciation, anemia, fatty liver, birth of weak young, and reduced milk production. Elliot (85) has suggested that ketosis, a relatively common disease of lactating dairy cattle, may be related to a metabolic deficiency of B_{12} which upsets the metabolism of propionate. Only a small proportion of the B_{12} synthesized in the rumen is absorbed (86) and, of the total vitamin B_{12} analogs produced in the rumen, only a small fraction have biological activity.

The literature reports no evidence that vitamin B_{12} given in large doses causes acute or chronic toxicity.

Folacin

Structure

The compound, monopteroylglutamic acid (folic acid) has the structure shown:

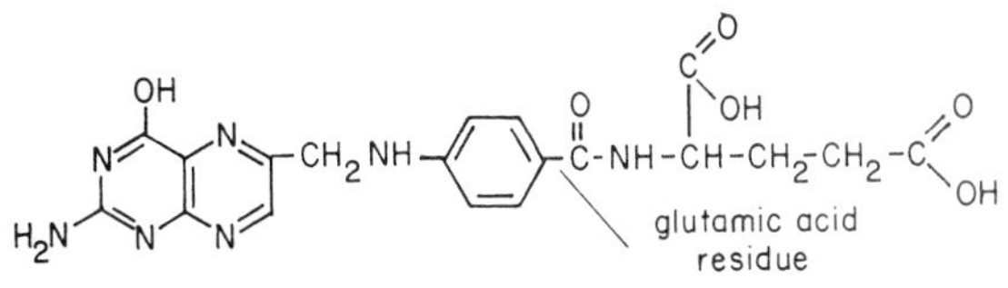

glutamic acid residue

Folacin

The term folacin should be used as the generic description for folic acid and related compounds with biological activity of folic acid (1). Folic acid derivatives with the glutamic acid residue combined through a peptide bond with one or more additional residues are designated folic acid derivatives with biological activity (1): tetrahydrofolic acid; 5-formyltetrahydrofolic acid; 10-formyltetrahydrofolic acid; or 5-methyltetrahydrofolic acid.

Functions

The relationship of folacin to B_{12} metabolism was emphasized in the previous section. Folic acid and its derivatives take part in a variety of metabolic reactions involving incorporation of Single-C units into large molecules. The metabolically active form of folacin is tetrahydrofolic acid which is a constituent of a coenzyme associated with metabolism of single-C fragments. A dietary source of folacin is converted by the body tissues to tetrahydrofolic acid which is known to be required for the biosynthesis of purines, pyrimidines, glycine, serine, and creatine (5). Folacin may also be concerned with synthesis of the enzymes, choline oxidase and xanthine oxidase (5), and is involved in choline and methionine metabolism (87). Other suggested roles of folacin in metabolism which are less well documented include: ascorbic acid biosynthesis in rats; porphyrin portions of metal porphyrin enzyme synthesis; and regulation of metal ions in the body (5).

Absorption, Storage and Excretion

Considerable intestinal synthesis of folic acid and its derivatives occurs, and man and animals may absorb significant quantities. Folacin is freely absorbed from the GIT and carried to all tissues of the body. Absorption is an active process (91). The liver contains high concentrations of folacin and is apparently the main site of conversion of folic acid into 5-formyltetrahydrofolic acid, along with the bone marrow (92). Vitamin B_{12} enhances the conversion (93).

Folic acid excretion occurs in urine, feces, and sweat. Sweat constitutes a major route of loss of folic acid in man, but in animals whose sweat glands are less well developed (pig, cow, sheep), most of the excretion occurs in feces and urine. Of course, much of the fecal excretion is of microbial origin in the GIT.

Deficiency Signs

The most prominent sign of folacin deficiency in man and animals — aside from reduced growth rate — is a macrocytic, hyperchromic anemia, leucopenia, and thrombocytopenia (3, 5). Resistance of animals to infections is affected by folacin; for example, monkeys show increased resistance to experimental poliomyelitis when suffering from chronic deficiency (88), and deficient rats show impairment of antibody response to murine typhus (89).

Folacin deficiency causes abnormal fetuses in pregnant rats. The central nervous system and the skeleton appear to be the most affected (90). Immature ruminants have been shown to require folacin in the diet, although a deficiency is unlikely because of the high content of most feedstuffs.

Folic acid antagonists such as aminopterin and amethopterin have been shown to inhibit cell division in normal and abnormal tissues, presumably because of their effect on purine and pyrimidine synthesis.

Because folacin is widely distributed in nature, it is unlikely to be deficient in common diets for birds and animals. Antagonists create the possibility for deficiency when they are present in the diet by accident or as antimicrobials, however.

Biotin

Biotin was first recognized as a growth factor for yeast. Later rats fed raw egg white developed skin lesions and loss of hair which were found to be cured by a protective factor in liver now known to be biotin (84). Biotin was first isolated from egg yolks (57) and later from liver (95, 96). The chemistry, isolation and biosyhesis of biotin have been reviewed (5, 57, 97).

Structure

The structure of biotin, formerly known as vitamin H or coenzyme R, is shown:

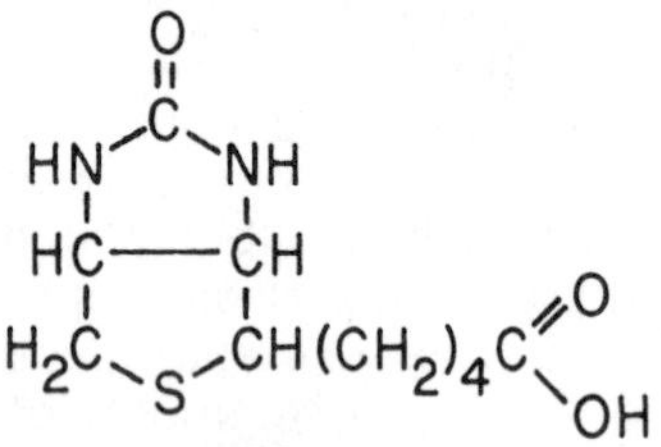

```
          O
          ‖
          C
        /   \
      HN     NH
       |      |
      HC ——— CH
       |      |           O
      H2C    CH(CH2)4C  ∥
         \  /            \
          S               OH
```

Biotin

Functions

Biotin is a component of several enzyme systems and, as such, participates in the following reactions: conversion of propionyl CoA to methylamalonyl-CoA; degradation of leucine; fat metabolism; and transcarboxylation reactions.

Biotin has been suggested to be involved in aspartic acid synthesis, in deamination of amino acids, and somehow involved in the activities of malic enzyme and ornithine transcarboxylase. The effects of biotin deficiency on protein, carbohydrate, and lipid metabolism can probably be explained on the basis of its involvement in the reactions listed above rather than on some general basis.

Deficiency Signs

A deficiency of biotin is unlikely under normal dietary conditions because the quantitative requirement of most animal species is low relative to the biotin content of common feedstuffs and because microbial synthesis in the GIT provides a source of biotin for ruminants and for simple-stomached animals that practice coprophagy.

Deficiency in rats is characterized by a progressive scaly dermatitis and alopecia (97). Biotin is needed for normal reproduction in female and male rats. Deficiency in females results in birth of young with abnormalities of the heart and circulatory system, and in deficient males development of the genital system is retarded. Dermatitis and perosis are the chief signs of biotin deficiency in chicks and turkey poults. Biotin deficiency is aggravated when pantothenic acid is also deficient in the diet of rats and chicks (98, 99). Biotin deficiency has been produced experimentally in young pigs (100), resulting in alopecia, seborrheic skin lesions, hind leg spasticity, and cracked hooves. Experimental biotin deficiency signs among healthy humans eating normal diets are unconvincing.

Scattered evidence indicates that biotin deficiency may decrease resistance to disease in some animals under some conditions, but much more work is needed to clarify this role of biotin (97).

Deficiency signs have been produced in chicks, rats, pigs, fish, hamsters, mink, guinea pigs, and other species fed raw egg white. A protein-like constituent of egg white, avidin, forms a stable complex with biotin in the GIT and renders the biotin unavailable for absorption. Avidin is destroyed by moist heat and the avidin-biotin complex is also destroyed by steaming for 30-60 minutes (102), so that biotin deficiency is not likely in animals fed cooked egg protein.

Biotin deficiency symptoms have been recently reported in swine under field conditions. Cunha (103) suggested that the increased use of slotted floors may be a factor, because coprophagy may be reduced. If so, we have an example of the effects of changing technology on the need for continued reappraisal of the nutrient requirements of animals under practical conditions of production.

There is no evidence for significant toxicity signs from oral or parental administration of massive doses of biotin in animals.

Choline

Choline was first isolated from hog bile in 1849 and its structure was reported in 1867. The chemistry and metabolism of choline have been reviewed recently by Griffin and Nye (104). Choline is widely distributed in animal tissues as free choline, acetylcholine, and as a component of phospholipids including lecithin and sphingomyelin. The structure of free choline is shown:

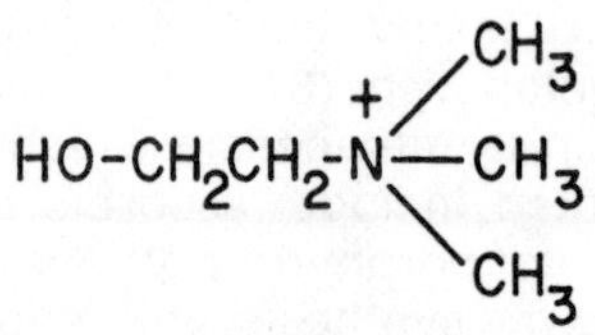

$$HO-CH_2CH_2-\overset{+}{N}(CH_3)_3$$

Choline

Choline added to feeds is usually supplied as choline chloride or choline dihydrogen citrate. Choline chloride is extremely hygroscopic and therefore is often added to the diet as a 70% aqueous solution.

Functions

Choline has the following broad functions: as a structural component of tissues (in lecithin, sphingomyelin); it is involved with transmission of nerve impulses as a component of acetylcholine; and it supplies biologically labile methyl groups. Unlike most other water-soluble vitamins, no good evidence exists for a role of choline or its derivatives as co-factors in enzymatic reactions.

The importance of choline as a component of phospholipids is clear from the discussion on lipid metabolism (Ch. 8). Its role as a component of acetylcholine is vital, as acetylcholine is the compound responsible for transmission of nerve impulses (104). The main physiologic effects of acetylcholine are peripheral vasodilation, contraction of skeletal muscle, and slowing of the heart rate. Acetylcholine is hydrolyzed to choline and acetic acid by cholinesterases. The acetylation of choline to acetylcholine is driven by an enzyme system requiring CoA.

Choline serves as a donor of methyl groups in transmethylation reactions. Labile methyl groups are attached to N or S which has or can acquire an additional positive charge. Labile methyl groups may come from choline or from its oxidation products, betaine aldehyde, betaine, and dimethylglycine.

Labile methyl groups, although for a time considered essential in the diet, are synthesized in body tissues from other one-C fragments. Folic acid is involved in the metabolism of these one-C fragments from which methyl groups are derived, and vitamin B_{12} is required in transfer of methyl groups from one metabolite to another once they are formed.

Deficiency Signs

The manifestations of choline deficiency are probably related to interference with all of its 3 distinct functions. As choline is a component of phospholipids, a deficiency is associated with fatty liver. Other signs of choline deficiency are hemorrhagic kidneys and other tissues of rats and other species, and perosis (slipped tendon) in chicks. The accumulation of lipids in the liver (fatty liver syndrome) in choline deficiency is common to all species studied. Liver fat accumulation is associated with a depressed level of serum lipids; Olsen et al (105) suggested that choline stimulates lipoprotein formation for lipid transport from liver to other tissues. Such an action of choline would explain the reduced serum lipid levels observed in choline deficiency.

Dietary methionine can completely replace choline for prevention of fatty liver in rats and pigs (106). McKittrick (107) had concluded earlier that the choline requirement of chicks consists of 2 parts — one that is indispensible and one for which methionine can be substituted. The same probably applies for other species as well. In young animals this can only occur if methionine is provided in excess of that needed for tissue protein synthesis; that available above the amount needed for tissue growth can supply labile methyl groups for choline synthesis from aminoethanol.

The choline requirement is reduced when the diet limits growth by a deficiency of another nutrient. Fat increases the requirement and caloric restriction reduces it. In fact, the fatty liver and hemorrhagic kidneys produced in rats fed a choline-deficient diet are prevented by restricting the fat and carbohydrate of the diet to reduce growth (104).

In slipped tendon (perosis) of choline-deficient chicks, the symptoms are the same as seen in Mn and biotin deficiencies.

In swine, choline deficiency causes an abnormal gait in growing pigs and reproductive failure in adult females in addition to the fatty liver and hemorrhages noted for other species (108).

No direct evidence exists for choline deficiency in man (109), but a considerable effort has been devoted to studying the possible relationship between fatty infiltration of the liver in alcholics and choline nutriture. The relationship between alcoholic liver cirrhosis and nutrition is a complex one in which choline appears to be only one of a number of factors.

Reports of choline toxicity have not been found in the literature.

Vitamin C [Ascorbic Acid]

A factor in citrus fruits capable of preventing scurvy in humans was recognized many years ago. The isolation of L-ascorbic acid and its characterization as a water-soluble vitamin have been reviewed by Hay et al (110).

Structure

The structure of L-ascorbic acid is shown:

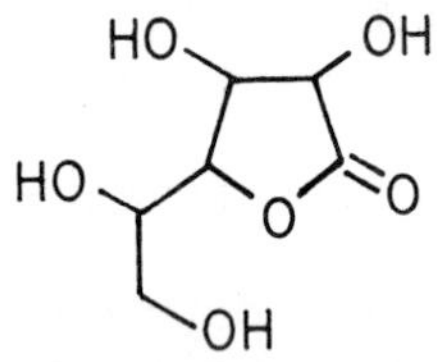

Ascorbic acid (reduced form)

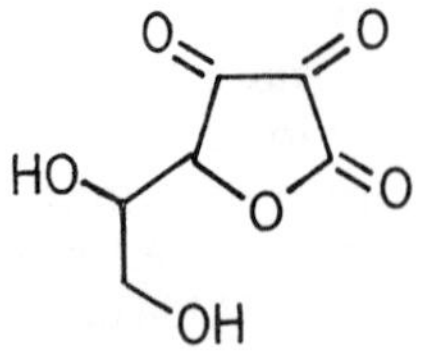

Dehydroascorbic acid (oxidized form)

Both the reduced and oxidized forms of L-ascorbic acid are physiologically active.

Functions

Ascorbic acid is involved in several important metabolic conversions which may be summarized as follows (111): it is directly involved with a number of enzymes catalyzing oxidation and reduction (electron transport) reactions. Ascorbic acid, itself easily oxidized and reversibly reduced, serves as a reducing agent. It is required to maintain normal tyrosine oxidation and for normal collagen metabolism. Specifically it is required for the formation of hydroxyproline from proline (111) and hydroxylysine from lysine (112). The hydroxylation reaction occurs only when the amino acids are in peptide linkage in collagen.

It is required as a co-substrate in certain mixed-function oxidations, as in the conversion of dopamine to norepinephrine (112). L-ascorbic acid can be replaced by D-ascorbic acid, isoascorbic acid, or glucoascorbic acid in this reaction (113). It is required along with ATP for the incorporation of plasma iron into ferritin. ATP, ascorbate, and iron appear to form an activated complex which allows release of ferric iron from a tight complex with plasma transferrin for incorporation into tissue ferritin.

Because ascorbic acid is not stored in appreciable quantities in body tissues, it must be supplied almost on a day-to-day basis for those species unable to synthesize it. Most mammalian and avian species are able to synthesize it from glucose in adequate amounts, but a growing list of species is known to develop a deficiency when dietary ascorbate is withheld. Man, other primates, and the guinea pig are the classical examples of animals unable to synthesize vitamin C (114). Other species now known to require a dietary source are the Indian fruit bat, red-vented bulbul, the flying fox, and some nonmammals such as rainbow trout, Coho salmon, two species of locust, and the silkworm. Although Braude et al (115) showed that the pig synthesizes ascorbic acid, more recent evidence (116, 117) suggests that a growth response may be obtained by supplemental vitamin C in growing-pig diets under some conditions.

Deficiency Symptoms

The first sign of vitamin C deficiency is a depletion of tissue concentrations of the vitamin (118). Early signs, known as scurvy in man, include edema, weight loss, emaciation, and diarrhea. Specific structural defects occur in bone, teeth, cartilage, connective tissues, and muscles. These defects can be explained largely by a failure in collagen formation in these tissues, including defective matrix formation in bones. Hemorrhages are commonly seen in muscles and in the gums as a result of increased capillary fragility. Hemorrhage, fatty infiltration, and necrosis may occur in the liver, and the spleen and kidney are damaged by hemorrhage. The adrenal gland is enlarged, congested, and infiltrated with fat. Adrenal hormone secretions are markedly affected. Biochemical changes in vitamin C deficiency include increased formation of mucopolysaccharides in connective tissue, increased hyaluronic acid in repair tissue of animals with scurvy, and decreased incorporation of sulfate into mucopolysaccharides. Vitamin C deficiency is associated with decreased serum protein concentration and anemia, prolonged blood clotting time, and delayed wound healing.

The role of ascorbic acid in Ca and P metabolism is apparently related to its association with bone matrix formation. Ca and P cannot be deposited normally in bone when matrix formation is impaired by vitamin C deficiency. Ascorbic acid enhances Fe absorption and has less predictable through generally beneficial effects on utilization of S, F, and I.

Deficiency signs in man resemble those reported in animals (119). Because of the generally recognized reduction in plasma ascorbic acid concentration and reduced urinary excretion of ascorbic acid in the presence of infectious disease, some have suggested that higher than recommended levels should be ingested in times of illness or stress. Pauling (120) has suggested that levels several times the NRC recommended levels are effective in prevention of the common cold. Solid support of this assertion seems unavailable. Although the toxicity level of ascorbic acid is apparently many-fold greater than the requirement, evidence suggests that chronic ingestion of 5-50X those levels recommended may increase susceptibility to scurvy when vitamin C is withdrawn (127). Furthermore, adverse effects of high levels in the diet of animals not considered to have a dietary ascorbic acid requirement may seriously reduce utilization of Cu and, perhaps, other trace elements (121). Therefore, indiscriminate ingestion of large doses of ascorbic acid in animals and man should be avoided, even though tissue shortage is limited and tolerance is high.

Other Water-Soluble Vitamins

Scattered reports have been made of a beneficial effect of the addition of inositol and para-aminobenzoic acid to purified diets for some animal species. No substantive evidence has been found that either of these compounds are essential dietary constituents for animals, however. Inositol is a common and abundant constituent of plant materials. Phytic acid, a major source of P in many cereal grains and other seeds, is the hexaphosphoric acid ester of inositol, but phytase enzymes needed to hydrolyze phytic acid to release free inositol are absent or limiting in normal digestive secretions. Nevertheless, inositol is present in abundant amounts in available form. Intestinal synthesis by microflora of the GIT also provides considerable inositol. Tissue synthesis of inositol has been demonstrated in the rat (122).

The chemistry, biogenesis and occurance, deficiency effects in animals, and pharmacology and toxicology of inositol have been recently reviewed (124-126). The final judgement as to whether inositol should be classed as a vitamin is still open.

Para-aminobenzoic acid (PABA) was first discovered as an essential growth factor for microorgansims. Although isolated reports have been made of a beneficial effect on animals by addition of PABA to the diet, no clear-cut evidence for a dietary requirement exists. Thus, PABA shares with inositol the status of being a questionable candidate for classification as a vitamin. It is abundantly synthesized in the intestine and the fact that it is a growth factor for some microorganisms suggests that a shortage in the diet of some animals may affect performance indirectly by limiting the synthesis of other vitamins.

The structure of inositol and PABA are shown:

Inositol (*myo*-inositol)

p-Aminobenzoic acid (PABA)

Although more than 25 years have passed since the discovery of the newest member of the water-soluble vitamin group (B_{12}), sporadic reports persist in the literature of unidentified factors that improve performance of animals and birds fed diets adequate in all known nutrients. Possibly one or more undiscovered organic compounds in nature are dietary essentials for animals and, will therefore be classified as vitamins. The nature of such compounds, their distribution in nature, and the amounts and conditions in which they are required are yet to be determined.

References Cited

1. UNS Committee on Nomenclature. 1971. J. Nutr. 101:133.
2. Williams, R.R. and J.K. Cline. 1936. J. Amer. Chem. Soc. 58:1504.
3. Follis, R.H., Jr. 1958. Deficiency Disease. C.C. Thomas Pub., Springfield, Ill.
4. Follis, R.H., Jr., M.H. Miller, M.M. Wintrobe and H.J. Stern. 1943. Amer. J. Path. 19:341.
5. Robinson, F.A. 1966. The Vitamin Co-Factors of Enzyme Systems. Pergamon Press Ltd., London.
6. Green, R.G. and C.A. Evans. 1940. Sci. 92:154.
7. Fujita, A. 1954. Advances Enzymol. 15:389.

8. Karrer, P., K Shopp and F. Benz. 1935. Helv. Chim. Acta 18:426.
9. Kuhn, R., K. Reinemund, F. Weygand and R. Strobele. 1935. Ber. deutch Chem. Gesellsch. 68:1765.
10. Wagner-Jauregg, T. 1972. In: The Vitamins. 2nd ed. Academic Press, N.Y.
11. Horwitt, M.K. 1972. In: The Vitamins. 2nd ed. Academic Press, N.Y.
12. Horwitt, M.K. and L.A. Witting. 1972. In: The Vitamins. 2nd ed. Academic Press, N.Y.
13. Wolbach, S.B. and O.A. Bessey. 1942. Physiol. Rev. 22:233.
14. Axelrod, A.E., B.B. Carter, R.H. McCoy and R. Geisinger. 1947. Proc. Soc. Exptl. Biol. Med. 66:237.
15. Pruzansky, J. and A.E. Axelrod. 1955. Proc. Soc. Exptl. Biol. Med. 89:323.
16. Harmon, B.G., et al. 1963. J. Nutr. 79:269.
17. Jusco, W.J. and G. Levy. 1967. J. Pharm. Sci. 56:58.
18. McCormick, D.B. 1972. Nutr. Rev. 30:75.
19. Yang, Chung-Shu and D.B. McCormick. 1967. J. Nutr. 93:445.
20. Mitchell, H.H., B.C. Johnson, T.S. Hamilton and W.T. Haines. 1950. J. Nutr. 41:317.
21. Unna, K. and J.G. Greslin. 1942. J. Pharmacol. Exp. Ther. 76:75.
22. Kaunitz, J., H. Wiesinger, F.C. Blodi, R.E. Johnson and C.A. Slanetz. 1954. J. Nutr. 52:467.
23. Hawkins, J. 1952. Biochem. J. 51:399.
24. Goldberger, J. and W.F. Tanner. 1925. U.S. Publ. Health Rep. 40:58.
25. Roe, D.A. 1973. A plague of corn: The social history of pellagra. Cornell Univ. Press. Ithaca, N.Y.
26. Maynard, L.A. and J.K. Loosli. 1969. Animal Nutrition. 6th ed. McGraw-Hill Book Co., N.Y.
27. Luecke, R.W., W.N. McMillan, F. Thorpe and C. Tull. 1947. J. Nutr. 36:417.
28. Kodicek, E., R. Braude, S.K. Kon and K.G. Mitchell. 1956. Br. J. Nutr. 10:51.
29. Chandhuri, D.K. and E. Kodicek. 1960. Br. J. Nutr. 41:35.
30. Luce, W.G., E.R. Peo and D.B. Hudman. 1966. J. Nutr. 88:39; 1967. J. Animal Sci. 26:76.
31. Hopper, J.H. and B.C. Johnson. 1955. J. Nutr. 56:303.
32. Firth, J. and B.C. Johnson. 1956. J. Nutr. 59:223.
33. Harmon, B.G., D.E. Becker, A.H. Jensen and D.H. Baker. 1969. J. Animal Sci. 28:848.
34. Badget, C.O., R.C. Provost, C.L. Ogg and C.F. Woodward. 1945. J. Amer. Chem. Soc. 67:1138; 1947. J. Amer. Chem. Soc. 69:2907.
35. Decker, R.H. and L.M. Henderson. 1959. J. Nutr. 68:17.
36. Woolley, D.W. 1945. J. Biol. Chem. 157:455; 1946. J. Biol. Chem. 162:179.
37. McDaniel, E.G., J.M. Hundley and W.H. Sebrell. 1955. J. Nutr. 55:623.
38. Williams, R.J. 1933. J. Amer. Chem. Soc. 55:2912.
39. Elvehjem, C.A. and C.J. Koehn. 1934. Nature 134:1007.
40. Lepkovsky, S. and T.H. Jukes. 1936. J. Biol. Chem. 114:109.
41. Oleson, J.J., C.A. Elvehjem and E.B. Hart. 1939. Proc. Soc. Exptl Biol. Med. 42:283.
42. Lunde, G. and H. Kringstad. 1939. Natureviss 27:755.
43. Draft, F.S., W.H. Sebrell, S.H. Babcock and T.H. Jukes. 1940. U.S. Publ. Health Req. 55:1333.
44. Salmon, W.D. and R.W. Engel. 1940. Proc. Soc. Exptl. Biol. Med. 45:621.
45. Melampy, R.M., D.W. Cheng and L.C. Northrup. 1951. Proc. Soc. Exptl. Biol. Med. 76:24.
46. Nelson, M.M. and H.M. Evans. 1946. J. Nutr. 31:497.
47. Gillis, M.B., G.F. Hensen and L.C. Norris. 1948. J. Nutr. 35:351.
48. Reid, M.E. and G.M. Briggs. 1954. J. Nutr. 52:507.
49. Scudi, J.V. and M. HAmlin. 1944. J. Nutr. 27:425.
50. Schaefer, A.E., J.M. McKibbin and C.A. Elvehjem. 1942. J. Biol. Chem. 143:321.
51. Follis, R.H., Jr. and M.M. Wintrobe. 1945. J. Exptl. Med. 81:539.
52. Teague, H.S., W.M. Palmer and A.P. Grifo. 1970. An. Sci. Mimeo 200. Ohio Agr. Exp. Sta., Wooster, Ohio.
53. Ludovici, P.P., A.E. Axelrod and B.B. Carter. 1951. Proc. Soc. Exptl. Biol. Med. 76:665.
54. Pruzansky, T. and A.E. Axelrod. 1955. Proc. Exptl. Biol. Med. 89:323.
55. Hodges, R.E., W.B. Bean, M.A. Ohlson and R.E. Bleiler. 1962. Amer. J. Clin. Nutr. 11:85.
56. Birch, T.W. and P. Gyorgy. 1936. Biochem. J. 30:304.
57. Harris, S.A. 1968. In: The Vitamins. Vol. II. Academic Press, N.Y.
58. Sauberlich, H.E. 1968. In: The vitamins. Vol. II. Academic Press, N.Y.
59. Illingworth, B., K. Kornfeld and D.H. Brown. 1960. Biochim. Biophys. Acta 42:486.
60. Rickert, D.A., B.Q. Pixley and M.P. Schulman. 1960. J. Nutr. 71:289.
61. Meites, J. and M.M. Nelson. 1960. Vitamins and Hormones 18:205.
62. Dasgupta, S.R., E.K. Killam and K.F. Killam. 1958. J. Pharmacol. Exptl. Therap. 122:16A.
63. Axelrod, A.E. and J. Pruzansky. 1955. Vitamins and Hormones 13:1.
64. Axelrod, A.E., S. Hopper and D.A. Long. 1961. J. Nutr. 74:58.
65. Weber, R., H. Wieser and O. Wiss. 1968. In: The Vitamins. Vol. II. Academic Press, N.Y.
66. Wintrobe, M.M., et al. 1943. Bul. Johns Hopkins Hosp. 72:1.
67. Unna, K.R. and G.R. Honig. 1968. In: The Vitamins. Vol. II. Academic Press, N.Y.
68. Snell, E.E. 1969. In: Vitamin Metabolism. Vol. II. Pergamon Press Ltd., London.

69. Smith, E.L. 1950-51. Nutr. Abstr. Rev. 20:795.
70. Moore, H.W. and K. Folkers. 1968. In: The Vitamins. Academic Press, N.Y.
71. Rickes, E.L., et al. 1948. Science 107:396.
72. Smith, E.L. and L.F.J. Parker. 1948. Biochem. J. 43:viii.
73. Barker, H.A. 1968. In: The Vitamins. Academic Press, N.Y.
74. Larrabee, A.R., S. Rosenthal, R.E. Cathose and J.M. Buchanan. 1961. J. Amer. Chem. Soc. 83:4094.
75. Reisner, E.H. 1968. In: The Vitamins. Academic Press, N.Y.
76. Coates, M.E. 1968. In: The Vitamins. Academic Press, N.Y.
77. Cooper, B.A. and W.B. Castle. 1969. J. Clin. Invest. 39:199.
78. Underwood, E.J. 1971. Trace Elements in Human and Animal Nutrition. Academic Press, N.Y.
79. Church, D.C., G.E. Smith, J.P. Fontenot and A.T. Ralston. 1971. Digestive Physiology and Nutrition of Ruminants. Vol. 2 - Nutrition. O & B Books, 1215 NW Kline Pl., Corvallis, Ore.
80. Ferguson, T.M., R.H. Rigden and J.R. Couch. 1955. Arch. Pathol. 60:393.
81. Alexander, W.F. 1957. In: Vitamin B_{12} and Intrinsic Factor. Enke, Stuttgart, Germany.
82. Neumann, A.L., J.L. Krider and B.C. Johnson. 1945. Proc. Soc. Exptl. Biol. Med. 69:513.
84. Anderson, G.C. and A.G. Hogan. 1950. J. Animal Sci. 9:646.
85. Frederick, G.L. and G.J. Brisson. 1961. Can. J. Animal Sci. 41:212.
85. Elliot, J.M. 1966. Proc. Cornell Nutr. Conf., p 73.
86. Kercher, C.J. and S.E. Smith. 1955. J. Animal Sci. 14:458.
87. Dalal, F.R., D.V. Rege and A. Sceenivasan. 1961. Nature 190:267.
88. Lickstein, H.C., K.B. McCall, C.A. Elvehjem and P.F. Clark. 1946. J. Bact. 52:105.
89. Wertman, K., F.D. Crisley and J.L. Sarandria. 1952. Proc. Soc. Exptl. Biol. Med. 80:404.
90. Nelson, M.M., H.V. Wright, C.W. Asling and H.M. Evans. 1955. J. Nutr. 56:349.
91. Bergess, A.S.V. and N.J. Goldberg. 1962. Br. J. Pharmacol. 19:313.
92. Nichol, C.A. 1953. Proc. Soc. Exptl. Biol. Med. 83:167.
93. Doctor, V.M., J.R. Couch and J.B. Trunnell. 1954. Proc. Soc. Exptl. Biol. Med. 87:228; 498; 1955. Proc. Soc. Exptl. Biol. Med. 90:251.
94. Boas-Fixsen, M.A. 1927. Biochem. J. 21:712.
95. Gyorgy, P., D.B. Melville, D. Burk and V. du Vigneud. 1940. Science 91:243.
96. du Vigneaud, V., D.B. Melville, P. Gyorgy and C.S. Rose. 1940. Science 91:243.
97. Gyorgy, P. and B.W. Langer. 1968. In: The Vitamins. Vol II. 2nd ed. Academic Press, N.Y.
98. Emerson, G.A. and C. Wentz. 1941. Proc. Soc. Exptl. Biol. Med. 57:47.
99. Robblee, A.R. and D.R. Clandinin. 1953. Poult. Sci. 32:576.
100. Lehrer, W.P., A.C. Wiese and P.R. Moore. 1952. J. Nutr. 47:203.
101. Sydensticker, V.P., et al. 1942. Science 95:176.
102. Gyorgy, P. and C.S. Rose. 1941. Science 94:261.
103. Cunha, T.J. 1971. Feedstuffs 43(9):20.
104. Griffith, W.H. and J.F. Nye. 1971. In: The Vitamins. Vol. II. 2nd ed. Academic Press, N.Y.
105. Olson, R.E., et al. 1958. Amer. J. Clin. Nutr. 6:310.
106. Kroening, G.H. and W.G. Pond. 1967. J. Animal Sci. 25:352.
107. McKittrick, D.S. 1948. Archives. Biochem. 15:133.
108. Ensminger, M.E., J.P. Bowland and T.J. Cunha. 1947. J. Animal Sci. 6:409.
109. Hartroft, W.S. and E.A. Porta. 1971. In: The Vitamins. Vol II. 2nd ed. Academic Press, N.Y.
110. Hay, G.W., B.A. Lewis and F. Smith. 1967. In: The Vitamins. Vol. II. 2nd ed. Academic Press, N.Y.
111. Mapson, L.W. 1967. In: The Vitamins. Vol. I. Academic Press, N.Y.
112. Barnes, M.J. and E. Kodicek. 1972. Vitamins and Hormones 30:1.
113. Levin, E.Y., B. Levenberg and S. Kaufman. 1960. J. Biol. Chem. 235:2080.
114. Burns, J.J. 1957. Nature 180:553.
115. Braude, R., S.K. Kon and J.W.G. Porter. 1950. Br. J. Nutr. 4:186.
116. Cromwell, G.L., V.W. Hays and J.R. Overfield. 1970. J. Animal Sci. 31:63.
117. Mahan, D.C., et al. 1967. J. Animal Sci. 25:1019.
118. Chatterjeee, G.C. 1967. In: The Vitamins. Vol. I. Academic Press, N.Y.
119. Vilter, R.W. 1967. In: The Vitamins. Vol. I. 2nd ed. Academic Press, N.Y.
120. Pauling, L. 1971. Vitamin C and the Common Cold. W.H. Freeman and Co., San Francisco.
121. Gipp, W.F., et al. 1974. J. Nutr. 104:532.
122. Halliday. J.W. and L. Anderson. 1955. J. Biol. Chem. 217:797.
123. Angyal, S.J. 1971. In: The Vitamins. Vol. III. 2nd ed. Academic Press. N.Y.
124. Alam, S.Q. 1971. In: The Vitamins. Vol. III. 2nd ed. Academic Press, N.Y.
125. Cunha, T.J. 1971. In: The Vitamins. Vol. III. 2nd ed. Academic Press, N.Y.
126. Milhorat, A.T. 1971. In: The Vitamins. Vol. III. 2nd ed. Academic Press, N.Y.
127. Sorensen, D.I., M.M. Devine and J.M. Rivers. 1974. J. Nutr. 104:1041.

Chapter 15 – Nonnutritive Feed Additives and Growth Stimulators

The term, **feed additive**, is usually defined to include drugs and other compounds of a non-nutritive nature, which do not supply necessary nutrients. Additives are used to stimulate growth or other types of performance, improve the efficiency of feed utilization, or improve the general health of the animal. Growth stimulators may be feed additives, but may also include such things as some hormones or hormone-like chemicals which may be administered subcutaneously or intermuscularly rather than orally.

Feed additives have been used extensively in the USA and in many other countries during the decades of the 1950's and 1960's, particularly following the discovery and commercial production of the antibiotics and sulfa drugs. In more recent years, a trend has developed for government regulatory agencies and law-making bodies to restrict usage of additives for a variety of reasons. In the USA, the Food and Drug Administration has control of feed additives usage. Currently, very extensive documentation is required on safety, efficacy, and tissue residues. Nevertheless, there is a trend towards reduced utilization of additives. For example, use of antibiotics as feed additives in most European countries is now severely limited and the trend is moving in that direction in the USA. Whether this trend is good or bad may be beside the point, as it appears to be more a matter of political decisions made on the basis of pressure by those who believe usage of antibiotics in animal feeds may be detrimental to man, although current evidence does not indicate a problem in this respect.

Types of Feed Additives

A wide Variety of feed additives have been tried out on domestic animals at one time or another, but many of them have not withstood the test of time and careful experimentation as well as practical usage. Additives have come and gone as a result of such factors as cost, tissue residues, toxicity, or what is more common, no beneficial response by the animal. Except for additives that are strictly medicinal, most of those in use give a response only in young animals. With the primary excretion of buffers and thyroprotein, mature animals do not respond with improved performance.

The nonnutritive additives in most common use in the USA at the present time are the antibacterial agents — antibiotics, arsenicals, nitrofurans, and sulfa drugs — and hormones or hormone-like compounds intended to stimulate gain and improve feed efficiency in young, rapidly growing animals. Other additives of a more specific nature may find usage in special situations.

For those readers interested in more information, a series of articles that might be recommended include those of Cravens and Holck (4), Kemp and Kiser (6), Wallace (9), Weston (10), and Jukes (5). Further detail on ruminant species is given in Ch. 5 of Church (3).

Antimicrobial Drugs

Antibiotics are compounds produced by one microorganism which inhibit the growth of another organism, and are the most widely used of the antimicrobial drugs. A list of those in common use in the USA as feed additives for domestic species is shown in Table 15-1. Chlortetracycline, oxytetracycline, penicillin, and bacitracin have enjoyed wider usage than most of the other antibiotics, although there are many others and new ones are continually being isolated and studied. Arsenicals and nitrofurans are used primarily for poultry and swine.

Antibiotics are still in use, where allowed by law, because they usually give a response in terms of growth, improved feed efficiency and, generally, better health. The response in growth and improved feed efficiency is apt to be variable from animal species to species, time to time, and place to place. For example, there is little or no response in new animal facilities, in very clean surroundings, or in germ-free animals raised in aseptic conditions. The response that may be expected in young pigs is shown in Table 15-2 and that for growing and fattening cattle in Table 15-3.

Table 15-1. Antibiotics and other antibacterial drugs commonly used as feed additives for growth promotion.

Name	Approved for use with	Approved dose level [a]		
		Concentration in feed		per day or per unit wt,
		mg/lb	mg/kg	mg
Arsanilic acid or Na arsanilate	poultry	22-45	48-99	
	swine	22-45	48-99	
Bacitracin*	poultry	2-25	4.4-55	
	swine	5-25	11-55	
	beef			35
Zn bacitracin*	feedlot cattle			35-70
Chlortetracycline* (Aureomycin)	poultry	5-25	11-55	
	swine	5-25	11-55	
	calves			25-70
	feedlot cattle			70
	lambs	10-25	22-55	
Erythromycin	chicks, poults	2.3-9	5-20	
	young pigs	5-35	11-77	
	growing pigs	5	11	
	feedlot cattle			37
Furazolidone	poultry	3.5-5	7.7-11	
	lactating sows	75	165	
	young pigs	50-100	110-220	
Oleandomycin*	broilers, turkeys	0.5-1.0	1.1-2.2	
	pigs	2.5-5.6	5.5-12.3	
Oxytetracycline* (Terramycin)	chicks, broilers	2.5-3.7	5.5-8.1	
	hens, turkeys	5-25	11-55	
	pigs, 10-30 lb	12-25	26-55	
	30-200 lb	3.7-5	8.1-11	
	calves, 0-12 wk			0.1 mg/lb body wt
	calf starters, milk replacers			0.5 mg/lb body wt
	calves under 400 lb			25-75 mg
	feedlot cattle			75 mg
	dairy cows (bloat)			75 mg
	fattening lambs			20-25 mg
Penicillin*	poultry	1.2-25	2.6-55	
	growing, fattening pigs	5-25	11-55	
	cattle (bloat)			75 mg
Roxarsone	poultry	11-22	24-48	
	swine	11-33	24-73	
Streptomycin*	poultry	6-25	13-55	
	pigs	3.7-25	8.1-55	
Tylosin*	poultry	2-25	4.4-55	
	pigs	5-50	11-110	

*Antibiotics

[a] Dose levels for medical or therapeutic purposes are usually substantially higher than those listed here for growth promotion.

Evidence indicates that most antibiotic-fed animals eat more than control animals receiving the same diet without antibiotics. Thus, this may largely account for the improved growth and efficiency. When animals are given the same amount of food with and without antibiotics, those fed antibiotics have usually not gained any faster. Evidence does indicate that antibiotics may have a sparing effect on dietary needs for some amino acids and B-complex vitamin in young chicks, pigs, or rats, the beneficial response being greater when diets contain submarginal or minimal levels of those nutrients.

It has been suggested that this nutrient-sparing effect may be a result of (a) stimulation of microorganisms of the GIT which favor nutrient synthesis, (b) suppression of organisms which compete for critical nutrients, or (c) improved nutrient absorption from the GIT as a result of thinner, healthier intestinal walls seen in antibiotic-fed animals (9). Another possibility is that growth may be enhanced as a result of stimulation of various enzyme systems. For example, Visek (8) suggested that antibiotics improve performance by their antiurease action which results in reduced ammonia production in the GIT. Ammonia is toxic to cells and may increase cell turnover in the GIT epithelium. A third effect proposed is that of disease control, particularly control of organisms causing subclinical or nonspecific diseases; this seems a likely factor since healthy animals respond very little to antibiotics. Note the response of runt pigs shown in Table 15-2.

Table 15-2. Effects of antibiotics on growth and feed efficiency in pigs of different weights.[a]

Initial wt, kg	No. of experiments	Growth response, %	Feed utilization response, %
<11.4	13	19.6	4.1
11.4-13.6	37	15.6	0.9
13.7-15.9	41	15.0	2.6
16.0-18.2	34	14.3	7.8
18.3-22.7	44	10.5	4.2
>22.7	46	8.7	4.1
runt pigs	12	82 (30-154)	11 (4-23)

[a] From Wallace (9) as adapted from Braude et al (2); a variety of antibiotics were used in these experiments.

Table 15-3. Effect of low-level feeding of anitbiotics to beef cattle.[a]

Item	Growing-wintering		Finishing		Finishing + stilbestrol	
	Control	Antibiotic[b]	Control	Antibiotic[b]	Control	Antibiotic[b]
Daily gain, kg	0.56	0.61	1.16	1.24	1.24	1.29
Feed/unit gain	12.70	12.02	9.85	9.37	9.33	8.94
Number of animals	5352		2354		3007	
Av. initial wt, kg	228		310		330	
Av. days on trial	112		117		124	
Gain response, %	8.9		6.7		3.7	
Feed conversion response, %	5.4		4.9		4.2	

[a] From Wallace (9).
[b] Continuous low level at about 70 mg/head/day. Antibiotics involved were chlortetracycline, oxytetracycline, and bacitracin.

In addition to some expected improvement in gain and feed efficiency, the feeding of antimicrobial drugs usually will reduce the incidence of diarrhea expected in young animals and especially in young mammals deprived of colostrum. Animals fed the drugs are also less prone to go off feed, and enterotoxemia and death loss are less apt to occur in lambs fed high-grain rations. In beef cattle, the incidence of abscessed livers is usually greatly reduced in antibiotic-fed cattle on high-grain rations.

Some comment on continued use of antibacterial drugs in animal feeds is required at this point. Individuals and organizations that object to the continuous use of low levels in animal feeds do so on the basis that resistant pathogenic strains of microorganisms might develop which could be harmful to humans. In the USA, antibiotics have been widely used for more than 20 years as feed additives. While it is true that microbial resistance to antibiotics was shown from the first (indeed some microorganisms may come to require an antibiotic), no convincing, documented instance has been shown where more virulent pathogens have been isolated as a result of feeding low levels of antibiotics to animals. As a matter of fact, the evidence shows that resistant strains are nearly always less virulent (6, 10). Those who object to antibiotic feeding ignore the considerable amount of favorable evidence for antibiotics and the experience and advice of respected scientists such as Thomas Jukes, a man who has spent a substantial amount of his professional career working in this area and who maintains that we have every reason to continue usage of low levels of antibiotics in animal feedstuffs (5). While it is true that antibiotics and other drugs are sometimes used in lieu of good management practices, the beneficial response to antibiotics in livestock enterprises attests to the assertion that the cost in productivity in animal agriculture in the USA would be considerable if routine use of antibiotics were not allowed (4, 5). The ratio of benefit:risk, a common measure used to evaluate drugs, is greatly in favor of continued use of antibiotics.

Hormones

Many different hormones have been fed or injected into animals at one time or another with the intent of increasing growth or milk production or to modify the normal fattening processes. This list includes growth hormone, natural or synthetic adrenal cortical hormones, natural and synthetic estrogens, androgens, progestogens, androgen-estrogen combinations, thyroid and antithyroid compounds.

At one time, diethylstilbestrol (DES) was used for "chemical" caponization of male birds, but its use is no longer allowed in the USA because of potential residue problems. With swine, hormones have not shown any consistent benefit. For practical usage with calves and feedlot cattle, compounds of interest would include oral or implanted DES and hexestrol, both synthetic estrogens; melengestrol acetate, a synthetic progestogen used orally for heifers; zeranol (Ralgro), said to be an anabolic agent, which is implanted; Synovex, a combination of estrogen and progesterone, used as an implant; and Rapid Gain, a combination of testosterone and estrogen, used as an implant. Most of these products also appear to work well for feedlot lambs, although much lower dosages are required. Other hormones such as cortisone and growth hormone have been tried, but there is little interest in them because of cost or lack of response.

In ruminants, the various natural or synthetic hormones appear to produce a response which results from increased N retention accompanied by an increased intake of feed. The result is, usually, an increased growth rate, an improvement in feed efficiency and, frequently, a reduced deposition of body fat which may, at times, result in a lower carcass grade for animals fed to the same weight as nontreated animals.

In the USA, far more DES has been used orally as subcutaneous implants in feedlot cattle than any other of the growth-stimulating hormones. Comparative studies generally show similar responses from the other products, but the cost nearly always favors DES. An example of the type of response and the savings to be expected is shown in Table 15-4.

DES usage was temporarily banned in the USA in 1973. The reason for the ban is that DES can, under some conditions, be considered a carcinogen (may cause cancer), although at much higher dose rates than used in domestic animals. Although only exceedingly small amounts of DES residue may be recovered in liver and kidney tissues of treated animals (from 0 to a few parts/billion), this was deemed sufficient to ban it from use. Action was taken because of a federal law (Delaney Clause), which specifies a zero tolerance in tissues of such drugs, a concept which has no sound basis in analytical chemistry.

Thyroid-active hormones have been used from time to time, particularly with cattle. It has been amply demonstrated that thyroprotein or iodinated casein, which produces the same physiological response, may be used over short periods of time to stimulate milk production, particularly when cows are past their peak of production. However, continued use during the remainder of a normal lactation is not apt to result in any improvement in milk production (7), and withdrawal will result in a prompt decrease in milk production. Thus, there is little current interest in use of such products. No beneficial response is likely in growing and fattening animals. Antithyroid compounds have also been used on the assumption that a decrease in thyroid activity might be beneficial in the fattening process. Some studies have shown favorable response, particularly with respect to improved feed efficiency, but others have not and there is no current interest in use of such products.

Many plants contain compounds producing estrogenic activity in animals. These compounds, known as phytoestrogens, may be found in relatively large concentrations in some leguminous plants, particularly alfalfa [*Medicago Sativa*] red [*Trifolium pratense*] and subterranean clover [*Trifolium subterraneum*]. Soybean seed [*Glycine max*], cottonseed [*Gossypium hirutum*], and flax [*Linum usitatissiumum*] may also contain appreciable levels of estrogenic activity. A limited amount of evidence indicates that such plant estrogens may produce some growth stimulation in ruminants. Conversely, plants such as red clover may have such high levels as to be detrimental to the normal ovarian cycle during the breeding season of sheep (1). Many plants also have growth inhibiting factors; for example, an antitrypsin factor and antithyroid compounds in soybeans and in seeds of other legumes, and gossypol in cottonseeds. These compounds often result in growth inhibition in some animal species.

Table 15-4. Econcomic benefit of using diethylstilbestrol in cattle.[a]

Item	Daily gain, kg	Days to gain 227 kg	Feed conversion	Feed to grain 227 kg
No hormone treatment	0.90	250	7.50	1,705
With hormone	1.00	223	6.75	1,535
Advantage of hormone	0.10	27	0.75	170
Savings				
27 days at 10¢/day (yardage)	=	2.70		
170 kg feed at 5.5¢/kg	=	9.35		
	Total =	$12.05		

[a] From Cravens and Holck (4). These calculations are based on the assumption of an improvement in gain of 12% and in feed conversion of 10%.

Other Feed Additives

A variety of feed additives are used from time to time for specific purposes which may or may not be related to stimulation of growth or other forms of production. For example, activated carbon has shown some promise for reducing absorption of certain pesticides that may be contaminants in the diet. A variety of anthelmintics are used routinely as medicinals in animals for control of stomach and intestinal worms. One of these, 2,2-dichlorovinyl dimethyl phosphate, apparently stimulates growth in cattle as well. Antioxidants have some value in reducing oxidation of nutrients such as vitamin E and unsaturated fats with the result that dietary requirements of vitamin A and E may be reduced. Sodium bentonite, a clay, is used as a pellet binder and shows some promise of improving N utilization in ruminants. A variety of surface-active compounds are used for prevention and treatment of bloat in ruminants. Various buffers such as $NaHCO_3$

aid in the prevention of indigestion following sudden ration changes in ruminants and may help to relieve the low-fat milk problem in dairy cows. Other buffers are useful in prevention and treatment of urinary calculi. Various feed flavors are used in different animal rations, although published research data are not overly encouraging, but suggest that the effect is mainly on preference when a choice of rations is given rather than on total feed consumption. Enzymes have been tried extensively for chicks, pigs, and ruminants, but results generally are not encouraging. Organic iodides have value for prevention of foot rot in cattle. Drugs which inhibit methane production may have some potential for ruminants if nontoxic ones can be developed. Live yeast cultures and dried rumen cultures are sometimes promoted, but with little factual evidence of their worth.

References Cited

1. Andrews, F.N. 1959. In: Reproductive Physiology and Protein Nutrition. Rutgers University Press, New Brunswick, N.J.
2. Braude, R., H.D. Wallace and T.J. Cunha. 1953. Antibiotics and Chemotherapy 4:864.
3. Church, D.C. (ed). 1972. Digestive Physiology and Nutrition of Ruminants. Vol. 3 - Practical Nutrition. O & B Books, 1215 NW Kline Pl., Corvallis, Ore.
4. Cravens, W.W. and G.L. Holck. 1970. J. Animal Sci. 31:1102.
5. Jukes, Thomas H. 1972. Bioscience 22:526; Oregon Agr. Expt. Sta. Sp. Rpt. 375, p 8.
6. Kemp, G and J. Kiser. 1970. J. Animal Sci. 31:1107.
7. Schmidt, G.H., R.G. Warner, H.F. Tyrrell, and W. Hansel. 1970. J. Dairy Sci. 54:481.
8. Visek, W.J. 1968. J. Dairy Sci. 51:286.
9. Wallace, H.D. 1970. J. Animal Sci. 31:1118.
10. Weston, J.K. 1970. J. Animal Sci. 31:1127.

Chapter 16 – Feeding Standards and the Productive Functions

Introduction

Feeding standards are statements or quantitative descriptions of the amounts of one or more nutrients needed by animals. Use of such standards dates back to the early 1800's (see Maynard and Loosli, 6). There has been a gradual development over the years to the point where nutrient requirements for farm animals may be specified with a high degree of accuracy, particularly for growing chicks and pigs; although there are still many situations where nutrient needs cannot be specified with great accuracy for animals, nutrient needs of animals are far better understood than for man. This is due to the simple fact that people do not lend themselves to the type of experimentation needed to collect good quantitative data.

In the USA, the most widely used standards are those published by the various committees of the National Research Council (NRC). These standards (see Appendix Tables 7-29) are continually revised and reissued at intervals of a few years. In England, the standards in use are put out by the Agricultural Research Council (ARC). Other countries have similar bodies which update information and make recommendations on animal nutrient requirements.

Terminology Used in Feeding Standards

Feeding standards are usually expressed in quantities of nutrients required/day or as % of a diet; the former being used for animals given exact quantities of a diet and the latter more commonly when rations are fed ad libitum. With respect to the various nutrients, most are expressed in weight units, % or ppm. Some vitamins — A, D, E — are given international units. Protein requirements are most frequently given in terms of digestible protein (DP), although crude protein is often used for ruminants, and amino acids can be substituted for DP in monogastric species when adequate information is at hand. Energy is expressed in a variety of different forms (see Ch. 9). The NRC uses ME for poultry, DE for swine, DE, ME and TDN for sheep, ME and TDN for beef cattle with alternative use of NE_m and NE_g in growing and fattening cattle, and, for dairy cattle, values are given for DE, ME, TDN, NE_m and NE_g with additional values as NE_{milk} for lactating cows (see Appendix Table 7-29). The ARC uses ME almost exclusively. Other European standards are based on starch equivalents, Scandinavian Feed Units, and so forth. Regardless of the units used, feeding standards are based on some estimate of animal needs and have been derived from data obtained from a great many experimental studies done under a wide variety of conditions with a diverse list of feedstuffs.

Some comments are in order regarding the use of NE_m and NE_g for beef cattle and growing dairy cattle and NE_{milk} for lactating cows by the NRC publications. If one looks at the tables on feed composition in the respective publications,it is apparent that values are given in this energy terminology for almost all of the feedstuffs listed except for mineral supplements. The reader should be aware that most of these NE values have been calculated from older data which were originally given as TDN as there is a wealth of older data expressed in this form; some values have also been derived from DE and ME. Only a few feedstuffs have been directly evaluated in terms of NE_m, NE_g or NE_{milk}. However applicable these values may be for these respective classes of cattle, recalculating from existing data does not necessarily improve the original data.

As pointed out in Ch. 9, energy (for adult animals) is required in relation to body weight$^{0.75}$ rather than in direct proportion to body weight. All current feeding standards recognize this approach, although not all emphasize it in their descriptions. For the other nutrients, Crampton (5) has suggested that energy should be used as the base value and that other nutrients should be specified in relation to the amount of available energy consumed. Crampton suggests that all animals have a requirement for maintenance of 19 g of digestible protein/100 Kcal of digestible energy. Although this approach is probably generally applicable, it has not been widely used in calculating protein requirements except for poultry where calorie:protein ratios have been shown to have practical significance. Because most animals tend to eat to meet energy needs, if we go from a moderate energy concentration to a high energy concentration in the ratio, feed consumption will decline; thus, more protein is needed/unit of ration, but the calorie:protein ratio may not change. With herbivores, however, DP is markedly influenced by dry matter intake. Thus, if DM intake is increased, DP declines as a result of abrasion in the lower GIT, but DE intake may remain essentially unchanged. As a result, relationships between caloric intake and protein requirements are more obscure when a wide range of ration types are considered.

The NRC poultry committee does express protein needs in relation to energy, but this approach is not used in the other NRC recommendations as protein needs are calculated on the basis of need/unit of $BW^{0.75}$, with additional amounts included for milk production.

In the case of mineral requirements, P need is sometimes computed in relation to protein intake (NRC beef). Ca need is based on body size with additional amounts for lactation or egg production of laying hens. Other mineral requirements are generally specified on the basis of metabolic size. Some vitamins, particularly thiamin and niacin, are known to be required in relation to energy needs, while riboflavin and B_6 needs are more directly related to protein intake.

Computation of nutrient needs in relation to energy seems to be a very sound approach for monogastric species. Certainly, if protein is required in direct relation to energy, then it should be feasible to relate the needs of most other nutrients to energy consumption. It is likely that in the future more of the NRC committees will recognize the value of relating nutrient needs to energy concentration of the diet as our nutritional information becomes more precise.

Methods of Estimating Nutrient Needs

Nutrient requirements of animals are evaluated and computed on the basis of many of different types of information. With respect to energy, animal needs may be based on calorimetric studies modified as necessary for feeding experiments under practical environmental conditions. Although calorimetric data provide useful information on fasting or basal metabolism or energy expenditure of closely confined animals, such data do not appear to give good estimates of the energy expenditure of grazing animals, as field studies indicate a considerably higher energy requirement than do calorimetric studies. In addition, the effect of environmental changes in temperature, solar radiation, wind, humidity and other stressing factors are difficult to quantify, so we are faced with the need to conduct feeding experiments under natural environmental and management conditions.

One limitation of feeding trials is that estimating energy gain or loss by the animal is difficult (see Ch. 9). We can easily calculate the energy value of milk or eggs produced and, in a long experiment, slight changes in body energy may not cause much of an error in estimating the energy needs of the animal, but some errors may be relatively large in short term studies. This problem can be partially overcome by using data obtained from slaughter experiments (see Ch. 9).

With protein and the major minerals, one method used at times to calculate requirements is designated as the factorial method. This method is based on back calculations of utilization using amounts excreted in urine and feces and, for lactating animals, amounts secreted in milk. The computations used in calculating protein requirements are too complex to justify a complete explanation here; the reader is referred to ARC (1) or McDonald et al (7) for more details, but a simple example will be given for Ca. In this method, the net requirement of Ca for maintenance and growth is calculated as the sum of the endogenous losses (fecal + urinary) and the quantity retained in the body. To determine the dietary requirement, the net requirement is divided by an average value of availability (expressed as a decimal). For example (McDonald et al, 7), a heifer at 700 lb gaining 1 lb/day might have endogenous losses of 5 g of Ca/day and retain 3 g/day; its net requirement would be 8 g Ca/day. For an animal of this weight, average availability of Ca would be about 40%, thus the animal's dietary requirement for Ca would be $8 \div 0.4 = 20$ g/day.

The factorial approach suffers from the fact that endogenous losses are difficult to estimate as endogenous losses are those originating from body tissues and must be estimated by feeding an animal a diet free of soluble Ca (or whatever nutrient is of interest) or by use of radioisotope procedures. In addition, availability of dietary nutrients is not a fixed biological reaction and may vary with age, source of the nutrient, and other dietary or environmental factors. Factorial estimates tend to be lower than those derived from feeding trials or balance studies.

With respect to vitamins, balance studies are of little value because excretory metabolites of some vitamins are difficult to measure or have not been identified; thus, data must be obtained from feeding trials or some measure of animal productivity. Data on blood levels, tissue storage, freedom from deficiency symptoms, and ability of the animal to produce at maximal levels are used to specify vitamin requirements.

Inaccuracies in Feeding Standards

For the nutritionist, feeding standards provide a useful base from which to formulate rations or estimate feed requirements of animals. They should not, however, be considered as the final answer on nutrient needs, but should be used as a guide. Current NRC recommendations are specified in terms believed by the committees to be minimum requirements for a population of animals of a given species, age, weight and productive status. Some of the earlier versions were called allowance and, as such, included a safety factor on top of what was believed to be required. It is well known that animal requirements vary considerably, even within a relatively uniform herd. For example, a protein intake that may be satisfactory for most animals in a given situation will probably not be sufficient for a few of the more rapid gainers or high producers; conversely, some of the herd will probably be overfed. With our present production methods, this usually is the most feasible basis of feeding. The poor producers would be culled and the high producers can be given extra allowances.

It is quite obvious from the literature that management and feeding methods may alter an animal's needs or efficiency of feed utilization apart from known breed differences in nutrient metabolism and requirements. In addition, current recommendations provide no basis for increasing intake in severe weather or reduction in mild climates. Nor is any allowance made for the effect of other stresses such as disease, parasitism, surgery, and so forth. Frequently, beneficial effects of additives or feed preparatory methods are not considered (or known). Thus, we may have many variables that may alter nutrient needs and nutrient utilization, and these variables are often difficult to include quantitatively in feeding standards, even when feed quality is well known.

The remainder of this chapter will be devoted to some general discussion relating the effect of various productive functions to nutrient requirements, followed by sections on individual species and classes of animals.

Nutritional Needs and the Productive Functions

Maintenance

Maintenance may be defined as the condition in which an animal is neither gaining or losing body energy (or other nutrients). With productive animals, there are only a few times when maintenance is desired; it is closely approximated or attained in adult male breeding animals other than during the breeding season and, perhaps, for a few days or weeks in adult females following the cessation of lactation and before pregnancy increases requirements substantially. However, as a reference point for evaluating nutritional needs, maintenance is a standard benchmark commonly used. Other things being equal, nutrient needs are minimal during maintenance. In field conditions during dry periods of the year, we may find that range animals may need to expend considerable energy just to obtain enough plant material for consumption as opposed to the amount of energy expended when forage growth is more lush, but this does not alter the fact that nutrient needs are less during maintenance than when an animal is performing some productive function.

Growth and Fattening

Growth, as measured by increase in body weight, is at its most rapid rate early in life. When expressed as % increase in body weight, the growth rate gradually declines until puberty, followed by an even slower rate until maturity. As animals grow, different tissues and organs develop at differential rates and it is quite obvious that the conformation of most newborn animals is different from that of adults; this differential development has, no doubt, some effect on changing nutrient requirements. Growth rate probably decreases because the biological urge to grow is lessened, because young animals cannot continue to eat as much/unit of metabolic size and, as measured by increase in body weight, because relatively more of the tissue of older animals is fat, which has a much higher caloric value than muscle tissue (see Ch. 9).

Nutrient requirements/unit of body weight or metabolic size are greatest for very young animals; these needs gradually taper off as the growth rate declines and as the animal approaches maturity. In young mammals, nutrient needs are so great that, because the capacity of the GIT is relatively limited in space and function, they must have milk (or a milk replacer) and additional highly digestible food to approach maximal growth rates. As the young mammal grows, quality of the diet generally decreases as more and more of its food is from non-milk sources with the result that digestibility is lower and the dry matter of food is used less efficiently.

Dry matter consumption for all young animals is usually far greater/unit of body weight during their early life than in later periods. This high level of feed consumption provides a large margin above maintenance needs, thus allowing a high proportion to be used for growth and development of the young animal. Due to differences in capability of the GIT for food utilization and because the rate, duration, and character of body growth vary with age and animal species, nutrient requirements may be quite different for different animal species. Nevertheless, it is characteristic for all species that nutrient requirements (in terms of nutrient concentration/unit of diet) are highest for the very young and then gradually decline as the animal matures. Naturally, total food and nutrient consumption are less for young animals because of their smaller size.

Nutrient deficiencies show up quite rapidly in young animals, particularly when the young are dependent in the early stages of life on tissue reserves obtained while in utero. With few exceptions, tissue reserves in newborn animals are low. Milk or other food may be an inadequate nutrient source so that deficiencies may frequently occur until the food supply changes or the young animal develops a capability to eat the existing food supply. The young pig is an example. Fe reserves are low and rapid growth soon depletes body reserves; because milk is a poor source of Fe, young pigs often become anemic unless supplemented with Fe.

It should be pointed out that young mammals are dependent upon an intake of colostrum early in life. In very early life, the intestinal tract is permeable to large molecules such as proteins. Colostrum has a large supply of globulins and other proteins that provide nutrition as well as a temporary supply of antibodies which greatly increase resistance to many diseases. In addition, colostrum is a rich source of most of the vitamins and trace minerals, and the young animal's tissues can rapidly be supplied with needed nutrients which may not have been adequately provided in utero.

From a production point of view, nutrient requirements/unit of gain are least and gross efficiency is greatest when animals grow at maximal rates, even though more fat may be deposited by fast gainers. However, net efficiency (nutrient needs over maintenance) may not be greatly altered. In a number of instances it may not be desirable or economical to attempt to achieve maximal gain. For example, if we want to market a milk-fed veal calf at an early age, maximal gain and fattening are desired. On the other hand, if the calf is being grown out for a herd replacement, then less than maximal gain will be just as satisfactory and considerably cheaper.

The biological urge to grow cannot be greatly suppressed in young animals without resultant permanent stunting. It is possible to maintain young animals for a period during which they do not increase body energy reserves, yet they will — if other nutrients are adequate — continue to increase in stature. Following a period of subnormal growth due to energy restriction, most young animals will gain weight at faster than normal rates when given adequate rations. This response is termed compensatory growth. This phenomenon has practical application when young calves are wintered at low to moderate levels (submaximal). When new grass is available in the spring or if they are put in the feedlot, weight gain occurs at a very rapid rate initially.

Efficiency for the total period and especially for a given amount of gain is greater, however, if the animal is fed at near maximal rates, but this practice may allow the use of much cheaper feedstuffs or deferred marketing and, as a result, be a profitable management procedure.

Work

Experimental studies with man and animals indicate that work results in an increased energy demand in proportion to the work done and the efficiency with which it is accomplished (see Brody, 3). Carbohydrates are said to be more efficient sources of energy for work than fats (8). With respect to protein, balance studies show little, if any, increase in N excretion in horses or man as a result of muscular exercise (3, 8). Although this evidence has been obtained in a number of studies, data on men indicate reduced quality and quantity of work when protein intake is on the low side, but still above maintenance levels. It is a common belief that hard workers need more protein, and it is a typical practice in competitive athletics to feed performers high levels of meat and other high-protein foods. Perhaps some of the effect on performance is psychological, but it seems to be real whatever the cause.

If sweating occurs, work may be expected to increase the need of Na and Cl, particularly. P intake should be increased during work, as it is a vital nutrient in many energy-yielding reactions. Likewise, the B-vitamins involved in energy metabolism, particularly thiamin, niacin and riboflavin, probably should be increased as work output increases, although data on this subject are not clear.

Reproduction

Although nutrient needs of animals for reproduction are generally considerably less critical than during rapid growth, they are certainly more critical than for an adequate level of maintenance. If nutrient deficiencies occur prior to breeding, they may render animals sterile, or result in low fertility, silent estrus or failure to establish or maintain pregnancy.

It has been amply demonstrated that underfeeding (energy, protein) during growth will result in delayed sexual maturity, and that both underfeeding and overfeeding (energy) will usually result in reduced fertility as compared to that of animals fed on a medium intake. Of the two, overfeeding is usually more detrimental to fertility.

Energy needs for most species during pregnancy are more critical during the last 1/3 of the term. This fact is illustrated by data on cattle shown in Table 16-1; daily energy deposits increase quite rapidly between day 200 and 280. Although the data in Table 16-1 show that only a relatively small amount of the cows' energy needs are needed for reproduction, other information indicates a somewhat greater need. Pregnant animals have a greater appetite and will spend more time grazing and searching for food than nonpregnant animals. Furthermore, the basal metabolic rate of pregnant animals is higher; the increase over maintenance is called the

Table 16-1. Deposition of various nutrients in the fetus, uterine tissues, and mammary gland of the cow at different stages of pregnancy.[a]

Days after conception	Uterine and fetal deposits (per day)				Mammary gland, g protein/day
	Energy, Kcal	Protein, g	Ca, g	P, g	
100	40	5			
150	100	14	0.1		
200	235	34	0.7	0.6	7
250	560	83	3.2	2.7	22
280	940	144	8.0	7.4	44
Approximate net daily maintenance requirement of 1000 lb cow	7000	300	8	12	

[a] From data of Moustgaard (9)

heat increment of gestation. By the end of pregnancy, the basal rate is about 1.5 fold that of similar nonpregnant cows.

Protein deposition in the products of conception follows the same trend as energy, but protein is relatively more critical for development of the fetus in the late stages than early in pregnancy as is true for Ca, P, and other minerals and vitamins.

Inadequate nutrition of the mother during pregnancy may have variable results, depending on the species of animal, the degree of malnutrition, the nutrient involved, and the stage of pregnancy. Nutrient deficiencies are usually more serious in late pregnancy although there may be exceptions to this. With a moderate deficiency, fetal tissues tend to have a priority over the mother's tissues; thus, body reserves of the mother may be withdrawn to nourish the fetus. A very severe deficiency, however, will usually result in partial depletion of the mother's tissues and such detrimental effects as resorption, abortion, malformed young, or birth of dead, weak, or undersized young with, sometimes, long-term effects on the mother. When the mother's tissues are depleted of critical nutrients, then tissue storage in the young animal is almost always low, nutrients secreted in colostrum are also low, milk production may be nil, and survival of the young animal is much less certain than when nutrition of the mother is at an adequate level.

Lactation

Heavy lactation probably results in more nutritional stress in mature animals than any other production situation, with the possible exception of heavy, sustained muscular excercise. During a year, high-producing cows or goats typically produce milk with a dry-matter content equivalent to 4-5 fold that in the animal's body and some animals reach production levels as high as 7X body DM (6). High-producing cows give so much milk that it is usually impossible for them to consume enough feed to prevent weight loss during peak periods of lactation.

Milk of most domestic species runs 80-88% water; thus, water is a critical nutrient needed to sustain lactation. All nutrient needs are increased during lactation, as milk components are either supplied directly via the blood or synthesized in the mammary gland and, thus, are derived from the animal's tissues or from food consumed; all recognized nutrients are secreted to some extent in milk. The major components of milk are fat, protein, and lactose with substantial amounts of ash, primarily Ca and P. Milk yield varies widely between and within species. In cows, peak yields usually occur between 60 and 90 days after parturition and then gradually taper off; thus the peak demand for nutrients follows the typical milk flow characteristic for the species concerned. Milk composition and quantity in ruminants may be altered by the type of ration, particularly the fat content and, to a lesser extent, the protein and lactose. In monogastric species, diet may affect fat, mineral and vitamin composition of milk.

Limiting water or energy intake of the lactating cow (and probably of any other species) results in a marked drop in milk production, whereas protein restriction has a less noticeable effect, particularly during a short period of time. Although deficiencies of minerals do not markedly affect milk composition, they will result in rapid depletion of the lactating animal's reserves. The needs of elements such as Cu, Fe and Se will be increased during lactation, even though they are found in very low concentrations in milk. The effects of marked nutrient deficiencies during lactation will often carry over into pregnancy and the next lactation.

Nutrition and the Productive Functions — Ruminants

In this section, space is set aside for some discussion of nutrition as affected by different productive functions for specific classes of domestic animals. Comments will not be made on each class within each species but, rather, only on those classes of animals in which nutritional problems are more critical in normal livestock production. Omissions of this nature are not meant to imply that nutrition may not be critical in all stages of the life cycle, but it is obvious that some stages of the life cycle are more critical than others, and these differences justify the selectivity used in the following discussions.

Young Ruminants

Young ruminant animals existing on milk or liquid milk substitutes are frequently referred to as preruminants because rumen function is not a dominant factor in food utilization. Rather, most of the liquid ingesta passes into the lower part of the stomach (abomasum) and on into the small intestine rather than into the rumen. This being the case, the nutrient requirements of these young animals are similar — both qualitatively and quantitatively — to monogastric species. As opposed to an older animal with a functioning rumen, preruminants require a dietary source of essential amino acids and fatty acids and of vitamin K and the various B-complex vitamins — dietary nutrients not normally needed by ruminants because the microbial population of the stomach and gut synthesizes adequate amounts of these organic nutrients. The point at which a preruminant animal develops a functioning rumen is dependent upon species and age, but is also markedly affected by the supply of milk and availability of palatable feedstuffs or forage. Young calves can be successfully weaned to dry feed at 3 wk of age and young lambs by 4 wk, providing the dry feed is palatable and has a satisfactory nutrient content (see Ch. 6 & 11 of Church et al, 4). More commonly, young animals begin to gradually consume solid feed; milk makes up less and less of their daily diet, and the rumen gradually develops in size and function.

Most of the nutritional data on preruminants has been developed on dairy calves taken off the cow at an early age. In order to feed these calves, milk replacers have been developed that allow satisfactory performance at a cost usually less than use of fluid milk. Milk replacers are also available for lambs and may be used to feed orphan lambs or extra lambs when a ewe has produced twins or triplets.

Milk replacers are usually based on use of relatively large amounts of dried skim milk supplemented with ingredients such as lard, starch, glucose, soy flour, dried whey or buttermilk, and with sources of the vitamins and minerals and feed additives such as antibiotics. Quality of protein is a critical factor as low quality proteins are poorly utilized and may cause digestive disturbances. It is also necessary to avoid too much carbohydrate in the form of starch or sucrose as the young animal does not have the gastrointestinal enzymes to handle much starch and has no sucrase for digestion of sucrose. High levels of whey result in diarrhea, either due to the excess lactose or high ash content. Fats from a variety of sources can be used, as can protein from sources such as soy flour, fish meal, liver meal, yeast, etc., although milk sources are preferred if cost is not prohibitive.

Except where veal calves with light-colored meat are being raised, all young animals on milk replacers should have supplementary dry feed available by 10 days of age, made up preferably of rolled grains and other appropriate supplementary feeds such as alfalfa hay, molasses, and soybean meal when cost allows. Young animals do not readily use urea as of source of N. Good to excellent quality forage should also be available at an early age.

Feedlot Cattle

The cattle feedlot industry in the USA has become a highly specialized segment of agriculture in the last decade, particularly in areas where there are large numbers of feeder cattle accompanied by surplus grain production. Many feedlots are in existence with capacities on the order of 30-40 thousand head and one or two with capacities of 100,000 head or more; with an annual turnover of about 2½ times capacity, these feedlots account for the majority of fat cattle marketed at the present time.

Cattle coming into these feedlots range in size from 200 to 1000 lb or more and in age from a few months to well over a year; thus, nutritional requirements and management problems vary tremendously. Because the cattle may come from many different farms and ranches, one of the biggest problems is the stresses imposed during marketing and transportation to the feedlot, exposure to new diseases and parasites, and adaptation of these cattle to their new environment. These various stresses can result in severe sickness and in death loss if management is not at a high level.

In addition, ruminant animals need time to adapt to unfamiliar feedstuffs. The time required is primarily because rumen microorganisms, which predigest the feed, develop into populations which reflect the composition of the feed. This takes time and means that animals should not suddenly be changed from one diet to quite a different one, particularly when going from a roughage diet to one high in readily available carbohydrates, which cause the most problems.

From a strictly nutritional point of view, except for the problem of adaptation, problems in the feedlot are more often of a management than of a strictly nutritional nature, as many of these large feedlots use nutrition consultants. Problems that crop up are concerned with pushing cattle too fast onto high grain rations, which may result in indigestion and founder;

ignoring mineral nutrition, which, in the case of Ca and P imbalance, may result in urinary calculi; and use of by-product or waste feedstuffs for which no good quantitative nutritional data are available.

As a result of the large numbers of animals involved, a high degree of mechanical handling of feed is required. With usual price relationships in the USA it is cheaper to feed rations high in grains and other concentrates. This type of ration is prone to cause digestive disturbances and disease and requires care in formulating rations and in the feeding management of these cattle.

The need for efficient feed conversion to minimize feed costs has resulted in the widespread use of grains processed in various ways (see Ch. 4 of Church et al, 4). Although processed grains are usually more digestible and produce more efficient gains, they are also more likely to result in digestive disturbances if improperly used in finishing rations.

Wintering Beef Cows

Beef cows are commonly wintered on straw, stover, grass hay, winter range or other roughages which are either of low quality or difficult to obtain. Dry beef cows can be wintered on such feedstuffs because their nutrient requirements are modest as compared to lactating dairy cows or growing animals. Nutrients that are usually of concern include protein, energy, minerals, and, at times, vitamin A. Although protein requirements are low, the protein content and digestibility in low-quality roughages are also low, so some supplementary feed is often needed, particularly in late gestation and lactation. Energy may or may not be of concern depending on quality of the roughage. Most low-quality roughages are deficient in P and often in Ca and Mg and some of the trace minerals such as I. Vitamin A may be of concern, particularly where cows have been on dry forage for several months.

During lactation, nutrient needs increase appreciably. If cows are underfed to a great extent, milk production will be markedly reduced and more problems may be expected with rebreeding. Better quality hays, silages or other roughage should be fed during lactation to avoid these problems. Young calves can often be profitably creep fed while cows are in winter quarters, particularly if feed supply for the cow is inadequate in quantity or quality.

Care is required in providing supplemental N for cows on poor quality roughage. Urea, a common supplementary form of N for cattle, must be fed along with some readily available carbohydrate (starch or sugars) in order to get good utilization and to prevent toxicity. Biuret, although more expensive, shows some promise in feeding in these situations.

Dairy Cows

The nutrition of lactating dairy cows is much more critical than that of beef cows because dairy cows are producing at a much higher level and, partly, because they have less time to recover between lactations.

Dairy cows are often prone to milk fever and ketosis, particularly cows in their 3rd and subsequent lactations. Milk fever is a result of abnormal Ca metabolism early in lactation. It can be controlled moderately well by careful management of the Ca:P ratio, taking care not to have an excess of Ca and adequate levels of P. Treatment prior to parturition with massive doses of vitamin D or its active metabolite, 25-hydroxycholecalciferol, is often helpful. Ketosis appears to be a result of faulty energy metabolism and it often occurs during peak milk flow. Its incidence can often be reduced by increasing energy intake during the latter stages of gestation, and, thus, having the cow adapted to a rapid increase in feed following parturition. Feeding of good-quality roughage and adequate total energy are factors that appear to reduce the incidence of ketosis.

Energy intake in lactating cows can be increased by increasing grain in the ration, but eventually this results in reduced roughage consumption. Reduced roughage intake, especially accompanied by feeding of heat-treated grains, usually causes a reduced milk-fat %. When milk is sold on the basis of its fat content, this is, obviously, undesirable. Consequently, the usual practice is to feed cows between 40 and 60% of their ration as roughage. Use of ground and pelleted roughage must also be restricted since milk fat % is apt to decline if they make up a high proportion of the ration.

Where the usual practice is to feed high-quality legume hay or excellent grass or legume silage, protein intake is usually not a problem in dairy cows. However, where other roughages such as corn silage are fed, protein intake can frequently be borderline to low. This is apt to lead to reduced milk production if continued for several weeks. Trace minerals, as well as some of the major minerals, must also be of more concern in dairy cows than in many other livestock classes.

Feedlot Lambs

Many fat lambs go to market directly off the ewe. Others that are late, small, or poor doers are often fattened on grass, wheat pasture in the Midwest, or in large commercial feedlots. Feedlot lambs have many of the same problems that cattle do; namely, they may be stressed considerably in the process of getting on feed in a new environment. Adaptation time when they are fed new rations is just as essential to lambs as for cattle. Fortunately, lambs will gain well and finish satisfactorily on considerably less grain than cattle, particularly if high roughage pelleted rations are feasible to use.

Lambs given sudden access to high grain rations, especially if large amounts of wheat are used, are subject to acute indigestion and often to enterotoxemia. Use of low levels of anitbiotics is helpful here. Lambs are also subject to urinary calculi if fed excess P in relation to Ca.

Energy consumption is not a problem in lambs provided rations are palatable and in a satisfactory physical form. In the author's (Church) opinion, lambs are often underfed on protein; they often respond to levels above those suggested by NRC and, in addition, are apt to have leaner carcasses, a desirable characteristic in terms of the human diet. Other than the factors mentioned, nutrition of feedlot lambs is not usually critical since the time factor is relatively short. If a lamb comes into the feedlot deficient in some nutrient, then this is something else.

Ewes

Nutrition problems of ewes, for the most part, are less likely than for most other ruminants. This is partly due to the fact that the usual management practice (at least in the USA) is to feed a much better quality forage to ewes than beef cows commonly receive.

Nutrition prior to and during the breeding season may have a substantial influence on the number of lambs born. If ewes are maintained in moderate flesh, but on an increasing plane of nutrition at breeding (called flushing), they will usually have more lambs than if managed in other ways. Overfeeding for too long a period will, however, reduce embryo survival and the number of lambs born.

A second critical period for ewes is in late gestation. Ewes carrying multiple fetuses have a markedly reduced stomach capacity so that food consumption drops off at a time when a greater nutrient supply is vital. This can easily result in pregnancy disease, a form of ketosis accompanied by acidosis. Pregnancy disease can largely be prevented by increasing the energy intake of such ewes by means of a gradually increasing concentrate intake during the last 6 wk of gestation.

Although lactating ewes do occasionally have milk fever, it is quite rare as compared to the incidence in lactating cows. Nutritional requirements increase substantially while the ewe is lactating, particularly so when more than one lamb is being nursed. Obviously, the ewe needs an increased intake of protein, energy, and major minerals found in milk.

Nutrition and the Productive Functions — Swine

Estimates of quantitative nutrient requirements for each phase of the life cycle have been tabulated by ARC (2) and NRC (10). As with other species, the nutrient requirements suggested must be considered only as guidelines as differences in genetic potential and environmental or climatic conditions may be expected to affect the requirement for each nutrient.

Gestation

The nutrient requirements for the sow during gestation are influenced by two separate productive functions — the need for maintaining the pregnant sow and the provision of an adequate nutrient supply for the developing pig fetuses. In general, nutrient requirements in early gestation are modest as compared to late gestation and, particularly, lactation.

It is possible, at least with some nutrients, to obtain normal reproduction through one reproductive cycle on a diet that is clearly inadequate for the dam. For example, results with sows fed protein-free diets during gestation show that the dam can draw on her own reserves to meet the needs of the fetuses for growth and survival (11). However, such a diet cannot be considered adequate since it does not satisfy the long-term requirements of both dam and fetuses.

Pregnant sows will voluntarily consume far more feed than required and will become obese if intake is not restricted to ca. 4-5 lb of a high concentrated diet/day (6600 Kcal of DE). The amount of restriction needed is dependent upon the size and condition of the animal, of course.

Lactation

The nutrient output in sow milk during a 5-wk lactation is much greater than the nutrient deposition in fetuses and placental membranes during a 114-day gestation period. Therefore, as with ruminants, the nutrient requirements of the sow for lactation are far more demanding than for gestation. While the sow can draw on her own body reserves for milk production, complete or partial lactation failure results if nutrient restriction is severe or prolonged. In general, a deficiency of a particular nutrient is manifested more by a reduction in total milk production than by a decreased concentration of that nutrient in the milk.

The most striking effect of level of intake on total milk production is with energy. If the lactating sow is not allowed to eat at or near ad libitum, milk production declines. In the well-fed sow, milk production increases as litter size increases, up to the point at which her genetic capacity for milk production is reached. Mature sows raising large litters will often consume 15-20 lb or more/day of a high-concentrate diet (3600 Kcal of ME/kg) at the peak of lactation (4-5 wk postpartum). Daily yield of milk, which has a dry matter content of ca. 20% (compared to 12% for bovines), may range from 12-20 lb or more during the 4th or 5th wk of lactation.

The energetic efficiency of lactation is higher when milk is produced by current energy intake than by dependence on body fat reserves. Therefore, in practical feeding the highest efficiency of energy utilization is acheived by controlled feeding during gestation to minimize mobilization of depot fats for milk production.

Inadequate protein (essential amino acids) intake will also result in reduced milk production, although the effect is less marked than for energy. Deficiencies of other required nutrients may also affect milk production, although to a lesser degree than for protein and energy.

Preweaning

The nutrient requirements of the suckling pig, except for Fe, are normally met by sow milk during the first 2-3 wk. After this time rapid growth of the pig, combined with the decline in milk yield after week 5, necessitates the provision of supplemental creep feed if maximum growth is to be attained. The protein content of the creep feed, as a % of diet, does not have to be as high as that of sow milk, as the energy concentration is less. The carbohydrate source must be palatable and low in fiber to encourage consumption. For this reason, cane sugar, dried molasses, or glucose is often added at a level of 5-10+% of the diet. Nonnutritive sweeteners, such as saccharin or monosodium glutamate, are also used, but at a much lower concentration. Although the energy concentration of the creep diet is not especially critical, the addition of 5-10% fat to the diet improves palatability and encourages early consumption.

The introduction of a dry, well-balanced diet early enough so that the suckling pig consumes enough for maximum weight gain is a major consideration during the preweaning period. Although consumption during this period is negligible as compared to later periods, a small amount may greatly increase growth of these young pigs.

Removal of the pig from the sow ealier than 5 wk of age can be considered early weaning. The younger the pig at weaning, the more critical are the dietary requirements. Pigs weaned at birth and deprived of colostrum must be fed a highly fortified diet, kept in a warm, sanitary environment and be given parenteral anitbody protection (porcine γ-globulin) for a reasonable chance of survival. Cow's milk can serve as a substitute for sow milk, although its lower caloric density does not allow maximum weight gain. Liquid sow milk substitutes of higher caloric density can be used. Composition of one such diet is (g/l. of finished milk): casein, 44.3; glucose, 44.1; lard, 33.0; soy lecithin, 2.0; minerals and vitamins. Baby pigs utilize casein 5-10% more efficiently than soybean protein, presumably due to limitations in digestive enzyme capacity in early life. Liquid diets must be fed at frequent intervals (minimum of 4X daily) to provide adequate intake for reasonable growth and to avoid digestive problems. If an adequate dry diet is available, young pigs can be weaned to dry diets at a few days of age.

Growing Period [Early Postweaning]

This stage of the life cycle of the pig is arbitrarily set as the period from weaning (usually 5-6 wk) to ca. 45 kg live weight (12-16 wk of age). During this period, nutrient requirements are less critical than at earlier stages, but more critical than during the finishing period. The changes in nutrient requirements as the pig matures are related to changes in growth rate and body composition. Fat concentration of the body increases rapidly at the expense of water and, although the % of protein remains rather stable, the calorie:protein ratio in the body increases steadily to market weight. Full-feeding of growing pigs on a high energy diet (1500 Kcal DE/lb) results in maximum growth rate and efficiency of feed utilization. Limited feeding may produce a leaner carcass, but the slower growth rate reduces energetic efficiency due to the higher proportion of daily energy intake needed for maintenance. Feeding a protein-deficient diet during the growing period results in a fatter carcass; feeding protein in excess of the requirement, however, does not result in a leaner carcass. Females and intact males have a higher carcass lean content than barrows and also require a higher % of protein in the diet.

In addition to nutritional factors, the physical form of the diet is important. Pelleting certain types of diets, especially those containing barley and other fibrous grains, may improve growth and efficiency of feed utilization. In addition, fineness of grind and dustiness of the feed contribute to variations in performance of growing pigs.

Finishing Period [45-90 kg]

The same principles and problems in meeting nutrient requirements during the growing period apply during the finishing period. However, the quantitative requirements for nutrients other than energy are less (as % of the diet) during the finishing period. The total daily feed requirement is considerably greater during the finishing period not only because of larger body size but also because of the higher feed requirement/unit of body weight gain; this is a reflection of increased fat disposition which requires considerably more energy/unit of gain.

Limiting feeding of finishing swine to 70-80% of ad libitum intake reduces carcass fatness, but also reduces rate and efficiency of gain due to the greater proportion of daily intake needed for maintenance. The system of feeding used for finishing will, therefore, be dictated by the economic relationship between carcass value of lean vs. fat pigs and the price of feed and labor.

The finishing pig, as well as the growing pig, tends to select proportions of grain and protein supplement appropriate to metabolic needs when the 2 feed sources are offered separately. It is more common, however, to feed complete diets as pellets, meal or in liquid or paste form.

While by far the greatest tonnage of feed in a swine enterprise is devoted to finishing pigs, this stage of the life cycle is the least critical in terms of meeting specific nutrient requirements of growing finishing-pigs.

Poultry

Since this segment of animal production accounts for at least 40% of all commercially mixed feeds, some general comments regarding it are necessary. It should be recognized that chickens and turkeys eat to satisfy their energy needs provided the ration allows them to do so. There are, of course, exceptions to this rule, particularly where heavy breed layers are concerned, when birds have a tendancy to overeat. Where this is a problem, it is the general practice to subject the birds to some degree of feed restriction, both during the growing and production periods.

In the poultry industry we are seeing greater concentrations of birds in the order of hundreds of thousands and up to millions in certain instances, all under the control of one operator. All of this can result in a multiplicity of problems confronting those interested in the field of nutrition as well as other disciplines.

The major ingredients that are integral parts of poultry rations at the present time in the USA are corn, as the primary source, and soybean meal as the major protein supplement. These major ingredients, usually available in plentiful supply, allow rapid growth or high egg production with very efficient feed conversion. Corn-soy rations however, are deficient in some nutrients* for chickens; these nutrients are normally supplied by alfalfa meal, other feedstuffs, or concentrated supplements.

A number of satisfactory substitutes may be available. For example, milo and wheat certainly may be used when economically priced. Similarly, cottonseed meal, fish meal, or meat and bone meal may be used as protein sources.

Note that the nutrient requirements of chickens and turkeys are more completely understood and quantitated with greater precision than is the case for other species.

* Amino acids: methionine; macrominerals: Ca, P, NaCl; trace minerals: I, Mn, Se, Zn; vitamins: all fat- and water-soluble vitamins except choline.

Since this is so and because a high % of poultry feed is commercially prepared, nutrient problems may (should) be less likely than for other species.

Broilers

Because of the very rapid growth of chicks, nutrient needs are more critical than for older birds. As with the young of other species, in newly hatched chicks body reserves of nutrients such as the various vitamins and minerals may be low. Thus, in order to obtain maximum growth rates and to avoid nutrient deficiencies, more attention must be paid to quality of protein and adequacy of the essential amino acids, especially methionine and lysine, and provision must be made for adequate supplementation of the necessary minerals and vitamins.

Probably in no other area does the concept that birds eat to satisfy their energy needs apply more than in the feeding of broiler chicks. Because of the increased demands made for efficiency of production, both energy and protein contents of the diet are most important. Accordingly, diets containing variable protein and energy levels, called multiple stage rations, are used at different ages during the 8-wk broiler growing period. It is generally recognized that protein requirements decrease with age and the indiscriminant use of these feeds without regard to protein level results in poor feed utilization and increased cost. Further, virtually all broiler feeds are now fed in pelleted or crumblized form and switching to mash feeds should be avoided. Supplemental fats from vegetables or animal sources are widely used as concentrated energy sources when the cost makes it feasible to include them in broiler rations. Keep in mind, however, that when the energy content of the ration is increased, feed consumption will usually be reduced; thus, high-energy diets require a greater concentration of all other nutrients, providing digestibility and absorption are not altered.

The importance of the diet as a carrier of nonnutritional factors cannot be underestimated. For example, the diet is the usual vehicle for administration of medicants such as antibiotics or coccidiostats. It is also the means for obtaining the yellow pigmented skin demanded by the consuming public. Skin pigmentation is related to the xanthophyll content of the feed. Both yellow corn and alfalfa meal are good sources of this substance and any dietary change that lowers xanthophyll intake must be corrected.

Laying Hens

In feeding layers, both energy and protein are important, but probably to a lesser degree than noted for broilers. In addition, protein quality is less important than for broilers. Under normal conditions, mash-type feeds may replace pelleted feeds used with broilers. Once pellets are used, however, they should be continued as hens prefer the physical texture and size of pelleted feeds. In general, rations now in use are considered complete diets from the standpoint of providing all nutrients required for egg production. Dietary protein intake is related to the daily protein requirement for egg production. Accordingly, because of the usual decrease in laying with age, rations are often formulated to meet these variable production rates. Higher protein levels are required during the earlier production periods when higher rates of lay are prevalent. Therefore, lower protein rations should not be used during early periods of production. This type of a feeding program is termed phase feeding.

Ca is frequently a critical nutrient in layer rations. The presence of Ca in adequate amounts insures good eggshell quality. Protein and P contents also may fall into this category. Any factor decreasing food intake can adversely affect shell quality. The presence of certain feedstuffs in rations may also influence egg quality. For example, the use of cottonseed meal can have a deleterious effect on internal egg quality as evidenced by the presence of olive-green yolks and pink albumens; its use should, therefore, be curtailed.

Turkeys

The feeding of turkeys utilizes most systems employed with chickens. Protein requirements are considerably greater than for chickens and the use of multiple stage rations is much more prevalent. With the recent increases in feed costs, interest in feeding for compensatory growth has occurred. By feeding poults low-protein of marginal diets during their early growth period, more efficient gains are possible during later growth stages with the end result that overall gains are obtained with less protein consumption.

Pelleted rations are often used for feeding

turkey breeders. In general, systems and practices are similar to those used with chickens. Breeder diets need to be especially fortified with the required vitamins (A, D, E, riboflavin, pantothenic acid, biotin and folic acid) and several trace minerals (Mn, Se, Zn) in order to insure the production of viable poults.

Nutrition and the Productive Functions — Horses

Before discussing nutrition and the productive functions of horses, it should be noted that horses in the USA are raised as pleasure rather than as meat or milk producing animals. Consequently, management goals are somewhat different from most livestock species in that horsemen are not so concerned with raising animals for their meat or milk producing abilities as they are with raising animals that can perform various athletic endeavors with competitive efficiency. The horse production areas of primary importance are: reproduction, lactation, growth, and work.

Reproduction [Conception-Gestation]

Reproductive efficiency in horses averages approximately 60-65% nationally, which is considerably lower than that of other livestock species. This poor performance can be attributed to a variety of causes such as updating the breeding season due to establishment of mandatory January 1 birthdates by many breed registries and to poor breeding farm management.

In regard to management improvement, researchers have recently established that reproductive efficiency of mares and stallions can be improved by increasing their plane of nutrition beginning about 60-80 days prior to breeding. The flushing of breeding animals which have been maintained in a slightly thrifty wintering condition tends to act with a catalytic effect on the mare's reproductive system. Many times the additional nutrients are all that is required for the mare's cycling pattern to become more regular and for normal ovulation to occur. Quite possibly, many wintering diets provide only for normal body maintenance (neglecting the reproductive processes in favor of essential body functions) causing reproductive activity to remain somewhat dormant. Diet mismanagement (over or under feeding) prior to any flushing period may reduce the beneficial effects of such a feeding program. In addition, it is known that over 60% of the foal fetus is produced during about the last 80-90 days of gestation; therefore, it is wise to increase a pregnant mare's plane of nutrition to meet this increased demand. These feeding practices will help to insure improved reproductive performance and production of foals by dams having adequate body reserves to meet the stresses of heavy lactation.

Lactation

As with most species, the lactation period (3-5 mo.) places the heaviest demands on a broodmare and results in a marked increase in nutritional requirements. Milk production by a 500 kg broodmare may range from 14 kg/day early in the lactation period to peak production of 17 kg/day at 2-3 mo. Mare's milk generally is lower in fat, protein and ash and higher in sugar than cow's milk. On a dry matter basis, mare's milk averages about 61.50%, 22.00%, 13.12% and 3.80% sugar, protein, fat, and ash, respectively.

In view of the fact that the mare is producing large quantities of milk for an extended time period, it is essential to increase her plane of nutrition. In the case of a 1200 lb mare, her daily feed intake should be increased from about 17.5 lb (maintenance) to 26.0 lb, and should provide an additional 14 Mcal of DE, 1.2 lb digestible protein, 30.8 g Ca and 26.9 g P in order to prevent the mare from drawing on her body stores during lactation.

Growth

Nutritional demands of immature horses for optimal growth follow general patterns similar to other livestock species. Of primary importance in feeding baby horses is the fact that foals have a limited digestive tract capacity. Hence, consideration must be given to feeding concentrated diets of high quality to insure intake of adequate quantities of required nutrients. Average birth weights of foals of light horse breeding run from 80-100 lb; average daily gains may be 2.2 lb/day for very young foals maturing at 1200 lb body weight and may range down to 1.5 lb/day for weanlings and to 0.8 lb.day and 0.4 lb/day for yearlings and two-year olds, respectively.

Mare's milk runs about 19-22% protein on a dry basis, and foals require a relatively high protein concentration in their diet (19-22% preweaned, 16% weanling, 13-14% yearling and

12% for two-year old). Studies of dietary amino acid requirements have indicated lysine as the most limiting in many foal diets. It appears that a minimum of 0.75% dietary lysine will support optimal growth in young horses.

Similar to protein requirements, foals require higher dietary concentrations of digestible energy during the early growing period with a diminished concentration required as they become older. This is partially due to the increasing capacity of the digestive tract and the establishment of an active bacterial population in the lower gut of older foals.

Minerals of primary importance in foal nutrition are Ca and P because of their involvement in the growing bones. Research has shown that foals require about 0.8% and 0.6% dietary Ca and P, respectively, with these values gradually declining with age. Ca and P fed in a 1.25-2:1 ratio will provide for optimum bone mineralization; extremely wide ratios (excessive amounts of P) ar associated with abnormal bone development in foals. Additional mineral needs will usually be met by normal dietary feedstuffs or by inclusion of 0.25% trace-mineralized salt in the diet.

Vitamin D has been shown to be important in absorption of dietary Ca. Other fat-soluble vitamins (A & E) also are generally added to foal rations. The mature horse is able to acquire much of its B-vitamin needs through microbial synthesis, but intestinal synthesis in the immature foal is not adequate to meet dietary needs. Therefore, it is essential that young horses receive dietary B-vitamin supplementation as well as vitamins A, D, & E. In most situations vitamin premixes or commercial vitamin supplements added to the diet will alleviate problems of vitamin deficiency.

Foals are not usually weaned until 4-5 mo. of age; however, many managers have discovered that providing nursing foals with a high quality creep feed allows for less milk demand from the dam and greater opportunity for foals to grow to their potential. In additon, creep-fed foals are less likely to be affected by the post-weaning slump observed in foals during the period of adapting from primarily a milk diet to a solid food diet. Recent research has shown that foals may be weaned from the mare at one month of age, providing they have access to adequate amounts of properly formulated milk replacer. Early-weaned or orphaned foals grow more slowly than their nonweaned comrades during the first year; however, they will usually compensate during their second year. Generally, early-weaned and normally weaned foals cannot be differentiated by two years of age if strict management practices are followed.

Work

In domestic livestock, athletic performance of work is one productive function which is rather unique to horses. Work by modern horses would essentially be classified as athletic endeavors such as carrying a rider for varying distances at various speeds over a variety of terrains. In other words, performing such activities as racing, jumping, trail riding, cutting, rodeo events or ranch work.

Unfortunately, most equine nutritionists are forced to classify nutrient requirements for work into 3 categories: light, moderate and strenuous. These categories are relics of the past when work by draft horses was defined in terms of weight pulled for a given duration at a given speed (usually at a walk). Today it is more difficult to establish a precise definition for work since most horses work for short time periods with their degree of exertion ranging from mild activity to near exhaustion. The preceding problems have most likely lent themselves to the development of "horse feeding" as an art with scientific application being made only in very recent years.

Many voids exist in discussing the nutrients required by horses for work. The current energy, vitamin and mineral requirements for work are in need of further investigation. It is assumed that protein requirements increase with work, however, not to the same degree as other nutrients (see Ch. 9). In most cases, the additional feed fed to meet increased energy demands will contain adequate protein. Contrary to many beliefs, mature performance horses do not require high protein diets to function at optimum capacities. B-vitamins are associated with energy metabolism; consequently, their requirement increases with incread energy consumption and utilization during work. Oxygen is involved in energy metabolsim; hence, the oxygen-carrying capacity of the blood is of importance in conditioning a performance horse. Vitamin E and Se have been associated with red blood cell fragility and muscle tissue capillary integrity. Since Fe also is associated with red blood cell production, it should receive due consideration in evaluating diets of performance horses.

There are a limited number of ways to evaluate the ability of a ration to provide adequate nutrients for strenuous athletic activity. Maintenance of body weight is an obvious positive sign and since O_2-carrying capacity is vital, the packed cell volume and red blood cell count are reasonably good indicators of the adequacy of a ration.

In general terms, rations for working horses should provide for increased DE, vitamin, and mineral concentrations with increased total daily protein, but not necessarily increased dietary protein concentration above that of maintenance (12%).

Many times coming two-year olds are placed in preparatory training for the race track, show ring or other activities during the later part of their yearling year. It is wise to insure that young horses in this situation are not only fed to meet their requirements for maintenance and growth but also fed to meet the additional nutrient needs for performing their required athletic functions. If these additional needs are not met, body development may be inhibited.

References Cited

1. ARC. 1965. The Nutrient Requirements of Farm Livestock. No. 2 Ruminants. Agricultural Research Council, London.
2. ARC. 1967. The Nutrient Requirements of Farm Livestock. No. 3 Pigs. Agricultural Research Council, London.
3. Brody, S. 1945. Bioenergetics and Growth. Reinhold Pub. Co.
4. Church, D.C., et al. 1972. Digestive Physiology and Nutrition of Ruminants. Vol. 3 - Practical Nutrition. O & B Books, 1215 NW Kline Pl., Corvallis, Ore.
5. Crampton, E.W. 1964. J. Nutr. 82:353.
6. Maynard, L.A. and J.K. Loosli. 1969. Animal Nutrition. 6th ed. McGraw-Hill Book Co.
7. McDonald, P., R.A. Edwards and J.F.D. Greenhalgh. 1966. Animal Nutrition. Oliver & Boyd, Edinburgh.
8. Mitchell, H.H. 1962. Comparative Nutrition of Man and Domestic Animals. Academic Press, N.Y.
9. Moustgaard, J. 1959. In: Reproduction in Domestic Animals. Academic Press, N.Y.
10. NRC. 1973. Nutrient Requirements of Swine. Nat. Acad. Sci. Pub. 1599. Washington, D.C.
11. Pond, W.G., W.C. Wagner, J.A. Dunn and E.F. Walker. 1968. J. Nutr. 94:309.

Chapter 17 – Factors Affecting Feed Consumption

Introduction

There is a considerable amount of interest in the various factors affecting feed intake by animals. This is understandable in view of the economic factors related to feed intake and cost of production. A great mass of evidence clearly shows that the gross efficiency of production can be greatly increased in growing, fattening, or lactating animals if their consumption can be maintained at a high level without any health problems. An example of this is shown in Table 17-1. Although these data are probably not applicable universally, they clearly illustrate the principle in question: that the maintenance requirement of animals gaining at a slow rate represents a much greater % of the total feed required than for animals gaining at more rapid rates. Other costs are also incurred at low levels of productivity, such as additional labor and additional time that money is tied up and less efficient use of facilities and equipment. High-level production is not without its costs, however, as animals are more prone to metabolic diseases and disorders which do not commonly occur in animals producing at more moderate rates; for example, milk fever, ketosis, and acute indigestion in cattle. In addition, higher quality feed ingredients and more costly rations are usually required to attain high levels of productivity; the net cost, however, is apt to be less for high rates of production. The need to attain high levels of productivity, thus, provides the impetus for studying and learning about factors that influence feed intake.

There are, however, situations where it is desirable to limit feed intake. For example, it is often convenient to self-feed different classes and species of animals, but often the desired or needed feed consumption is considerably less than ad libitum intake. The need of wintering beef cows for supplemental protein is a good example here. The same situation could easily apply to rations designed for any species where the animal is primarily in a maintenance situation, i.e., adult animals under no productive stress. Consequently, more information is needed on factors that tend to stimulate or inhibit food consumption.

Table 17-1. Energy required to produce 700 lb of gain on 500 steers at three rates of gain. [a]

Daily gain, lb	Days on feed	Av. wt, lb	Net energy required for maintenance, Mcal	Net energy required for gain, Mcal	Corn equivalent, bushels
1.0	700	750	2,156,000	708,500	62,784
2.0	350	750	1,078,000	740,250	43,419
3.0	234	750	720,720	782,730	37,595

[a] From McCullough (22)

Palatability and Appetite

Palatability is a term which is frequently misused. A quick and simple definition would be the overall acceptance and relish with which an animal consumes any given feedstuff or ration; obviously, this is not a very quantitative measure. Palatability is, essentially, the result of a summation of many different factors sensed by the animal in the process of locating and consuming food and depends upon appearance, odor, taste, texture, temperature, and, in some cases, auditory properties of the food. These various factors are affected by the physical and chemical nature of the food, and the effect on individual animals may be modified by physiological or psychological differences. **Appetite**, on the other hand, generally refers to internal factors (physiological or psychological) which stimulate hunger in the animal. The effects of different variables on appetite or palatability are discussed in succeeding sections.

Taste

In taste research, the basic tastes are described as sweet, sour, salty, and bitter along with what is called the common chemical sense, which means that animals may detect certain chemicals that do not fall in one of these four classes. Most of the taste research has been done with pure chemicals in water solutions or when added to feedstuffs. However, in the usual environment, the animal is only occasionally exposed to a pure chemical and most tastes are a result of complex mixtures of organic and inorganic compounds which frequently defy description. In addition, odor frequently has a pronounced effect on taste perception.

Much research on taste of humans and animals such as the rat has been conducted, but only a handful of papers are available on the domestic farm species * ; thus, only a limited amount of information is available on them. Research with domestic animals clearly shows that their responses are not typical of human responses, leading to the assumption that sensations perceived by man for a given chemical may be quite different than for many animals, although there is no way to conclusively prove this theory.

Animals are able to taste chemicals, which must be at least partly dissolved, because of taste buds located primarily on the tongue, but also on the palate, pharynx, and other parts of the oral cavity. Lower animals may have external taste buds on antennae, feet, and other appendages. On the tongue, different areas are sensitive to different tastes. In man, for example, the tongue is particularly sensitive to sweet and salty flavors at the tip and front edges. On the sides, it is more sensitive to sour, and at the back, bitter. The number of taste buds varies greatly between species. It is said that averages are: chicken, 24; dog, 1,700; man, 9,000; pig and goat, 15,000; and cattle, 25,000. The number of taste buds does not necessarily reflect taste sensitivity as the chicken will reject certain flavored solutions that are apparently inperceptible to cattle.

With tastants in water solutions, sheep show a positive preference only for a few sugar in water solutions; cattle have a strong preference for sweets and a moderate preference for sour flavors; deer show a strong preference for sweets and moderate to weak preference for sour and bitter flavors; and goats tend to show a preference for all four taste classes. The concentration of the solution has a marked effect on preference or rejection. Other information indicates that sheep and cattle prefer grasses with high organic acid content, and Arnold (4) has shown that a variety of organic compounds common to some plants resulted in a reduced intake by sheep. Salt-deficient sheep, however, show a marked preference for various Na salts, and it has been shown that ruminant animals will choose to graze on grasses that have been fertilized with P and N in preference to those not so fertilized. Pigs show a pronounced liking for sweets, and chickens show a positive response to a wide variety of plant or animal tissue components, but little or no liking for some sugars, although they will readily consume xylose, which may be toxic to them.

Man's taste has clearly been shown to be affected by a variety of factors such as age, sex, physiological condition (pregnancy, for example), and disease. Information of this type is almost nonexistent in domestic animals, although some is available on rats and dogs. It has been demonstrated that buck deer show a stronger preference for sour compounds than does, and bucks show a preference for bitter compounds such as quinine, although does do not. Very likely, domestic species may show the same type of response.

In man, it is well known that there are great differences in the ability of different individuals to detect low concentrations of different tastants and, when detected, in the response — that likes and dislikes for foods and fluids vary widely. The same principle probably applies to animals. An example of the variability to be expected in similar sheep given different test solutions is shown in Table 17-2. Note the tremendous range in acceptibility of these different solutions, a situation that is typical of animal response to such tests.

A variety of different flavoring agents, which usually have moderate to strong odors also, are sold for use in commercial feeds. Many times they are added to animal feeds on the assumption that if it smells good to men it ought to taste and smell good to animals. It is very probable that a substantial amount of money is wasted as a result (13), although a limited amount of published information indicates that some commercial flavoring agents have resulted in increased feed consumption in some situations. Flavors (and odors) are likely to be of less consequence when animals do not have a choice of feedstuffs than when a variety of feedstuffs are available at one time.

* See bibliography covering years from 1556-1966 in Kare and Maller [16].

Table 17-2. An example of variation in response of sheep to different chemical solutions when given a choice between the chemical solution and tap water.[a]

Chemical in test solution	Consumption of test solution, % of total					
	Sheep no.					
	1	2	3	4	5	Av.
Sucrose, 15%	95	24	30	82	8	48
Lactose, 1.3%	73	54	42	46	68	56
Maltose, 2%	8	61	33	44	21	33
Glucose, 5%	69	88	27	71	90	69
Saccharin, 0.037%*	36	34	7	28	57	32
Sodium chloride, 2.5%*	8	23	14	22	29	19
Quinine HCl, 0.0063%*	51	80	23	27	43	45
Urea, 0.16%*	20	46	60	64	48	48

[a] From Goatcher (14)
* Those identified with * are a second set of sheep.

Odors

Odors are produced by volatile compounds and, as the reader will know, there are a tremendous variety that may be produced by food and feed. At the present time, there is little agreement as to a single broad classification of odors such as we find in taste.

It is generally conceded that most animals have a keener sense of smell than man does, but quantitative information at this time is not adequate to predict the response of an animal. Some of the best research with domestic species (sheep) has been reported by Arnold (3, 4). When various senses — smell, taste or touch — were impaired by surgery, Arnold found that loss of the sense of smell did not statistically increase consumption of any of 5 plant species and intake of 2 was decreased. Sheep were less apt to consume flowering heads, a fact also noted by Tribe (28). When various odoriferous compounds were added to feed, Arnold found that the response depended on whether the animals had a choice or not, and the response (increased or decreased total consumption) was not predictable. Of 6 different compounds, only butyric acid increased consumption with and without a choice of feeds. Tribe (28) has shown that a wide variety of odoriferous compounds may be objectionable to sheep at first, with the sheep reacting strongly, but they eventually overcame their objections and ate the feed even when other uncontaminated feed was available. These reports would indicate that odor may serve as an attractant, but may not have a great deal of influence on total or eventual consumption.

Sight

It is well recognized that many, although not all, animal species have better vision than man. However, the importance of sight as it affects food consumption in animals is not fully understood. Pangborn (26) points out that experimental evidence on man clearly indicates that sight has a pronounced effect on taste, because individuals tend to associate different colors, shapes, and other visual cues with known flavors and odors. Sight in animals appears to be used more for orientation and location of food. Research with cattle and sheep shows no effect of coloring feeds red, green, or blue — indicating that they may be color blind.

Texture-Physical Factors

It is well known that the texture and particle size of feedstuffs may influence acceptability. Pelleted feeds are a good example, as most domestic and some wild ruminant species will readily take to pelleted feeds even though they may be completely unfamiliar with them. Particle size and texture may also have some effect as evidenced by the fact that many animals will take more readily to rolled or cracked grains than to whole grains. Feed preparatory methods that reduce dustiness usually result in an increased feed intake, as almost all animals discriminate against dust if given a chance to do so; this is probably one reason why succulent feeds are readily consumed as compared to dry feed.

Appetite and Regulation of Feed Intake

Appetite has been defined as the desire of an animal to eat and **satiety** is the lack of desire to eat. **Hunger** may be defined as the physiological state that results from the deprivation of food of a general or specific type and is abolished by the ingestion of these foods (5). Appetite is frequently quantitated by measuring the intake of food in a limited time span. An example of equipment used in appetite research is shown in Fig. 17-1.

All animals unquestionably have physiological means of regulating food intake, both in short-term and in long-term situations. This is clearly shown by the fact that, when adequate food of an acceptable nature is available, wild animals do not starve or overeat to a harmful extent, although man and some domestic animals are exceptions. This seems particularly true for wild species which seldom accumulate the amount of body fat seen in tame species.

It has been clearly demonstrated that hunger and satiety centers are located in the hypothalamus (midbrain). Studies with experimental animals have shown that lesions in the appropriate location will cause temporary loss of appetite and lack of thirst. By the same token, appropriate lesions in the hypothalamus may cause an animal to overeat. Electrodes placed in these areas can be used to pin-point electrical stimuli to cause the same effects so that the animal will eat during stimulation and quit eating as soon as the electrical stimulus is stopped (see Fig. 17-2). The specific mechanisms which actuate these appetite centers have not been clearly identified by any means, particularly with respect to long-term control of appetite. Some of the various theories concerning appetite control are discussed briefly in succeeding sections.

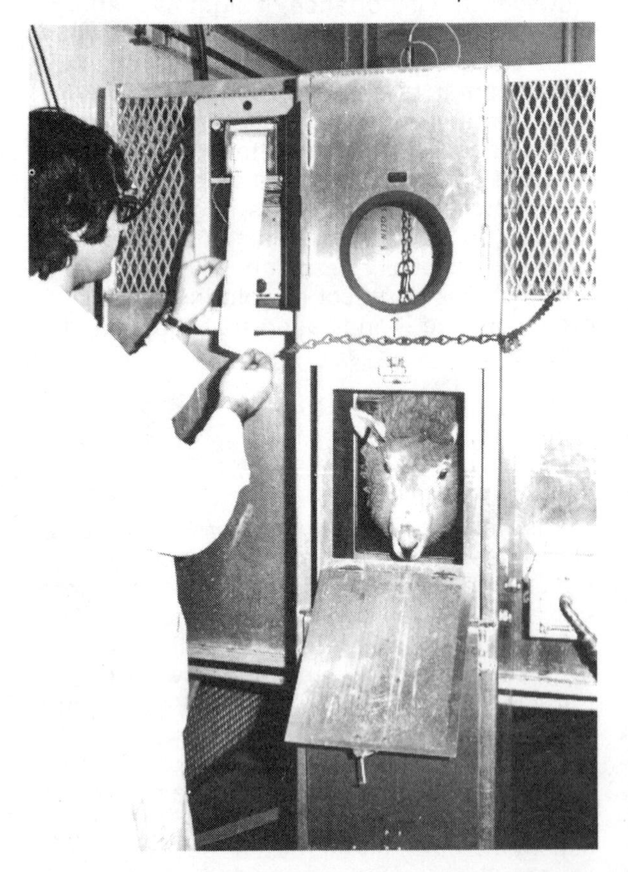

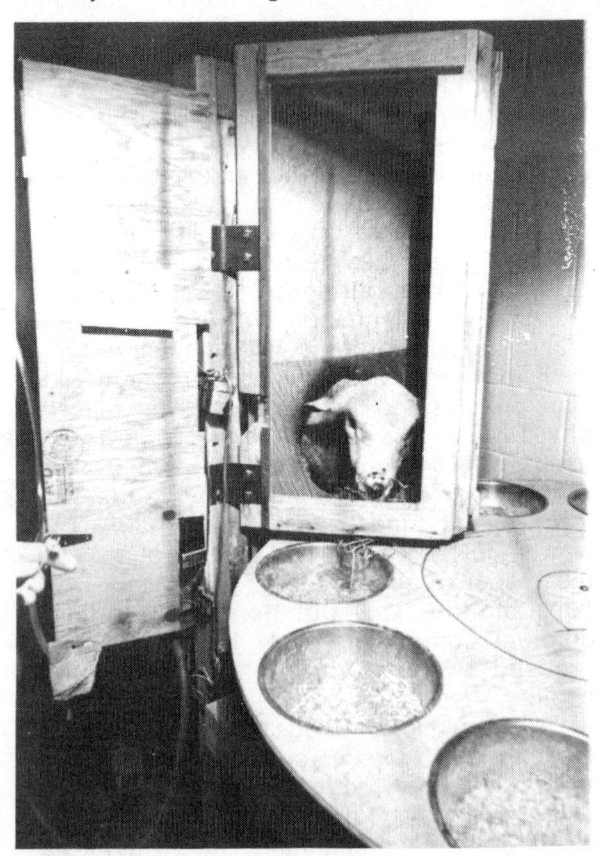

Figure 17-1. Left. Equipment used in appetite research. Feed intake recording unit used for the continuous monitoring of feed intake by sheep or goats. Courtesy of L.A. Muir, Merck Institute for Therapeutic Research, Rahway, N.J. Right. Apparatus illustrating how feed consumption can be measured in timed intervals to study periodic intake of feed. Courtesy of P.J. Wangsness, Pa. State Univ.

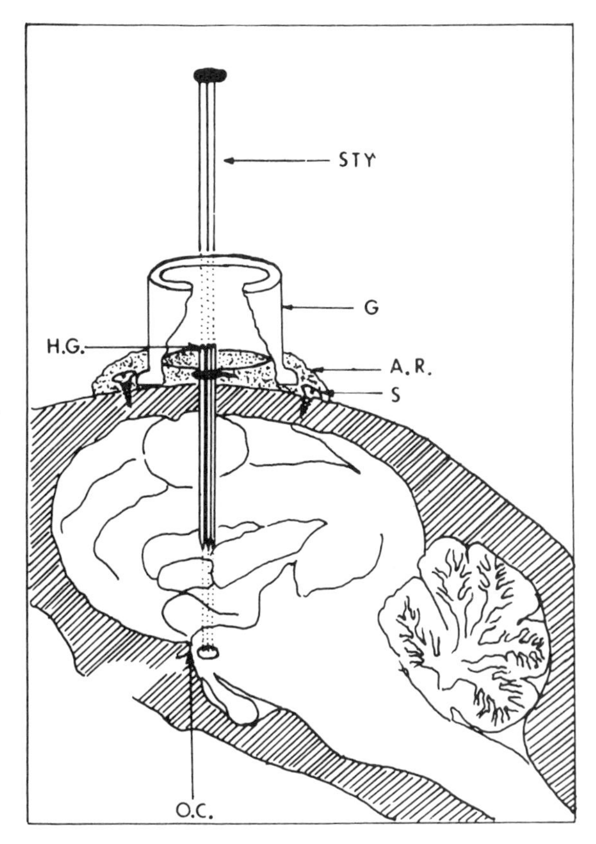

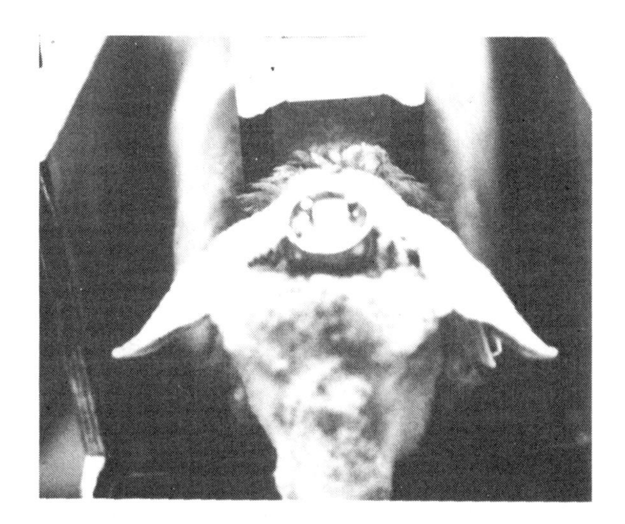

Figure 17-2. Left. A diagram illustrating the apparatus and placement of cannulas into the brain of a sheep. Abbreviations are: STY-stylet; G-gard [syringe barrel]; HG-hypothalamic guides; AR-acrylic resin; S-stainless steel screw; OC-optic chiasm. Courtesy of C.A. Baile, Smith Kline Research Laboratories. Right. Head of a sheep showing implanted apparatus. Courtesy of C.A. Baile.

Thermostatic Regulation of Appetite

This theory is based on the observation that animals generally eat more when they are cold and markedly reduce intake when stressed by heat (8). Studies with experimental animals also indicate that cooling or heating the hypothalamus with probes inserted into it may cause the same response. Further evidence is required to validate the effect, although research data do show that body and environmental temperature have some effect on appetite and satiety.

Lipostatic Regulation of Appetite

This theory suggests that the amount of body adipose tissue (fatty tissues) may serve in some way to increase or decrease intake as the amount of body fat decreases or increases. Little evidence to support or refute it exists at this time, other than some data showing a negative correlation between blood free fatty acids and appetite.

Chemostatic Regulation of Appetite

In most monogastric species there is evidence to show that blood glucose concentration is negatively related to feed intake over the short-term, and that hunger contractions of the stomach are more pronounced when blood glucose is low. This is the glucostatic theory of appetite control (17, 20). In ruminant species, however, such relationships do not hold as blood glucose concentration has little, if any, relationship to feed intake. Recent data on ruminants indicate that blood insulin levels increase after feeding and that the level of insulin in plasma is influenced by the energy level of the diet and/or amount of food consumed (19). Some evidence shows that blood propionate levels are negatively related to feed intake as are rumen volatile fatty acids; other unidentified chemical factors are also likely to have some affect on appetite in ruminants (9).

Caloric Density and Physical Limitations of the GIT on Appetite.

Most adult animals apparently are capable of maintaining a relatively stable body weight over long periods of time. Likewise, young animals of a given species tend to grow at uniform rates. Both adults and growing young do this in spite of marked variation in physical activity and energy expenditure, indicating that the animal is able to adjust energy intake to energy expenditure by some unknown means of appetite control. If no other problems interfere — such as nutrient deficiencies, disease, and so on — animals eat to meet their caloric needs. If the diet is diluted by water, then a much greater volume will be consumed. By the same token, if the diet is diluted with undigestible ingredients, then the animal will eat more up to a point at which its GIT can no longer handle the bulk in the diet. This principle is illustrated in Table 17-3 with chickens and graphically in Fig. 17-3. In Table 17-3, note that feed consumption gradually declines as energy content of the diet is increased over a rather narrow range, although total energy intake increased in this example. Feed conversion also was improved as energy concentration in the diet increased. These principles are well understood in the nutrition of monogastric species, and diets are adjusted to provide more or less optimum caloric density for a given production function. For example, if one wishes to control energy intake of pregnant sows, it can be done by diluting the ration with ground alfalfa hay.

In ruminant species, where a large proportion of the diet is composed of roughages — pasture, hay, silages — the relation is less clearly understood by most producers, probably because so many different factors affect the optimum caloric density for a given class and species of animal. Factors such as physical density of the feed, particle size, amount of indigestible residue, rapidity of rumen fermentation, and level and frequency of feeding influence rate of passage through the GIT. This, in turn, influences the amount of space in the stomach and gut for the next meal. For example, it has been demonstrated many times that consumption of low-quality roughages such as straw and poor hay can be markedly increased by supplementation with protein supplements and, sometimes, with P or molasses. Such forage is usually deficient in both N and P for adequate rumen digestion and small amounts of readily available carbohydrate tend to stimulate cellulose digestion in the rumen. Other nutrients required by rumen microorganisms would be expected to have the same effect (Mg and S, for example). By way of another example, pelleting of low-quality roughages will almost always greatly increase consumption due to more rapid digestion and passage out of the rumen.

As a result of these factors, pin-pointing a precise value at which increasing caloric density of a ration would result in a reduced intake of dry matter or energy would be most difficult. Age is a factor also, as young lambs are less able to handle high-roughage diets than older animals. With respect to dairy cows, Conrad et al (11) concluded that 67% digestible dry matter was the lowest ration digestibility at which lactating cows could regulate energy intake (no longer physically limiting),

Table 17-3. Effect of energy level on performance of finishing broilers from 4.5 to 8 weeks of age.[a]

Ration energy, ME, Mcal/kg	Consumption of Feed, kg	Energy, Mcal	Feed conversion	Daily gain, g
2.81	2.30	6.46	2.80	34.2
2.98	2.31	6.87	2.65	36.3
3.10	2.31	7.17	2.45	39.3
3.14	2.22	6.98	2.39	38.8
3.18	2.20	7.02	2.35	39.1
3.31	2.18	7.20	2.32	39.1
3.79	2.09	7.90	2.14	40.6

[a] From Combs and Nicholson (10)

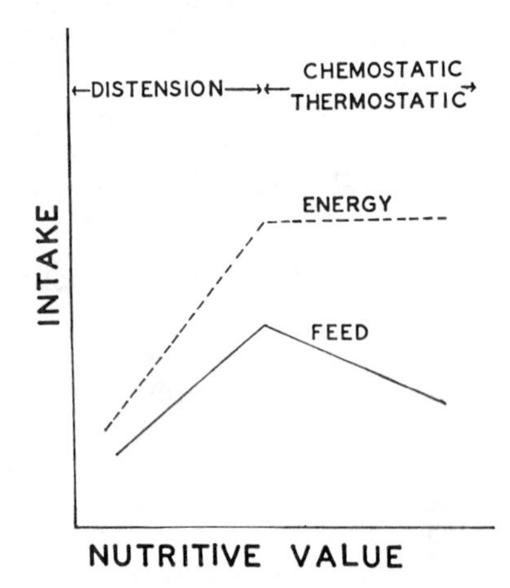

Figure 17-3. A graphic example of probable relationships between energy and feed intake and some controlling mechanisms. From Montgomery and Baumgardt [23].

whereas Montgomery and Baumgardt (23) found that energy intake of dairy heifers was maintained when the ration (pelleted) was above 56% digestible dry matter.

Two sets of data showing the effects of caloric density on energy intake are shown in Table 17-4. The data on steers clearly illustrate the curvilinear effect on energy intake as the caloric density goes from low to high. At the low end, physical limitations of the GIT are, most likely, limiting intake to less then would be consumed otherwise. At the 2 highest energy concentrations, feed and energy intake were reduced considerably, probably because of abnormal rumen fermentation or appetite inhibition by some chemical factor produced in or present in the GIT. In the example with lambs, dry matter was almost the same with each of these rations which were composed of varying amounts of barley, oats and oat husks. Estimated ME intake and carcass gain were highest with the most concentrated ration (79.5% DE), however. Note also, that dry matter found in the fore-stomach and gut after slaughter increased progressively as the caloric density of the rations decreased. These data illustrate the physical limitation effect of a roughage (oat hulls) which is of low digestibility. With respect to these lambs, the high energy ration had not, apparently, reached the point of diminishing returns as it did with the steers.

The point (in terms of caloric density) at which an animal will have satisfied its energy demand may be expected to be quite variable with different classes and species of animals. Baumgardt (6) points out that energy needs will vary with productive functions. Thus, a lactating cow will have a greater demand than a pregnant, nonlactating cow, which, in turn, will have a higher requirement than an animal that is neither pregnant nor lactating. Rapid growth will stimulate a greater need for energy as will various environmental factors and stressing situations.

Inhibition of Appetite

As the reader has probably concluded, control of appetite is a complex mechanism with inputs from a variety of different sources. In addition to the various theorized mechanisms, appetite is generally affected by factors that interfere with the normal functions of the GIT or by many different diseases that affect non-GIT tissues and organs.

Nutritionally, it has been known for many, many years that a high-protein meal (in man, at least) tends to dull the appetite. This is believed

Table 17-4. Effect of caloric density on feed and energy intake of ruminant species.

Item	Experimental Data										
	Steers [a]										
Roughage in ration, %	100	90	80	70	60	50	40	30	20	10	0
DE, Kcal/g	2.50	2.62	2.81	2.92	3.02	3.08	3.21	3.25	3.36	3.46	3.60
Dry matter intake, g/kg $BW^{0.75}$	94	91	98	97	91	97	98	87	85	71	46
DE intake, Kcal/kg $BW^{0.75}$	235	242	276	282	276	299	316	282	288	245	166

Item	Lambs [b]				
DE, %	65.2	71.6	74.9	76.6	79.5
Calculated ME, Mcal/kg	2.36	2.59	2.71	2.73	2.82
Dry matter intake, kg/day	1.06	1.03	1.00	1.00	1.02
ME intake, Mcal/day	2.50	2.66	2.71	2.73	2.88
Carcass gain, g/day	105	118	121	125	143
Dry matter in forestomach, g	1287	1302	818	816	503
Dry matter in hind gut, g	265	236	226	220	188

[a] Data from Parrott et al (27)
[b] Data from Andrews et al (1)

to be related to the prolonged and relatively high heat increment resulting from the metabolism of amino acids. Fat is also inhibitory, presumably because a high-fat meal does not pass out of the stomach rapidly as fat entering the duodenum triggers hormonal mechanisms that may cause restriction of the pylorus, resulting in a slow emptying of the stomach.

In human nutrition and dietetics there is a great interest in appetite inhibitors (called anorectics) as there are so many over-weight people, at least in the USA. A wide variety of compounds have been used at one time or another, but most of those currently in use are amines similar to or derived from amphetamine; these compounds are stimulatory to the central nervous system, but, at the same time, result in some inhibition of appetite. Undesirable side effects frequently occur and, in addition, many people develop tolerance to the effect on appetite after a period of a few weeks. Generally speaking, these chemical inhibitors have not proven to be very successful and there are many conditions where their use is contraindicated.

In the nutrition of animals, we are more concerned usually with obtaining maximum feed intake, although in some situations it is desirable to limit feed intake. Feed intake is more commonly limited by caloric dilution — dilution of rations with feedstuffs of low digestibility — or by restricting intake by reducing the daily allowance. Feed intake can be partially restricted in other ways. For example, if feed is provided in a physical form not preferred — such as a dusty meal — this will nearly always reduce feed intake. Unpalatable ingredients can be used, for example diammonium phosphate or quinine. With ruminant species, salt (NaCl) has been used successfully to restrict intake, but this would be hazardous with birds and pigs as they are relatively susceptible to salt toxicity. Research in the early 1950's clearly demonstrated that salt, when mixed with protein supplements such as cottonseed meal, can be fed for several months to cattle or sheep without detrimental results if water is not restricted. Salt may be needed in a concentration of 20-30% of the mixture to restrict protein supplement consumption to about 1 kg/day for cattle. The exact amount depends on availability of other feedstuffs and the age of the animal. Fat has also been used to restrict intake of concentrates and, nutritionally, is probably preferable to dilution with salt as the excess salt, even though it may not be harmful, is wasted from a nutritional point of view and may increase urinary loss of other electrolytes.

Factors Affecting Feed Intake

Some knowledge of expected feed intake is required in any situation where ad libitum intake is allowed, i.e., for high producing or fast gaining animals. If we assume that suggested nutrient allowances are reasonably accurate, then we must know how much feed is likely to be consumed in order to arrive at the desired concentration of nutrients in any ration provided to an animal. To put it another way, if we assume that a cow requires 3 lb of protein/day, we must know how much feed is consumed in order to arrive at a desired % of protein in the diet if we do not want to over- or under-feed the cow. As the reader may have concluded, several factors may affect the amount of feed consumed by an animal. These are discussed briefly in subsequent sections.

Body Weight

As indicated in Ch. 9, energy requirements of adult animals are related, more or less, to body weight raised to the 0.75 power. Thus, an increase in body size from, let us say, 1000 lb to 1100 lb, does not increase energy requirements by 10%, but by a lesser amount, 7.4%*. Body weight, by itself, however, is not always a good measure of feed intake, particularly when we are dealing with fattening animals. For example, beef animals just starting on feed are apt to have relatively little body fat (10%+) and may be expected to consume on the order of 2.5-2.75% of body weight/day. When they are approaching market condition (USDA Choice grade) with, perhaps, 25-30% of body tissues being fat, feed consumption is apt to be more nearly 2.2% of body weight/day. This is a reflection of the fact that energy need (and feed consumption) is more nearly related to lean body mass than to total body weight; also, more fat has been deposited in the abdominal cavity, thus reducing the amount of space that may be occupied by the GIT and the feed that it contains. Furthermore, the exces fat may inhibit appetite (lipostatic control?) as well as causing a physical limitation on feed intake.

* $1000^{0.75}$ = 177.8; $1100^{0.75}$ = 191.0; the difference, 13.2 = 7.4% increase over 177.8.

Individuality of Animals

Anyone who has had any experience with individually-fed animals realizes that they do not all eat alike and may readily show pronounced likes and dislikes when they have the opportunity to express themselves. It is also well known that hormonal differences may result in hyperexcitable or phlegmatic animals, with resultant effects on activity and feed consumption. Such differences make it difficult to predict how much some of these animals may consume. In modern day agriculture, catering to such individual animal variation may not be profitable, even though we should be aware that it exists.

Type and Level of Production

All young animals have what we might call a 'biological urge' to grow and growth cannot be greatly impeded without an adverse effect on eventual size and/or productivity. Almost invariably, those animals having the most rapid growth rates have the best appetites. Ample evidence also shows that pregnancy and lactation result in a stimulation of appetite. With respect to lactating animals, appetite is not completely correlated with production. Some cows, for example, have the ability to lactate at very high levels, yet do not have the appetite to go along with it, with the result that they may lose a considerable amount of body weight during peak lactation. On the other hand, some animals will not greatly deplete body reserves in order to produce milk. Considerable variation among animals is found in this respect.

Miscellaneous Factors

Hot ambient temperatures, especially along with high humidity, have an inhibitory effect on appetite. On the other hand, cold temperatures usually stimulate appetite. Ration characteristics can modify these influences somewhat, although not greatly.

Health of the animal is certainly a factor of great consequence. Most infectious diseases result in a reduced feed intake, more or less related to the severity of the infection. Likewise, parasites such as stomach worms will usually result in reduced feed consumption, often for prolonged periods of time. Metabolic problems such as ketosis, bloat, and diarrhea result in restricted feed consumption.

Stresses, (in addition to disease or parasites) will usually reduce consumption. Such stresses as crowding (lack of adequate space), noise and disturbances, and excessive handling tend to keep animals excited and reduce feed consumption.

Proper design for feed bunks, mangers, and water supplies can encourage increased intake. Cleanliness of feed and water containers may have a pronounced effect, also, as most animals object to dirt, molds, and manure in or near their feed or water. Inadequate water supplies, particularly if the water is contaminated or foul-tasting, will reduce feed consumption.

Expected Feed Intake

Expected Feed Intake of Cattle

The total feed consumption of cattle, as with other ruminants, is highly dependent upon the quality of the roughage being consumed. As an approximate guide, the information in Table 17-5 can be used, bearing in mind that many different factors may modify the values given. For example, it is clear that young, rapidly growing animals will consume more/unit of size than adult animals, particularly if fed high quality roughages. Also, adult animals with high physiological needs, such as lactating cows, may consume forage at the upper limits or even exceed the values shown in the table, but those with low requirements will not consume nearly as much. Feed preparation, such as pelleting, may greatly increase consumption as will proper supplementation of low quality roughages. Silages are only rarely consumed at levels approaching that of high-quality hay, even though the digestibility of the silage may be quite good; the high water content may be partially responsible as well as the high acid content.

Conrad et al (11) maintain that consumption by dairy cows of rations of low digestibility is in direct proportion to body weight rather than in proportion to metabolic size ($BW^{0.75}$). Equations developed by Conrad (see McCullough, 21) can be used to predict maximum feed intake and minimum allowable digestibility in the rations. These equations are:

Maximum feed intake (values in lb)
$= 10.7\ (BW/1000) + 0.058\ BW^{0.75} + 0.33$ milk production $+ 0.53$

Minimum allowable digestibility (values in lb) $= (0.058\ BW^{0.75} + 0.33$ milk production $+ 0.53)$/maximum feed intake

Table 17-5. Expected roughage consumption by cattle.

Feedstuff	Usual range in energy		Dry matter intake,
	TDN, %	DE, Mcal/kg	% body wt/day
Lush, young legume, grass pasture; barn-dried grass	70 +	3.10 +	2.75-3.5
Well-eared corn silage; high quality sorghum silage	70	3.10	2.0-2.5*
Moderate quality, actively growing pasture	60-65	2.65-2.87	2.5-3.2
Grass and grass-legume silage of good quality	55-60	2.43-2.65	2.0-2.5*
Good quality legume hays	50-55	2.20-2.43	2.5-3.0
Grass hay from mature plants; re-growth pasture	45-50	1.98-2.20	1.5-2.0
Poor quality grass hays; dormant pasture	40-45	1.76-1.98	1.0-1.5
Cereal and grass straws	35-40	1.54-1.76	1.0 or less

* Cattle will only rarely eat as much dry matter from silage as when the same crop is preserved in other forms.

where BW = body weight. According to McCullough (21) these 2 formulas have worked out well in practice on Georgia dairy farms.

Figure 17-4 illustrates the range in forage consumption that may be expected by dairy cows given various levels of concentrates (12). As indicated in this figure, small amounts of concentrates may not depress forage consumption; they may even increase consumption of low-quality roughages; however, as a general rule, increased concentrate consumption will result in a gradual reduction in roughage intake and in total dry-matter consumption, although energy intake will nearly always increase as concentrates are added. For dairy cows, most nutritionists recommend 35-55% concentrate in the total ration for

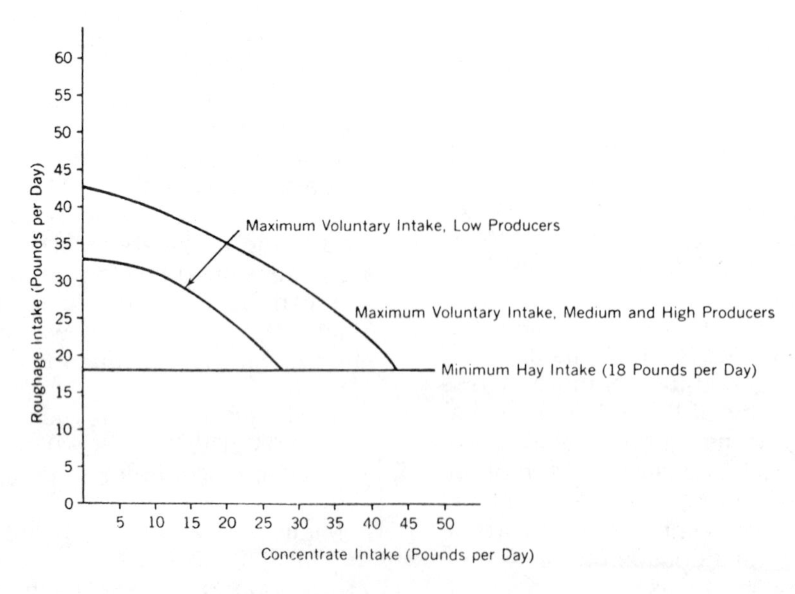

Figure 17-4. Estimated maximum voluntary intake of roughages [90% dry matter] by 1200 lb cows at varying concentrate intakes. From Dean et al [12]. By permission of Lea & Febiger.

optimum utilization of roughage and near-maximum energy intake. For finishing beef cattle, the amount of concentrate fed in the USA in current practice is apt to be about 70-90%, depending on the quality of the roughage and on feed preparation methods used on both roughage and concentrates.

With respect to total feed consumption, young animals would be expected to consume relatively more than older animals. Young Holstein calves may be expected to consume 3.2-3.4% of body weight at 2 mo. of age of a complete feed and to consume about 3% at 6 mo. of age. After this their feed intake will gradually decline (18). Data on beef calves (25; see Appendix tables) indicate a lower intake than this, more on the order of 2.7% of body weight for calves weighing about 200-300 kg (441-661 lb). With finishing steer calves, NRC gives values equivalent to 2.5% (at 200 kg) ranging down to 2.1% at 450 kg (992 lb), and for finishing yearlings, values ranging from 2.9% at 250 kg (551 lb) down to 2.3% at 500 kg (1100 lb). In the author's (Church) experience, steer calves fed moderate energy rations (ca. 70% TDN) will average 2.1-2.2% of body weight as dry matter intake from 600 to 1050-1060 lb (ca. 480 kg).

Expected Feed Intake of Sheep

Experiments carried out under calorimetric conditions indicate that sheep will consume roughages at about the same level/unit of metabolic size ($BW^{0.75}$) as cattle (7). However, since their actual size is considerably smaller, the amount, when expressed as % of body weight, will be greater. NRC values (see Appendix tables) indicate consumption of total ration by finishing lambs to be about 4.3% of body weight at 30 kg (66 lb) ranging down to 3.5% at 55 kg (121 lb). This level of consumption may easily be exceeded when lambs are fed high-roughage, pelleted rations. NRC values for other sheep classes are given in the appropriate tables.

Expected Feed Intake of Swine

Feed consumption of growing pigs from weaning to market weight (40-220 lb; 18-100 kg) may be expected to be about 4.5-7 lb/day (2.0-3.2 kg). Consumption level will be markedly affected by physical and caloric density as well as other factors previously discussed. A graphic presentation of typical feed consumption is shown in Fig. 17-5. Note that average daily consumption is about 5% of body weight, although it is higher for young pigs and gradually decreases as the pigs get older and fatter. Mature animals, except for lactating sows, are usually limit fed at about 4-5 lb daily as they will usually become too fat unless a low-energy ration is used. Lactating sows may be expected to consume 3-4% of body weight/day of a moderate energy ration.

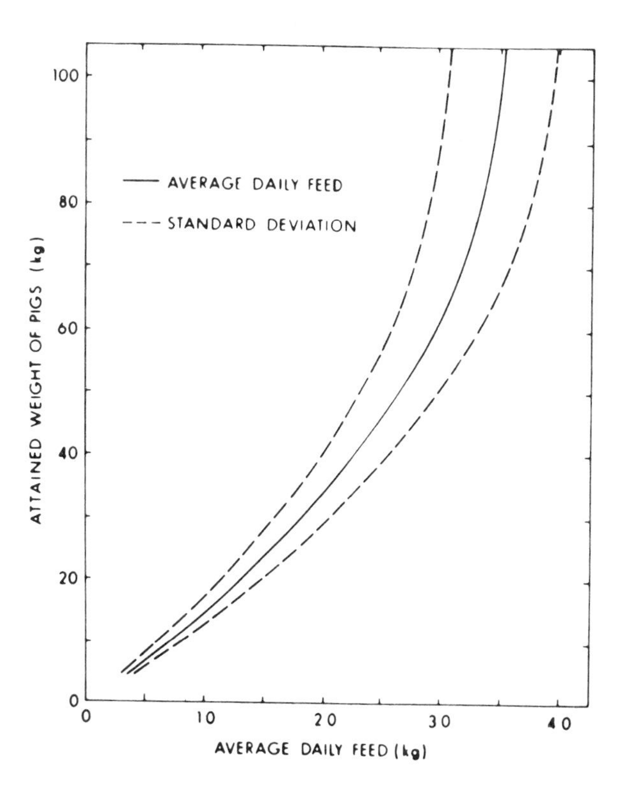

Figure 17-5. Expected average daily feed intake and range for full-fed pigs when fed according to NRC standards. From NRC [24].

Expected Consumption for Chickens

Expected feed consumption of mixed sexes of broiler chicks on typical rations and with favorable conditions is illustrated in Table 17-6. As with other species, growth rate declines as does feed consumption/unit of body weight as animals age. With respect to laying hens, feed consumption is markedly affected by size of the hen and by her level of egg production. With 5 lb birds producing at a level of 60-70%, daily feed consumption will be about 24-25 lb (10.9-11.3 kg)/100 birds.

Table 17-6. Expected feed consumption of broiler chicks, mixed sexes. [a]

Age in weeks	Av. body weight lb	g	Feed/broiler/week lb	g
1	0.26	118	0.29	132
2	0.57	259	0.43	195
3	1.02	463	0.72	327
4	1.53	695	0.94	427
5	2.06	935	1.07	486
6	2.67	1212	1.26	572
7	3.33	1512	1.45	658
8	3.98	1807	1.62	735

[a] Anonymous (2)

References Cited

1. Andrews, R.P., M. Kay and E.R. Orskov. 1969. Animal Prod. 11:173.
2. Anonymous. 1972. Poultry Management and Business Analysis Manual. Maine Extension Service Bul. 566.
3. Arnold, G.W. 1966. Austral. J. Agr. Res. 17:531.
4. Arnold, G.W. 1970. In: Physiology of Digestion and Metabolism in the Ruminant. Oriel Press, Newcastle upon Tyne, England.
5. Balch, C.C. and R.C. Campling. 1962. Nutr. Abstr. Rev. 32:669.
6. Baumgardt, B.R. 1969. In: Animal Growth and Nutrition. Lea & Febiger, Philadelphia.
7. Blaxter, K.L., F.W. Wainman and J.L. Davidson. 1966. Animal Prod. 8:75. Acad. Press, N.Y.
8. Brobeck, J.R. 1960. Hormone Research. Academic Press.
9. Church, D.C., G.E. Smith, J.P. Fontenot and A.T. Ralston. 1971. Digestive Physiology and Nutrition of Ruminants. Vol. 2 - Nutrition. O & B Books, 1215 NW Kline Pl., Corvallis, Ore.
10. Combs, G.F. and J.L. Nicholson. 1964. Feedstuffs 36(34):17.
11. Conrad, H.R., A.D. Pratt and J.W. Hibbs. 1964. J. Dairy Sci. 47:54.
12. Dean, G.W., D.L. Bath and S. Olayide. 1969. J. Dairy Sci. 52:1008.
13. Ewing, W.R. 1963. Poultry Nutrition. 5th Ed. The Ray Ewing Co., Pasadena, California.
14. Goatcher, W.D. 1969. M.S. Thesis, Oregon State University, Corvallis.
15. Hervey, G.R. 1959. J. Physiol. 145:336.
16. Kare, M.R. and O. Maller (eds.). 1967. The Chemical Senses and Nutrition. The John Hopkins Press, Baltimore.
17. Janowitz, H.D. and M.I. Grossman, 1949. Amer. J. Physiol. 159:143.
18. Kliewer, R.H., W.H. Kennick and D.C. Church. 1970. J. Dairy Sci. 53:1766.
19. Lofgren, P.A. and R.G. Warner. 1972. J. Animal Sci. 35:1239.
20. Mayer, J. 1955. Ann. N.Y. Acad. Sci. 63:15.
21. McCullough, M.E. 1969. Optimum Feeding of Dairy Animals. Univ. of Georgia Press, Athens, Ga.
22. McCullough, M.E. 1973. Feedstuffs. 45(7):34.
23. Montgomery, M.J. and B.R. Baumgardt. 1965. J. Dairy Sci. 48:569.
24. NRC. 1973. Nutrient Requirements of Swine. National Acad. of Sci., Washington, D.C.
25. NRC. 1970. Nutrient Requirements of Beef Cattle. National Acad. of Sci., Washington D.C.
26. Pangborn, R.M. 1967. In: The Chemical Senses and Nutrition. The John Hopkins Press, Baltimore.
27. Parrott, C., H. Loughhead, W.H. Hales and C.B. Theurer. 1968. Arizona Cattle Feeders Day Report. Univ. of Arizona, Tucson.
28. Tribe, D.C. 1949. J. Agri. Sci. 39:309.

Chapter 18 – Feedstuffs for Animals

In the study of animal nutrition and feeding, we need to be concerned with feedstuffs (or feeds) because they are the raw material that is essential for animal production. With domestic animals used for production of food or fiber, we are concerned with the efficient conversion of feedstuffs to useful products for man's use or enjoyment. Thus, some understanding of the chemical and nutritional composition of important classes of feedstuffs will provide a better understanding of applied animal nutrition.

A tremendous variety of feedstuffs is used for animal feeding throughout the world, the variety in a given location depending on the local products that are grown or harvested and the class and species of animals involved. Well over 2000 different products have been characterized to some extent for animal feeds, not counting varietal differences in various forages and grains.

The desirability of a given feedstuff is dependent upon a number of different factors. Cost is an important item and, generally, products fed to animals are either those that are not edible for humans or those that are produced in surplus to human needs in a given location or country. With our imperfect systems of distribution, a grain that may be surplus in the USA might well be in high demand for human use in some other area of the world. Thus, it is only in relatively recent times that livestock have been fed large quantities of edible food grains. Other factors affecting value of a feedstuff will include acceptability by the animal, ability of a given animal species or class to utilize a given product, the nutritional content, and the handling and milling properties of the product.

The number of feedstuffs is so great that it is not feasible to cover many individual feeds with any detail in this book. Rather, we will deal with major differences of feedstuffs in the various classes (next section) with occasional emphasis on individual items. Readers desiring more detailed information, can refer to NRC (30) or to older publications such as Morrison (28) or Schneider (34), the latter primarily dealing with digestibility.

Classification of Feedstuffs

A **feedstuff** may be defined as any component of a ration that serves some useful function. Most feedstuffs provide a source of one or more nutrients, but ingredients may also be included to provide bulk, reduce oxidation of readily oxidized nutrients, to emulsify fats, provide flavor, color, or other factors related to acceptability, rather than serving strictly as a source of nutrients. Generally speaking, medicinal compounds are usually excluded. The usual classification of feedstuffs, essentially as given by NRC (30) but with a number of added items, is as follows:

- Roughages
  - Pasture, range plants, and plants fed green
    - Grazing plants
      - Growing
      - Dormant
    - Soilage or green crop
    - Cannery and food crop residues
  - Dry forages and roughages
    - Hay
      - Legume
      - Nonlegume
    - Straw
    - Fodder
    - Other products with >18% crude fiber
      - Corn cobs
      - Hulls
      - Shells
      - Sugar-cane bagasse
      - Paper and wood
  - Silages
    - Corn
    - Sorghum
    - Grass
    - Legume
    - Miscellaneous
  - Concentrates
    - Energy feeds
    - Cereal grains
    - Milling by-products (primarily from

cereal grains)
Molasses
Seed and mill screenings
Animal and vegetable fats
Miscellaneous
Brewery by-products
Waste from food processing plants
Garbage, manures, and sewage
Roots and tubers
Fruits and nuts

Protein concentrates
Animal
Marine
Avian
Seeds from plants
Dehydrated legumes
Single-cell sources (bacteria, yeast, algae)
Nonprotein nitrogen (urea, etc.)

Mineral supplements
Vitamin supplements
Nonnutritive additives
Antibiotics
Antioxidants
Buffers
Colors & flavors
Emulsifying agents
Enzymes
Hormones
Medicines

Roughages

Roughages are the natural feeds for all herbivorous animals existing under natural conditions and such food provides the major portion of their diet for most if not all of the year. Harvested and stored roughages (hays, silages and other forms) are utilized by man to increase animal productivity under conditions that would not allow it otherwise in nature. Thus, roughages are of primary interest for domestic ruminants, horses, and other herbivores. Although other species such as swine can survive on roughage, productivity with no other source of feed would be too low to be economical in our current economy.

Nature of Roughages

To most livestock feeders, a **roughage** is a bulky feed that has a low weight/unit of volume. This is probably the best means of classifying a feedstuff as a roughage, but any means of classifying roughages has its limitations since, due to the nature of the products we are dealing with, there is a great variability in physical and chemical composition. Most feedstuffs classed as roughages have a high crude fiber (CF) content and a low digestibility of nutrients such as crude protein and energy. If we attempt, as does NRC (30), to classify all feedstuffs as roughages that have >18% CF and/or with low digestibility, we immediately find exceptions. Corn silage is a good example; it nearly always has >18% CF, but the TDN content of well-eared corn silage is about 70% on a dry basis. Lush young grass is another example. Although its weight/unit volume may be relatively low and fiber content relatively high, its digestibility is quite high. Soybean hulls are another exception for ruminants.

Most roughages have a high content of cell-wall material (see Ch. 2). The cell-wall fraction may have a highly variable composition, but contains appreciable amounts of lignin, cellulose, hemicellulose, pectin, polyuronides, silica, and other components. In contrast, roughages are generally low in readily available carbohydrates as compared to cereal grains.

The amount of lignin is a critical factor with respect to digestibility. Lignin is an amorphous material which is closely associated with the fibrous carbohydrates of the cell wall of plant tissue. It greatly limits fiber digestibility, probably because of the physical barrier between digestive enzymes and the carbohydrate in question. Removal of lignin with chemical methods greatly increases digestibility by rumen microorganisms and, probably, by cecal organisms. Lignin content of plant tissue gradually increases with maturity of the plant and a high negative correlation exists between lignin content and digestibility, particularly for grasses, although somewhat less for legumes (see Ch. 2 of Church et al, 8). There is also evidence that the silica content of plant tissue is negatively related to fiber digestibility.

The protein, mineral and vitamin contents of roughages are highly variable. Legumes may have 20% or more crude protein content, although a third or more may be in the form of nonprotein N. Other roughages, such as straw, may have only 3-4% crude protein. Most others fall between these two extremes.

Mineral content may be exceedingly variable; most roughages are relatively good sources of Ca and Mg, particularly legumes. P content is apt to be moderate to low, and K content high; the trace minerals vary greatly depending on plant species, soil, and fertilization practices.

In overall nutritional terms roughages may range from very good nutrient sources (lush young grass, legumes, high-quality silage) to very poor feeds (straws, hulls, some browse). The nutritional value of the very poor can often be improved considerably by proper supplementation or by some feed preparatory methods. The feeder must, however, use some wisdom in selecting the appropriate roughage for a given class and species of animal.

Factors Affecting Roughage Composition

A number of factors may affect roughage composition and nutritive value, if we ignore the generally poor-quality roughages. Maturity at harvest (or grazing) has one of the more pronounced effects. One relatively detailed example of stage of maturity on composition and digestibility of orchard grass [*Dactylis glomerata*] is shown in Table 18-1. Many other examples are available in the literature. In Table 18-1 and Fig. 18-1, note the rapid decline in crude protein, particularly. A gradual decline occurs also in ash and soluble carbohydrates and an increase in lignin, cellulose and CF, all during a 6 wk period. Digestibility declines, also, as these changes occur in plant composition. The changes that occur will depend on the plant species and on the environment in which the plant is grown. For example, if the growing season progresses rather rapidly from cool spring weather to hot summer weather, changes in plant composition will be more rapid than where the weather remains cool during plant maturation, especially in a cool season grass. In plants such as alfalfa [*Medicago sativa*], which has quite different growing habits than grass, rapid changes take place as the plant matures and blooms. Crude portein contents given by NRC publications indicate the following values (% on dry basis) for second cuttings: immature, 21.5; prebloom, 19.4; early bloom, 18.4; mid-bloom, 17.1; full bloom, 15.9; and mature, 13.6. Corresponding changes in TDN range from 63 to 55%. Part of these differences are due to the

Table 18-1. Effect of stage of maturity on composition and digestibility of orchard grass.[a]

Item	Stage of maturity			
Composition, %	6-7" high cut 5/19 pasture	8-10" high cut 5/31 late pasture	10-12" high cut 6/14 early hay	12-14" high cut 6/27 mature hay
Composition, %				
Crude protein	24.8	15.8	13.0	12.4
Ash	9.3	6.8	7.1	7.2
Ether extract	4.0	3.5	3.9	4.2
Organic acids	6.3	6.0	5.4	5.0
Total carbohydrates*	49.9	63.0	64.4	63.1
Sugars	2.1	9.5	5.4	2.4
Starch	1.2	9.5	0.8	0.9
Alpha cellulose	19.5	19.8	19.1	27.7
Beta & gamma cellulose	3.4	5.4	3.8	2.5
Pentosans	15.1	15.8	16.8	18.1
Nitrogen-free-extract	35.0	45.7	44.2	41.2
Lignin	5.7	5.0	6.2	8.1
Crude fiber	26.9	28.2	31.8	35.0
Digestibility, %				
Dry matter	73	74	69	66
Crude protein	67	63	59	59
Crude fiber	81	77	71	68

[a] From Ely et al (12)
* Total carbohydrates = (crude fiber + NFE) - (lignin + organic acids)

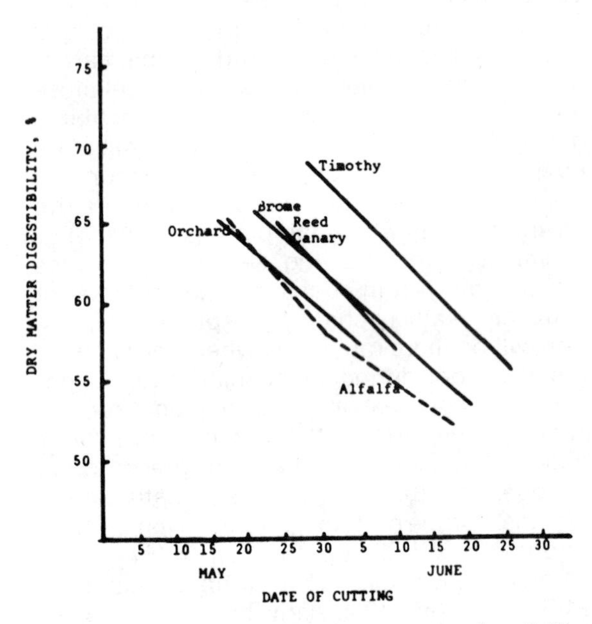

Figure 18-1. Effect of maturity on dry matter digestibility of first cutting forages. R.R. Johnson, Oklahoma State University, unpublished data.

fact that the plant loses leaves as it matures, and leaves have a higher nutrient value than the remainder of the plant. As plants mature, there is also a decline in concentration of Ca, K, and P and for most of the various trace minerals (33).

For many years, soil fertility and fertilization practices have been known to have a pronounced effect on quality of forage and crops produced. In addition, some alterations in plant composition may occur as a result of these factors, although the differences are much less dramatic than those changes associated with increased maturity, and results given in the literature show many discrepancies in responses to fertilizer. In pastures with mixed plant species, one obvious change that may occur as a result of fertilization is an alteration in the vegetative composition as some plants respond more to fertilizer than others. If a grass-legume mixture is fertilized with high levels of N, for instance, this practice is apt to kill out the legume or to stimulate the grass much more than the legume. Fertilization of grasses with N tends to increase total, nonprotein, and nitrate N of the plants. Nitrate levels usually drop off rapidly after fertilization, however. K content and, perhaps, some other minerals may increase in response to N; however, the marked increase in plant growth that may be obtained by high levels of N may be expected to result in some dilution of most mineral elements, particularly during the first few days or weeks or rapid growth after fertilization (33).

Digestible protein is apt to be increased and, in some instances, palatability and dry matter intake may increase in response to fertilization with N, although not all data agree on some of these points. Fertilization studies have shown, in general, that plant concentration of most of the mineral elements may be increased by fertilization with the element in question. P use may increase palatability when used alone or in combination with N. Thus, results to date indicate that soil fertility and/or fertilization practices may alter nutrient concentration and consumption of forage plants.

Harvesting and storage methods may have some effect on nutritive value of roughages, particularly when forage cut for hay is unduly bleached by the sun or damaged by dew and/or rain. Sun bleaching results in a rapid loss of carotenes, although this is of less consequence than it used to be because of the low cost of synthetic vitamin A. Leaching by rain will result in losses in soluble carbohydrates and N, and added handling required may result in loss of leaves. Any harvesting method used for legumes that tends to reduce leaf loss will affect nutritive value. With respect to storage, hays may lose a considerable amount of original nutritive value if stored too wet with resultant heating and mold formation. Otherwise, there is relatively little nutritive loss if the hay is stored in dry conditions over a period of several years. The same comment applies to silages, as well-preserved silages will keep for a number of years without much effect on nutritive value.

Latitude, along with accompanying effects of temperature and light intensity, may have an appreciable effect on plant composition and nutritive value. Not a great deal of information is available on this subject, but data do indicate that nutritive value of forage cut on a given date is apt to be higher as we go away from the equator toward the poles (northern or southern).

Pasture and Grazed Forages

A wide variety of vegetation is utilized by herbivorous animals. With respect to agricultural production, herbage is usually divided into native and cultivated species, the latter being utilized to improve productivity or versatility of crop production. Herbage may also be devided into the following classes:

- Grasses — members of the family Gramineae (5000 + species)
  - Cool season grasses — grasses that make their best growth in the spring and fall
  - Warm season grasses — grasses that grow slowly in the early spring and grow most actively in early summer, setting seed in summer or fall
- Legumes — members of the family Leguminosae (11000 + species)
- Forbs — primarily broadleaf, nonwoody, plants
- Browse — woody plants consumed in some degree by ruminants, particularly selective eaters such as deer and antelope

Grasses

Grasses are by far the most important plant that man is concerned with agriculturally, because the grass family not only includes all of the wild and cultivated species used for grazing but also the cultivated cereal grains and sorghum species. In this section, however, we will be concerned only with grass as a forage, and will not consider it from an agronomic point of view. The discussion will also be very general because of the tremendous numbers of grasses of importance.

As a food for grazing animals, grass has many advantages. Most grass species are quite palatable when immature and only a few are highly toxic for any appreciable part of the grazing season. Grasses of one type or another have the ability to grow in most environments in which ruminants can survive, arctic regions being one notable exception. Furthermore, nutrients supplied in grasses roughly provide needed amounts that more or less parallel animal needs during a yearly life cycle of reproduction and production, except during midwinter in cold climates.

Some more or less typical values for chemical composition are illustrated in Table 18-1. Early in the growing season, grasses — especially cool season species — have a very high water content and an excess of protein for ruminant animals. The result is that animals may have diarrhea and, because of the low dry matter content, may have difficulty in obtaining a maximum intake of energy. Grass proteins are usually high in the amino acid, arginine, and also contain appreciable amounts of glutamic acid and lysine. If N application is liberal, and particularly if S is deficient, there is apt to be a high level of NPN in the form of amino acids and amides and a relative deficiency of the S-containing amino acids. High nitrate may also result from fertilization. Toxic symptoms may occur at levels of about 0.07% nitrate N in dry matter and amounts on the order of 0.22% may be fatal to ruminants; however, if ruminants are adapted and fed on high-nitrate grasses continuously, toxicity is less likely as rumen microorganisms are capable of reducing nitrate to ammonia which is well utilized.

In comparison with legumes such as alfalfa, the protein content of grasses is nearly always lower, particularly in mature plants. Digestible energy is very good in young grass, but declines rapidly with maturity. Mature plants, especially those that are weather leached, will be low in digestible energy and protein, as well as in other solubles such as carbohydrates and some of the mineral and in carotenes; thus, this type of plant material may not meet the animal's needs, even when productive requirements are quite low.

Mineral content of grasses may vary considerably, depending on species and soil fertility, particularly. Of those elements of concern to ruminants, grasses are usually adequate in Ca, Mg, and K, but are apt to be borderline or deficient in P. Trace minerals also vary considerably; recent evidence from Canada (26) indicates that 80% or more of forage species measured contained less than levels generally considered adequate for ruminants. Ranges and typical values to be expected are shown in Table 18-2.

The desirability of different grasses depends on the local environment and their growing habits and on animal needs. Considerable differences may exist in the composition of grasses that fall into the cool- or warm-season classes. Generally, cool-season grasses mature at slower rates and their quality deteriorates less rapidly than do warm-season grasses. Lush, young grass is usually quite palatable, but palatability usually declines as the plants mature, and most animals object to the seed heads of many grass species. Quality differences between species of grasses become more evident with maturity. Furthermore, regrowth of grass in the fall is usually not as nutritious as spring grass, partly due to the

Table 18-2. Range and typical mineral concentrations for pasture grasses and alfalfa plants.

Mineral element	Grasses			Alfalfa		
	Low	Typical	High	Low	Typical	High
Major elements, % of dry matter						
Ca	< 0.3	0.4-0.8	> 1.0	< 0.60	1.2-2.3	> 2.5
Mg	< 0.1	0.12-0.26	> 0.3	< 0.1	0.3-0.4	> 0.6
K	< 1.0	1.2-2.8	> 3.0	< 0.4	1.5-2.2	> 3.0
P	< 0.2	0.2-0.3	> 0.4	< 0.15	0.2-0.3	> 0.7
S	< 0.1	0.15-0.25	> 0.3	< 0.2	0.3-0.4	> 0.7
Trace elements, ppm of dry matter						
Fe	< 45	50-100	> 200	< 30	50-200	> 300
Co	< 0.08	0.08-0.25	> 0.30	< 0.08	0.08-0.25	> 0.3
Cu	< 3	4-8	> 10	< 4	6-12	> 15
Mn	< 30	40-200	> 250	< 20	25-45	> 100
Mo	< 0.4	0.5-3.0	> 5	< 0.2	0.5-3.0	> 5
Se	< 0.04	0.08-0.01	> 5	< 0.04	0.08-0.1	> 5
Zn	< 15	20-80	> 100	< 10	12-35	> 50

lower concentration of soluble carbohydrates.

Cultivated grasses held in high esteem include perennial ryegrass [*Lolium perenne*], Italian ryegrass [*Lolium multiforum*], orchard grass [*Dactylis glomerata*], blue grass [*Poa spp*], and smooth brome grass [*Bromus inermus*]. Others considered less desirable include Bermuda grass [*Cynodon dactylon*], foxtail [*Alopecurus pratensis*], bent [*Agrostis spp*] and tall fescue [*Festuca arundinacea*]. In addition to varying in growth habits, these species may vary in nutrient composition, palatability, and digestibility.

In many regions some of the cereals are used for pasture, particularly winter wheat, with lesser use of barley, oats, and rye. These plants can be pastured during the winter and early spring with little or no effect on grain yield, provided soil conditions permit. The forage of these plants is quite high in readily available carbohydrates (50% +) and crude protein is high. Extensive use of such pasture is made in the southwestern states of the USA, particularly for growing calves and lambs.

Several sorghum species are also used for pasture or harvested forage. Sudan grass [*Sorghum vulgare sudanense*] is one of the more common ones used in the USA, but others, such as Johnson grass [*Sorghum halepense*], find some use. Sudan-sorghum hybirds have also been developed. These species are often utilized because they can be sown during early summer in temperate areas and will produce late summer and fall pastures. They are prone to have high levels of glycosides which can be converted to prussic acid (hydrocyanic acid), particularly following drouth or frost damage, so care must be used if these conditions occur while pasture is in use.

Legumes

A wide variety of legumes are utilized by grazing animals, although the cultivated legumes comprise a much smaller group than do cultivated grasses. In overall usage, alfalfa [*Medicago sativa*] is by far the most common legume used for pasture, hay-crop silage, and hay in most temperate climates. As a point of reference, alfalfa is known as lucerne in most English-speaking areas other than in North America.

Other legumes that find extensive usage for pasture or hay include clovers such as ladino or red [*Trifolium pratense*], white [*T. repens*] and subterranean [*T. subterraneum*], as well as common lespedeza [*Lespedeza striata*], lupines [*Lupinus spp*] and vetches [*Vicia spp*].

Legumes have higher protein contents than grasses, particularly in more mature plants. Fiber in the stems tends to be particularly high and soluble carbohydrates relatively low. The leaves are rich sources of nutrients but stems are of much less value, especially in mature plants. Changes in plant composition with maturity are largely from lignification and increased fiber in the stems and to a reduced leaf:stem ratio. Compared to grasses, legumes have characteristically high concentrations of Ca, Mg, and S (Table 18-2) and, frequently, Cu. They tend to be lower in Mn and Zn than grasses. On the whole, legumes are palatable, although most are bitter and may require some adaptation before they are readily consumed by cattle.

Some legumes, particularly alfalfa and white, ladino, and red clover are prone to cause bloat in grazing ruminants, especially cattle. Bloat is caused primarily by foam-producing compounds from the plant, of which cytoplasmic proteins and, perhaps, pectins are the most important. Foam in the rumen (see Church, 7) causes entrapment of normal rumen gases which cannot be gotten rid of, resulting in a gradual increase in rumen pressures and, if not relieved, eventual suffocation of the animal. Recent evidence indicates that alfalfa plants probably can be selected to have a lower content of the proteins that are involved in bloat production (27).

Native Pastures and Range

Pastures and rangeland comprised of uncultivated native forage plants account for many millions of acres of land in areas of the world where the environment, soil, or topography rule out intensive agricultural methods. These areas are apt to contain a wide range of grasses, sedges, forbs, and browse. The nutrient properties of these various plants vary widely and, in addition, there are apt to be distinct seasonal patterns of use by different grazing animals. The subject is too complex to discuss here; for more information, the reader is referred to Stoddart and Smith (35).

Miscellaneous Forage Plants

In some areas and for some specific seasonal usage, plants such as the cabbage family [*Brassica spp*] are used extensively. Kale, cabbage, and rape are included in this group. Rape, for example, is often planted for use as fall pasture by sheep. The tops of root crops such as beets and turnips, are also frequently used as forage. As a part of the total resource, however, these crops account for a very small percentage of the total.

Harvested Dry Roughages

Roughages stored in the dried form are the most common type used for feeding during the time of year when grazing is not available or for feeding of confined animals. Roughages harvested in the form of long hay or bales require a relatively high labor input and present difficulties in mechanical handling, both during harvest and feeding, with the result that the cost of nutrients is increased as compared to cereal grains in intensive livestock operations. Fortunately, machinery currently available allows rather complete mechanization in operations where it is financially feasible, (see Fig. 18-2, 18-3 and 18-4), and continual improvement is being made in this respect.

Hays

Hay — from grasses or legumes — is one crop that is grown and harvested almost exclusively for animal usage. Haymaking has been practiced for many centuries and much information has been accumulated on the nutritive value of hays as affected by many different factors too numerous to discuss in detail here. Although haymaking is the most common method of conserving green crops, its relative importance has declined some in recent years with the development of newer methods of forage preservation.

Figure 18-2. Equipment designed for ease and convenience of handling loose hay. Courtesy of the Hesston Corp.

Figure 18-3. Equipment designed for convenience in feeding loose hay. Courtesy of the Hesston Corp.

Figure 18-4. A hay baler. This particular model throws the bales into a trailing wagon. Courtesy of Deere & co.

The intent in haymaking is to harvest the crop at a more or less optimum stage of maturity which will provide a maximal yield of nutrients/unit of land without damage to the next crop. To make good hay, the water content of the plant material must be reduced to a point low enough to allow storage without marked nutritional changes. Moisture content of green herbage may range from 65 to 85% or more, depending on maturity and the plant species. For hay to keep satisfactorily in storage, the water content must be reduced to about 15% or less.

Losses in Haymaking

It is impossible to cut, dry and move hay into storage without losses occurring in the process, although it may be possible to harvest more units of nutrients/unit of land than could be obtained by grazing due to trampling and feed refusal resulting from contamination by dung and urine and to selective grazing of some species of plants. Both the quality and quantity of field-cured hay that can be harvested depends on such factors as maturity when cut, method of handling, moisture content, and weather conditions during harvest; for example, rain on freshly cut hay will cause little damage; however, when hay is partially dried, rain is very damaging. Early cut hay in many areas is often of low quality because of rain and resultant spoilage, leaching, and leaf loss. One report indicates dry matter losses ranging from about 6% for artificially dehydrated hay up to about 33% for rain-damaged field-cured hay. Another report (25) indicated the following losses: plant respiration losses (before plant is dry), 3.5% in 24 hr; leaching by rain, 5-14%; and leaf shattering in legume hays, 3-35%. Thus, very substantial amounts of dry matter may be lost in haymaking under adverse conditions; an average loss of 15-20% is not abnormal for legume hays.

Changes During Drying

Ample evidence shows that rapid drying, provided it is not accompanied by excessively hot temperatures, results in the least changes in chemical components of forage. Machines such as crimpers have been developed to crush the stems of plants like alfalfa and speed up the drying process. If drying is slow in the field, stack, or bale, appreciable changes may occur as a result of plant enzymes, microorganisms, or oxidation. After the plant is cut, the cells continue to function for a time with the result that soluble carbohydrates may be oxidized. Oxidative reactions may continue for some time, depending on the temperature and how the hay is stored. The most obvious change is a loss in pigmentation as plants lose carotenes by oxidation. Proteins may also be modified as some hydrolysis occurs, resulting in relatively greater amounts of NPN, primarily amino

Table 18-3. Composition and digestibility of green ryegrass and material from the same field which was dried artificially or made into wilted or unwilted silage.[a]

Item	Fresh grass	Dried grass	Wilted silage	Unwilted silage
Composition of dry matter				
Organic matter, %	90.8	92.0	92.2	91.7
Total water-soluble carbohydrate, %	9.2	8.4	trace	trace
Cellulose, %	24.2	24.3	25.0	26.8
Hemicellulose, %	14.0	13.3	12.9	13.1
Total N, %	2.85	2.99	3.09	3.08
Gross energy, Kcal/g	4.59	4.55	4.46	4.89
Digestibility, %				
Energy	67.4	68.1	67.5	72.0
Cellulose	75.2	75.5	76.5	80.6
Hemicellulose	59.4	57.7	59.9	63.2
N	75.2	71.0	76.5	76.4

[a] From Beever et al (5). In this experiment the fresh grass was quick-frozen so that it could be fed at the same time as the other forms.

acids. Slow drying is almost always accompanied by excessive mold growth which reduces the palatability and nutritive value of hay.

Hay stored in the stack or bale while too wet to dry rapidly may undergo enough fermentation to result in marked temperature increases, which may cause browning and, sometimes, spontaneous combustion. Newly baled alfalfa hay should have no more than 35% moisture to avoid these problems. Excess heating (25) or molding results in a marked reduction in digestibility of protein and energy.

If drying is not complicated by weather factors, relatively little change in the composition takes place between green plant and hay (5, 14) nor is there any pronounced effect on nutrient utilization (see Table 18-3). Animals generally ingest dry matter from fresh herbage at a slower rate than for hay, and some slight differences may occur in rumen fermentation, digestibility, and site of digestion in ruminants, but the differences appear to be inconsequential.

Thus, we see that drying may not have any great effect on forage utilization. In practice, however, we must expect some loss of leaves in legumes and reduced soluble carbohydrates. Hay, if made from moderately mature plants, will have lower protein and digestible energy than young herbage, but is apt to be better than very mature herbage. Nutritive properties of hays are, then, similar to those of forages, but with slightly to greatly reduced values depending on freedom from weather damage and method of harvesting.

Figure 18-5. Round bales of hay, a form sometimes left directly in the field for later consumption by animals. Surprisingly, wastage is relatively low. Photo by R. Bogart.

Artificially Dried Forage

Rapid drying is required for good haymaking. At one time in the USA, particularly in areas where weather was a problem, considerable interest was shown in barn drying. This is accomplished by circulating air though the hay after storage. Although a very good product can be produced, interest has declined with greater usage of silage and recent developments in bale-handling machinery.

Dehydration of herbage with appropriate machinery is a viable industry both in the USA and some areas of northern Europe. In the USA, alfalfa is the primary crop that is dehydrated, but in Europe, grasses or grass-clover mixtures are more commonly used. In making dehydrated alfalfa, the herbage is cut at a prebloom stage, dried quickly at hot temperatures, ground, and sometimes pelleted. Because carotene pigments are important, the product is often stored using inert gases such as N_2 to reduce oxidation. Herbage so processed is high in crude protein and quite digestible, moderately low in fiber, and has a high carotene content. The relatively low fiber content makes such feedstuffs more suitable for monogastric species such as poultry and swine and, in the USA, a high proportion of dehydrated alfalfa goes into commercial rations for these species. For poultry, the carotenes and xanthophylls serve to increase pigmentation of the skin of broilers or the yolk of eggs, as well as provide protein and other nutrients. For swine, particularly sows, nutrients of interest are the vitamins, Ca and trace minerals. These feeds serve very well for horses or ruminants, but generally are more expensive/unit of nutrients provided as compared to other roughage sources.

Straws and Chaff

An appreciable amount of straw and chaff is available for animal feed in most farming areas, although much less than in the days when stationary threshing machines were used for harvesting cereal grains. Straw consists mainly of the stems and variable amounts of leaves of plants that remain after the removal of the seeds. Chaff consists of the small particles removed from the seed head along with limited amounts of small or broken grains.

Figure 18-6 A. One type of forage harvester utilized in making silage or for green chop. Courtesy of Deere & co.

The primary supply of straw and chaff comes from the small cereal grains — wheat, barley, rye, rice, oats — but, in some areas, substantial amounts may be available from the grass or legume seed industry and from various miscellaneous crops. As a whole, straws are very low in digestible protein, very high in fiber and, usually, lignin, and are poor feed, although some are of less value than others. For example, values given from NRC (30) for winter wheat straw (dry basis) are: crude protein, 3.2%; crude fiber, 40.4%; and digestible energy (cattle), 1.92 Mcal/kg or 47% TDN. Of the various cereal straws, that from oats is regarded as the best, partly because the grain is often harvested before it is fully ripe. As a feedstuff, straws are best used as a diluent in high-concentrate rations or as the basal feed for wintering cattle when properly supplemented with deficient nutrients — protein, vitamin A, minerals. Even though straws are low in metabolizable energy, the energy derived from that which is digested and from the heat increment (see Ch. 9) provides energy that can be used for animals, such as pregnant cows, which have a low productive requirement.

Miscellaneous

In the USA nearly all of the grain from corn and sorghum is now field harvested and often the cobs or threshed seed head are left in the field, along with a substantial amount of grain at times. The nutritive problems here are the same as with straws, and some supplementary feeding is nearly always required to supply needed minerals, carotene, and protein. Protein may be less critical than with straws.

Figure 18-6 B. A second type of forage harvester used for windrowed crops. Courtesy of Western Farmers Assoc.

Harvested High-Moisture Roughages

Green Chop [Soilage]

Green chop (or soilage) is herbage that has been cut and chopped in the field and then fed to livestock in confinement. Plants used in this manner include the forage grasses, legumes, sudan grass, the corn plant and, at times, residues of food crops used for human consumption.

Although this is one of the simplest means of harvesting herbage (Fig. 18-6), it requires constant attention to animal needs as opposed to other methods of harvesting herbage. A major advantage of use of green chop is that more usable nutrients can be salvaged/unit of land than with other methods such as pasturing, haymaking, or ensiling (24). Thus, it is often feasible to harvest in this manner rather than using such crops in other ways, provided land productivity is relatively high. When herbage growth outruns daily need, the excess can be made into hay or silage before it becomes too mature for efficient usage. Weather, of course, is less of a factor than in haymaking.

Data, in general, indicate that beef cattle will gain as well as when pastured using intensive systems such as short-term rotations or strip grazing (24) and that dairy cows do equally well when fed soilage as when fed alfalfa preserved in other ways (4). Because harvestable nutrients can be increased, the result in recent years has been a steady increase in use of green chop, particularly for lactating cows. Practical experience has indicated that optimal usage is obtained when green chop is fed along with hay or silage rather than by itself, partly because total intake tends to be greater.

Silage

Silage has been used for feeding animals, primarily ruminants, for many years (Fig. 18-7). It is the material produced by controlled fermentation of high-moisture herbage. When such material is stored under anaerobic conditions (in the absence of oxygen) and, if the supply of fermentable carbohydrates is adequate, sufficient lactic acid is produced to stabilize the mass so that no more fermentation occurs. If undisturbed, silage will keep for an indefinite period. Alternate methods, primarily used in Europe, require the addition of strong acids which lower the pH, thus preventing fermentation.

Good silage is a very palatable product which is well utilized (see Table 18-4), and excellent results may be obtained with high-producing animals such as lactating cows. In addition to any advantages in harvesting or in nutritive properties, the fermentation that occurs usually will greatly reduce the nitrate content, if nitrate is present, and other toxic materials, such as prussic acid, will be reduced in amount.

Most silage in the USA is made from the whole corn plant [*Zea mays*] or from any of a number of sorghum varieties in areas where rainfall is insufficient for growing corn. Grass, grass-legume, or legume silages are also used extensively as technology has developed in recent years so that very good silage can usually be produced from these crops. For that matter, palatable and nutritious silage can be produced from a wide variety of herbage, even from products such as potato tubers. Some of the nutritional characteristics of principal silage crops will be discussed in later sections.

Chemical Changes During Fermentation

Chemical changes during silage fermentation are rather complex and a complete discussion is beyond the scope of this book. For further information, the reader is referred to sources such as Barnett (3), Watson and Nash (39), McDonald et al (21) or McCullough (20).

The chemical changes are the result of plant enzyme activity and action of microbes present on the herbage or which find their way into it from other routes. The plant enzymes continue to be active for the first few days after cutting while oxygen is available, resulting in some metabolism of soluble carbohydrates to CO_2 and water and the production of heat. Optimum temperatures during fermentation are said to be between 80° and 100°F. Excessive heat is objectionable, but is not likely if the silage is well packed to exclude air.

Plant proteins are partially broken down by cellular enzymes, resulting in an increase in NPN compounds such as amino acids. Anaerobic microorganisms rapidly multiply, using

Figure 18-7. A view of the face of a bunker [pit, trench] silo, a type now being commonly used in many areas for storage of large amounts of silage. This type facilitates equipment usage in both filling and unloading. Courtesy of Bill Fleming, Beef Magazine.

sugars and starches as primary energy sources and producing mainly lactic acid with lesser amounts of acetic acid and others such as formic, propionic and butyric, although little butyric acid is present in well-preserved silage. Continued action occurs on N-containing compounds with further solubilization and production of ammonia and other NPN compounds (Table 18-4). The level of lactic acid rises in well-preserved silage, eventually reaching levels of 7-8%, and the pH drops to about 4.0, depending on buffering capacity and dry matter content of the crop in question. If the silage is too wet or the supply of soluble carbohydrates too low, the pH will not go this low, allowing the development of clostridia bacterial species, relatively large amounts of butyric acid, and further fermentation of NPN compounds, resulting in production of amines such as tryptamine, histamine and others. These amines have an undesirable odor and they may be toxic. On the other hand, if the mass is too dry or poorly packed, excess heating may occur, and molds may develop, producing unpalatable and, sometimes, toxic silage.

For most crops, dry matter contents of 25-35% and a soluble carbohydrate content of 6-8% (dry basis) are near optimal for silage making. Consequently, if grass or legume silage is to be made, it is usually wilted some before ensiling. If direct-cut herbage is used, it will usually be too wet for good silage, allowing clostridia bacteria to multiply. When herbage such as grass or legumes is direct cut and ensiled, preservatives or sterilants are often added. Added dry matter is sometimes used to soak up some of the moisture and soluble carbohydrates may be added in the form of grain or molasses; these provide useful insurance when making grass-legume silage. Silage sterilants such as formic acid, sulfur dioxide, or sodium metabisulfite may be advantageous where moisture content is high, but are of doubtful value in silage made from wilted grasses or legumes or from the corn or sorghum plant.

Table 18-4. Composition and nutritive value of 3rd cutting alfalfa harvested as hay, low-moisture and high-moisture silage. [a]

Item	Hay	Wilted forage*	Low-moisture silage	Green chop*	High-moisture silage
Dry matter, %	92.6	58.1	59.0	28.8	28.1
Others on DM basis					
Crude protein, %	20.6	18.1	20.4	19.8	21.6
Cellulose, %	34.0	33.3	37.4	30.3	37.9
Soluble carbohydrates, %	7.8	5.2	2.5	3.8	3.7
Total N, %	3.3	2.9	3.3	3.2	3.4
Soluble N, % of total N	31.8	37.2	51.1	32.6	67.0
NPN, % of total N	26.0	28.4	44.6	22.6	62.0
Volatile fatty acids, %					
acetic			0.4		4.2
propionic			0.05		0.14
butyric			0.002		0.11
pH			4.7		4.7
Dry matter intake, g/day/kg body wt	25.3		21.8		25.8
Digestibility, %					
Dry matter	64.5		59.0		59.5
Nitrogen	77.5		63.0		72.8
Cellulose	58.8		60.4		64.3

[a] Taken from Sutton and Vetter (36)

* Wilted herbage used to make low-moisture silage and the green chop was used for the high-moisture silage.

Losses from Ensiling

Losses occur during ensiling because of fermentative activity and resultant heat produced. Obviously, these may be quite variable. Gaseous losses are said to range from 5-30% of original dry matter. Most of the losses originate from soluble and highly digestible nutrients with the result that silage (Table 18-4) is apt to contain a higher % of fibrous and insoluble ingredients.

Very substantial amounts of seepage may occur in high-moisture silage, particularly where the dry matter content is less than 30% (20). Above this level such losses are moderate to low. Seepage moisture contains many soluble nutrients, and such losses are to be avoided when possible. Further losses occur from molding in all but air-tight silos. Molding nearly always takes place around the perimeter and on top of the silage. Recent data on alfalfa silage (37) indicate spoilage losses on the order of 4-12% of original dry matter ensiled. Such material is not only unpalatable but may also be toxic. Obviously, overall losses may be quite variable, but, when field losses are included, total losses may be expected to be about 20-25% of herbage dry matter present in the field.

Nutritive Properties of Silages

One of the nutritive problems associated with feeding silages is that consumption of dry matter in the form of silage is nearly always lower than when the same crop is fed as hay, and this seems to apply whether the crop in question is legume, grass, or other herbage. Intake of silage is usually greater as the dry-matter content increases (see Table 18-5). Data covering 70 different silages (40) indicate that ad libitum intake was positively correlated with the dry matter, N, and with lactic acid as a % of total acids. Intake was negatively correlated with acetic acid content and ammonia as a % of total N. Intake was positively related to digestibility for legumes, but negatively for grasses other than ryegrass. This information would indicate that maximum intake should be achieved with silage containing just enough moisture to allow for preservation with minimum production of ammonia and acetic acid, which tend to increase at higher moisture contents (see Table 18-4, 18-5).

Differences in digestibility and in animal performance between direct cut or wilted silage and hay have been inconsistant (37). Some investigations (ibid) indicate that digestibility of energy is improved enough by ensiling to compensate for the reduced intake. Unfortunately, errors in analyzing silage have been included in many experimental reports; if silage is typically dried in an oven, substantial amounts of volatile materials are lost, thus giving a low value for nutrient content of silage as fed and an underestimate of organic matter intake and digestibility.

Grass-Legume Silages

Grass or legume silages are, normally, high in crude protein (20% +) and carotene, but only moderate in digestible energy. This combination is too high in protein for most ruminants and more nearly optimum results may be expected by supplementing grass silage with some form of energy (grains) or by diluting the protein by feeding some other form of low-protein roughage.

Low-moisture silage, often called haylage, is a very palatable feed, probably because it tends to have relatively less acetic acid and less N in the form of ammonia and the othr NPN compounds (Tables 18-4, 18-5). It is used to a great extent in the USA for dairy cows. This type of silage is best made in air-tight silos now in common use in many areas (Fig. 18-8).

Corn and Sorghum Silage

Corn silage is the most popular silage in the USA in areas where the corn plant grows well. It is becoming more popular in other areas of the world because, except for sugar cane, maximum yields of digestible nutrients/unit of land can be harvested from this crop. In addition, the corn plant can also be easily handled mechanically at a convenient time of the year. A host of research reports have appeared on this subject; two relatively recent reviews include those of Coppock (10) and Hillman (16).

Well-made corn silage is a very palatable product with a moderate to high content of digestible energy, but it is usually low to moderate in digestible protein, particularly for the amount of energy contained. Although corn silage contains much grain (up to 50% in well-eared crops), maximum growth rates or milk yields cannot be obtained without energy and protein supplementation.

The effect of stage of maturity on nutritive value of corn silage is shown in Table 18-6. Note that the crude fiber declines and NFE increases as the corn becomes more mature. Digestibility of energy was quite constant due to the offsetting effect of increasing energy

Table 18-5. Effect of dry matter of alfalfa silage on intake and utilization by sheep.[a]

Item	Dry matter of silage, %			
	22	40	45	80
Silage pH	4.48	4.49	4.52	5.87
Acid content, % of DM				
Lactic	5.06	4.11	3.19	0.19
Acetic	4.91	3.21	2.88	0.62
Total	10.46	7.53	6.26	1.00
Dry matter intake, g/kg body wt $^{0.75}$	49.1	57.3	58.8	63.2
Dry matter digestibility, %	58.2	60.1	60.0	61.1
N intake, g/day	14.9	16.2	18.3	19.6
N retention, g/day	-0.9	1.0	2.0	2.9
N digestibility, %	70.5	72.2	72.0	70.0
Water intake, g/g DM	3.5	2.4	2.4	2.2

[a] From Hawkins et al (15)

from grain vs. the lower digestibility of the more mature stalk. No information was given in this paper on total yield, but the data clearly show that intake and digestibility were greater for the medium-to-hard dough stage.

Research in recent years has shown that some additives may improve corn silage. Treatment at ensiling time with limestone (0.5-1%) or other Ca salts tends to buffer the acids produced during fermentation and results in a very substantial increase in lactic acid production and, usually, an improvement in intake and animal performance. Likewise, addition of urea (0.5-1%) or other NPN sources to corn silage will increase the crude protein content and has worked out well for dairy cows, particularly (17). The value of the added urea will depend, somewhat, on fertilization practices used on the crop as heavy fertilization with N will increase the N content of the plant.

These same general comments apply to sorghum silage. As a rule, sorghum silage has a somewhat lower nutritive value than corn silage. This is a result of lower digestibility and lower sugar content of the stalk, and the passage through the GIT of the small seeds if not broken up during ensiling. When the seeds have been partially broken, then sorghum silage is comparable to corn silage if the grain content is comparable.

Figure 18-8. An illustration of gas-tight silos which have become widely used in recent years. These silos are also useful for storage of high-moisture grain. Courtesy of the A.O. Smith Co.

Miscellaneous Silages

A wide variety of other herbaceous material has been used to make silage. Waste from

Table 18-6. Effect of maturity on nutritive value of corn silage as fed.[a]

Item	Stage of maturity			
	Soft dough	Med.-hard dough	Early dent	Glazed & frosted
Dry matter, % as fed	24	27	32	39
Crude protein, % of DM	8.0	7.7	7.8	8.0
Crude fiber, % of DM	24.0	20.9	17.8	16.6
NFE, % of DM	58.4	62.9	65.9	68.3
Dry matter intake, % body wt/day,	1.66	2.02	1.90	1.71
Gross energy digestibility, %	67.7	68.1	65.7	65.0
Protein digestibility, %	53.7	53.9	53.8	54.6
Energy balance, Kcal/day	962	3285	2870	2341

[a] Data from Colovos et al (9). Silage fed ad libitum along with a small amount of molasses-urea supplement.

canneries processing food crops such as sweet corn, grean beans, green peas, and various root or vegetable residues have been used successfully. Residues of this type are difficult to use on a fresh basis because of a variable daily supply or because they are available for only short periods of time. Ensiling is advantageous in that it tends to result in a more uniform feed and a known supply provides for more efficient planning. Preliminary information indicates that some low-quality roughages such as grass straws can be ensiled and, when additives such as molasses and urea are added, a reasonably palatable and digestible product can be produced.

Estimated Digestibility of Roughages

As indicated previously, roughages are apt to be quite variable in composition and nutritive value. Furthermore, the nutritive value of a given batch of roughage may be different for sheep than for cattle or may be influenced by level of feeding, nature of other feedstuffs, the environment, and so on. Even so, one is faced with making some estimate of value in practical feeding situations. We can go to appropriate tables and use listed digestion coefficients to calculate digestible protein or TDN, etc.; when no analytical information is available, this is about all that can be done. However, if a minimal amount of analytical information is at hand for the roughage of interest, then formulas are available for estimating digestible protein (DP) and TDN content. Some formulas (shown below) being relatively widely used in the USA were derived by Swift and coworkers at Penn. State University, as well as at other laboratories. In these formulas DP and CF (crude fiber) are on a dry matter basis.

Digestible Protein (DP)
- All forages DP=CP x 0.929 - 3.48
- Ensiled crops below 61% moisture, multiply DP x 0.86

Total Digestible Nutrients (TDN)
- Legumes (including peavines, soybeans forage)
  TDN=74.43 + 0.35 CP - 0.73 CF
- Grasses and corn stover (no ears)
  TDN=50.41 + 1.04 CP - 0.07 CF
- Mixed hay crop and forages of unknown origin
  TDN=65.14 + 0.45 CP - 0.38 CF
- Annuals other than corn
  TDN=90.36 - 0.29 CP - 0.86 CF
- Corn silage
  TDN=77.07 - 0.75 CP - 0.07 CF
- Corn fodder silage subtract 6% from TDN if DM less than 37%

Estimated Net Energy (ENE)
- ENE = (TDN x 1.393) - 34.63 where ENE expressed as Mcal/100 lb

California research workers have developed a method of predicting DP, TDN, and ENE for alfalfa based on a modified CF analysis. The formulas used are:

DP = 30.75 - 0.54 MCF
TDN = 1.58 DP + 32.59
or = 81.05 - 0.86 MCF
or 78.7 - (CF x 0.8)

For those who wish more information, appropriate tables are given in Calif. Agric. Extension publication AXT-290.

A formula for corn silage developed by Goodrich (Minnesota) is:

TDN = 72.1 - (CF x 0.34)

The author has not attempted to verify the accuracy of these various formulas. Some of them are being used by forage analytical laboratories as a means of putting a meaningful energy value on forage. It is to be expected that more complex and, perhaps, more reliable formulas will be developed as feedstuffs analyses become a more common practice.

High-Energy Feedstuffs

Feedstuffs included in this class are those that are fed or added to a ration primarily for the purpose of increasing energy intake or to increase energy density of the ration. Included are the various cereal grains and some of their milling by-products and liguid feeds — primarily molasses and mixtures in which molasses predominates — and fats and oils. High energy feedstuffs generally have low to moderate levels of protein. However, several of the high-protein meals could be included on the basis of available energy. Furthermore, on the basis of energy/unit of dry matter we could include some of the roots and tubers; because of their high water content they are usually considered separately. The same comments apply to fluid milk.

Energy from high-energy feedstuffs is supplied primarily either by readily available carbohydrates — sugars and/or starches — or by fat. Depending upon the type of diet and the class of animal involved, feedstuffs in this class may make up a substantial percentage of an animal's total diet. As such, other nutrients provided by these base feeds — amino acids, minerals, vitamins — must be considered also, but quantitatively these nutrients are of less concern than the available energy.

Cereal Grains

Cereal grains are produced by members of the grass family (Gramineae) grown primarily for their seeds. A tremendous tonnage is produced in the USA primarily for feed as shown in Table 18-7. Of course, some of these grains go into human food. For example, corn may be consumed as popped corn, corn flakes, corn flour, corn starch, and corn syrup, but the amount used in this manner is much less than goes into animal feed. Wheat and rice are grown primarily for human consumption, although substantial amounts of wheat may go into animal feed in the USA when price and supply allow it. Barley and oats, although good feeds, are becoming relatively less important because they do not usually yield as well as some of the other feed grain plants. Other grains such as millet and rye find only limited use. A wheat-rye cross, triticale, appears to show some promise as a feed grain.

Table 18-7. Feed grains produced in the USA in 1971.

Grain	Millions of tons
Corn (maize)	5400
Grain sorghum	892
Oats	884
Barley	470
Rye	51
Wheat*	1628

* Primarily used for human food although substantial amounts of milling by-products are consumed by animals.

Average values for some nutrients in cereal grains are illustrated in Table 18-8. Note here that relatively small differences occur between grains and identifying them on the basis of their chemical composition would be difficult. Although grains are usually said to be less variable in composition than roughages, many factors influence nutrient composition and, thus, feeding value for a given grain. For example, factors such as soil fertility and fertilization, variety, weather and rainfall, and insects and disease may all affect plant growth and seed production so that average book values may not be meaningful. With wheat, for example, if a hot, dry period occurs while the grain is ripening, shriveled, small seeds may result; although they may have a relatively high protein content, the starch content is apt to be much lower than usual, the weight per bushel will be low, and the feeding value appreciably lower.

Although crude protein content of feed grains is relatively low, ranging from 8-12% for most grains, we may find some lower than this

Table 18-8. Average composition of the major cereal grains, dry basis. [a]

Item	Corn, dent	Opaque-2 corn	Wheat		Rice, with hulls	Rye	Barley	Oats	Sorghum (Milo)
			hard winter	soft white					
Crude protein, %	10.4	12.6	14.2	11.7	8.0	13.4	13.3	12.8	12.4
Ether extract, %	4.6	5.4	1.7	1.8	1.7	1.8	2.0	4.7	3.2
Crude fiber, %	2.5	3.2	2.3	2.1	8.8	2.6	6.3	12.2	2.7
Ash, %	1.4	1.8	2.0	1.8	5.4	2.1	2.7	3.7	2.1
NFE, %	81.3	76.9	79.8	82.6	75.6	80.1	75.7	66.6	79.6
Total sugars, %	1.9		2.9	4.1		4.5	2.5	1.5	1.5
Starch, %	72.2		63.4	67.2		63.8	64.6	41.2	70.8
Essential amino acids, % of DM									
Arginine	0.45	0.86	0.76	0.64	0.63	0.6	0.6	0.8	0.4
Histidine	0.18	0.44	0.39	0.30	0.10	0.3	0.3	0.2	0.3
Isoleucine	0.45	0.40	0.67	0.44	0.35	0.6	0.6	0.6	0.6
Leucine	0.99	1.06	1.20	0.86	0.60	0.8	0.9	1.0	1.6
Lysine	0.18	0.53	0.43	0.37	0.31	0.5	0.6	0.4	0.3
Phenylalanine	0.45	0.56	0.92	0.57	0.35	0.7	0.7	0.7	0.5
Threonine	0.36	0.41	0.48	0.37	0.25	0.4	0.4	0.4	0.3
Tryptophan	0.09	0.16	0.20		0.12	0.1	0.2	0.2	0.1
Valine	0.36	0.62	0.79	0.56	0.50	0.7	0.7	0.7	0.6
Methionine	0.09	0.17	0.21	0.19	0.20	0.2	0.2	0.2	0.1
Cystine	0.09	0.22	0.29	0.34	0.11	0.2	0.2	0.2	0.2
Minerals, % of DM									
Calcium	0.02		0.06	0.09	0.06	0.07	0.06	0.07	0.04
Phosphorus	0.33		0.45	0.34	0.45	0.38	0.35	0.30	0.33
Potassium	0.33		0.57	0.44	0.25	0.52	0.63	0.42	0.39
Magnesium	0.12		0.11	0.11	0.11	0.13	0.14	0.19	0.22

[a] Analytical data taken from NRC publications.

and, particularly with wheat, some may be much higher, sometimes as high as 22% crude protein. Sorghum grains are also quite variable. Of the nitrogenous compounds, 85-90% is in the form of proteins; most cereal grains are moderately low to deficient for monogastic species in lysine and often in tryptophan (corn) and threonine (sorghum and rice) and in methionine for poultry, whose requirement is higher than that of pigs.

The fat content may vary greatly, ranging from less than 1% to more than 6% with oats usually having the most and wheat the least. Most of the fat is found in the seed embryo. Seed oils are high in linoleic and oleic acids, both unsaturated fatty acids that tend to become rancid quickly, particularly after the grain is processed.

The carbohydrates in grains, with the exception of the hulls, are primarily starch with small amounts of sugars. The starch, which makes up most of the endosperm, is highly digestible, providing hull permeability allows access of digestive juices. Starch from the different grains has specific physical characteristics that can be identified by microscopic examination. Minor differences are also noted in chemical characteristics, some of which have not been well understood with respect to animal utilization.

Hulls of seeds have a substantial effect on feeding value. Most hulls (or seed coats) must be broken to some extent before feeding for efficient utilization, particularly for ruminant animals. Barley and oats are sometimes known as rough grains because of the very fibrous hulls that are relatively resistant to digestion. Rice hulls are almost totally indigestible.

Corn [*Zea mays*]

Corn is frequently called King Corn, a tribute to the fact that, in areas where it grows well, corn will produce more digestible nutrients/unit of land than any other grain crop (Fig. 18-9). Yields of corn on small acreages have exceeded 300 bushels/acre (8.4 + tons). In addition, corn is a very digestible and palatable feed, being relished by all domestic animals.

The chemical composition of corn has been studied in detail (31). Zein, a protein in the endosperm, makes up about half the total protein in the kernal of most varieties. This protein is low in many of the essential amino acids but particularly so for lysine and tryptophan; the total protein of corn is deficient in these amino acids for monogastric animal species and requires supplementation for adequate performance. The low tryptophan (which is a precursor for niacin) plus the low niacin content will lead, eventually, to a niacin deficiency and pellagra in monogastric animals depending on corn as a major dietary constituent (see Ch. 14). N fertilization has been shown to increase protein content and decrease protein quality, due primarily to an increase in the zein fraction.

The energy value of corn is generally used as a standard of comparison for other grains (Table 18-9). Thus, if the relative energy value of corn is taken at 100, the value of other cereal grains is usually lower. This is because of the low fiber content of the corn kernel and the high digestibility of its starch. Wheat compares favorably with corn, but rations composed of a high % of wheat often promote lower feed intake, particularly by ruminants.

White and yellow corn have similar compositions except that yellow corn has a much higher content of carotene and xanthophylls, vitamin A precursors. Both white and yellow corn are fair sources of vitamin E, but low in vitamin D and the B-vitamins. Corn is very deficient in Ca. Although the P content is relatively high, much of it is in the form of phytic acid P which has a low availability to most monogastric animals. Trace elements are relatively low to deficient.

Several genetic mutants of corn have been isolated and are being developed. One of these known as opaque-2 (22, 23) is of particular interest because it has a high level of lysine and increased levels of most other essential amino acids (see Table 18-8). This change results from a reduction in zein and an increase in glutelin, a protein found both in the endosperm and germ. The result is an improvement in the quality of the corn protein. Performance of monogastric species may be considerably improved by the use of this mutant. It is questionable if opaque-2 is of added value for ruminants. Yields of opaque-2 are not equivalent to common corn nor are prices proportionately higher, so interest of corn producers has not resulted in a marked increase in production of this mutant.

A second mutant of interest is known as floury-2. Although it also has higher levels of lysine and some of the other essential amino acids, comparative experiments do not indicate it to be as good as the opaque-2 strain. Other experimental mutants include those designated as high-fat, high-amylose, and brown midrib.

Grain Sorghum [*Sorghum vulgare*]

Sorghum is a hardy plant that is able to withstand heat and drought better than most grain crops. In addition, it is resistant to pests such as root worm and the corn borer and is adaptable to a wide variety of soil types. Consequently, sorghum is grown in many areas where corn does not do well. Sorghum yields less than corn, where corn thrives. The seed from all varities is small and relatively hard and usually requires some processing for good animal utilization.

A wide range of sorghum varieties [all *Sorghum vulgare*] are used for seed production. These include milo, various kaffirs, sorgo, sumac, hegari, darso, and feterita. Milo is a favorite in drier areas because it is a short plant adaptable to harvesting with grain combines (Fig. 18-10). Development of hybrid varieties which have higher yields has greatly increased production in recent years in the USA.

Chemically, grain sorghums are similar to corn. Protein content averages about 11%, but

Figure 18-9. A field of corn [maize], a crop used to feed millions of animals when harvested in different ways. Photo courtesy of R.W. Henderson.

Table 18-9. Relative values (dry basis) of different cereal grains as given by NRC publications.

Grain	Digestible protein,	Energy				
		TDN		ME	NE_m	NE_g
	cattle	cattle	swine	chickens	cattle	cattle
Corn, #2 dent	100	100	100	100	100	100
Barley	131	91	88	77	93	95
Milo	95	88	96	103	81	83
Oats	132	84	79	74	76	77
Wheat, hard red winter	152	97	99	90	95	96

is apt to be quite variable; lysine and threonine are said to be the most limiting amino acids. Content of other nutrients is similar to corn. Feeding trials indicate that sorghum grains are usually worth somewhat less than corn (Table 18-9), although some data indicate a higher value. Some bird-resistant varieties, whose seed coats are high in tannin, are not well liked by most animals.

Wheat [*Triticum spp*]

Wheat is rarely grown intentionally for animal feed and all commonly grown varieties were devoloped with flour milling qualities in mind rather than feeding values. The hard winter wheats are high in protein, averaging 13-15%, but the soft white wheats have less protein (Table 18-8). The amino acid distribution is better than that of most cereal grains, and wheat is a very palatable and digestible feed, having a relative value equal to or better than corn for most animals. Feeding wheat to ruminants requires some caution as wheat is more apt than other grains to cause acute indigestion in animals unadapted to it. Some processing (grinding, rolling) is required for optimal utilization. When available for feed, it can be substituted equally for corn on the basis of digestible energy.

Barley [*Hordeum vulgare or H. distichon*]

Barley is widely grown in Europe and in the cooler climates of North America and Asia. Although a small amount goes into human food and a substantial amount is used in the brewing industry in the form of malt, most of it is used for animal feeding.

Barley contains more total protein and higher levels of lysine, tryptophan, methionine, and cystine than corn, but its feeding value for ruminants is appreciably less in most cases, due to the relatively high fiber content of the hull and the lower digestible energy. Barley is a very palatable feed for ruminants, particularly when rolled before feeding, and few digestive problems result from its use. Hulless varieties are roughly comparable to corn and, thus, more suitable for swine and poultry feeding; however, not much hulless barley is produced. A new high-lysine barley containing 25% more lysine than common barley shows promise as an energy source for monogastric species; this would allow a saving in supplemental protein (29).

Oats [*Avena sativa*]

Oats represent only about 4.5% of the total world production of cereal grains, most of the production being concentrated in northern Europe and America. A substantial amount of the production goes into human food. The protein content of oats is relatively high and the amino acid distribution is more favorable than for corn, but oats are not widely used for feeding of swine or poultry because of the hull which makes up about 1/3 of the seed. The hull is quite fibrous and poorly digested. Even when ground, the result is a very bulky feed. Including a high percentage in the ration does not allow optimal feed intake for growing poultry or swine, although oats have some value in protecting young pigs from stomach ulcers. For ruminants and horses, oats is a favored feed for breeding stock, one well liked by animals. For feedlot cattle, however, oats do not supply sufficient energy for most rapid gains. Oat groats (whole seed minus the hull) have a feeding value comparable to corn, but the price is usually not favorable for animal use.

Triticale

A relatively new cereal grain, triticale, shows promise as a swine feed. Triticale is a hybrid

Figure 18-10. One type of grain combine used for harvesting a variety of different types of crops such as milo or, in this example, corn. This machine shells the corn in the process. Courtesy of Deere & Co.

cereal derived from a cross between wheat [*Triticum duriem*] and rye [*Secale cereale*]. Its feeding value as an energy source is comparable to that of corn and other cereal grains for swine. Triticale can replace all of the grain sorghum and part of the soybean meal in diets for growing pigs. Total protein content tends to be higher than that of corn and grain sorghum and similar to that of wheat. Current knowledge of triticale as a livestock feed was recently brought together at an International Symposium on Triticale (18).

Milling By-Products of Cereal Grains

The milling of cereal grains for production of flour, particularly wheat, results in production of a number of by-products that are used

Figure 18-11. Grain combines being used to harvest wheat in the Pacific Northwest. Photo by R.W. Henderson.

primarily by the commercial feed trade. Details of milling procedures, seed composition, and more complete descriptions of by-products may be found in other references (2, 13, 28).

Milling by-products from wheat account for about 25% of the kernel. They are classified and named on the basis of decreasing fiber as bran, middlings (or mill run), shorts, red dog, and feed flour. These are relatively bulky and laxative feeds, particularly bran, but are quite palatable to animals. The bran and middlings are from the outer layers of the seed and contain more protein than the grain. Protein quality is usually somewhat improved, although these products are apt to be deficient in lysine and methionine as well as some other essential amino acids. These outer layers of the seed are relatively good sources of most of the water-soluble vitamins, except for niacin, which is entirely unavailable. These feedstuffs are low in Ca and high in P and Mg. The bulk of the production of wheat by-products is used in swine and poultry rations; although when bran is available, it is favored feed in rations for dairy cows, all breeding classes of ruminants, and for horses.

Corn milling by-products are somewhat different than wheat by-products as corn is often milled for purposes other than flour production. When milled for corn meal, hominy feed is produced, a product that consists of corn bran (hulls), corn germ, and part of the endosperm. It is similar to corn meal but higher in protein and fiber. Some wet-milling processes are used for various purposes resulting in by-products such as corn gluten meal, germ meal, corn solubles, and bran. The gluten meal is a high-protein product. Of these by-products, hominy feed is most valuable as an energy feed and it is used in a wide variety of rations. It is quite digestible but of variable energy value depending on its fat content.

Milling by-products of barley, sorghum, rye, and oats are similar, more or less, to those of wheat or corn and of comparable nutritional value. With rice, the bran may be of good quality, but is apt to be quite variable because of inclusion of hulls. The bran tends to become rancid rapidly because of its relatively high content of unsaturated fats.

Liquid Energy Sources

Molasses

Molasses is a major by-product of sugar production, the bulk of it coming from sugar cane. About 25-50 kg of molasses results from production of 100 kg of refined sugar. Molasses is also produced from sugar beets and other products as shown in Table 18-10, and hemicellulose extract (wood molasses, Masonex) is a similar product.

Molasses is essentially an energy source and the main constituents are sugars. Cane molasses contains 25-40% sucrose and 12-35% reducing sugars with total sugar content of 50-60% or more. Crude protein content is usually quite low (3 ± %) and variable, and the ash content ranges from 8-10%, largely made up of K, Ca, Cl, and sulfate salts. Molasses is usually a good source of the trace elements but has only a moderate to low vitamin content. In commercial use molasses is usually adjusted to about 25% water content, but may be dried for mixing into dry diets.

One of the problems in molasses feeding is that such products (except for corn molasses) are quite variable in composition. Age, type, and quality of the cane, soil fertility, and system of collection and processing have a bearing on composition of molasses. As an example, Buitrago (cited in 31) noted that the Ca content of 11 samples of cane molasses ranged from 0.3 to 1.68%.

Molasses is widely utilized as a feedstuff, particularly for ruminants. In the USA alone, more than 2.5 million tons are used annually and large amounts are used in Europe and in other areas where it is produced. The sweet taste makes it appealing to most species. In addition, molasses is of value in reducing dust, as a pellet binder, as a vehicle for feeding medicants or other additives, and as a liquid protein supplement when fortified with a N source. The cost is often attractive as compared to grains. Most molasses products are limited in use, however, because of milling problems (sticky consistency of molasses) or because levels exceeding 15-25% of the ration are apt to result in digestive disturbances, diarrhea, and inefficient animal performance. The diarrhea is largely a result of the high level of various mineral salts in most molasses products. The problem is not due to sugar content, as pigs or ruminants can utilize comparable amounts of sugar supplied in other forms. Preston and Willis (32) have shown that high test molasses, which has a lower ash content, can be fed at very high levels to either pigs or cattle without any particular problem, but not much high-test molasses is available for animal feeding.

Table 18-10. Analytical and TDN values of different sources of molasses.[a]

Item	Source of molasses					
	Cane*	Beet	Citrus	Corn	Sorghum	Refiners
Standard Brix, degrees	79.5	79.5	71.0	78.0	78.0	79.5
Total solids, %	75	76	65	73	73	73
Crude protein, %	3.0	6.0	7.0	0.5	0.3	.30
Ash, %	8.1	9.0	6.0	8.0	4.0	8.2
Total sugars, %	48-54	48-52	41-43	50	50	48-50
TDN, %	72	61	54	63	63	72

[a] From Anonymous (1)
* Also known as backstrap molasses.

Other Liquid Feeds

Other liquids utilized for animals include distillers solubles, fish solubles, corn steep liquor (corn fermentation solubles), glutamic acid fermentation liquor, and propylene glycol. The fermentation solubles tend to be quite variable, although they are usually a good source of vitamins and so-called unidentified growth factors. These products are used in the commercial trade as additives to other feedstuffs. An experimental process of preparing juices from grasses and legumes with hydraulic presses is being studied in Great Britain and the USA. It shows promise as a source of B-vitamins, trace elements and high-quality protein.

Other High-Carbohydrate Feedstuffs

Root Crops

Root crops used in feeding animals, particularly in northern Europe (21), include turnips, mangolds, swedes, fodder beets, carrots, and parsnips. These crops are frequently dug and left lying in the field to be consumed as desired when used as animal feed. The bulky nature of these feeds limits their use for swine and poultry, so most are fed to cattle and sheep.

Root crops are characterized by their high water (75-90 + %), moderately low fiber (5-11% dry basis), and crude protein (4-12%) contents. These crops tend to be low in Ca and P and high in K. The carbohydrates range from 50-75% of dry matter and are mainly sucrose which is highly digestible by ruminants and nonruminants. Animals (sheep, cattle) not adapted to beets or mangolds (both *Beta vulgaris*) tend to be subject to digestive upsets, probably due to the high sucrose content.

A tropical root crop, [*Manihot esculenta*] called cassava, yucca, manioc, tapioca or mandioca is of great potential importance as a livestock feed. It is 9th in world production of all crops and 5th among tropical crops, and has been shown in experimental plots to be capable of yielding 75-80 tons/hectare/year (92 million Kcal of digestible energy) which is many times greater than can be produced by rice, corn or other grains adapted to the tropics (31). Although it is strictly a tropical plant, significant amounts of cassava are now used in the USA and Europe for livestock feeding. Cassava root contains approximately 65% water, 1-2% protein, 1.5% crude fiber, 0.3% fat, 1.4% ash and 3% NFE. Thus, its dry matter is largely readily available carbohydrates. Dried cassava is equal in energy value to other root crops and tubers and can be used to replace all of the grain portion of the diet for growing-finishing pigs if the amount of supplemental protein is increased to compensate for the low protein content of cassava. It can also be used as the main energy source in diets of gestating and lactating swine. The stalk and leaf portion of the plant is well utilized by ruminants, but is too high in fiber for monogastrics.

Freshly harvested cassava roots and leaves may be high in prussic acid (hydrocyanic acid). Oven-drying at 70-80°C, boiling in water or sun-drying are effective in reducing the HCN content of freshly harvested cassava. As improved harvesting and processing methods for cassava are developed, large-scale commercial production of cassava can be expected to make a significant contribution to the world energy feed supply for animals.

Tubers

Surplus or cull white potatoes [*Solanum tuberosum*] are often used for feeding cattle or sheep in areas where commercial potato production occurs. Pigs and chickens do poorly on raw potatoes, but cooking improves digestibility of the starch so that it is comparable to corn starch. Potatoes are high in digestible energy (dry basis) which is derived almost entirely from starch. Water content is about 78-80%; crude protein content is low, and the quality of the protein is poor. The Ca content is usually low. Potatoes and particularly potato sprouts contain a toxic compound, solanin, which may cause problems if potatoes are fed raw or unsiled. In cattle finishing rations, cull potatoes are frequently fed at a level to provide about ½ of the dry matter intake.

Various by-products of potato processing are available in some areas as high percentages of white potatoes are partially processed before entering the retail trade in the USA. Potato meal is the dried raw meal of potato residue left from processing plants. Potato flakes are residues remaining after cooking, mashing, and drying; potato slices are the residue after raw slices are dried with heat; and potato pulp is the by-product remaining after extraction of starch with cold water. The raw meals have about the same relative nutrition values as raw cull potatoes. A product called dried potato by-product meal (tater meal) is produced in some areas; it contains the residue of food production such as off-color french fries, whole potatoes, peelings, potato chips, and potato pulp. These are mixed, limestone added, and the mixture dried with heat. The composition and digestibility by ruminants is shown in Table 18-11. Potato slurry is a high-moisture product remaining after processing for human food; it contains a high amount of peel. Generally, the value of these potato products is roughly comparable to that of raw or cooked cull potatoes, depending on how the by-product is dried. However, residues of the potato chip processing may have much higher levels of fat, so the energy value may be increased accordingly. For swine, cooked potato products are usually restricted to 30% or less of the ration.

Dried Bakery Product

This is a product produced from reclaimed (unused) bakery products. While relatively little is available, it is an excellent feed as most of the energy is derived from starch, sucrose and fat. It is well utilized by pigs and is a preferred ingredient in starter rations. A wide variety of unconsumed bakery products make very satisfactory feeds for swine or ruminants.

Dried Beet and Citrus Pulp

Dried beet pulp is the residue remaining after extraction of sugar from sugar beets. It frequently has molasses added before drying and may be sold in shredded or pelleted form. The physical nature and high palatability make it a favored feed for cattle or sheep and it has a feeding value comparable to cereal grains (6).

Citrus pulp is a similar product remaining after the juice is extracted from citrus fruits; it includes peel, pulp, and seeds. Citrus pulps are used primarily as feedstuffs for cattle or sheep.

Distillery and Brewery By-products

By-products of the distilling and brewing industries find some use as animal feed. The main distillery products are wet or dried distillers grains, dried distillers solubles, or mixtures thereof. The protein content is generally higher than that of the original grain and energy values are similar to barley grain. Distillers dried solubles is an excellent source of B-vitamins and trace elements. Dried brewers' grains has relatively high crude protein content (26%), but is low in starch and digestible energy. Dried distillers grain by-products are used more for ruminants, but the solubles are used in all animal feeds when available.

Fats and Oils

Surplus animal fats and, occasionally, vegetable oils are frequently used in commercial feed formulas, depending on relative prices. A variety of different products are available and these have been described as to composition and source by the American Assoc. of Feed Control Officials. For the most part, feeding fats are animals fats derived by rendering of beef, swine, sheep, or poultry tissues. Vegetable oils generally command better prices for use in producing margarine, soap, paint, and other industrial products so that they are usually priced out of use in animal feeding.

Although most animals need a source of the essential fattly acids (see Ch. 8), these are usually supplied in sufficient amounts in natural feedstuffs and supplementation is not required except when low-fat energy sources are used. Fats are added to rations for several reasons. As a source of energy, fats are highly digestible, and digestible fat supplies about 2.25X that of digestible starch or sugar; thus, they have a high caloric value and can be used to increase energy density of a ration. Fats often tend to improve rations by reducing

Table 18-11. Composition and digestibility of dried potato by-product meal.[a]

Item	Composition, % air dry	Digestibility, %		Digestible nutrients by cattle, % dry
		cattle	sheep	
Water	8.64			
Protein	7.38	0.0	7.3	0.0
Fat	4.91	83.7	75.0	9.2
Fiber	4.64	37.5	28.2	1.7
NFE	58.47	73.4	64.8	42.9
Ash	15.96			
TDN				53.9
Energy		56.2	64.8	

[a] After Dickey et al (11)

dustiness and increasing palatability. There is some evidence that fats will reduce bloat in ruminants. In additon, the lubrication value on milling machinery is often of interest. Fats are subject to oxidation (see Ch. 8) with development of rancidity which reduces palatability and may cause some digestive and nutritional problems. Feeding fats usually have antioxidants added, especially if the fats are not to be mixed into the ration and fed immediately.

Adding fat at low to moderate levels can sometimes increase total energy intake through improved palatability, although animals usually consume enough energy to meet their demand when it is physically possible. In swine and poultry rations 5-10% fat is often added to creep diets for pigs or to broiler rations. Amounts above 10-12% will usually cause a sharp reduction in feed consumption, so concentration of other nutrients may need to be increased in order to obtain the desired consumption of other nutrients.

For ruminants, high levels of fats are used in milk replacers; depending on the purpose of the replacer, it may contain 15-30% added fat. Ruminants on dry feed, however, are less tolerant of high fat levels than are monogastrics. Concentrations of more than 7-8% are apt to cause digestive disturbances, diarrhea, and greatly reduced feed intake. In practice, 2-4% fat is commonly added in finishing rations for cattle, and some fat is occasionally added to rations for lactating dairy cows.

Protein Concentrates

Protein is one of the critical nutrients, particularly for young, rapidly growing animals and high-producing adults, although it may be secondary to energy or other nutrients at times. In addition, protein supplements are usually more expensive than energy feeds, so optimal use is a must in any practical feeding system. Most energy sources, except for fat, starch or refined sugar, supply some protein, but usually not enough to meet total needs except for adult animals. Furthermore, for monogastric animals the quality of protein from a given source is rarely adequate to sustain maximal production with minimal amounts of protein.

Protein supplements (less than 20% crude protein) are available from a wide variety of animal and plant sources, and nonprotein N sources such as urea and biuret are available from synthetic chemical manufacturing processed. It is beyond the scope of this book to detail all of the available sources; refer to Pond and Maner (31) for more detailed discussion and data on amino acid composition of some of the less well known protein sources. Table 18-12 contains data on composition of a few common protein supplements.

Protein Supplements of Animal Origin

Protein supplements derived from animal tissues are obtained primarily from inedible tissues from meat packing or rendering plants, from surplus milk or milk by-products, or from marine sources. Those of animal origin include meat meal, meat and bone meal, blood meal, and feather meal; milk products include dried skim and whole milk, liquid or dried whey, and dried buttermilk; fish products include whole fish meal made from a variety of species, meals

Table 18-12. Composition of several important protein supplements.

Item	Dried skim milk	Meat meal	Fish meal, herring	Cottonseed meal, sol. ext.	Soybean meal, sol. ext.	Yeast, dried brewers
Dry matter, %	94.3	88.5	93.0	91.0	89.1	93.7
Other components, %*						
Crude protein	36.0	55.0	77.4	45.5	52.4	47.8
Fat	1.1	8.0	13.6	1.0	1.3	1.0
Fiber	0.3	2.5	0.6	14.2	5.9	2.9
Ash	8.5	21.0	11.5	7.0	6.6	7.1
Ca	1.35	8.0	2.2	0.2	0.3	0.1
P	1.09	4.0	1.7	1.1	0.7	1.6
Essential amino acids						
Arginine	1.23	3.0	4.5	4.6	3.8	2.3
Cystine	0.48	0.4	0.9	0.7	0.8	0.5
Histidine	0.96	0.9	1.6	1.1	1.4	1.2
Isoleucine	2.45	1.7	3.5	1.3	2.8	2.2
Leucine	3.51	3.2	5.7	2.4	4.3	3.4
Lysine	2.73	2.6	6.2	1.7	3.4	3.2
Methionine	0.96	0.8	2.3	0.5	0.7	0.7
Phenylalanine	1.60	1.8	3.1	2.2	2.8	1.9
Threonine	1.49	1.8	3.2	1.3	2.2	2.2
Tryptophan	0.45	0.5	0.9	0.5	0.7	0.5
Valine	2.34	2.2	4.1	1.9	2.8	2.5

* Composition on dry basis. Data from NRC publications.

made from residues of fish or other seafood and fish protein concentrate.

Meat Meal, Meat and Bone Meal, Blood Meal

These animal by-products are used almost exclusively for swine, poultry and pet rations. Palatability is moderate to good. However, the quality of meat or meat and bone meal varies considerably depending upon dilution with bone and tendonous tissues and on methods and temperatures used in processing. Biological value of the proteins is lower, generally, than for fish or soybean meal. Growth studies with pigs indicate that these products are best used as a part rather than the total source of supplemental protein. Blood meal is a high-protein source (80-85% CP, dry basis), but it is quite deficient in isoleucine and is best used as a partial supplementary protein source, also.

Milk Products

Dried skim or whole milk, although excellent protein sources, are usually quite expensive as compared to other feedstuffs. These products are used primarily in milk replacers or in starter rations for young pigs or ruminants. The quality of dried milk can be impaired by overheating during the drying process (drum drying), therefore spray-dried milk is preferred. Poor quality milk, when used in milk replacers, is apt to lead to diarrhea and digestive distrubances. Whey, a by-product of cheese or casein production, is used similarly to dried skim milk. The high lactose and mineral content may be laxative, so the amount that can be used is limited. Whey is relatively low in protein (13-17% dry basis), although the protein is of high biological value. It is an excellent source of B-vitamins. Delactosed whey is available in some areas.

Marine Protein Sources

Fish meals are primarily of two types — those from fish caught for making meal and those made from fish residues processed for human food or other industrial purposes. Herring or related species provide much of the raw material for fish meals. These fish have a high body oil content, much of which is removed in preparation of the finished fish meal.

Good quality fish meals are excellent sources of proteins and amino acids, ranking close to milk proteins. The protein content is high (see Table 18-12), and it is highly digestible. Fish meals are especially high in essential amino acids, including lysine, which are deficient in the cereal grains. In addition, fish meal is usually a good source of the B-vitamins and most of the mineral elements (19). As a result, fish meal is a highly favored ingredient for swine and poultry feeds, although the cost of such meal is usually considerably higher than for other protein sources except milk. Some fish meal is used for ruminants in Europe, but very little is used elsewhere. Some use of fat extracted meals appears feasible in milk replacers.

The quality of fish may be variable due to factors such as partial decomposition before processing or overheating during processing. Excess oil may lead to rancidity and inadequate drying may allow molding. Even with good quality meal, high levels, when fed to swine, may result in fishy-flavored pork if extraction of fat from the meal is incomplete.

Plant Protein Concentrates

The most important commercial sources of plant-protein concentrates are derived from cottonseed and soybeans with lesser amounts from peanuts, flax (linseed), sunflower, sesame, safflower, rapeseed, various legume seeds and other miscellaneous sources of lesser importance. Meals made from the seeds mentioned are called oilseed meals because these seeds are all high in oils that have a number of important commercial uses. Originally, the meals were by-products of little interest, but the meals have become much more important in recent years.

Three primary processes are used for removing the oil from these seed crops. These are known as expeller (screw press), prepress-solvent and solvent extraction. In the expeller process, the seed, after cracking and drying, is cooked for 15-20 minutes, then extruded through dies by means of a variable pitch screw. This results in rather high temperatures which may cause reduced biological value of the proteins. The solvent-extracted meals are extracted with hexane or other solvents, usually at low temperatures. In the prepress-solvent process the seed oils are partially removed with a modified expeller process and then extracted with solvents; this is usually done with seeds containing more than 35% oil as they are not commercially suitable for direct solvent extraction.

As a group the oilseed meals are high in crude protein, most being over 40%, and they are usually standardized before marketing by dilution with hulls or other material. A high percentage of the N is present as true protein (95 ± %) which is usually highly digestible and of moderate to good biological value, although of usually lower value than good animal protein sources. Most meals are low in cystine and methionine and have a variable and usually low lysine content; soybean meal being an exception in lysine content. The energy content varies greatly, depending on processing methods; solvent extraction leaves less fat and, thus, reduces the energy value. Ca content is usually low, but most are high in P content although ½ or more is present as phytin P, a form poorly utilized by monogastrics. These meals contain low to moderate levels of the B-vitamins and are low in carotene and vitamin E.

Soybean Meal

Whole soybeans [*Glycine max*] contain 15-21% oil which is usually removed by solvent extraction. In processing, the meal is toasted, a process which improves the biological value of its protein; the protein content is standardized at 44% or 50% by dilution with soybean hulls. Soybean meal is produced in large amounts in the USA and is a highly favored feed ingredient as it is quite palatable, highly digestible, of high energy value, and results in excellent performance when used for different animal species. Methionine is the most limiting amino acid for monogastric species and the B-vitamin content is low. In overall value, however, soybean meal is probably the best plant protein source available in any quantity.

As with most other oil seeds, soybeans have a number of toxic, stimulatory, and inhibitory substances. For example, a goitrogenic material is found in the meal and its long term use may result in goiter in some animal species. Of major concern in monogastric species is a trypsin-inhibitor material which inhibits digestibility of protein. Fortunately, this inhibitor and other factors (saponins and a hemagglutinin) are inactivated by proper heat treatment during processing. Soybeans also contain genistein, a plant estrogen which may account, in some cases, for its high growth-inducing properties.

Currently, there is interest in feeding whole soybeans after appropriate heat processing to inactivate the trypsin inhibitors (110°C for 3 minutes). This product is known in the feed trade as full fat soybean meal; it contains about 38% CP, 18% fat and 5% CF. Heating-extruding equipment has been developed for on the farm processing and its use appears feasible in relatively small operations. Such meal has found some favor in dairy cow rations and, in moderate amounts, in rations for swine and poultry. Heat-treated soybeans can be used to replace all of the soybean meal in corn-soy diets for growing-finishing pigs.

Cottonseed Meal

Due to the growth habits of cotton [*Gossypium spp*], cottonseed meal (CSM) is available in many areas where soybeans do not grow, particularly in some areas in South America, northern Africa, and Asia. CSM protein is of good although variable quality. Most meals are standardized (in the USA) at 41% crude protein. The protein is low in cystine, methionine, and lysine, and the meal is low in Ca and carotene. Although palatable for ruminants, CSM is less well liked by swine and poultry; nevertheless, it finds widespread use in animal feeds, although its use is much more limited by various problems than that of soybean meal.

The cotton seed and CSM contains a yellow pigment, gossypol, which is relatively toxic for monogastric species, particularly young pigs and chicks. In addition, CSM results in poor egg quality as gossypol tends to result in green egg yolks. Sterculic acid, which is found in CSM, can cause egg whites to turn pink.

Gossypol is found bound to free amino groups in the seed protein or in a free form which can be extracted with solvents. The free gossypol is the toxic form. Gossypol toxicity can largely be prevented by addition of ferrous sulfate and other iron salts (31). Choice of an appropriate processing method such as prepress-solvent extraction or more complicated extraction methods can remove most of the free gossypol. Further improvement can be made by plant breeding. Gossypol is produced in glands that can be reduced in size or removed by genetic changes in the plant. Meals from glandless seeds have resulted in good performance in poultry, although not for young pigs. Biologically tested meals that are screened by feeding to hens are also available in some areas.

Other Oilseed Meals

Linseed meal is made from flax seed [*Linum usitatissimum*] which is produced for the drying oils it contains. Linseed meal accounts for only a small part of the total plant proteins produced in the USA. The crude protein content is relatively low (35%) and is deficient in lysine. Although favored in ruminant diets, linseed meal is used sparingly in monogastric diets.

A substantial amount of peanuts or ground nuts [*Arachis hypogaea*] is grown worldwide (2nd to soybeans amoung seed legumes), mostly for human consumption. Peanut meals are quite deficient in lysine and the protein is low in digestibility. Peanuts contain a trypsin inhibitor, as do many beans, and they may be contaminated with molds such as *Aspergillus flavus* which produces potent toxins (aflatoxins) that are detrimental, particularly to young animals.

Safflower [*Carthamus tinctorius*] is a plant grown in limited amounts for its oil. The meal is high in fiber and low in protein unless the hulls are removed. The protein is deficient in S-containing amino acids and lysine. This meal finds most use in ruminant rations because of the fiber content and inferior amino acid content.

Sunflowers [*Helianthus annuus*] are produced for oil and seeds, primarily in northern Europe and Russia as they will grow in relatively cool climates. The meals, although high in protein, are also high in fiber and have not produced performance in swine comparable to that obtained with other protein sources.

Rapeseed meal [from *Brassica napus* or *B. campestris*) is of interest because it grows in more northern climates than most oilseed plants. Most rapeseed meals are about 40% crude protein with an amino acid pattern similar to other oilseed meals. Use of rapeseed meal is limited, at least in monogastric species, by the content of goitrogenic substances common to members of the mustard family. These meals also tend to be unpalatable.

Other Plant Protein Sources

Coconut or copra meal, the residue after extraction and drying the meat of the coconut, is available in many areas of the world. The crude protein content is low (20-26%) and of variable digestibility. It can be used to partially supply protein needs of most species.

Field beans and peas of a number of species are sometimes available for animal feeding, although they may be grown primarily for human food. These seeds generally contain about 22-26% protein and the proteins tend to be deficient in S-containing amino acids as well as tryptophan. Some seeds contain toxic factors, such as trypsin inhibitors, as well as other toxins. Because of the toxins and poor protein quality, use in feed for monogastric species is often limited, although they make useful supplementary protein for ruminant animals. Some of these grain legumes hold promise as complete energy-protein feeds for swine in tropical areas of the world (31). Their favorable chemical composition is illustrated by data shown in Table 18-13. Considerable research effort is currently being expended to identify high-yielding varieties and to develop economical methods of destroying inhibitors and toxins.

New and Miscellaneous Protein Sources

As the world protein shortage becomes more acute, efforts are underway to identify and develop additional sources for use in livestock feeding. Potential sources include animal wastes, seaweed, and single cell protein products such as algea, bacteria and yeasts.

Animal and poultry wastes (feces and urine) have potential value as animal feed. To date it appears that ruminants are best able to utilize them and until technological advances provide the knowledge needed to convert these waste products to feed useful to monogastrics, cattle and sheep will probably be the main avenue of utilization. Current research efforts are aimed at possible utilization of microbial populations grown on animal wastes as a means of improving the nutritive value of these wastes for monogastric species.

Feather meal, a principal by-product from poultry processing, contains about 86% protein, is high in cystine, threonine and arginine, but the protein is very poorly digested by monogastrics and must be hydrolyzed for good utilization. No more than 3-5% of hydrolyzed feather meal should be used in swine diets for optimum performance.

Algae is an attractive possibility as a protein source. Preliminary results with cultivated fresh water algae indicate a potential for about 10 times as much protein/unit of land area as soybeans. Algae contains approximately 50% protein, 6-7% crude fiber, 5-6% fat and 6% ash. Because of its bitter taste and protein of low biological value, it should not be used at levels exceeding 10% of the diet of growing swine.

Seaweed is plentiful in many areas of the world but most research work indicates that it is not a good source of either energy or protein and should be used mainly as a mineral supplement and should not exceed 10% of the diet. It contains about 2% Ca, 0.4-0.5% P and is a good source of Fe and extremely high in I.

Certain yeasts can use the carbon from petroleum by-products as a source of energy. A culture of yeast is propagated by blending a N source such as ammonia, an appropriate mixture of inorganic minerals, and petroleum hydrocarbons under controlled conditions (38). The yeast culture multiplies rapidly and produces large yields. The nutritive content varies according to production methods and type of yeast. Methionine-cystine is generally limiting, but feeding trials with pigs fed diets containing up to 70% yeast replacing fish meal produced good performance, indicating a high biological value of hydrocarbon-yeast protein. It contains approximately 54% protein, 7% fat, 8% crude fiber and 7% ash and is high in P (2%) but low in Ca (0.14%). As technology improves, large-scale production of yeast protein may become a significant source for animal feeding.

Table 18-13. Chemical composition of some grain legumes.

Item	Moisture	Protein	Ether extract	Fiber	Ash	Nitrogen-free extract
	%	%	%	%	%	%
Chickpeas (forage)	11.8	17.3	3.5	8.4	2.8	56.2
Cowpeas	11.0	23.4	1.8	4.3	3.5	56.0
Dry beans	11.0	23.9	1.3	4.2	3.4	56.2
Field beans	11.0	23.4	2.0	7.8	3.4	52.4
Field peas	10.1	25.9	1.5	6.0	2.6	53.9
Pigeon peas	11.0	20.9	1.7	8.0	3.5	54.9

[a] See Pond and Maner (31)

Nonprotein-Nitrogen [NPN]

NPN includes any compounds that contain N but are not present in the polypeptide form of precipitable protein. Organic NPN compounds would include ammonia, amides, amines, amino acids, and some peptides. Inorganic NPN compounds would include a variety of salts such as ammonium chloride and ammonium sulfate. Although some feedstuffs, particularly some forages, contain substantial amounts of NPN, from a practical point of view, NPN in formula feeds usually refers to urea or, to a lesser extent, such compounds as biuret and diammonium phosphate, mainly because other NPN compounds are often too costly to use for feeding of animals at the present time, and those such as ammonium sulfate are not suitable sources of N.

NPN, especially urea, is primarily of interest for feeding of animals with a functioning rumen. The reason for this is that urea is rapidly broken down to ammonia which is then incorporated into amino acids and microbial proteins by rumen bacteria which are utilized later by the host. Thus, the animal, itself, does not utilize urea directly.

A variety of factors must be considered in utilizing urea in feeds, but complete discussion of the subject is too complex for this chapter (see references 7 and 8). Urea, if consumed too rapidly, may be toxic or lethal. Furthermore, urea must be fed with some readily available carbohydrate for good utilization and to prevent toxicity, and some adaptation time is required for the animal. Where livestock management is good and feed is formulated and mixed properly, urea can provide a substantial amount of the supplemental N required, even to the point where it supplies all of the N in purified diets. In practical rations current recommendations are that not more than 1/3 of the total N be supplied by urea (8). Most states require labeling of feed tags with maximal amounts of urea as a protective measure for the buyer. With the current high costs of plant and animal protein sources, we are likely to see more and more urea being utilized every year as supplementary N for ruminants.

In monogastric species, the only microorganisms that can convert urea to protein are found in the lower intestinal tract at a point where absorption of amino acids, peptides and proteins is believed to be rather low or nonexistent. Research with pigs, poultry, horses, and other species indicates that some ^{15}N from urea can be recovered in amino acids and proteins from body tissues, probably as a result of its incorporation into tissue amino acids by various known pathways of amination and transamination. However, little if any net benefit is to be expected from feeding urea to monogastric species. The most likely situation in which it might be useful would be when the supply of essential amino acids was adequate and the nonessentials in short supply; then supplementing with urea might be of some slight benefit.

Mineral Supplements

Although minerals make up only a relatively small amount of the diet of animals, they are vital to the animal and, in most situations, some diet supplementation is required to meet needs. Of course, all of the required mineral elements are needed in an animal's diet, but needed supplementary minerals will vary according to the animal species, age, production, diet, and mineral content of soils and crops in the area where grown. Generally, those minerals of concern include common salt (NaCl), Ca, P, Mg, and sometimes S of the macrominerals and, of the trace elements, Cu, Fe, I, Mn and Zn and, in some places, Co and Se. Most energy and nitrogen sources — fat and urea being 2 marked exceptions — provide minerals in addition to the basic organic nutrients, but the flexibility needed in formulating rations usually requires more concentrated sources of one or more mineral elements. Some of these sources are discussed briefly.

Mineral Sources

Common salt (NaCl) is usually required for good animal production except in areas where the soil and/or water are quite saline. It is a common practice to add 0.5% salt to most commercial feed formulas, although this is probably more than is actually required for most species. Salt is often fed ad libitum, particularly to ruminants and horses, as their requirement is probably higher than for swine or poultry, and different feeding methods lend themselves to this practice. Excess salt may be a problem for all species, but particularly for swine and poultry as they are much more susceptible to toxicity than other domestic species, especially if water consumption is restricted.

In many cases salt is used as a carrier for some of the trace elements (see Table 18-14).

Additional sources of the trace elements can be added to feedstuffs as inorganic salts in areas where they are known to be needed, or feedstuffs high in a needed mineral can be used in the feed formula. Se is the only required element that cannot be legally added to all animal diets in the USA, but legislation is approaching the point of approval at the time this was written. Current regulations (spring, 1974) allow Se to be added to diets only for poultry and swine.

Ca and P are required in large amounts by the animal body and many feedstuffs are deficient in one or both of these elements. Some of the common supplements are shown in Table 18-15. Most sources of Ca are well utilized by different animal species. Although net digestibility may be low, particularly in older animals, there is little difference between Ca sources. This statement does not apply to P, however, because sources differ greatly in availability. In plants, about ½ of the P is bound to phytic acid and P in the product, phytin, is poorly utilized by monogastric species. The usual recommendation is to consider only ½ of plant P available to monogastric species although ruminants utilize it very well. Marked differences also exist in the biological availability of inorganic sources. Phosphoric acid and the mono-, di-, and tri-calcium phosphates are well utilized. Sources such as Curacao Island and colloidal (soft) phosphates are utilized less well by most animals. Some sources are high in fluoride and must be defluorinated to prevent toxicity.

Of course, some differences exist in utilization of some of the inorganic salts and oxides of different elements. For example, Mg in magnesite is used very poorly; Fe from ferric oxide is almost completely unavailable, and I from some organic compounds such as diiodosalicylic acid, although it is absrobed, is also almost all excreted. Se from plant sources is more biologically available than that from inorganic sources. Legume forage and animal and fish proteins are usually excellent sources of the trace elements.

Currently, there is interest in utilization of chelated trace minerals. Chelates are compounds in which the mineral atom is bound to an organic complex (hemoglobin for example). Feeding of chelated minerals is proposed on the basis that chelates will prevent the formation of insoluble complexes in the GIT and reduce the amount of the particular mineral that will be required in the diet. Limited evidence indicates that this works sometimes (with Zn in poultry diets) but not in all, and the cost/unit of mineral is appreciably higher. Further data are required to know how well chelated trace minerals will work out for all species.

Table 18-14. Composition of a typical trace-mineralized salt and its contribution to the diet when included at a level of 0.5%.

Mineral[a]	Amount in salt mixture, %	Amount provided in complete diet when 0.5% included, ppm
NaCl	Not less than 97.000 or more than 99.000	5000
Cobalt	Not less than 0.015	0.70
Copper	0.023	1.15
Iodine	0.07	3.50
Iron	0.117	5.4
Manganese	0.225	11.25
Sulfur	0.040	2.0
Zinc	0.008	0.40[b]

[a] Ingredients are: Salt, cobalt carbonate, copper oxide, calcium iodate, iron carbonate, manganese oxide, solium sulfate, zinc oxide; yellow prussate of soda (sodium ferrocyanide) added as an anti-caking agent.

[b] This amount supplies only about 1% of the needs of swine.

Vitamin Sources

Nearly all feedstuffs contain some vitamins, but vitamin concentration varies tremendously because it is affected by harvesting, processing, and storage conditions as well as plant species and plant part (seed, leaf, stalk). As a rule, vitamins are easily destroyed by heat, sunlight, oxidizing conditions, and storage conditions that allow mold growth. Thus, if any question arises of adequacy of a diet, it is frequently better to err on the positive side than to have a diet that is deficient.

As a general rule, vitamins likely to be limiting in natural diets are: vitamins A, D, E, riboflavin, pantothenic acid, niacin, choline and B_{12}, depending on the species and class of animal we are concerned with. For adult ruminants, only A, D, and E are of concern with A being the most likely to be deficient. Vitamin K and biotin can be a problem for both swine and poultry under some conditions, in addition to those vitamins listed previously.

Fat-Soluble Vitamins

The best sources of carotene and vitamin A are green plants and organ tissues such as liver meal. Commercially, most vitamin A is now produced synthetically at such low prices that the cost of adding it to rations is negligible and there is no economic reason not to add it where any likelihood of a need exists. Sun-cured forages are good sources of vitamin E; very high activity is obtained from fish liver oils or meals and irradiated yeast or animal sterols are common sources. Vitamin E is found in highest concentrations in the germ or germ oils of plants and in moderate concentrations in green plants or hays. Vitamin K is available in synthetic form (see Ch. 13) and, where gut synthesis is not adequate, it can be added at reasonable costs.

Water-Soluble Vitamins

Animal and fish products, green forages, fermentation products, oil seed meals, and some seed parts are usually good sources of the water-soluble vitamins. The bran layers of cereal grains are fair to moderate sources, and roots and tubers are poor to fair sources. Commercially, some of the crystalline vitamins or mixtures used for humans are prepared from liver and yeast or other fermentation products and some are produced synthetically. Thus, these various sources can be used to meet vitamin needs in animal rations. Vitamin B_{12} is produced only by microorganisms, thus is found only in animal tissues or some type of fermentation product or in animal manures. Animal and fish products are good sources.

Table 18-15. Some common supplements for Ca, P and Mg.

Mineral source	Ca, %	P, %	Mg, %
Bone meal, steamed	24-29	12-14	0.3
Bone black, spent	27	12	
Dicalcium phosphate	23-26	18-21	
Tricalcium phosphate	38-39	19-20	
Defluorinated rock phosphate*	31-34	13-17	
Raw rock phosphate**	24-29	13-15	
Sodium phosphate		22-23	
Soft phosphate	18	9	
Curacao phosphate	35	15	
Diammonium phosphate		20	
Oyster shell	38		
Limestone***	38		
Calcite, high grade	34		
Gypsum	22		
Magnesium oxide			60-61

* Usually less than 0.2% F.
** 2-4% F.
*** Dolomite limestone contains up to 10% or more Mg.

References Cited

1. Anonymous. 1969. Liquid supplements for livestock feeding. Chas. Pfizer & Co.
2. Anonymous. 1972. Millfeed Manual. Millers' National Federation, Chicago, Ill.
3. Barnett, A.J.G. 1954. Silage Fermentation. Butterworth, London.
4. Baxter, H.D., et al. 1973. J. Dairy Sci. 56:119.
5. Beever, D.E., D.J. Thomson, E. Pfeffer and D.G. Armstrong. 1971. Br. J. Nutr. 26:123.
6. Bhattacharya, A.N. and F.T. Sleiman. 1971. J. Dairy Sci. 54:89.
7. Church, D.C. 1969. Digestive Physiology and Nutrition of Ruminants. Vol. 1 - Digestive Physiology. O & B Books, 1215 NW Kline Pl., Corvallis, Ore.
8. Church, D.C. et al. 1972. Digestive Physiology and Nutrition of Ruminants. Vol. 3 - Practical Nutrition. O & B Books, 1215 NW Kline Pl., Corvallis, Ore.
9. Colovos, N.F. et al. 1970. J. Animal Sci. 30:819.
10. Coppock, C.E. 1969. J. Dairy Sci. 52:848.
11. Dickey, H.C., H.A. Leonard, S.D. Musgrave and P.S. Young. 1971. J. Dairy Sci. 54:876.
12. Ely, R.E., E.A. Kane, W.C. Jacobson and L.A. Moore. 1953. J. Dairy Sci. 36:334.
13. Ewing, W.R. 1963. Poultry Nutrition. 5th ed. Ray Ewing Co., Pasadena, Cal.
14. Graham, N. McC. 1964. Austral. J. Agric. Res. 15:974.
15. Hawkins, D.R., H.E. Henderson and D.B. Purser. 1970. J. Animal Sci. 31:617.
16. Hillman, D. 1969. J. Dairy Sci. 52:859.
17. Huber, J.T. and J.W. Thomas. 1971. J. Dairy Sci. 54:224.
18. International Symposium on Triticale. 1973. Texas Tech. Univ., Lubbock, Tex.
19. Kifer, R.R., W.L. Payne, D. Miller and M.E. Ambrose. 1968. Feedstuffs. 40(20):36.
20. McCullough, M.E. 1969. Optimum Feeding of Dairy Animals. Univ. of Georgia Press, Athens, Ga.
21. McDonald, P., R.A. Edwards and J.F.D. Greenhalgh. 1966. Animal Nutrition. Oliver & Boyd, Edinburgh.
22. Mertz, E.T. 1963. Proc. 18th An. Hybrid Corn Research-Industry Conf. Pub. 18. pp 7-12.
23. Mertz, E.T. 1968. Agric. Sci. Rev. Third Quarter, pp 1-9.
24. Meyer, J.H., G.P. Lofgreen and N.R. Ittner. 1956. J. Animal Sci. 15:64.
25. Miller, L.G., D.C. Clanton, L.F. Nelson and O.E. Hoehne. 1967. J. Animal Sci. 26:1369.
26. Miltimore, J.E., J.L. Mason and D.L. Ashby. 1970. Can. J. Animal Sci. 50:293.
27. Miltimore, J.E., J.M. McArthur, J.L. Mason and D.L. Ashby. 1970. Can. J. Animal Sci. 50:61.
28. Morrison, F.B. 1956. Feeds and Feeding. 22nd ed. The Morrison Pub. Co., Clairmont, Alberta, Canada.
29. Munck, L., K.I. Karlsson, A. Hagberg and B.O. Eggum. 1970. Science. 168:985.
30. NRC. 1969. United States-Canadian Tables of Feed Consumption. NRC Pub. 1684, Washington, D.C.
31. Pond, W.G. and J. Maner. 1974. Swine Production in Temperate and Tropical Environments. W.H. Freeman and Co.
32. Preston, T.R. and M.B. Willis. 1970. Feedstuffs. 42(13):20.
33. Reid, R.L., E.K. Odhuba and G.A. Jung. 1967. Agron. J. 59:265.
34. Schneider, B.H. 1947. Feeds of the World. West Virginia Agr. Expt. Sta., Morgantown, W. Va.
35. Stoddart, L.A. and A.D. Smith. 1955. Range Management. 2nd ed. McGraw-Hill Book Co.
36. Sutton, A.L. and R.L. Vetter. 1971. J. Animal Sci. 32:1256.
37. Thomas, J.W. et al. 1969. J. Dairy Sci. 52:195.
38. Waldroup, P.W. 1972. Proc. Distillers Res. Council Conf. 27:34.
39. Watson, S.J. and M.J. Nash. 1960. The Conservation of Grass and Forage Crops. Oliver & Boyd, Edinburgh.
40. Wilkins, R.J., K.J. Hutchinson, R.F. Wilson and C.E. Harris. 1971. J. Agric. Sci. 77:531.

Chapter 19 – Feed Preparation and Processing

Feed represents the major cost in animal production. Even with sheep, which typically consume more forage than other domestic species, feed may represent 55+% of total production costs; with poultry, a value of 75-80% might be more appropriate. Thus, it is imperative to supply an adequate diet — in terms of nutrient content — and to prepare the ration in a manner that will encourage consumption without waste and allow high efficiency of feed utilization.

Feed processing may be accomplished by physical, chemical, thermal, bacterial or other alteration of a feed ingredient before it is fed. Feeds may be processed to alter the physical form or particle size, to preserve, to isolate specific parts, to improve palatability or digestibility, to alter nutrient composition, or to detoxify.

Generally speaking, feed preparatory methods become more important as level of feeding increases and as maximum production is desired. This is so because heavily fed animals become more selective and because, in ruminants, digestibility tends to decrease as level of feeding increases, the latter primarily because feed does not remain in the GIT long enough for maximal effect of the various digestive processes. Feed preparation may also become more important as we go to larger animal production units with greater mechanization. This applies particularly to roughage because long or baled hay is less convenient to handle than chopped or pelleted hay.

Feed preparatory methods that are used for swine and poultry are relatively simple as compared to the variety that are available and in use for ruminant feeds. Thus, the bulk of the discussion in this chapter will deal with ruminants with brief sections on swine and poultry. Also, the discussion will deal primarily with roughages and grains and not with methods that are used for processing of oil seed meals, grain by-products, and similar feedstuffs. More detailed discussion of feed processing and its effect on nutritive values for swine may be found elsewhere (7).

Grain Processing for Ruminants

Grain processing is done primarily to improve digestibility and efficiency of utilization as grain is already in a form that can easily be handled mechanically. Improvement in utilization can usually be obtained by various means that break up the hull or waxy seed coat and improve utilization of the starch in the endosperm. Some methods may also provide a more favorable particle size and/or density which facilitates more optimal passage through the rumen and which may, also, improve palatability. As indicated, grain (or roughage) processing is expected to give greater returns/unit of cost when feed intake of grain is high. Animals on a maintenance diet would not normally be fed much grain and any improvement in efficiency might not return the added cost.

In rations typically in use at this time in the USA, grain and other concentrates in finishing rations for cattle may account for 70-95% of the total intake. With this type of ration the need and benefits of grain processing can clearly be shown in most instances. Recent research evidence (see Ch. 4 of reference 1) indicates that whole corn grain may be an exception to this general rule when corn makes up a high percentage of the ration. It appears that the size and density of the corn grain is a factor; furthermore, cattle apparently masticate whole corn well enough so that much of it will be partly broken before swallowing. Usually, as the percentage of roughage in the diet increases, the physical nature of the concentrate becomes less important, although processing may still influence digestibility and efficiency. Generally, sheep are believed to masticate grain more efficiently than cattle with the result that grinding or other preparatory methods are of less relative value for sheep.

Grain Processing Methods

Grain processing methods are conveniently divided into dry and wet processes as shown below. Heat is an essential part of some methods, but it is not utilized at all in others. Examples of milo processed in different ways are shown (Fig. 19-1).

Dry Processing

Grinding	Extruding
Dry rolling or cracking	Micronizing
	Roasting
Popping	Pelleting

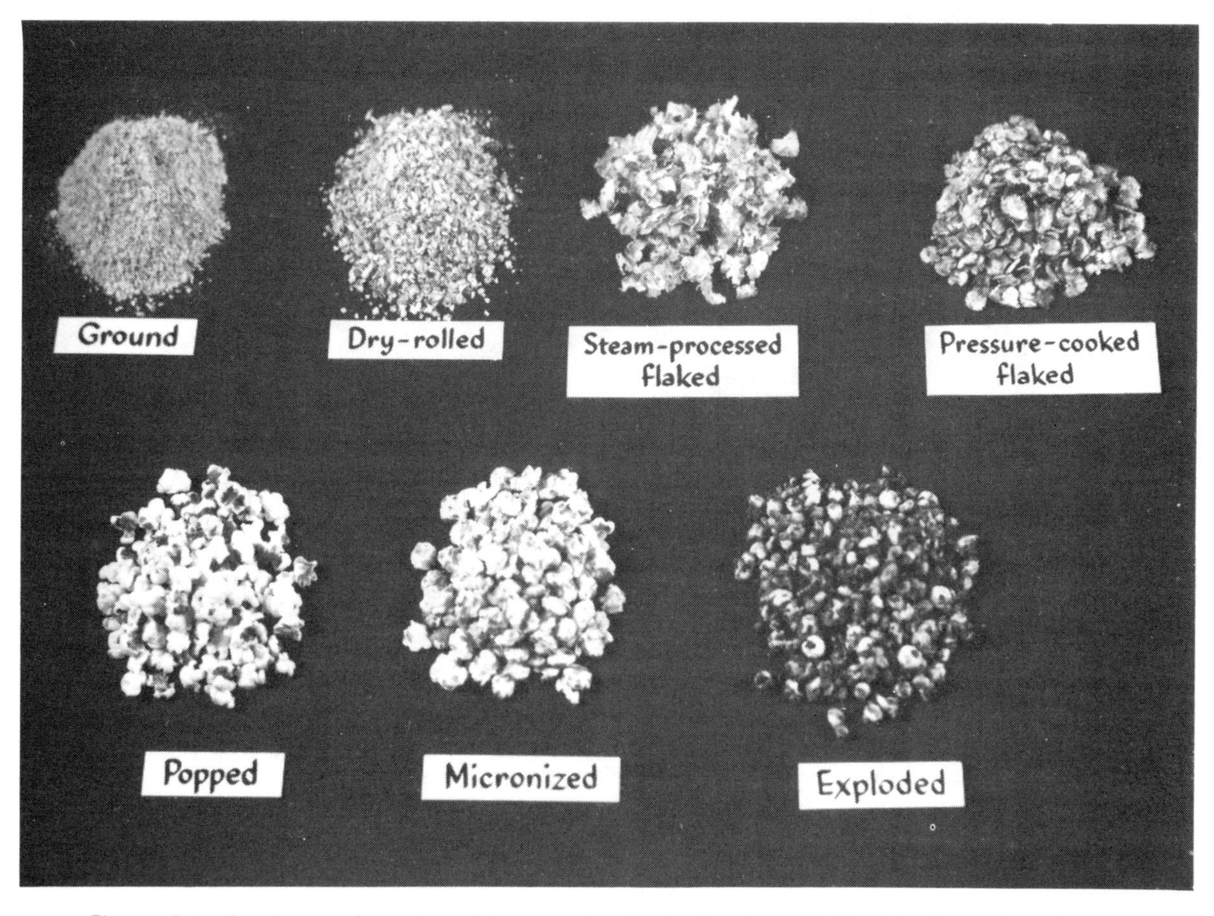

Figure 19-1. Sorghum grain processed by several methods. Courtesy of W.H. Hale, University of Arizona.

Wet Processing

Soaking	Exploding
Steam rolling	Reconstitution
Steam processing and flaking	Ensiling at high moisture content
Pressure cooking	

Grinding

Grinding is probably the most common method of feed processing and, other than soaking, is the cheapest and most simple process. A wide variety of equipment is available on the market and all of it allows some control of particle size of the finished product. Grinding generally improves digestibility of all small, hard seeds such as sorghum grain. Coarsely ground grains are preferred for ruminants, because they dislike finely ground meals, particularly when the meals are dusty. Grinding is just as satisfactory as other more expensive methods when grain intake is relatively low.

Dry Rolling or Cracking

Rolled grain is similar to grain coarsely ground in a hammer mill and the physical nature is usually attractive to ruminants. Particle size can be varied considerably, however. Rolled grain is well utilized and is very satisfactory when roughage intake is relatively low.

Popping and Micronizing

Most readers are familiar with popped corn which is produced by action of dry heat, causing a sudden expansion of the grain which ruptures the endosperm. This process increases rumen starch utilization, but results in a low-density feed. Consequently, it is usually rolled before feeding to reduce bulk. Use of popped grains is quite new because commercial equipment has been developed in only the last 2-3 years. Research data are scarce but results look promising for high-grain rations

Figure 19-2. One example of a pellet mill. Pellet mills can be used to pellet a wide variety of feedstuffs, although roughage and cereal grains must usually be ground prior to pelleting. Courtesy of Sprout-Waldron.

and the equipment is relatively foolproof as compared to some other types. Micronizing is essentially the same as popping, except that heat is furnished in the form of infrared energy.

Pelleting

Pelleting is accomplished by grinding the material and then forcing it through a thick die. Feedstuffs are usually, but not always, steamed to some extent prior to pelleting. Pellets can be made in different diameters, lengths, and hardnesses and have been commercially available for many years. Animals generally like the physical nature of pellets, particularly as compared to meals. Results with ruminants on high-grain diets, however, have not been particularly favorable because of decreased feed intake even though feed efficiency is usually improved over other methods. Pelleting the ration fines (finely ground portions of a ration) frequently is desirable, however, as the fines will often be refused otherwise. Supplemental feeds such as protein concentrates are often pelleted so they can then be fed on the ground or in windy areas with little loss. One example of a pellet mill is shown (Fig. 19-2).

Roasting

Roasting is accomplished by passing the grain through a flame, resulting in heating and some expansion of the grain which produces a palatable product. Limited data on corn indicate a good response with cattle in terms of daily gain and feed efficiency. Roasting may also be used with whole soybeans to destroy heat-labile inhibitors and, thus, improve nutritive value for poultry and swine.

Extruding

Extruded grains are prepared by passing the grain through a machine with a spiral screw which forces the grain through a tapered head. In the process the grain is ground and heated, producing a ribbon-like product. Results with cattle on high-grain rations are similar to those with other processing methods.

Steam-Rolled and Steam-Processed Flaked Grains

Steam-rolled grains have been used for many years, partly to kill weed seeds. The steaming is accomplished by passing steam up through a tower above the roller mill. Grains are subjected to steam for only a short time (3-5 min.) prior to rolling. Most results indicate little if any improvement in animal performance as compared to dry rolling.

Steam-processed flaked grains, as defined by Hale (Ch. 4 in reference 1) are prepared in a similar manner but with relatively rigid quality controls. Grain is subjected to high-moisture steam for a sufficient time to raise the water content to 18-20%, and the grain is then rolled to produce a rather flat flake. Feedlot data with cattle indicate that best response is produced with thin flakes which apparently allow more efficient rupture of starch granules and produce a more desirable physical texture. Corn, barley, and sorghum usually give a good response in terms of increased gain; although feed efficiency is improved with corn and sorghum, barley shows little improvement.

Pressure Cooking

Pressure cooking is accomplished by cooking the grain under steam pressure and then rolling it, the product being similar to steam-processed flaked grain. One advantage is that only 1-2 min. are required for cooking as compared to 15-30 min. for the steam-processed grain. Flakes from pressure-cooked grain are less apt to break up on handling, but the cooked grain usually requires cooling before flaking and maintenance costs of equipment are relatively high. One example is shown of equipment used for pressure cooking and then exploded (Fig. 19-3).

Soaked Grain

Grain soaked for 12-24 hr in water has long been used by livestock feeders. The soaking, sometimes with heat, softens the grain which swells during the process, making a palatable product that should be rolled before using in finishing rations. Research results do not show any marked improvement as compared to other methods. Space requirements, problems in handling, and potential souring have discouraged large scale use.

Reconstitution

Reconstitution is similar to soaking and involves adding water to mature dry grain to raise the moisture content to 25-30% and storage of the wet grain in an oxygen-limiting unit for 14-21 days prior to feeding. This procedure has worked well with sorghum and corn, resulting in improved gain and feed conversion (Table 19-1) in cattle fed high-concentrate rations when whole grain was used.

Figure 19-3. An example of equipment used for exploding grains. The grain is subjected to steam under pressure for a short period and then the pressure is suddenly released causing expansion of the grain kernal. Courtesy of Triplett Exploders, Inc.

High-Moisture Grain

This term refers to grain harvested at a high moisture content (20-35%) and stored in a silo to preserve the moisture content. It is usually ground or rolled when fed. This is a particularly useful procedure when weather conditions do not allow normal drying in the field and it obviates the need to dry the grain artificially. Storage costs may be relatively high, but high moisture grain produces good feedlot results, comparable to some of the processing methods discussed. Feed conversion, particularly, is improved.

Acid Preservation of High Moisture Grains

With the energy shortage has come increased interest in eliminating artificial drying of newly harvested cereal grains. British and Scandinavian data with barley for pigs, Canadian work with corn for pigs, and research in the USA with corn for beef cattle shows promise for the use of acids to preserve high-moisture grains (3, 5, 6, 9). Thorough mixing of 1-1.5% propionic acid or mixtures of acetic-propionic acid, or formic-propionic acid into

Table 19-1. Examples of effect of grain processing on steer performance.[a]

Item	Av. daily gain, lb	Daily feed intake, lb	Feed/lb of gain
Milo, dry rolled	2.83	22.7	8.02
vs. steam-processed flaked	3.10	23.7	7.64
Barley, dry rolled	2.88	20.8	7.22
vs. steam-processed flaked	3.10	22.7	7.32
Corn, dry rolled	3.03	20.2	6.70
vs. pressure cooked	3.35	21.0	6.33
Whole corn, reconstituted	2.61	20.5	7.85
vs. steam-processed flaked	2.65	20.3	7.66
Sorghum grain, dry rolled	2.20	18.2	8.21
vs. reconstituted	2.64	18.5	7.02

[a] Data excerpted from Church et al (1)

high-moisture (20-30%) whole corn or other cereal grains retards molding and spoilage with out appreciably affecting animal performance compared with that obtained with dried grains.

Roughage Processing for Ruminants

Baled Roughage

Baling is still one of the most common methods of handling roughage, particularly where it is apt to be sold or transported some distance. Baling has a considerable advantage over loose hay stacked in the field or roughage in other less dense forms. Although baled hay can now largely be handled mechanically, it still requires considerably more hand labor than many other feedstuffs. Furthermore, considerable waste may occur in feeding, depending on how it is fed (feed bunks, on the ground) and on the level of feeding. Heavily fed animals such as dairy cows may be quite selective so that stems will not be consumed. Thus, nearly always a high loss occurs in feeding baled hays. Consumption is not adequate for high levels of performance (Table 19-2) when it provides the only feed for ruminants.

Chopped and Ground Roughage

Chopping or grinding puts roughage in a physical form in which it can readily be handled by some mechanical equipment, tends to provide a more uniform product for consumption, and usually reduces feed refusal and waste. However, additional expense may be incurred by grinding and loss of dust may be appreciable from grinding in hammer mills. This dust loss is sometimes reduced in commercial mills by spraying fat on bales before they are ground. Ground hays are, as a rule, quite dusty and may not be consumed readily. Adding molasses, fat, or water will usually improve intake. Chopping produces a physical texture of a more desirable nature than grinding, but chopped hay does not lend itself as well to incorporation into mixed feeds as does ground hay.

Pelleting

Pelleted roughages are usually readily consumed by ruminants as the particle size and physical texture are of a desirable nature. Roughages such as long hay must be ground before pelleting — a slow, costly process compared to similar treatment of grains. Thus, cost of processing is a bigger item than for most other feed processing methods. Pelleting usually gives the greatest relative increase in performance for low-quality roughages. This appears to result from an increase in density with more rapid passage through the GIT and not to any great improvement in digestibility. Pelleted roughages are also metabolized somewhat differently; as a result of more rapid

Table 19-2. Effect of roughage processing on animal performance of cattle.

Item	Av. daily gain, kg	(lb)	Daily dry matter intake, kg	(lb)	Feed/unit of gain
Costal Bermuda grass [a]					
long	0.33	(0.73)	4.67	(10.30)*	17.0
vs. ground	0.50	(1.11)	6.07	(13.38)*	13.9
vs. pelleted	0.66	(1.45)	6.53	(14.4) *	11.3
Alfalfa hay [b]					
baled	0.29	(0.63)	4.31	(9.5)	15.1
chopped	0.28	(0.62)	4.22	(9.3)	15.1
pelleted	0.78	(1.73)	6.49	(14.3)	8.3
Alfalfa hay [c]					
baled	0.67	(1.48)	6.14	(13.5)	9.1
cubes	0.86	(1.90)	6.65	(14.7)	7.8
haylage	0.77	(1.70)	6.79	(15.0)	8.9

[abc] Data from Cullison (2), Webb and Cmarik (8) and Kercher et al (4), respectively.
* Feed intake as fed.

passage out of the rumen, less cellulose is digested and relatively less acetic acid is produced with relatively greater digestion in the intestines. Utilization of metabolizable energy is usually more efficient. Pelleted high-quality roughages will produce performance (gain in wt) in young cattle (see Table 19-2) or lambs almost comparable to high grain feeding. An example is shown in Fig. 19-4 of the change in density achieved with pelleting and cubing.

Cubed Roughages

Cubing is a relatively new process in which dry hay is forced through dies that produce a square product (ca. 3 cm in size) of varying lengths. Grinding before cubing is not required, but usually water is sprayed on the dry hay as it is cubed. Field cubers have been developed (Fig. 19-5), and stationary cubers are also used which use hay from stacks or bales. Alfalfa hay produces the best cubes. Research data (1) indicate that cubes will produce satisfactory performance in cattle provided they are not too hard. Cubes are widely used now for dairy cattle and usage is likely to increase in the future as cubes lend themselves to mechanization of feeding.

Feed Processing for Dairy Cows

Feed processing may result in somewhat different responses in dairy cows than for growing or finishing cattle or lambs. Generally, feeding lactating cows high grain rations,

Figure 19-4. An example of the densification of alfalfa hay that can be achieved by baling, grinding and pelleting. Each sample contains 5 lb of alfalfa hay.

Figure 19-5. A portable hay cuber. These machines are being used more and more to facilitate handling, transportation and feeding of roughages. Courtesy of Deere & Co.

particularly heat-treated grains, or all of their roughage in ground or pelleted form results in reduced rumen acetate production and lower milk-fat %. Total milk-fat production may not be decreased as these feedstuffs are apt to result in increased milk production. When milk is sold on the basis of its fat %, however, then use of heat-treated grains or pelleted roughages may need to be restricted.

Feed Processing for Swine and Poultry

Grinding and pelleting are the most common means of preparing feed for these two classes of animals. Generally, results show that grinding to a medium to moderately fine texture results in better performance than when rations are finely ground. Feed particles of different ingredients should be of a similar size so that animals do not sort out the coarse particles and leave the fines. Digestibility of grains by swine is generally improved by grinding, and finely ground (0.16 cm screen size) grain promotes improved efficiency as compared with coarse grinding (1.27 cm screen or larger). Finely ground feed, however, is associated with an increased incidence of stomach ulcers in swine.

Pelleting usually results in an improvement in performance, partly because animals tend to eat more in a given period. Efficiency is often improved, sometimes because of less feed wastage with pellets. A common practice, particularly for poultry, is to pellet meal rations and then roll them, producing a product called crumbles. The texture of crumbles is well liked, particularly if pellets are quite hard. It might be noted that birds fed pellets tend to exhibit more cannibalism.

Research data with swine (4) indicate that pelleting generally improves gain and feed efficiency (5-10%) of the feed grains that are low in fiber — corn, wheat, sorghum — but it has less effect on barley and oats. Although oats are improved considerably by medium-to-fine grinding, pelleting may improve efficiency with little or no effect on gain.

Other feed preparatory methods have been used with varying degrees of success for both pigs and poultry. With barley, soaking and treatment with different enzymes have resulted in some improvement, but little use is currently being made of these methods, partly because little barley is fed to poultry, and the improvement for swine probably does not justify the expense. Steam-flaking and reconstitution may improve digestibility for pigs, but have little effect on gain.

References Cited

1. Church, D.C. (ed.). 1972. Digestive Physiology and Nutrition of Ruminants. Vol. 3 - Practical Nutrition. O & B Books, 1215 NW Kline Pl., Corvallis, Ore.
2. Cullison, A.E. 1961. J. Animal Sci. 20:478.
3. English, P.R., J.H. Topps and D.G. Dempster. 1973. Animal Prod. 17:75.
4. Kercher, C.J., W. Smith and L. Paules. 1971. Proc. West. Sec. Amer. Soc. An. Sci. 22:33.
5. Madsen, A. et al. 1973. Nat. Agr. Lab. of Denmark Bul. No. 407, Copenhagen, Denmark.
6. Miller, J.I., J.B. Robertson and R. Logan. 1972. Proc. Cornell Nutr. Conf., Dept. An. Sci., Cornell University.
7. Pond, W.G. and J. Maner. 1974. Swine Production in Temperate and Tropical Environments. W.H. Freeman & Co., San Francisco.
8. Webb, R.J. and C.F. Cmarik. 1957. Univ. of Illinois rpt. 15-40-329, Dixon Springs Sta.
9. Young, L.G., R.G. Brown and B.A. Sharp. 1970. Can. J. Animal Sci. 50:711.

Chapter 20 – Ration Formulation

Ration formulation has been left until now because it is a complicated subject, although the mathematical manipulations required are relatively simple. A great deal has been written on this subject. Unfortunately, much of the literature is either poorly written, too incomplete, inadequately illustrated, or is obtuse to the point of being worthless for students and novices wishing to understand the subject. We will hope that such statements do not apply to this chapter.

In ration formulation, the intent is to translate knowledge about nutrients, feedstuffs, and animals into nutritionally adequate rations that will be eaten in sufficient amounts to provide the level of production desired. This, obviously, requires more knowledge and experience than may be expected of beginners, and space in this book does not allow complete treatment for all of the different species and classes of animals we may be concerned with in the field.

Information Needed

Before we can begin any mathematical manipulations, several different types of information are required for an organized approach to ration formulation for a given situation. These are discussed briefly.

Nutrient Requirements of the Animal

The first step is to arrive at some estimate of the daily nutrient requirements of the animal for which the ration is intended. In the USA, the NRC (National Research Council) recommendations are most commonly used as a guide and tables from these publications are given in Appendix Tables 7-29. The reader should understand that these estimates are not necessarily the best or last word and they should be subject to modification when the formulator has knowledge indicating a change should be made.

Depending on the situation, we may not be concerned with all known nutrients as some are usually more critical than others. For example, if we were interested in formulating a protein supplement for cows on the range, we would normally consider protein first and possibly the energy content and 1 or 2 of the macrominerals such as P and Ca. We would probably disregard most other nutrients, with the possible exception of vitamin A. However, if we are formulating a broiler ration where the chick has access only to the feed we provide, then we need to be concerned with a much more detailed list of nutients.

We may need to adjust the NRC recommendations in situations where it is likely that NRC values may not be satisfactory. For example, high-producing dairy cows may need more liberal nutrient allowances or, if we are in a cold climate, nutrient needs, especially energy, will be increased. If we are in an area where one of the trace minerals in native forage is quite low, then we would certainly consider it whereas it might normally be ignored in routine ration formulation.

Feedstuffs

The next step is to list available feedstuffs (as in Table 20-1) which are suitable for the particular animal in question. In many cases only a limited number of locally grown feeds are available at competitive prices; although we might contrive a long list of by-product and supplementary feeds, this is usually unnecessary.

The contribution by each feed or critical nutrients should be listed. Analytical data on roughages and grains are preferred where available; if not, then average composition data can be used from NRC tables (Appendix tables) or other available information.

The list of feeds we need to consider can be greatly reduced by careful consideration. Is this feed suitable and economical for its intended use? For each species and class of animal some feeds are more useful than others. For adult ruminants we would usually want to include urea where protein is of concern, but on the basis of present knowledge we would not consider urea at all for any of the monogastric species or for milk-replacer formulas. Meat and fish meals, on the other hand, may be favored feeds for poultry and pigs, but are usually too high priced for ruminant feeds.

Furthermore, we must consider if the feed should be processed and, if so, in what manner

Table 20-1. Composition and cost of feedstuffs used in formulation illustrations.

Feedstuff		DM, %	Nutrients					Cost/lb of feed, ¢	Cost/unit, ¢/lb		
			CP, %	DE, Mcal/lb	TDN, %	Ca, %	P, %		CP	DE	TDN
Alfalfa hay,	dry	100	17.1	1.12	56	1.35	0.22				
chopped	as fed	89.2	15.2	1.00	50	1.20	0.20	2.00	13.2	2.00	4.00
Barley, P.C.,	dry	100	10.9	1.64	82	0.07	0.45				
rolled	as fed	89.0	9.7	1.46	73	0.06	0.40	3.00	30.9	2.05	4.11
Beet pulp, Mol.	dry	100	9.9	1.48	74	0.61	0.11				
dried	as fed	92.0	9.1	1.36	68	0.56	0.10	3.10	34.1	2.28	4.56
Corn, #2, cracked	dry	100	10.0	1.82	91	0.02	0.35				
	as fed	89.0	8.9	1.62	81	0.02	0.31	3.25	36.5	2.01	4.01
Cottonseed, meal,	dry	100	44.8	1.50	75	0.17	1.31				
41%	as fed	91.5	41.0	1.37	69	0.16	1.20	5.00	12.2	3.64	7.25
Fat (tallow)	dry	100		3.90	195						
	as fed	99.5		3.88	194			8.00		2.06	4.12
Molasses, cane	dry	100	4.3	1.82	91	1.19	0.11				
	as fed	75.0	3.2	1.36	68	0.89	0.08	2.50	78.1	1.84	3.68
Oat straw	dry	100	4.4	1.04	52	0.33	0.10				
	as fed	90.1	4.0	0.94	47	0.30	0.09	1.00	25.0	1.06	2.12
Urea	as fed		2.80					4.50	1.60		
Minerals											
Dicalcium phosphate						24.0	18.0	5.20			
Limestone						38.0		1.15			
Salt								2.50			

and at what cost? What effect will the processing have on animal production? Is it a palatable feed or will the mixture be palatable? Does it present problems in handling, mixing, or storage? Are feed additives required and, if so, what additives and at what concentrations?

Obviously, it takes a fund of knowledge and experience to answer these questions. The ability to answer them correctly is a must for a practicing nutritionist. For our purposes here, the beginner can ignore such questions as posed in the last paragraph as time and experience will help provide the answers which cannot be covered in this basic book.

Type of Ration

The type of ration has a great deal to do with its needed composition and nutrient content. That is, is it a complete feed, a finishing grain mix to feed along with roughage, or is it a supplemental feed formulated primarily for its protein, vitamin, or mineral content? If a complete feed, is it intended to be fed on a restricted or ad libitum basis? If we are dealing with herbivorous animals such as ruminants, we would normally first consider roughage as the base feed and then determine what nutrients are needed to supplement the roughage. Sometimes, however, roughage may be added only as a diluent to control intake or produce a desired physical texture for the ration.

Expected Feed Consumption

As explained in Ch. 17, rations should be designed so that animals will consume a desired amount because the required concentration of a nutrient in a ration depends on consumption. For example, if we want a steer to consume 500 g of protein/day, the feed needs to contain only 10% protein if the steer eats 5 kg of feed; if it will only eat 4 kg, then 12.5% protein is needed to achieve the desired protein intake. Energy concentration greatly affects feed intake (see Ch. 17) as do such factors as physical density, deficiency of some nutrients or presence of unpalatable ingredients. Pelleted hay, for example, will usually be consumed in much larger amounts by ruminants than long hay, thus the concentration of some nutrients can be reduced with pelleted hay.

Guidelines and Thumb Rules for Ration Formulation

Note in the appendix tables that feed composition data may be given either on the basis of % of dry matter or on an as-fed basis depending on which publication the information was taken from. Therefore, some recalculation may be required before ration formulation is commenced because rations should be formulated on a dry basis. If laboratory data are available on dry matter, then this should be used; if not, then use appropriate values from the appendix tables or other sources. For example, if the dry matter content is 90%, then multiply each nutrient % by 0.9 to get the correct concentration in the feed (see Table 20-1).

Formulation can be done on the basis of daily needs, although this is hardly ever done in practice. Rather, it is more common to formulate on the basis of a given weight unit — 100 lb, 1000 lb, 1 ton. Use of % units is the simplest means as the final values can easily be converted to any final weight unit.

If formulas are worked out to exact specifications, we may often come up with fractional units (g, lb, kg) which are undesirable in commercial feed mill usage. Where lb units are used to make up ton batches, these should be rounded off to at least 10 lb increments, except for additives included in the ration in very small amounts. As a rule, the errors caused by this procedure will be small.

A rule of thumb is that simple nutrient needs can be met adequately by simple feed formulas. Thus, when nutrient needs are simple, complex formulas do not guarantee improved performance. Also, the more complex the nutrient specifications, the more complex a formula is required to meet all specifications without having an excess of some nutrient(s). One individual feedstuff rarely will suffice to supply all needed nutrients without one or more being in excess. For example, if we wish to feed a roughage to cows to supply 12% crude protein, the feeding of alfalfa hay, which usually has 15% or more protein, will be wasteful of protein although it may have other nutrient properties of considerable value. Therefore, we could dilute the alfalfa with some other roughage which has a lower protein content, such as grass straw, without having excess protein consumption. In some instances it may be cheaper to use the feed with excess nutrients rather than dilute it with another ingredient.

Many commercial concentrate mixes frequently contain a wide variety of feedstuffs, even in situations where nutrient requirements are relatively simple. There are two reasons why they may be included. First, it is felt that a variety of feedstuffs in a mixture may be more palatable for heavily fed animals — perhaps this is true where the same mix is fed for a long period of time. The second reason is that a mixture of several energy or protein sources may provide some insurance against trace nutrient deficiencies. This may or may not be true, depending on the feed ingredients chosen for the mixture.

As seasonal changes occur in availability or cost of feeds, many times it is desirable to alter formulas, particularly for the feed grains and protein concentrates. Some substitution can take place with relatively little effect; for example, corn, sorghum and wheat have about the same energy value for most animals. For ruminants, barley and oats can be substituted more liberally than for monogastric species. If drastic changes are in order, then formulas should be recalculated. With respect to protein sources, most nutritionists suggest using protein for ruminants on the basis of least cost/unit of protein. Factors other than protein content may affect value (plant hormones, Se in linseed meal), but they are poorly quantified. Some meals, such as rapeseed, may not be very palatable, so some judgement and knowledge of the specific feedstuffs is required when liberal substitution is practiced. For monogastric species, the amino acid content of protein substitutes must be watched, as well as other factors such as gossypol content of cottonseed meal.

Attention should be called to a few other simple guidelines. Salt (NaCl) can be adequately supplied for herbivorous species by providing it ad libitum in a separate container from other feed. In complete feeds or concentrate mixtures, usually 0.25-0.5% salt is included for poultry and swine and 0.5-1% for ruminants. When fat is added, it is the usual practice to add not more than 5% for swine and poultry and more on the order of 2-3% in finishing rations for cattle and, sometimes, in dairy rations. Feeds with high levels of fat do not store well as they are apt to become rancid. With molasses or other similar liquid feeds, the amount commonly added in mixed feeds is usually restricted to 7-8% because of handling and mixing problems, although most species can utilize more than this amount.

Mathematics of Ration Formulation

A variety of different methods may be used to formulate rations, some more appropriate in some situations than in others. These generally include: (a) use of Pearson's square, (b) trial and error or 'cut and try' method, (c) use of weighted averages, and (d) linear programming with computers, which will be covered in a separate section.

Pearson's Square

Use of Pearson's square is a simple procedure illustrated below. Suppose we have some protein concentrate such as cottonseed meal (CSM) with 40% crude protein (CP) as fed and some grain with 10% CP as fed and we wish to have a mixture with 18% CP.

Using the square

```
CP % of CSM    40     8      parts of CSM
                 \   /
                  18
                desired
                 /   \
CP % of grain  10     22     parts of grain

                      30     total parts
```

Answer:

8 parts CSM and 22 parts grain
or % of CSM in mix = 100 x 8 ÷ 30 = 26.7%
% of grain = 100 x 22 ÷ 30 = 73.3%

Check for CP
26.7% CSM x 40% CP = 10.68%
73.3% grain x 10% CP = 7.33%
for a total of 18.01%

As shown in the illustration, we compare the CP % of each feed on the left with the desired % in the middle of the square, the lesser value is subtracted from the greater value and the answer, in parts of a mixture rather than in %, is recorded diagonally. We would usually want to calculate the % of the final mixture; although this is not necessary for this illustration, working with percentages is easier than with fractions. The same procedure can be used for energy, minerals, and so on, but keep in mind that one feed (or mixture) must have a value higher and the other must be lower than the desired solution. Furthermore, calories or ppm can be used as well as %. The Pearson square is not utilized nearly to the extent that it could be by many formulators. Further illustration of its use will be shown in conjunction with other problems in subsequent paragraphs.

As a matter of information, the check run on CP (Pearson square example) illustrates the use of a weighted average. Where we have two or more ration ingredients with different nutrient concentrations, the total is obtained by multiplying the amount of each ration ingredient by its nutrient concentration and summing the answers for all ingredients.

Simple Substitutions

Frequently, in simple rations the most convenient way to meet ration specifications is to make some simple substitutions. For example, with a ration as shown,

Original formula

Feeds	Amount	CP, %	CP, lb
Roughage	50	10	5.0
Barley	45	10	4.5
CSM	5	41	2.05
		Total	11.55

suppose we wish to increase the CP content to 14% by substituting CSM for barley. For each lb (or kg other unit) of CSM substituted, we add 0.41 lb of CP and take away 0.10 in barley, so the substitution value is 0.31 (0.41 - 0.10). Because we want to increase from 11.55 to 14% CP, we need 2.45 lb (14 ÷ 11.55) of additional protein. This can be added by supplying 7.9 lb (2.45 ÷ 0.31) of CSM. If we round to 8% added CSM, we now have 14.03% CP as shown above. If we try to do this by 'cut and try' methods, several calculations are apt to be required rather than just one. Note that a 4th feedstuff could have been introduced into the formula rather than substituting CSM for barley. For example, some urea could have been substituted for CSM or barley.

Revised formula

Feeds	Amount	CP, %	CP, lb
Roughage	50	10	5.0
Barley	37	10	3.7
CSM	13	41	5.33
		Total	14.03

Example of a Simple Problem

One of the simplier situations encountered is where one feedstuff is available and we want to determine how much is needed for the animal in question. Suppose some alfalfa hay (Table 20-1) is to be fed and we want to know how much a 500-kg dry pregnant beef cow needs. According to NRC (Appendix Table 6) the needs and expected consumption are as shown:

	Feed consumption, kg		Total protein			TDN		
	DM	Hay as fed	kg	% of DM	% as fed	kg	% of DM	% as fed
Cows' requirements	7.6	8.5*	0.44	5.9	5.2	3.8	50	44.7
Nutrient content of hay, %		89.2			15.2			50
Nutrient intake if cow consumes maximum DM, kg	8.5		1.29			4.25		
Excess intake, kg			0.85			0.45		
Hay required to fill nutrient needs, kg			2.89			7.6		

* 7.6 ÷ 89.2% = 8.5

From the information shown, the protein needs of the cow can be met by feeding only 2.89 kg of hay (0.44 ÷ 15.2% x 100) and her energy needs can be met by feeding 7.6 kg (3.8 ÷ 50% x 100), well within the expected consumption. If the cow is fed just enough for energy needs, she will still overconsume protein (7.6 - 2.89 x 15.2% or 0.72 kg excess); this overconsumption is not harmful, but it may be more costly than need be if other roughage is available for feeding.

Since oat straw is available (Table 20-1), we can calculate with the square what mixture of alfalfa and straw will meet her needs. Assuming that the cow will still eat 7.6 kg of DM and that the CP should be 5.2% as fed, we go through the square as shown:

Alfalfa hay 15.2 → 1.2 parts alfalfa 10.7%

5.2

Oat straw 4.0 → 10.0 parts straw 89.3%

Check for		CP
Alfalfa	15.2% x 10.7	1.63
Oat straw	4.0% x 89.3	3.57
Total		5.20

Check for		TDN
Alfalfa	50% x 10.7	5.35
Oat straw	47% x 89.3	41.97
Total		47.32

The cow needs to consume 8.03 kg (3.8 ÷ 47.3% x 100) of this mixture to meet her TDN needs; because it is below expected consumption, this combination should be satisfactory for most cows although in practice we might want to increase alfalfa to about 15% of the mixture to insure adequate protein intake.

Lamb Finishing Ration

We will now go a step farther and put together a simple ration for fattening lambs. From the feedstuffs in Table 20-1, corn and alfalfa will be used as previous experience shows that lambs do well on these two feeds. The nutrient concentration (NRC) and the amount of nutrients are shown above along with data on alfalfa and corn. The alfalfa will be fed in pelleted form. NRC does not give a value for pelleted alfalfa, but we know that lambs will readily consume more pellets than hay so

Table 20-2. Formulating a simple finishing ration for lambs weighing 35 kg (77 lb).

Line	Lamb requirements	DM	CP	DP	DE	Ca	P
1	Nutrient concentration		11%	6.7%	1.34 Mcal/lb	0.34%	0.21%
2	Quantity of nutrients	3.1 lb	0.34 lb	0.21 lb	4.15 Mcal	0.011 lb	0.0065 lb
3	Alfalfa hay pellets	89.2%	15.2%	12.1%	1.10*Mcal/lb	1.20%	0.20%
4	Corn, #2, cracked	89.0	8.9	6.7	1.62	0.02	0.31
	Try 50-50 mix of alfalfa						
5	pellets and corn, av.	89.1	12.05	9.4	1.36	0.61	0.26
6	Deficiency or excess		+1.05	+2.7	+0.02	+0.27	+0.05
7	Amount of mixture to exactly meet nutrient needs, lb	3.47	2.82	2.23	3.05	1.80	2.50

* Increased by 10% over value for alfalfa hay shown in Table 10-1.

increasing the DE value by 10%** is justified. When lamb needs (line 1) are compared to the nutrients furnished from alfalfa (line 3), it is apparent that alfalfa provides excess CP, DP, and Ca, but not enough DE and that P is borderline. Next, we arbitrarily (cut and try) choose a mixture of 50-50 alfalfa and corn and calculate the composition of the mix (line 5). The mix has an excess of protein and Ca, but DE and P are right on the mark. Line 7 shows how much feed would be required to exactly meet the requirement for each nutrient shown. The DM is well within the range suggested by NRC, so the lambs should easily eat this much. The ration is close enough for the various nutrients, but it could be cheapened, no doubt, by including some feedstuff with less protein. Had we worked the solution out with Pearson's square, a mixture of 33.3% alfalfa pellets and 66.7% corn would satisfy the CP and one of 53.8% alfalfa and 46.2% corn would suffice for DE. Cut and try methods worked out easily in this example, but they do not work well with more complex rations with a wide variety of feedstuffs, particularly where specifications need to be more exact.

Steer Finishing Ration

In this example we will go to a relatively complicated problem and will approach it more on the order of a commercial solution. The ration to be formulated will be a finishing ration for cattle in which the specifications call for exactly 12% CP and 74% TDN as fed and with at least 10% alfalfa hay, 5% molasses, and 2% tallow in a ration which should be adjusted, if required, for Ca and P and with some thought to cost of the various feed ingredients. With more complicated ration specifications such as would be encountered with swine or poultry (amino acids, vitamins), we might have a relatively long list of required ingredients.

First step. Calculate the amount of nutrients furnished by the required feedstuffs. At this point we will ignore Ca and P. We will also leave some space in the ration, designated as 'slack,' for manipulation of Ca and P supplements, salt, or other additives. In the total ration (100 lb units),

	Amount	CP, %*	TDN, %	CP, lb	TDN, lb
Alfalfa	10	15	50	1.50	5.0
Molasses	5	3	68	0.15	3.4
Tallow	2		194		3.9
Slack	1				
	18			1.65	12.3

* For simplicity CP values are rounded to nearest whole number

** Although digestibility is not likely to be increased, consumption is enough greater to provide at least a 10% increase in energy intake.

we need 12 lb of CP and 74 lb of TDN, so in the remaining 82 lb (100-alfalfa, molasses, tallow, slack), we need (2nd step):

CP 12 ÷ 1.65 = 10.35 lb   or   10.35 ÷ 82 x 100 = 12.62%

TDN 74 ÷ 12.3 = 61.7 lb   61.7 ÷ 82 x 100 = 75.24%

Third step. The easiest way to solve a problem of this type is to make up two mixtures using Pearson's square. We can solve first for either CP or TDN. If we choose CP, then make up one mixture with TDN > 75.24 and one < 75.24, both with 12.62% CP. For the low TDN mixture we will choose alfalfa and barley as they are moderately priced, and for the high TDN we will choose corn and urea because they are inexpensive sources of energy and CP, respectively.

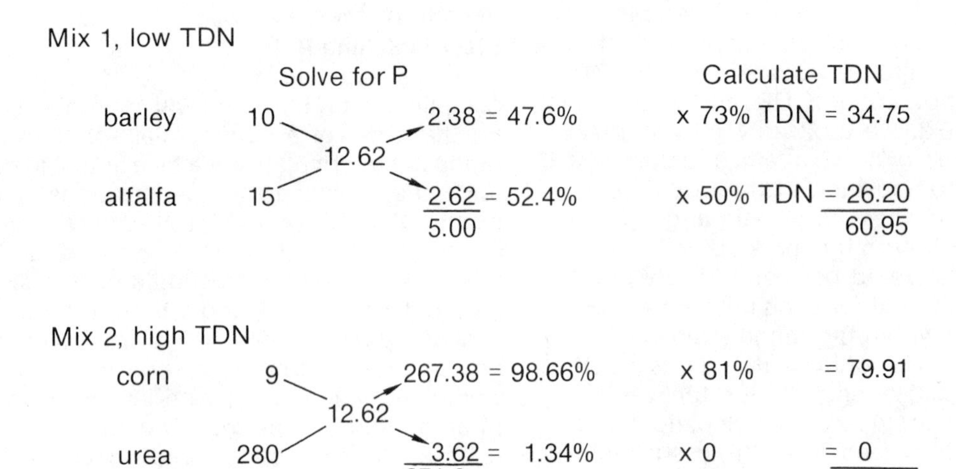

Step 4. Now, put these two mixes through the square and solve for TDN.

Mix 1	60.95		4.67 = 24.63%
		75.24	
Mix 2	79.91		14.29 = 75.37%
			18.96

These two mixtures are going to be used to provide 82% (or lb) of the final ration, so the final ration will contain 20.20 lb of mix 1 (24.63% x 82) and 61.80 lb of mix 2 (75.37% x 82). The amount of each feed ingredient in the two mixes is calculated as shown (step 5):

Mix 1
- barley 47.6% x 20.20 = 9.62 lb
- alfalfa 52.4% x 20.20 = 10.58 lb

Mix 2
- corn 98.66% x 61.80 = 60.97 lb
- urea 1.34% x 61.80 = 0.83 lb

Thus, the final mixture, except for adjustment in the slack, is as shown:

		Amount	CP, lb	TDN, lb	Ca, lb	P, lb	Cost, $
Alfalfa	10 required +						
	10.58 from Mix 1	20.58	3.10	10.32	0.2470	.0412	0.412
Molasses	5 required	5	0.15	3.40	.0445	.0040	0.125
Tallow	2 required	2		3.88			0.16
Barley	9.62 from Mix 1	9.62	0.96	7.07	.0058	.0385	0.289
Corn	60.97 from Mix 2	60.97	5.48	49.28	.0122	.1890	1.982
Urea	0.83 from Mix 2	0.83	2.32				0.037
					.3095	.2727	
Slack (1%) made up of*							
Limestone (38% Ca)		0.24			.0912		0.003
Salt		0.76					0.019
Totals		100.00	12.01	73.98	0.4007	0.2727	3.027

*See following discussion

The mixture comes out with a very slight overage on CP and a slight deficiency of TDN (rounding errors). Ca and P have been calculated in the same manner as CP and TDN and the subtotals for Ca and P are shown before addition of limestone (step 6). The P content is at a satisfactory level, but Ca is a bit low, particularly where fat is included in the ration. Therefore, the slack has been used to adjust Ca to 0.4% by addition of limestone (step 7). For 0.4% Ca in 100 lb we need 0.4 lb of mix. We have 0.309, so 0.091 lb (0.4 − 0.309) are needed. Limestone is the cheapest source available with 38% Ca. The amount of limestone needed is 0.24 lb (0.091 ÷ 0.38). The remainder of the slack space can be filled up with salt, antibiotic, vitamin A premix, or one of the feedstuffs already used. For simplicity we will add 0.76% salt as some salt should be included anyway. With the cost figures shown in Table 20-1, this ration comes to $3.03/cwt or $60.54/ton.

Formulating with NE_m and NE_g values

NRC has adopted (or suggested) the use of NE_m and NE_g values developed by California workers. Use of this system adds a slight complication in that two different energy values must be used, both values do not apply to a given unit of feed. As pointed out in Ch. 16 the animal does not use one feed only for maintenance and the other for gain. One example will be worked out for steer calves using kg weight units and ignoring nutrient requirements other than dry matter, CP, and energy.

Energy requirements for finishing steer calves of 400-kg size are shown for calves gaining 1.3 kg/day (Table 20-3). Animals of this weight should eat about 8.8 kg of DM/day. If the steers will eat 8.8 kg of DM, this is roughly equivalent to 9.9 kg of a ration as fed. Only a short list of feedstuffs will be used (as fed basis):

	DM, %	CP, %	NE_m	NE_g
			Mcal/kg	
Alfalfa hay	90	16.6	1.22	0.44
Barley	89	11.6	1.90	1.25
Corn	89	8.9	2.03	1.32

We will specify that the total ration (as fed) should contain between 15 and 20% alfalfa or between 1.485 and 1.980 kg (15 or 20% of 9.9). To avoid likely recalculation, we will start with about 18% of expected feed consumption, or 1.8 kg of alfalfa. This ration works out so that we do not need any added protein supplement such as urea or a high-energy feedstuff such as tallow, and it requires less than the expected feed and dry matter consumption to meet requirements. On a percentage basis, the ration is 18.55% alfalfa, 25.47% barley, and 55.98% corn.

Note that Pearson's square procedures could be used to solve this problem just as well as with previous examples, although we would need to use the square to solve the NE_m needs and then the NE_g part of the ration. Also, we could work out this solution in the manner shown by using

Table 20-3. Calculation of a ration for 400 kg finishing steers gaining 1.3 kg/day.

Step	Item	Feed kg	NE_m Mcal/kg		NE_g Mcal/kg	DM, kg	CP, kg
1	Animal requirements	9.9	6.89		7.17	8.8	0.98
2	From alfalfa	1.8	2.196	or*	0.792	1.620	0.299
3	NE_m		4.694				
4	Barley needed to fill NE_m requirement	2.471[a]	4.695			2.199	0.287
5	Corn required for NE_g	5.423[b]			7.170	4.834	0.483
		9.703	6.891		7.170	8.653	1.069

[a] = 4.694 ÷ 1.90, [b] 7.17 ÷ 1.32

* Both values cannot be used for the same unit (lb, kg) of feed.

a mixture of alfalfa and barley for maintenance needs and then continue with the corn, or we could simply use a complete mixture and calculate how much of it would be required for this level of production. If we use the latter approach the problem could be solved as shown, ignoring DM and CP. We will arbitrarily use 40% alfalfa hay.

	NE_m	NE_g
	Mcal	
Requirement for 400 kg steer gaining 1.3 kg/day	6.89	7.17
Energy in 1 kg of alfalfa	1.22	0.44
Energy in 1 kg of corn	2.03	1.32
Energy in 1 kg of mixture with 60% corn and 40% alfalfa	1.706	0.968

Amount needed for maintenance = 6.89 ÷ 1.706 = 4.039 kg
Amount needed for gain = 7.17 ÷ 0.968 = 7.407 kg
Total feed required = 4.039 + 7.407 = 11.446 kg

This example works out to require somewhat more feed than some steers might consume, but it clearly illustrates the problem of packaging the needed nutrients in the right sized package as compared to the previous example.

Use of Tables to Calculate Expected Gain

Tabular data on TDN requirements for gain of beef cattle are shown in Table 20-4 as estimated by Winchester's formula (3). These data are included because, in the author's (Church) experience, they give a better estimate of gain by finishing steers than do the NE_m and NE_g values shown in Appendix Table 9. The author's comparisons were made in a moderate climate, but the experimental data used to develop NE_m and NE_g values were collected in a much hotter climate, and the animals likely were subjected to enough heat stress to reduce the NE values of the feedstuffs utilized. Published NE_m and NE_g values for feedstuffs appear to work well in the warmer areas in the USA, but it may be questionable if they are applicable in cooler areas.

At any rate, these tables can be used to calculate expected gain when animals are consuming a known amount of a given ration and when the average weight of the animals on feed is known. For example, if we have a pen of steers averaging 900 lb and they are consuming 25 lb/day of a ration which should average 74% TDN, how much should they be gaining? TDN consumed/day = 25 x 74% = 18.5 lb of TDN. We now look in the line for 900 lb animals until we find the number closest to 18.5 (18.48), and then look up at the gain value at the top of the table. It is 3.2 lb/day in this example. By making these calculations and comparing them to known gain, it gives us a reasonably good idea if animals are performing up to

Table 20-4. Estimated TDN (total digestible nutrients) required by beef cattle gaining at varying rates.[a]

Ave. wt	Expected average daily gain																			
	.2	.4	.6	.8	1.0	1.2	1.4	1.6	1.8	2.0	2.2	2.4	2.6	2.8	3.0	3.2	3.4	3.6	3.8	4.0
300	2.88	3.28	3.68	4.08	4.48	4.88	5.28	5.68	6.08	6.48	6.88	7.28	7.68							
325	3.04	3.46	3.88	4.30	4.73	5.15	5.57	5.99	6.41	6.84	7.26	7.68	8.10							
350	3.20	3.64	4.08	4.52	4.97	5.41	5.85	6.30	6.74	7.18	7.63	8.07	8.50	8.96						
375	3.34	3.81	4.27	4.74	5.20	5.66	6.13	6.59	7.06	7.52	7.98	8.45	8.91	9.37						
400	3.49	3.98	4.46	4.94	5.43	5.91	6.40	6.88	7.37	7.85	8.34	8.82	9.30	9.79	10.27					
425	3.64	4.14	4.64	5.15	5.65	6.16	6.66	7.17	7.67	8.17	8.68	9.18	9.69	10.19	10.70					
450	3.78	4.30	4.83	5.35	5.87	6.40	6.92	7.45	7.97	8.49	9.02	9.54	10.06	10.59	11.11	11.64				
475	3.92	4.46	5.00	5.55	6.09	6.63	7.17	7.72	8.26	8.80	9.35	9.89	10.43	10.98	11.52	12.06				
500	4.05	4.61	5.17	5.74	6.30	6.86	7.42	7.98	8.54	9.11	9.67	10.23	10.79	11.35	11.91	12.48				
525	4.19	4.77	5.35	5.93	6.51	7.09	7.67	8.25	8.83	9.41	9.99	10.58	11.16	11.74	12.32	12.90	15.48			
550	4.33	4.93	5.52	6.13	6.73	7.33	7.92	8.53	9.13	9.73	10.33	10.93	11.53	12.13	12.73	13.33	13.93	14.53		
575	4.45	5.07	5.68	6.30	6.92	7.53	8.15	8.77	9.38	10.00	10.62	11.23	11.85	12.45	13.09	13.70	14.32	14.94		
600	4.58	5.21	5.85	6.48	7.12	7.75	8.38	9.02	9.65	10.29	10.92	11.56	12.19	12.83	13.46	14.10	14.73	15.37	16.00	
625	4.70	5.36	6.01	6.66	7.31	7.96	8.62	9.27	9.92	10.57	11.23	11.88	12.53	13.18	13.83	14.49	15.14	15.79	16.44	
650	4.83	5.50	6.17	6.84	7.51	8.18	8.85	9.52	10.19	10.85	11.52	12.19	12.86	13.53	14.20	14.87	15.54	16.21	16.88	17.55
675	4.95	5.64	6.32	7.01	7.70	8.38	9.07	9.76	10.44	11.13	11.82	12.50	13.19	13.88	14.56	15.25	15.94	16.62	17.31	18.00
700	5.07	5.78	6.48	7.18	7.89	8.59	9.29	10.00	10.70	11.40	12.11	12.81	13.51	14.22	14.92	15.62	16.33	17.03	17.73	18.44
725	5.19	5.91	6.63	7.35	8.07	8.79	9.51	10.23	10.95	11.67	12.39	13.11	13.83	14.56	15.28	16.00	16.72	17.44	18.16	18.88
750	5.31	6.05	6.78	7.52	8.26	8.99	9.73	10.47	11.20	11.94	12.68	13.41	14.15	14.89	15.62	16.36	17.10	17.83	18.57	19.31
775	5.43	6.18	6.93	7.69	8.44	9.19	9.95	10.70	11.45	12.20	12.96	13.71	14.46	15.22	15.97	16.72	17.47	18.23	18.98	19.73
800	5.54	6.31	7.08	7.85	8.62	9.39	10.16	10.93	11.70	12.47	13.23	14.00	14.77	15.54	16.31	17.08	17.85	18.62	19.39	20.15
825	5.66	6.44	7.23	8.01	8.80	9.58	10.37	11.15	11.94	12.72	13.51	14.29	15.08	15.86	16.65	17.43	18.22	19.00	19.79	20.57
850	5.77	6.58	7.38	8.18	8.98	9.78	10.58	11.38	12.18	12.98	13.78	14.58	15.38	16.19	16.97	17.79	18.59	19.39	20.19	20.99
875	5.89	6.70	7.52	8.34	9.15	9.97	10.78	11.60	12.42	13.23	14.05	14.87	15.68	16.50	17.31	18.18	18.95	19.76	20.58	21.40
900	6.00	6.83	7.66	8.49	9.33	10.16	10.99	11.82	12.65	13.49	14.32	15.15	15.98	16.81	17.65	18.48	19.31	20.14	20.97	21.80
925	6.11	6.96	7.80	8.65	9.50	10.35	11.19	12.04	12.89	13.73	14.58	15.43	16.28	17.12	17.97	18.92	19.66	20.51	21.36	22.21
950	6.22	7.08	7.94	8.81	9.67	10.53	11.39	12.25	13.12	13.98	14.84	15.70	16.57	17.43	18.29	19.15	20.01	20.88	21.74	22.60
975	6.33	7.20	8.08	8.96	9.84	10.71	11.59	12.47	13.35	14.22	15.10	15.98	16.86	17.73	18.61	19.50	20.38	21.23	22.12	23.00
1000	6.42	7.31	8.20	9.09	9.98	10.87	11.76	12.65	13.54	14.43	15.32	16.21	17.10	17.99	18.88	19.78	10.67	21.56	22.45	23.34
1025	6.55	7.46	8.36	9.27	10.18	11.09	12.00	12.90	13.81	14.72	15.63	16.54	17.44	18.35	19.26	20.17	21.08	21.98	22.89	23.80
1050	6.65	7.57	8.50	9.42	10.34	11.26	12.19	13.11	14.03	14.95	15.87	16.80	17.71	18.64	19.56	20.49	21.41	22.33	23.25	
1075	6.75	7.69	8.63	9.56	10.50	11.44	12.37	13.31	14.25	15.18	16.12	17.06	18.00	18.93	19.87	20.91	21.74	22.67	23.62	
1100	6.86	7.81	8.76	9.71	10.66	11.61	12.56	13.51	14.46	15.41	16.37	17.32	18.27	19.22	20.17	21.12	22.07	23.02		
1125	6.96	7.93	8.89	9.86	10.82	11.79	12.75	13.72	14.68	15.65	16.61	17.58	18.54	19.51	20.47	21.44	22.40	23.37		
1150	7.06	8.04	9.02	10.00	10.98	11.96	12.94	13.92	14.90	15.88	16.86	17.84	18.81	19.79	20.77	21.75	22.73			
1175	7.16	8.15	9.14	10.14	11.13	12.12	13.12	14.08	15.10	16.09	17.09	18.08	19.07	20.06	21.06	22.05	23.04			
1200	7.27	8.28	9.29	10.30	11.31	12.32	13.33	14.33	15.34	16.35	17.36	18.37	19.38	20.39	21.39	22.40				

[a]From Winchester (1953). TDN = $0.0553\ BW^{2/3}(1 + 0.805\ \text{gain})$, where gain is in pounds.

expectation. If gains are considerably less than expected, we might find that management of the animals is very poor, that they are unduly stressed by the environment, that disease is a problem, or that the ration is not what it should be. The same type of calculations serve as a useful check on factors other than nutritional input.

Linear Programming

Linear programming (LP) is a mathematical technique for maximizing or minimizing some function subject to constraints. In nutrition, it is nearly always used to formulate rations to a given set of specifications at the least possible cost, thus the alternate name of least-cost ration formulation. The mathematics are complex enough so that almost all LP is done with electronic computers (Fig. 20-1), and several different computer programs are available. LP can be used to most advantage when a very wide range of feedstuffs is available at price likely to be feasible, when prices change rapidly, or when ration specifications are quite complex. Often only a limited list of feed ingredients are economically feasible and, if ration specifications are not too complex, a perfectly adequate job can be done using the approach shown earlier with the steer finishing ration.

A thorough coverage of the subject cannot be justified in this book. For more detailed presentations, the reader is referred to Ch. 3 of Church et al (1) or Ch. 15 of Foley et al (2). Other detailed presentations are available for interested readers. In this chapter, two examples will be given for illustrative purposes.

With LP it is feasible to specify minimums, maximums, ranges, ratios, or exact amounts of each individual nutrient that needs to be considered; in addition, the same type of constraints may be used with individual feedstuffs. The greater the number of constraints, the higher the cost is likely to be, however, as reduced cost is the primary objective and each specification restricts the computer solution. Formulas can be produced on almost any basis — percentage, 100 lb units, tons, and so forth.

First Example — Steer Finishing Ration

This example is one of a rather simple set of specifications. The computer was given a list of feedstuffs with data only on DP, DE, RAC (readily available carbohydrates), and cost, and ration specifications were built around these items with specific instructions on the amount of molasses and fat that were required, with minimums on alfalfa hay and beet pulp, and a maximum on urea. The solution is shown. Note that several of the feedstuffs did not come into the solution. Had the specifications been more detailed, they likely would have been included, although the requirement for a high level of RAC at a relatively low energy level probably kept wheat out of the solution.

Second Example — Broiler Ration

The second LP example (shown below) is a formula devised by a computer for a broiler starter ration. In this particular example, a rather modest number of constraints have been

Item	Specifications	Solution, lb	Feed cost ¢/lb
Feedstuffs included			
Alfalfa hay pellets	≥100	182.910	2.75
Barley		1319.154	2.80
Beet pulp	≥100	265.133	3.45
Molasses	=150	150.000	1.95
Oats			2.80
Wheat			2.75
Fat	= 60	60.000	8.00
Cottonseed meal			5.75
Urea	≤ 5	2.803	4.30

Weight	1980 (20 lb slack for additives)
Cost, $	58.96
Digestible protein	140 (7% in a ton)
Digestible energy	3040 Mcal
Readily available carbohydrates	1040 (54% in a ton)

used. Limits were placed on the amount of only three feedstuffs (fish meal, phosphate and salt) and specific requirements are shown for a trace mineral premix. With the nutrients, there are requirements for crude protein, lysine, methionine + cystine, calcium and added phosphorus.

This number of constraints allows the computer more freedom than if additional requirements were used for such items as additional amino acids, other minerals, fiber, ash, and additional constraints on individual feedstuffs. Generally, most commercial LP programs use more constraints than have been used in this example. However, remember that additional constraints reduce the flexibility of

Figure 20-1. Computers, long used in many areas, are becoming more frequently utilized for ration formulation as indicated here. Photo by R.W. Henderson.

Item	Cost, $/cwt	Opportunity price		Restrictions		Solution, lb/ton
		Low	High	Min., lb/t	Max., lb/t	
Ingredients used						
Corn grain, gr.	5.63	5.27	7.65			1133.1
Soybean meal	11.25	9.50	11.60			595.7
Meat & bone scrap	10.55	9.77	10.83			129.1
Fat, poultry	16.75	10.06	17.86			76.1
Fish meal, anchovy	16.75	16.53		40	100	40.0
Defluorinated phos.	9.14			10		10.0
Salt	2.10			7	7	7.0
Vitamin premix	45.00			5	5	5.0
Methionine, DL	123.00	14.36	141.81			2.0
Trace mineral mix	7.15			2	2	2.0
Ingredients not used						
Wheat middlings	5.53	3.42				
Alfalfa meal	5.97	1.78				
Limestone	0.88					
Nutrient Specifications						
Weight, lb				2000	2000	2000
Metabolizable energy, lb				1450	1475	1450
Crude protein, %				24	25	24
Arginine, %						1.70
Lysine, %				1.25		1.40
Methionine & cystine, %				0.86		0.86
Crude fat, %						6.98
Crude fiber, %						2.18
Moisture, %						9.80
Ash, %						5.49
Calcium, %				0.90	1.00	1.00
Phosphorus, added, %				0.50	0.60	0.50
Phosphorus, total, %						0.72
Cost, $/cwt						8.490

[a] Courtesy of Leo Jensen, Univ. of Georgia

the computer and nearly always result in higher costs and more complex formulas.

Most current computer programs will give a printout showing opportunity prices as illustrated in the broiler ration. The printout may show a low range price and a high range price. Normally, if the cost of the particular feedstuff is higher than the low range (example wheat middlings), the computer will not select that feedstuff unless it is forced to. In some circumstances opportunity prices can be used as buying guides. However, they may change with only minor changes in the computer program and this fact makes it more difficult to utilize opportunity prices as buying guides.

References Cited

1. Church, D.C. et al. 1972. Digestive Physiology and Nutrition of Ruminants. Vol. 3 - Practical Nutrition. O & B Books, 1215 NW Kline Pl., Corvallis, Ore.
2. Foley, R.D., D.L. Bath, F.N. Dickinson and H.A. Tucker. 1972. Dairy Cattle: Principles, Practices, Problems, Profits. Lea & Febiger, Philadelphia, Pa.
3. Winchester, C.F. 1953. U.S.D.A. Tech. Bul. 1071.

APPENDICES

Appendix Table 1. Composition of feedstuffs commonly fed to ruminant animals and horses. Data primarily from NRC publications.

Feed class and ingredient name	NRC reference number	Typical dry matter	Crude protein	Crude fiber	Ca	P	Mg	Digest. protein	DE	ME	NE_m	NE_g	NE_{milk}	TDN	Digest. protein	DE	TDN
		Composition, dry basis						Digestible protein and available energy, dry basis — Ruminants							Horses		
		%	%	%	%	%	%	%	Mcal/kg	Mcal/kg	Mcal/kg	Mcal/kg	Mcal/kg	%	%	Mcal/kg	%
Forage and Roughage																	
Pasture Grasses																	
Bluegrass, Kentucky, immature	2-00-778	30.5	17.3	25.1	.56	.47	.18	12.6	3.18	2.60	1.59	1.02	1.77	72	12.2	2.44	62
Bluestem, immature	2-00-821	31.6	11.0	28.9	.63	.17	---	7.2	2.82	2.31	1.38	.79	---	64	5.6	---	----
Brome, immature	2-00-892	32.5	20.3	23.9	.59	.37	.18	15.1	3.00	2.46	1.52	.95	1.63	68			
	2-00-956	32.5	22.1	22.4	.62	.57	---	---	---	---	---	---	---	----	16.3	2.36	60
Fescue, meadow, immature	2-08-420	25.0	17.6	23.6	.76	.48	---	12.9	3.01	2.47	---	---	---	68	---	---	----
Orchard grass, immature	2-03-440	23.8	18.4	23.6	.58	.55	.31	13.5	2.87	2.35	1.41	.82	1.64	65	13.2	2.36	60
Ryegrass, Italian, fresh	2-04-073	23.8	15.5	22.5	.65	.41	.35	11.0	2.58	2.11	1.33	.72	1.52	62	---	---	----
Ryegrass, Italian, early bloom	2-04-071	35.3	5.8	30.1	---	---	---	2.8	2.58	2.12	---	---	---	58	2.5	---	----
Timothy, immature	2-04-901	26.1	15.7	22.7	.47	.43	.18	11.2	2.92	2.39	---	---	---	66	10.9	---	----
Wheatgrass, crested, immature	2-05-420	30.8	23.6	22.2	.46	.35	.28	18.0	3.26	2.68	1.65	1.08	---	74	17.5	2.36	60
Mixed grasses (English data)	------------																
Very leafy	------------	18	22.2	20.0	---	---	---	18.3		2.58	---	---	---	70	---	---	----
Leafy	------------	19	17.4	23.7	---	---	---	13.2		2.56	---	---	---	69	---	---	----
Early flowering	------------	21	14.3	25.7	---	---	---	10.0		2.51	---	---	---	68	---	---	----
Flowering	------------	23	10.4	27.0	---	---	---	7.0		2.36	---	---	---	64	---	---	----
Mature	------------	25	8.4	29.6	---	---	---	5.2		2.29	---	---	---	62	---	----	----
Green Legumes																	
Alfalfa, prebloom	2-00-181	20.9	21.2	23.6	2.26	.35	.25	15.9	2.74	2.25	---	---	---	62	14.3	2.36	60
Clover, ladino, fresh	2-01-383	18.0	14.1	24.7	1.27	.42	.48	19.5	3.31	2.71	---	---	---	75	18.5	---	----
Clover, red, early bloom	2-01-428	19.7	19.4	23.3	2.26	.38	---	13.9	3.05	2.50	1.56	.99	1.83	69	14.0	---	----
Clover, white, fresh	2-01-468	17.7	28.2	15.7	1.40	.51	.45	19.8	2.94	2.41	---	---	---	67	---	---	----
Lespedeza, annual, prebloom	2-08-453	25.0	16.4	32.0	1.12	.28	.24	11.8	2.54	2.08	---	---	---	58	11.5	---	----
Trefoil, birdsfoot, fresh	2-07-998	22.5	21.4	20.7	1.76	.20	---	16.1	3.33	2.73	---	---	---	75	15.7	----	----
Other Pasture Crops																	
Rape, fresh	2-03-867	16.9	17.6	14.7	1.47	.43	.06	12.9	3.37	2.76	---	---	---	76	12.5	---	----
Sudangrass, fresh	2-04-489	20.8	14.1	27.5	.49	.44	.35	9.9	3.04	2.49	---	---	---	69	9.5	---	----
Wheat, fresh	2-08-078	25.9	16.0	22.9	.31	.31	.30	11.7	3.50	2.87	---	---	---	79	11.1	---	----
Silages																	
Alfalfa, early bloom, <30% DM	3-08-149	28.3	18.6	28.9	1.40	.32	.36	11.2	2.34	1.92	1.12	.39	1.12	53	---	---	----
Alfalfa, early bloom, >50% DM	3-08-151	55.0	17.9	32.4	1.61	.38	---	10.7	2.29	1.83	1.10	.35	1.07	52	---	---	----
Alfalfa, wilted	3-00-221	36.2	17.8	30.2	1.40	.32	.33	11.9	2.56	2.08	1.31	.69	1.28	58	---	---	----
Alfalfa, molasses added	3-00-238	32.2	17.5	28.8	1.74	.31	.34	12.2	2.60	2.12	1.27	.63	1.31	59	---	---	----
Corn, well eared, <30% DM	3-08-154	27.9	8.4	26.3	.28	.21	.18	4.9	3.09	2.57	1.56	.99	1.70	70	---	---	----
Corn, cannery waste	3-02-837	22.5	9.3	23.9	---	---	---	5.2	3.23	2.65	---	---	---	72	---	---	----
Grass-legume	3-02-303	29.3	11.8	31.4	.78	.28	---	6.0	2.47	2.00	1.19	.50	1.21	56	---	---	----
Sorghum, grain variety	3-07-962	29.4	7.3	26.3	.25	.18	---	2.0	2.51	2.03	1.22	.57	1.24	57	---	---	----
Grass Hays																	
Bermudagrass, coastal, s-c	1-00-716	91.5	9.5	30.5	.46	.18	.17	5.1	2.36	1.89	1.14	.42	1.13	51	4.5	1.94	44
Bluegrass, Kentucky, s-c	1-00-776	88.9	10.2	30.0	.45	.30	.21	5.8	2.78	2.28	---	---	---	63	6.2	---	----
Brome, s-c	1-00-890	89.7	11.8	32.0	---	---	---	5.0	2.38	1.91	1.15	.44	1.14	54	6.2	1.91	44
Fescue, tall, s-c	1-01-912	88.5	10.5	31.2	.50	.36	.50	6.0	2.73	2.24	1.33	.72	1.41	62	5.3	1.96	49
Meadow, intermountain, s-c	1-03-181	93.5	9.1	30.1	.58	.16	.24	2.9	2.02	1.66	1.00	.11	---	46	4.2	1.63	40

Orchard grass, s-c	1-03-438	88.3	9.7	34.0	.45	.37	.32	5.8	2.51	2.03	1.22	.55	1.25	57	4.2	1.85	46
Prairie, midbloom, s-c	1-07-956	91.0	8.1	32.1	.34	.21	---	4.1	2.20	1.81	1.07	.28	---	50	---	---	---
Sudangrass, s-c	1-04-480	88.9	12.7	28.9	.56	.31	.40	5.5	2.60	2.12	1.26	.63	1.31	59	5.8	---	---
Timothy, s-c, mid bloom	1-04-883	88.4	8.5	33.5	.41	.19	.16	4.6	2.56	2.08	1.24	.59	1.28	59	3.6	1.94	49
Legume Hays																	
Alfalfa, s-c, immature	1-00-050	89.1	21.5	26.3	2.12	.30	.26	15.0	2.78	2.29	1.36	.76	1.46	63	---	---	---
Alfalfa, s-c, early bloom	1-00-059	90.0	18.4	29.8	1.25	.23	.30	12.7	2.51	2.03	1.20	.52	1.25	57	12.3	2.33	59
Alfalfa, s-c, mid-bloom	1-00-063	89.2	17.1	30.9	1.35	.22	.35	12.1	2.47	2.00	1.18	.49	1.21	56	11.0	2.25	57
Alfalfa, s-c, mature	1-00-071	91.2	13.6	37.5	---	---	---	9.5	2.42	1.95	1.17	.47	1.17	55	7.7	1.94	49
Clover, Alsike, s-c	1-01-313	87.9	14.7	29.4	1.31	.25	.45	9.3	2.65	2.16	1.29	.66	1.36	60	8.7	2.21	56
Clover, crimson, s-c	1-01-328	87.4	16.9	32.2	1.42	.18	.27	11.8	2.65	2.16	1.29	.66	1.36	60	10.9	2.18	55
Clover, ladino, s-c	1-01-378	91.2	23.0	19.2	1.38	.40	.50	14.5	2.69	2.20	1.31	.69	1.39	61	---	---	---
Clover, red, s-c	1-01-415	87.7	14.9	30.1	1.61	.22	.45	8.9	2.60	2.12	1.26	.62	1.31	59	8.3	2.16	55
Lespedeza, s-c, mid-bloom	1-02-511	93.0	15.7	30.7	1.19	.26	.27	7.1	2.20	1.75	1.07	.27	1.00	50	---	---	---
Trefoil, birdsfoot, s-c	1-05-044	91.2	15.6	29.6	1.75	.22	---	10.7	2.69	2.20	1.31	.69	1.39	61	---	---	---
Vetch, s-c	1-05-106	88.2	20.0	28.5	1.36	.34	.27	13.2	2.73	2.24	1.33	.73	1.41	62	---	---	---
Straw																	
Barley straw	1-00-498	88.2	4.1	42.4	.34	.09	.19	.5	2.16	1.71	1.05	.23	.97	49	.5	1.56	38
Oats straw	1-03-283	90.1	4.4	41.0	.33	.10	.18	1.4	2.29	1.83	1.11	.35	1.07	52	---	---	---
Ryegrass straw	1-04-059	89.1	4.3	36.9	---	---	---	.7	2.35	1.93	---	---	---	53	1.2	---	---
Wheat straw	1-05-175	90.1	3.6	41.5	.17	.08	.12	.4	2.03	1.59	.99	.10	.88	46	.2	1.45	35
Miscellaneous																	
Corn cobs, ground	1-02-782	90.4	2.8	35.8	.12	.04	.07	.0	2.07	1.62	1.06	.25	.90	47	.0	1.23	29
Corn stover, no ears, mature	1-02-776	87.2	5.9	37.1	.49	.09	---	2.2	2.60	2.12	1.21	.55	1.31	59	---	---	---
Cottonseed hulls	1-01-599	90.3	4.3	47.5	.16	.10	.14	.2	1.90	1.47	.94	.03	.77	43	---	---	---
Rice hulls	1-08-075	92.4	3.1	44.5	.09	.08	---	.2	.66	.30	---	---	---	11	---	---	---
Energy Feeds																	
Cereal grains and by-products w < 20% CP																	
Barley	4-00-530	89.0	13.0	5.6	.09	.47	.14	9.8	3.75	3.08	2.04	1.36	2.38	85	8.2	3.66	83
Barley, Pac. Coast	4-07-939	89.0	10.9	7.0	.07	.45	---	8.2	3.70	3.03	1.99	1.33	2.34	84	6.2	3.62	82
Corn, dent	4-02-931	89.0	10.0	2.2	.02	.35	---	7.5	3.92	3.21	2.18	1.43	2.52	89	5.3	4.01	91
Corn, flaked		89.0	11.0	1.7	---	---	---	10.6	4.14	3.39	2.39	1.53	2.71	94	---	---	---
Corn and cob meal	4-02-849	87.0	9.3	9.2	.05	.31	.17	4.6	3.70	3.03	1.99	1.33	2.34	84	4.6	3.44	78
Corn, hominy feed, 5% fat	4-02-887	90.6	11.8	5.5	.06	.58	.26	7.9	4.19	3.60	2.45	1.55	2.56	95	---	---	---
Oats	4-03-309	89.0	13.2	12.4	.11	.39	.19	9.9	3.35	2.82	1.73	1.14	1.90	76	8.3	3.09	70
Oats, Pac. Coast	4-07-999	91.2	9.9	12.1	.10	.36	---	7.4	3.40	2.86	1.76	1.16	1.94	77	5.2	3.09	70
Oat groats	4-03-331	91.0	18.4	3.3	.08	.47	.10	12.9	4.10	3.51	2.36	1.52	2.48	93	13.3	3.70	84
Rice, polished	4-03-942	89.0	8.2	.4	.03	.14	.02	3.5	3.70	3.14	2.00	1.33	2.18	84	---	---	---
Rice bran	4-03-928	91.0	14.8	12.1	.07	2.00	1.04	9.6	2.91	2.41	1.43	.85	1.55	66	9.9	2.87	65
Sorghum grain, milo	4-04-444	88.9	12.2	2.5	.03	.31	.15	7.0	3.54	2.90	1.85	1.23	2.05	80	7.6	3.53	80
Wheat	4-05-211	89.0	14.3	3.4	.06	.41	.18	11.2	3.88	3.31	2.15	1.42	2.32	88	9.4	3.88	88
Wheat, Pac, Coast	4-08-142	89.2	11.1	3.0	.14	.34	---	8.6	3.92	3.21	2.18	1.43	2.52	89	6.3	3.88	88
Wheat bran	4-05-190	89.0	18.0	11.2	.16	1.32	.62	14.0	3.09	2.57	1.53	.96	1.70	70	12.9	2.57	65
Wheat mill run	4-05-206	90.0	17.0	8.9	.10	1.13	.57	11.6	3.57	3.02	1.89	1.26	2.07	81	---	---	---
Liquid Feeds																	
Animal fat, feed grade		95.0*	---	---	---	---	---	---	8.41	6.90	---	---	6.10	191	---	---	---
Molasses, cane	4-04-696	75.0	4.3	---	1.19	.11	.47	2.4	4.01	3.43	2.27	1.48	2.42	91	.0	3.18	72
Molasses, citrus	4-01-241	65.0	10.9	---	2.01	.25	.22	5.6	3.40	2.86	1.97	1.32	1.94	77	---	---	---
Molasses, sugar beet	4-00-668	77.0	8.7	---	.21	.04	.30	5.0	3.88	3.31	2.15	1.42	2.32	88	5.0	3.92	89

* % total fatty matter.

Appendix Table 1. Composition of feedstuffs commonly fed to ruminant animals and horses — Continued.

Feed class and ingredient name	NRC reference number	Typical dry matter	Composition, dry basis: Crude protein	Crude fiber	Ca	P	Mg	Digestible protein and available energy, dry basis — Ruminants: Digest. protein	DE	ME	NE_m	NE_g	NE_{milk}	TDN	Horses: Digest. protein	DE	TDN
		%	%	%	%	%	%	%	Mcal/kg	Mcal/kg	Mcal/kg	Mcal/kg	Mcal/kg	%	%	Mcal/kg	%
Energy Feeds continued																	
Roots and Tubers																	
Beet, mangels	4-00-637	10.6	13.2	8.3	.19	.19	.19	9.8	3.44	2.90	1.80	1.20	1.97	78	---	---	---
Potatoes, fresh, cull	2-03-787	23.1	9.6	2.4	.05	.24	.14	5.3	3.53	2.90	1.95	1.30	2.15	80	6.0	---	---
Potato peelings	4-03-774	23.0	9.9	3.4	.14	.19	---	5.1	3.65	3.00	---	---	---	83	---	---	---
Potato waste, dried	4-03-775	88.6	8.7	6.9	.23	.24	---	4.0	3.93	3.22	---	---	---	89	5.2	---	---
Miscellaneous																	
Beet pulp, dried w molasses	4-00-672	92.0	9.9	17.4	.61	.11	.14	6.5	3.75	3.08	2.04	1.36	2.38	85	8.2	3.66	83
Citrus pulp, dried	4-01-237	90.0	7.3	14.4	2.18	.13	.18	3.9	3.40	2.86	1.97	1.32	1.94	77	2.9	3.09	70
Grass seed screenings		88.0	8.0	20.0	---	---	---	6.5	2.86	2.35	---	---	---	65	---	---	---
Protein Supplements w >20% CP																	
Oil seed meals																	
Coconut, solv-extd	5-01-573	92.0	23.1	16.3	.18	.66	---	18.7	3.26	2.73	1.66	1.08	1.83	74	---	---	---
Cottonseed, mech-extd	5-01-615	94.0	43.6	12.8	.17	1.28	.60	35.3	3.44	2.90	1.81	1.20	1.97	78	---	---	---
Cottonseed, solv-extd	5-01-621	91.5	44.8	13.1	.17	1.31	.61	36.3	3.31	2.78	1.69	1.11	1.87	75	38.4	3.31	75
Linseed, solv-extd	5-02-048	91.0	38.6	9.9	.44	.91	.66	34.0	3.35	2.82	1.73	1.14	1.91	76	32.5	3.35	76
Peanut, solv-extd	5-03-650	92.0	51.5	14.1	.22	.71	.04	46.4	3.40	2.86	1.76	1.16	1.94	77	44.7	3.40	77
Rape, solv-extd	5-03-871	90.3	43.6	15.3	.44	1.00	---	37.5	3.04	2.53	1.51	.94	1.66	69	---	---	---
Safflower, w hulls, solv-extd	5-04-110	91.8	23.3	35.2	.37	.92	---	18.7	2.42	1.95	1.17	.48	1.17	55	---	---	---
Safflower, wo hulls, solv-extd	5-07-959	90.5	49.1	9.4	.26	1.83	---	41.3	3.35	2.82	1.56	.99	1.90	76	---	---	---
Sesame, mech-extd	5-04-220	93.0	51.5	5.4	2.18	1.39	---	41.2	3.31	2.78	1.69	1.11	1.87	75	---	---	---
Soybean, solv-extd	5-04-604	89.0	51.5	6.7	.36	.75	.30	43.8	3.57	3.02	1.93	1.29	2.07	81	44.7	3.53	80
Sunflower, wo hulls, solv-extd	5-04-739	93.0	50.3	11.8	---	---	---	44.8	2.87	2.37	1.41	.83	1.53	65	---	---	---
Grain by-products																	
Brewer's dried grains	5-02-141	92.0	28.1	16.3	.29	.54	.15	20.8	2.91	2.41	1.42	.83	1.55	66	22.5	2.25	51
Corn distillers grains, dehy.	5-02-842	92.0	29.5	13.0	.10	.40	.07	23.1	3.70	3.14	1.99	1.33	2.18	84	23.8	3.70	84
Corn distillers solubles, dehy.	5-02-844	93.0	28.9	4.3	.38	1.47	.69	22.6	3.88	3.31	2.15	1.42	2.32	88	---	---	---
Corn gluten feed	5-02-903	90.0	28.1	8.9	.51	.86	.32	24.2	3.62	3.07	1.93	1.29	2.11	82	---	---	---
Corn gluten meal	5-02-900	91.0	47.1	4.4	.18	.44	.05	39.2	3.70	3.14	1.99	1.33	2.18	84	40.5	3.70	84
Wheat germ meal	5-05-218	90.0	29.1	3.3	.08	1.16	---	27.4	4.19	3.60	2.44	1.55	2.56	95	---	---	---
Legume Seeds																	
Alfalfa seed screenings	5-08-326	90.3	34.4	12.3	---	---	---	---	3.63	2.98	---	---	---	82	---	---	---
Bean seed, navy	5-00-623	90.0	25.4	4.7	.17	.63	---	22.4	3.66	3.10	1.96	1.31	2.14	83	---	---	---
Clover, red, seed screenings	5-08-005	90.5	31.2	11.3	---	---	---	---	3.45	2.83	---	---	---	83	---	---	---
Pea seeds, cull	5-08-480	91.6	24.0	6.4	.19	.35	---	---	3.65	3.00	---	---	---	83	---	---	---
Animal and Fish By-products																	
Fishmeal, menhaden	5-02-009	92.0	66.6	1.1	5.97	3.05	---	53.5	3.27	2.68	---	---	---	74	---	---	---
Meat meal	5-00-385	93.5	57.1	2.5	8.49	4.31	.29	52.0	3.35	2.82	1.73	1.14	1.90	76	---	---	---
Meat and bone meal	5-00-388	94.0	53.8	2.3	11.25	5.39	1.20	49.0	3.18	2.66	1.61	1.03	1.77	72	---	---	---

Milk and Milk By-products																	
Buttermilk, dried	5-01-160	93.0	34.2	---	1.43	1.02	.52	---	3.77	3.09	2.05	1.36	2.23	86	----	----	----
Milk, skim	5-01-170	9.6	28.5	---	1.26	1.03	.10	27.4	4.10	3.51	2.32	1.50	2.48	93	----	----	----
Milk, skim, dried	5-01-175	94.0	35.6	---	1.34	1.10	.12	32.4	3.62	2.97	2.07	1.37	---	82	----	3.79	86
Milk, whole, fresh	5-01-168	72.0	25.8	---	.93	.75	.07	24.8	5.73	5.03	4.59	2.01	3.76	130	----	----	----
Milk, whole, dried	5-01-167	93.7	26.5	---	.94	.74	---	---	5.05	4.14	---	---	---	114	----	----	----
Whey, dried	4-01-182	93.2	16.0	---	.98	.81	.14	10.7	2.66	2.18	1.29	.67	1.36	60	----	----	----
Non-Protein-Nitrogen																	
Biuret			250.0														
Diammonium phosphate			132.0														
Monoammonium phosphate			76.0														
Urea			281.0														
Miscellaneous																	
Yeast, dried Brewer's	7-05-527	93.0	47.9	3.2	.14	1.54	.25	44.1	3.44	2.90	1.77	1.17	1.97.	78	41.3	3.09	70
Yeast, dried Torula	7-05-534	93.0	51.9	2.2	.61	1.81	.14	47.2	3.53	2.98	1.86	1.24	2.05	80	----	----	----

Appendix Table 2. Composition of feedstuffs commonly fed to poultry and swine. Data from NRC publications.

Feed class and ingredient name	NRC reference number	Composition, as fed					As fed basis				
							Poultry	Swine			
		Dry matter	Crude protein	Crude fiber	Ca	P	ME	Digest. protein	DE	ME	TDN
		%					Kcal/kg	%	Kcal/kg		%
Roughage											
Alfalfa, dehy, mn 15% CP	1-00-022	93.1	15.2	26.4	1.23	.22	1587	7.0	1436	1331	32
Alfalfa, dehy, mn 17% CP	1-00-023	93.0	17.9	24.3	1.33	.24	1653	8.3	1435	1322	32
Alfalfa, dehy, mn 20% CP	1-00-024	93.1	20.6	20.2	1.52	.27	1720	12.6	2217	2029	50
Alfalfa, dehy, mn 22% CP	1-00-851	92.9	22.5	18.5	1.48	.28	1764	13.7	2253	2052	51
Alfalfal hay, s-c, gnd	1-00-111	92.2	16.7	25.8	---	---	----	7.7	1382	1276	31
Alfalfa leaf meal	1-00-246	88.8	21.3	14.6	2.11	.26	1580	13.0	2192	2000	50
Pasture grass, closely grazed		20.0	5.2	3.4	---	---	----	3.5	517	----	12
Energy Sources (< 20% CP)											
Animal fat		99.5	---	---	---	---	7900	---	8790	7900	199
Barley grain	4-00-530	89.0	11.6	5.0	.08	.42	2646	8.2	3080	2876	70
Buckwheat grain	4-00-994	88.0	11.1	9.0	.11	.33	2712	8.0	3026	2829	69
Corn germ meal	5-02-898	93.0	18.0	12.0	.10	.40	1700	---	----	----	--
Corn grain	4-02-935	86.0	8.8	2.0	.03	.27	3417	7.0	3488	3275	79
Corn hominy feed	4-02-887	90.6	10.7	5.0	.05	.53	2866	8.5	3595	3365	82
Millet grain	4-03-098	90.0	12.0	8.0	.05	.28	----	8.8	2897	2703	66
Molasses, beet	4-00-668	77.0	6.7	---	.16	.03	1962	---	----	----	--
Molasses, cane	4-04-696	75.0	3.2	---	.89	.08	1962	---	2464	2343	56
Oats, grain	4-03-309	89.0	11.8	11.0	.10	.35	2535	9.9	2860	2668	65
Oats, groats	4-03-331	91.0	16.7	3.0	.07	.43	3549	14.0	3250	2999	74
Potatoes, cooked	4-03-784	22.5	2.2	.7	.01	.05	----	1.6	863	811	20
Potato meal	4-07-850	90.3	5.9	1.4	.07	.20	3527	5.0	3345	3168	76
Rice bran	4-03-928	91.0	13.5	11.0	.06	1.82	1630	10.2	3256	3028	74
Rice grain w hulls, grnd	4-03-938	89.0	7.3	9.0	.04	.26	2668	5.5	2511	2367	57
Rye grain	4-04-047	89.0	11.9	2.0	.06	.34	2888	9.6	3300	3079	75
Sorghum grain, milo	4-04-444	89.0	11.0	2.0	.04	.29	3250	7.8	3453	3229	78
Wheat bran	4-05-190	89.0	16.0	10.0	.14	1.17	1146	12.2	2512	2321	57
Wheat grain	4-05-211	89.0	12.7	3.0	.05	.36	3071	11.7	3520	3277	80
Wheat middlings	4-05-203	89.0	18.0	2.0	.08	.52	2756	16.0	3212	2952	73
Wheat mill run	4-05-206	90.0	15.3	8.0	.09	1.02	1764	12.2	3168	2934	72
Wheat shorts	4-05-201	90.0	18.4	5.0	.11	.76	2646	15.4	3168	2912	72
Whey, dried	4-01-182	94.0	13.8	---	.87	.79	1852	12.6	3432	3191	78

Plant Protein Sources (> 20% CP)											
Barley malt sprounts	5-00-545	93.0	26.2	14.0	.22	.73	1411	20.7	1558	1406	35
Brewers dried grains	5-02-141	92.0	25.9	15.0	.27	.50	2513	20.4	1892	1708	43
Coconut meal, solv. extd	5-01-573	92.0	21.3	15.0	.17	.61	1540	15.5	3123	2852	71
Corn distillers grains											
w solubles, dehy.	5-02-843	92.0	27.4	9.0	.09	.37	2425	---	----	----	--
Corn dist. sol., dehy	5-02-844	93.0	26.9	4.0	.35	1.37	2932	16.1	3300	2976	75
Corn gluten meal	5-02-900	91.0	42.9	4.0	.16	.40	3307	---	----	----	--
Cottonseed meal,											
pre-press solv. extd.	5-07-874	92.5	50.0	8.5	.16	1.01	2150	45.0	3018	2569	68
Pea seed, grnd	5-03-598	91.0	22.5	9.0	.17	.50	2601	19.3	3531	3213	80
Peanut meal, solv. extd.	5-04-650	92.0	47.4	13.0	.20	.65	2205	44.5	3408	2920	77
Rapeseed meal, solv. extd.	5-03-871	90.3	39.4	13.8	.40	.90	----	32.3	2747	2396	62
Soybean meal, solv. extd.	5-04-604	89.0	45.8	6.0	.32	.67	2249	41.7	3300	2825	75
Soybean meal, dehulled,											
solv. extd.	5-04-612	89.8	50.9	2.8	.26	.62	2425	46.3	3405	2881	77
Sunflower meal, solv. extd.	5-04-739	93.0	46.8	11.0	.40	1.00	1760	42.1	3034	2604	69
Wheat germ meal	5-05-218	90.0	26.2	3.0	.07	1.04	3086	23.6	3770	3397	86
Yeast, Brewers dried	7-05-527	93.0	44.6	3.0	.13	1.43	2425	39.2	3076	2654	70
Animal and Fish Protein Sources											
Blood meal	5-00-380	91.0	79.9	1.0	.28	.22	2844	62.3	2684	2101	61
Blood flour	5-00-381	91.0	82.2	1.0	.45	.37	----	64.1	2608	2029	59
Buttermilk, dried	5-01-160	93.0	32.0	---	1.34	.94	2756	29.8	3388	3015	77
Casein, dried	5-01-162	90.0	81.8	---	.61	.99	4120	76.0	3532	2740	80
Fish meal, anchovy	5-02-985	93.0	66.0	1.0	4.50	2.85	2900	60.7	2994	2446	68
Fish meal, herring	5-02-000	92.0	70.6	1.0	2.94	2.20	2976	66.3	3650	2938	83
Fish meal, menhaden	5-02-009	92.0	61.3	1.0	5.49	2.81	2866	56.4	3123	2580	71
Fish solubles, dried	5-01-971	92.0	62.8	1.0	---	---	2866	60.3	3408	2801	77
Liver meal	5-00-389	92.6	66.5	1.3	.50	1.25	----	64.4	3920	3195	89
Meat meal	5-00-385	93.5	53.4	2.4	7.94	4.03	1984	47.5	3010	2543	68
Meat meal tankage	5-00-386	92.0	59.8	2.0	5.94	3.17	2646	37.1	2475	2052	56
Meat and bone meal	5-00-388	94.0	50.6	2.2	10.57	5.07	1984	45.0	2859	2434	65
Milk, dried skim	5-01-175	94.0	33.5	---	1.26	1.03	2513	32.8	3784	3360	86

Appendix Table 3. Amino acid composition of selected feedstuffs.

	Crude protein	Arginine	Cystine	Glycine	Histidine	Isoleucine	Leucine	Lysine	Methionine	Phenylalanine	Threonine	Tryptophan	Tyrosine	Valine
	Amino acids, as fed basis													
	%													
Forage - Roughage														
Alfalfa, dehy, 15% CP	15.2	.60	.17	.70	.30	.68	1.10	.60	.20	.80	.60	.40	.40	.70
Alfalfa, dehy, 20% CP	20.6	.90	---	1.00	.40	.80	1.50	.90	.30	1.10	.90	.50	.70	1.19
Alfalfa leaf meal, s-c	21.3	.90	.34	.90	.33	.90	1.25	.95	.30	.80	.70	.25	.60	.90
Grass, dehy	14.8	.99	.19	.72	.46	1.38	1.98	1.06	.31	1.30	.89	.31	[illegible]	1.57
Energy Feeds														
Barley grain	11.6	.53	.18	.36	.27	.53	.80	.53	.18	.62	.36	.18	.36	.62
Corn hominy feed	10.7	.50	.18	.50	.20	.40	.80	.40	.18	.30	.40	.10	.50	.50
Corn germ meal	18.0	1.20	.32	---	---	---	1.70	.90	.35	.80	.90	.30	1.50	1.30
Corn grain	8.8	.50	.09	.43	.20	.40	1.10	.20	.17	.50	.40	.10	---	.40
Millet grain	12.0	.35	.08	---	.23	1.23	.49	.25	.30	.59	.44	.17	---	.62
Oats grain	11.8	.71	.18	---	.18	.53	.89	.36	.18	.62	.36	.18	.53	.62
Potato meal	8.2	.43	---	---	.11	.48	.30	.47	.07	.29	.21	.15	---	.39
Rice grain w hulls	7.3	.53	.10	.80	.09	.27	.53	.27	.17	.27	.18	.10	.60	.51
Rye grain	11.9	.53	.18	---	.27	.53	.71	.45	.18	.62	.36	.09	.27	.62
Sorghum grain, milo	11.0	.36	.18	.40	.27	.53	1.42	.27	.09	.45	.27	.09	.36	.53
Wheat grain	12.7	.71	.18	.89	.27	.53	.89	.45	.18	.62	.36	.18	.45	.53
Wheat shorts	18.4	.95	.20	.40	.32	.70	1.20	.70	.18	.70	.50	.20	.40	.77
Whey, dried	13.8	.40	.30	.30	.20	.90	1.40	1.10	.20	.40	.80	.20	.30	.70
Plant Protein Sources														
Brewers dried grains	25.9	1.30	---	---	.50	1.50	2.30	.90	.40	1.30	.90	.40	1.20	1.60
Corn dist. sol., dehy	26.9	1.00	.60	1.10	.70	1.50	2.10	.90	.60	1.50	1.00	.20	.70	1.50
Corn gluten meal	42.9	1.40	.60	1.50	1.00	2.30	7.60	.80	1.00	2.90	1.40	.20	1.00	2.20
Cottonseed meal, solv.	50.0	4.75	1.00	2.35	1.25	1.85	2.80	2.10	.80	2.75	1.70	.70	.80	2.05
Peanut meal, solv.	47.4	4.69	---	---	1.00	2.00	3.10	1.30	.60	2.30	1.40	.50	---	2.20
Rapeseed meal, solv.	39.4	2.16	---	1.88	1.05	1.43	2.63	2.09	.76	1.49	1.65	.48	.83	1.90
Soybean meal, solv.	45.8	3.20	.67	2.10	1.10	2.50	3.40	2.90	.60	2.20	1.70	.60	1.40	2.40
Sunflower meal, solv.	46.8	3.50	.70	2.70	1.10	2.10	2.60	1.70	1.50	2.20	1.50	.50	---	2.30
Yeast, Brewers dried	44.6	2.20	.50	1.70	1.10	2.10	3.20	3.00	.70	1.80	2.10	.50	1.50	2.30

Animal and Fish Protein Sources														
Blood meal	79.9	3.50	1.40	3.40	4.20	1.00	10.30	6.90	.90	6.10	3.70	1.10	1.80	6.50
Buttermilk, dried	32.0	1.10	.40	.60	.90	2.70	3.40	2.40	.70	1.50	1.60	.50	1.00	2.80
Casein, dried	81.8	3.40	.30	1.50	2.50	5.70	8.60	7.00	2.70	4.60	3.80	1.00	4.70	6.80
Fish meal, anchovy	66.0	4.46	1.00	5.10	1.84	3.40	7.01	5.40	2.19	2.48	3.04	.80	1.77	3.54
Fish meal, herring	70.6	4.00	1.60	5.00	1.30	3.20	5.10	7.30	2.00	2.60	2.60	.90	2.10	3.20
Fish meal, menhaden	61.3	4.00	.94	4.40	1.60	4.10	5.00	5.30	1.80	2.70	2.90	.60	1.60	3.60
Liver meal	66.5	4.10	.90	5.60	1.50	3.40	5.40	4.80	1.30	2.90	2.60	.60	1.70	4.20
Meat meal	53.4	3.70	.60	2.20	1.10	1.90	3.50	3.80	.80	1.90	1.80	.30	.90	2.60
Meat and bone meal	50.6	4.00	.60	6.60	.90	1.70	3.10	3.50	.70	1.80	1.80	.20	.80	2.40
Meat meal tankage	59.8	3.60	---	---	1.90	1.90	5.10	4.00	.80	2.70	2.40	.70	----	4.20
Milk, dried skim	33.5	1.20	.50	.20	.90	2.30	3.30	2.80	.80	1.50	1.40	.40	1.30	2.20

Appendix Table 4. Vitamin content of selected feedstuffs, fresh basis.

	Carotene	Vitamin E	Choline	Niacin	Pantothenic acid	Riboflavin	Thiamin	Vitamin B_6	Vitamin B_{12}
	ppm								
Plant sources									
Alfalfa, dehy., 15% CP	102	98	1550	42	21	11	3.0	6.5	---
Alfalfa leaf meal, s-c	62	--	1600	55	33	15	--	11	---
Barley grain	--	11	1030	57	6.5	2.0	5.1	2.9	---
Brewers dried grains	--	--	1587	43	8.6	1.5	.7	.7	---
Corn dist. sol., dehy.	1	55	4818	115	21	17	6.8	10	---
Corn grain	4	22	537	23	5	1.1	4.0	7.2	---
Cottonseed meal, solv., 41% CP	--	15	2860	40	14	5.0	6.5	6.4	---
Oats grain	--	36	1073	16	13	1.6	6.2	1.2	---
Peanut meal, solv.	--	3	2000	170	53	11	7.3	10	---
Rice grain w hulls	--	14	800	30	3.3	1.1	2.8	---	---
Rye grain	--	15	---	1.2	6.9	1.6	3.9	---	---
Sorghum grain, milo	--	12	678	43	11	1.2	3.9	4.1	---
Soybean meal, solv., 45% CP	--	3	2743	27	14	3.3	6.6	8.0	---
Wheat grain	--	34	830	57	12	1.2	4.9	---	---
Wheat middlings	--	58	1100	53	14	1.5	19	11	---
Yeast, brewers dried	--	--	3885	447	110	35	92	43	---
Animal sources									
Buttermilk, dried	--	6	1808	9	30	31	3.5	2.4	.02
Fish meal, herring	--	27	4004	89	11	9.0	---	3.7	219
Fish meal, menhaden	--	9	3080	56	9	4.8	.7	---	.1
Meat meal	--	1	1955	57	4.8	5.3	.2	3.0	51
Meat and bone meal	--	1	2189	48	3.7	4.4	1.1	2.5	45
Liver meal	--	--	----	204	45	46	.2	---	501
Milk, cow's, dried skim	--	9	1426	11	34	20	3.5	3.9	42
Whey, dried	--	--	20	11	48	30	3.7	2.5	.03

Appendix Table 5 Mineral composition of selected salts and mineral supplements used in livestock rations, as fed basis, feed grade purity.+

Name	Basic chemical formula	Elemental composition, % feed grade products										Acid-base reaction
		Ca	Cl	F	K	Mg	Mn	N	Na	P	S	
Ammonium chloride	NH_4Cl		66.28					26.18				Acid pH 4.5-5.5
Ammonium phosphate												
monobasic	$(NH_4)H_2PO_4$			.0025*				12.18		26.93		Acid pH 4.5
dibasic	$(NH_4)_2HPO_4$			.0025*				21.21		23.48		Basic pH 8.0
Animal bone charcoal		27.1				.5		1.4		12.7		
Animal bone meal, steamed		29.0				.6	trace	1.9	.5	13.6		
Calcium												
Clacium carbonate	$CaCO_3$	40.04										Basic
Limestone	$CaCO_3$	35.8				2.0	trace			trace		Basic
Limestone, dolomitic	$CaCO_3$	22.3			.4	10.0	trace					Basic
Oyster shell	$CaCO_3$	38.0			.1	.3	trace		.21	.07		Basic
Calcium phosphates												
Monocalcium phosphate	$CaH_4(PO_4)_2$	15.9		.0025*						24.6		Acid pH 3.9
Dicalcium phosphate	$CaHPO_4$	23.35		.02						18.21		Acid pH 5.5
Triclacium phosphate	$Ca_3(PO_4)_2$, CaO	38.76		.18*						19.97		sl. Acid pH 6.9
Defluorinated rock phosphates		32.0		.18*	.09				4.0	18.0		
Gypsum	$CaSO_4$	29.44									23.55	
Magnesium carbonate	$MgCO_3$	.02				25.2						sl. Basic
Magnesium oxide	MgO					60.3			.5			sl. Basic
Manganese sulfate	$MnSO_4 \cdot H_2O$ MnO						28.7				14.7	
Potassium chloride	KCl		47.6		52.4							Neutral pH 7.00
Sodium chloride	NaCl		60.66						39.34			Neutral
Sodium phosphate												
Monobasic	$NaH_2PO_4 \cdot H_2O$			.0025*					16.6	22.40		Acid pH 4.4
Dibasic	Na_2HPO_4			.0025*					32.39	21.80		Basic pH 9.0
Tripoly	$Na_5P_3O_{10}$			.0025*					24.94	30.85		Basic pH 9.9

+Elemental content varies according to source; use guaranteed analysis where available.

*Maximum.

Appendix Table 6. Nutrient requirement of beef cattle as suggested by NRC (1970).

Body wt.		Average daily gain		Daily dry matter/animal[a]		Nutrient concentration in ration dry matter							
						Total protein, %	Digestible protein, %	Energy: ME,[b] Mcal/kg	Energy: TDN,[c] %	Ca, %	P, %	Carotene, mg/kg	Vitamin A, IU x 10^3/kg
kg	lb	kg	lb	kg	lb								
Finishing Steer Calves													
150	331	0.90	1.98	3.5	7.7	12.8	8.6	2.82	78	0.60	0.43	5.5	2.2
200	441	1.00	2.20	5.0	11.0	12.2	8.1	2.67	74	0.46	0.34	5.5	2.2
300	661	1.10	2.43	7.1	15.7	12.2	8.1	2.67	74	0.37	0.27	5.5	2.2
400	882	1.10	2.43	8.8	19.4	11.1	7.1	2.67	74	0.28	0.23	5.5	2.2
450	992	1.05	2.31	9.4	20.7	11.1	7.1	2.67	74	0.22	0.22	5.5	2.2
Finishing Yearling Steers													
250	551	1.30	2.87	7.2	15.9	11.1	7.1	2.61	72	0.40	0.28	5.5	2.2
300	661	1.30	2.87	8.3	18.3	11.1	7.1	2.61	72	0.35	0.25	5.5	2.2
400	882	1.30	2.87	10.3	22.7	11.1	7.1	2.61	72	0.27	0.22	5.5	2.2
500	1,102	1.20	2.65	11.5	25.4	11.1	7.1	2.61	72	0.23	0.22	5.5	2.2
Finishing Two-Year-Old Steers													
350	772	1.40	3.09	10.3	22.7	11.1	7.1	2.56	71	0.29	0.22	5.5	2.2
400	882	1.40	3.09	11.3	24.9	11.1	7.1	2.56	71	0.27	0.22	5.5	2.2
500	1,102	1.40	3.09	13.4	29.5	11.1	7.1	2.56	71	0.22	0.22	5.5	2.2
550	1,213	1.30	2.87	13.7	30.2	11.1	7.1	2.56	71	0.22	0.22	5.5	2.2
Finishing Heifer Calves													
150	331	0.80	1.76	3.5	7.7	12.8	8.6	2.82	78	0.51	0.37	5.5	2.2
200	441	0.90	1.98	5.0	11.0	12.2	8.1	2.67	74	0.42	0.30	5.5	2.2
300	661	1.00	2.20	7.3	16.1	12.2	8.1	2.67	74	0.31	0.25	5.5	2.2
400	882	0.95	2.09	8.7	19.2	11.1	7.1	2.67	74	0.26	0.22	5.5	2.2
Finishing Yearling Heifers													
250	551	1.20	2.65	7.6	16.8	11.1	7.1	2.61	72	0.36	0.26	5.5	2.2
300	661	1.20	2.65	8.6	19.0	11.1	7.1	2.61	72	0.31	0.23	5.5	2.2
400	882	1.20	2.65	10.7	23.6	11.1	7.1	2.61	72	0.28	0.22	5.5	2.2
450	992	1.10	2.43	11.0	24.3	11.1	7.1	2.61	72	0.22	0.22	5.5	2.2
Growing Steers													
150	331	0.00	0.00	2.7	6.0	7.8	4.2	2.06	57	0.19	0.19	5.5	2.2
		0.25	0.55	3.1	6.8	11.1	7.1	2.28	63	0.26	0.23	5.5	2.2
		0.50	1.10	3.2	7.1	12.2	8.1	2.61	72	0.38	0.31	5.5	2.2
		0.75	1.65	3.2	7.1	13.3	9.0	2.82	78	0.53	0.41	5.5	2.2
200	441	0.00	0.00	3.3	7.3	7.8	4.2	2.06	57	0.18	0.18	5.5	2.2
		0.25	0.55	4.5	9.9	10.0	6.1	2.06	57	0.18	0.18	5.5	2.2
		0.50	1.10	4.9	10.8	11.1	7.1	2.28	63	0.27	0.20	5.5	2.2
		0.75	1.65	5.0	11.0	11.1	7.1	2.50	69	0.36	0.28	5.5	2.2
300	661	0.00	0.00	4.5	9.9	7.8	4.2	2.06	57	0.18	0.18	5.5	2.2
		0.25	0.55	6.1	13.4	8.9	5.2	2.06	57	0.18	0.18	5.5	2.2
		0.50	1.10	7.7	17.0	10.0	6.1	2.06	57	0.18	0.18	5.5	2.2
		0.75	1.65	8.0	17.6	11.1	7.1	2.28	63	0.21	0.18	5.5	2.2
400	882	0.00	0.00	5.6	12.3	7.8	4.2	2.06	57	0.18	0.18	5.5	2.2
		0.25	0.55	7.7	17.0	8.3	4.6	2.06	57	0.18	0.18	5.5	2.2
		0.50	1.10	9.7	21.4	8.9	5.2	2.06	57	0.18	0.18	5.5	2.2
		0.75	1.65	9.9	21.8	8.9	5.2	2.28	63	0.18	0.18	5.5	2.2

Growing Heifers													
150	331	0.00	0.00	2.7	6.0	7.8	4.2	2.06	57	0.19	0.19	5.5	2.2
		0.25	0.55	3.2	7.1	11.1	7.1	2.28	63	0.25	0.22	5.5	2.2
		0.50	1.10	3.2	7.1	12.2	8.1	2.61	72	0.38	0.31	5.5	2.2
		0.75	1.65	3.3	7.3	13.3	9.0	2.82	78	0.52	0.39	5.5	2.2
200	441	0.00	0.00	3.3	7.3	7.8	4.2	2.06	57	0.18	0.18	5.5	2.2
		0.25	0.55	4.6	10.1	10.0	6.1	2.06	57	0.18	0.18	5.5	2.2
		0.50	1.10	5.0	11.0	11.1	7.1	2.28	63	0.26	0.20	5.5	2.2
		0.75	1.65	5.4	11.9	11.1	7.1	2.50	69	0.33	0.26	5.5	2.2
300	661	0.00	0.00	4.5	9.9	7.8	4.2	2.06	57	0.18	0.18	5.5	2.2
		0.25	0.55	6.2	13.7	8.9	5.2	2.06	57	0.18	0.18	5.5	2.2
		0.50	1.10	8.2	18.1	10.0	6.1	2.06	57	0.18	0.18	5.5	2.2
		0.75	1.65	8.6	19.0	11.1	7.1	2.28	63	0.20	0.18	5.5	2.2
400	882	0.00	0.00	5.6	12.3	7.8	4.2	2.06	57	0.18	0.18	5.5	2.2
		0.25	0.55	7.7	17.0	8.3	4.6	2.06	57	0.18	0.18	5.5	2.2
		0.50	1.10	10.2	22.5	8.9	5.2	2.06	57	0.18	0.18	5.5	2.2
		0.75	1.65	10.6	23.4	8.9	5.2	2.28	63	0.18	0.18	5.5	2.2
Dry Pregnant Mature Cows													
350	772	–	–	5.8	12.8	5.9	2.8	1.80	50	0.16	0.16	6.1	2.4
400	882	–	–	6.4	14.1	5.9	2.8	1.80	50	0.16	0.16	6.1	2.4
450	992	–	–	6.8	15.0	5.9	2.8	1.80	50	0.16	0.16	6.1	2.4
500	1102	–	–	7.6	16.8	5.9	2.8	1.80	50	0.16	0.16	6.1	2.4
550	1213	–	–	8.0	17.6	5.9	2.8	1.80	50	0.16	0.16	6.1	2.4
600	1323	–	–	8.6	19.0	5.9	2.8	1.80	50	0.16	0.16	6.1	2.4
Cows Nursing Calves, First 3-4 Months Pospartum													
350	772	–	–	8.6	19.0	9.2	5.4	2.06	57	0.29	0.23	9.7	3.9
400	882	–	–	9.3	20.5	9.2	5.4	2.06	57	0.28	0.23	9.7	3.9
450	992	–	–	9.9	21.8	9.2	5.4	2.06	57	0.28	0.22	9.7	3.9
500	1102	–	–	10.5	23.1	9.2	5.4	2.06	57	0.27	0.22	9.7	3.9
Bulls, Growth and Maintenance (Moderate Activity)													
300	661	1.00	2.20	8.7	19.2	13.9	9.6	2.35	65	0.26	0.21	9.7	3.9
400	882	0.90	1.98	10.0	22.0	13.3	9.0	2.35	65	0.19	0.18	9.7	3.9
500	1102	0.70	1.54	12.0	26.5	13.3	9.0	2.15	60	0.18	0.18	9.7	3.9
600	1323	0.50	1.10	11.6	25.6	12.2	8.2	2.15	60	0.18	0.18	9.7	3.9
700	1543	0.30	0.66	12.7	28.0	11.1	7.1	2.06	57	0.18	0.18	9.7	3.9
899	1984	0.00	0.00	9.9	21.8	10.0	6.1	2.06	57	0.18	0.18	9.7	3.9
900	1984	0.00	0.00	10.7	23.6	10.0	6.1	2.06	57	0.18	0.18	9.7	3.9

[a]Feed intake was calculated from the NE requirements and average NE values for the kind of ration fed.

[b]ME requirements for growing and finishing cattle were calculated from the NE_m and NE_{gain} requirements for weights and rates of gain.

[c]TDN was calculated from ME by assuming 3.6155 kcal of ME per g of TDN.

Appendix Table 7. Net energy requirements for growing and finishing steers and heifers (NRC, 1970).

Daily gain, kg	Bodyweight, kg								
	100	150	200	250	300	350	400	450	500
	NE_m required, Mcal/day								
	2.43	3.30	4.10	4.84	5.55	6.24	6.89	7.52	8.14
Steers	Daily Gain – NE_g required, Mcal/day								
0.1	0.17	0.23	0.28	0.34	0.39	0.43	0.48	0.52	0.56
0.2	0.34	0.46	0.57	0.68	0.78	0.88	0.97	1.06	1.14
0.3	0.52	0.70	0.87	1.03	1.18	1.33	1.47	1.61	1.74
0.4	0.70	0.95	1.18	1.40	1.60	1.80	1.99	2.17	2.34
0.5	0.89	1.20	1.49	1.77	2.02	2.27	2.51	2.74	2.97
0.6	1.08	1.46	1.81	2.15	2.46	2.76	3.05	3.33	3.60
0.7	1.27	1.73	2.14	2.53	2.90	3.26	3.60	3.93	4.25
0.8	1.47	2.00	2.47	2.93	3.36	3.77	4.17	4.55	4.92
0.9	1.67	2.27	2.81	3.34	3.82	4.29	4.74	5.18	5.60
1.0	1.88	2.55	3.16	3.75	4.29	4.82	5.33	5.82	6.29
1.1	2.09	2.84	3.52	4.17	4.78	5.37	5.93	6.47	7.00
1.2	2.31	3.13	3.88	4.60	5.27	5.92	6.55	7.14	7.73
1.3	2.53	3.43	4.26	5.04	5.77	6.49	7.17	7.82	8.46
1.4	2.76	3.74	4.63	5.49	6.29	7.06	7.81	8.52	9.22
1.5	2.99	4.05	5.02	5.95	6.81	7.65	8.46	9.23	9.98
Heifers									
0.1	0.18	0.25	0.30	0.36	0.41	0.46	0.51	0.56	0.61
0.2	0.37	0.50	0.62	0.74	0.84	0.95	1.05	1.14	1.24
0.3	0.57	0.77	0.95	1.13	1.29	1.45	1.60	1.75	1.90
0.4	0.77	1.05	1.30	1.54	1.76	1.98	2.18	2.39	2.58
0.5	0.99	1.34	1.66	1.96	2.25	2.53	2.79	3.05	3.30
0.6	1.21	1.64	2.03	2.40	2.75	3.09	3.41	3.73	4.03
0.7	1.44	1.95	2.42	2.85	3.27	3.68	4.06	4.44	4.80
0.8	1.67	2.28	2.81	3.33	3.82	4.28	4.73	5.17	5.59
0.9	1.92	2.60	3.23	3.81	4.37	4.91	5.42	5.93	6.41
1.0	2.17	2.94	3.65	4.32	4.95	5.56	6.14	6.71	7.26
1.1	2.43	3.30	4.09	4.84	5.55	6.23	6.88	7.52	8.13
1.2	2.70	3.66	4.55	5.37	6.16	6.92	7.64	8.35	9.03
1.3	2.98	4.04	5.01	5.92	6.79	7.63	8.42	9.21	9.96
1.4	3.26	4.42	5.49	6.49	7.44	8.36	9.23	10.09	10.91
1.5	3.56	4.82	5.98	7.07	8.11	9.11	10.06	11.00	11.90

Appendix Table 8. Daily nutrient requirements of **dairy cattle** (NRC, 1971).

Body weight kg	Daily gain g	Dry feed kg	Protein Total g	Protein Digestible g	Energy NE_m Mcal	Energy NE_{gain} Mcal	Energy DE Mcal	Energy ME Mcal	TDN kg	Ca g	P g	Carotene mg	Vitamin A 1000 IU	Vitamin D IU
Growing Heifers (Large Breeds)														
40	200	0.5[a]	110	100	0.9	0.4	2.2	1.8	0.5	2.2	1.7	4.2	1.7	265
45	300	0.6[a]	135	120	1.1	0.5	2.6	2.1	0.6	3.2	2.5	4.8	1.9	300
55(5)[b]	400	1.2	180	145	1.3	0.6	4.0	3.3	0.9	4.5	3.5	5.8	2.3	360
75(10)	750	2.1	330	245	1.5	0.9	6.6	5.4	1.5	9.1	7.0	7.9	3.2	495
100(15)	750	2.9	370	260	2.0	1.1	8.8	7.2	2.0	10.9	8.4	11	4	660
150(24)	750	4.1	435	295	3.1	1.5	11.9	9.8	2.7	15	12	16	6	990
200(34)	750	5.3	500	330	4.1	1.8	15.0	12.3	3.4	18	14	21	8	1320
250(43)	750	6.5	570	365	4.8	2.2	17.6	14.4	4.0	21	16	26	10	–
300(53)	750	7.5	640	395	5.6	2.5	19.8	16.2	4.5	24	18	32	13	–
350(62)	750	8.4	715	430	6.2	2.8	21.6	17.7	4.9	25	19	37	15	–
400(72)	750	9.3	800	465	6.9	3.1	22.9	18.8	5.2	26	20	42	17	–
450(82)	700	9.5	885	495	7.5	3.1	23.4	19.2	5.3	27	21	48	19	–
500(93)	600	9.5	935	505	8.1	2.9	23.4	19.2	5.3	27	21	53	21	–
550(107)	400	8.9	915	475	8.7	2.0	22.0	18.0	5.0	26	20	58	23	–
600(133)	150	8.6	810	405	9.3	0.7	19.0	15.5	4.3	24	18	64	26	–
Growing Heifers (Small Breeds)														
20	100	0.3[a]	65	60	0.6	0.2	1.3	1.1	0.3	1.1	0.8	2.1	0.8	130
25	150	0.4[a]	90	80	0.8	0.3	1.8	1.5	0.4	1.5	1.1	2.6	1.0	165
35(5)[b]	300	0.8	135	110	0.9	0.5	2.6	2.1	0.6	3.2	2.5	3.7	1.5	230
50(10)	500	1.2	215	160	1.0	0.9	4.0	3.3	0.9	4.9	3.8	5.3	2.1	330
75(17)	550	1.7	275	190	1.5	1.0	5.3	4.3	1.2	7	5.4	7.9	3.2	495
100(23)	550	2.4	320	210	2.1	1.1	7.1	5.8	1.6	9	7	11	4	660
150(36)	550	3.6	390	245	3.7	1.3	10.1	8.3	2.3	12	9	16	6	990
200(49)	550	4.8	465	280	4.1	1.6	12.8	10.5	2.9	15	11	21	8	1320
250(62)	550	6.1	550	320	4.8	1.9	15.4	12.6	3.5	17	13	26	10	–
300(76)	500	6.8	590	330	5.6	2.0	16.7	13.7	3.8	19	14	32	13	–
350(93)	350	6.6	585	315	6.2	1.5	16.3	13.4	3.7	19	14	37	15	–
400(121)	150	6.4	555	290	6.9	0.7	15.9	13.0	3.6	19	14	42	17	–
450(192)	50	6.1	580	290	7.5	0.5	15.0	12.3	3.4	19	14	48	19	–
Growing Bulls (Large Breeds)														
40	200	0.5[a]	110	100	0.9	0.4	2.2	1.8	0.5	2.2	1.7	4.2	1.7	265
45	300	0.6[a]	135	120	1.1	0.5	2.6	2.1	0.6	3.2	2.5	4.8	1.9	300
55(5)[b]	400	1.2	180	145	1.3	0.6	4.0	3.3	0.9	4.5	3.5	5.8	2.3	360
75(9)	800	2.1	345	255	1.6	1.0	6.6	5.4	1.5	9.7	7.5	7.9	3.2	495
100(13)	1000	3.2	455	320	2.1	1.3	9.7	8.0	2.2	13	10	11.0	4.0	660
150(20)	1000	4.5	520	355	3.2	1.8	13.2	10.8	3.0	18	14	16	6	990
200(27)	1000	5.9	595	390	4.5	2.2	16.7	13.7	3.8	21	16	21	8	1320
250(34)	1000	7.3	670	430	6.0	2.7	19.8	16.3	4.5	24	18	26	10	–
300(41)	1000	8.7	745	465	7.2	3.0	22.9	18.8	5.2	27	20	32	13	–
350(49)	1000	10.2	830	500	8.1	3.4	26.0	21.3	5.9	29	22	37	15	–
400(56)	1000	11.8	930	540	9.0	3.8	29.1	23.8	6.6	30	23	42	17	–
450(63)	1000	12.5	1055	590	9.8	4.1	30.8	25.3	7.0	30	23	48	19	–
500(70)	900	13.0	1110	610	10.6	4.0	32.2	26.4	7.3	30	23	53	21	–
550(79)	800	13.8	1160	625	11.4	3.8	33.9	27.8	7.7	30	23	58	23	–
600(88)	700	13.8	1190	630	12.1	3.5	33.9	27.8	7.7	30	23	64	26	–
650(99)	600	13.6	1220	635	12.9	3.2	33.5	27.5	7.6	30	23	69	28	–
700(112)	500	13.4	1235	630	13.6	2.8	33.1	27.1	7.5	30	23	74	30	–
750(128)	400	13.2	1240	620	14.4	2.3	32.6	26.8	7.4	30	23	79	32	–
800	250	12.7	1165	570	15.1	1.4	31.3	25.7	7.1	30	23	85	34	–
850	100	12.1	1060	510	15.7	0.6	30.0	24.5	6.8	30	23	90	36	–

Appendix Table 8. — Continued.

Body weight kg	Daily gain g	Dry feed kg	Protein: Total g	Protein: Digestible g	Energy: NE_m Mcal	Energy: NE_{gain} Mcal	Energy: DE Mcal	Energy: ME Mcal	TDN kg	Ca g	P g	Carotene mg	Vitamin A 1000 IU	Vitamin D IU
Growing Bulls (Small Breeds)														
20	100	0.3[a]	65	60	0.5	0.2	1.3	1.1	0.3	1.1	0.8	2.1	0.8	130
25	150	0.4[a]	90	80	0.6	0.3	1.8	1.4	0.4	1.5	1.1	2.6	1.0	165
35(4)[b]	300	0.8	135	110	0.7	0.5	2.6	2.2	0.6	3.2	2.5	3.7	1.5	230
50(8)	650	1.4	265	200	1.0	1.1	4.4	3.6	1.0	6.5	5.0	5.3	2.1	330
75(13)	750	2.0	345	240	1.5	1.3	6.2	5.1	1.4	8.4	6.5	7.9	3.2	495
100(18)	750	2.8	390	255	2.1	1.6	8.4	6.9	1.9	11	8	11	4	660
150(28)	750	4.3	460	295	3.1	1.9	11.9	9.8	2.7	15	11	16	6	990
200(37)	750	5.7	530	330	4.5	2.3	15.0	12.3	3.4	18	14	21	8	1320
250(47)	750	7.0	610	365	6.0	2.7	17.6	14.5	4.0	21	16	26	10	–
300(56)	750	8.2	680	395	7.2	3.1	20.3	16.6	4.6	23	17	32	13	–
350(66)	750	9.3	760	430	8.1	3.4	22.9	18.8	5.2	24	18	37	15	–
400(76)	700	10.2	820	450	8.9	3.6	25.1	20.6	5.7	25	19	42	17	–
450(88)	600	10.4	875	465	9.8	3.3	25.6	20.9	5.8	26	20	48	19	–
500(106)	400	10.0	885	455	10.6	2.3	24.7	20.2	5.6	26	20	53	21	–
550(134)	250	10.0	845	420	11.4	1.4	24.7	20.2	5.6	25	19	58	23	–
600	100	9.8	800	385	12.1	0.6	24.2	19.9	5.5	24	18	64	26	–
Veal Calves														
35	500	0.7[a]	155	130	1.0	0.8	3.1	2.5	0.7	3.0	2.3	3.7	1.5	230
40	800	1.1[a]	240	205	1.5	1.4	4.8	4.0	1.1	4.8	3.7	5.3	2.1	330
75	1000	1.4[a]	310	260	1.9	1.8	6.2	5.1	1.4	7.9	5.9	7.9	3.2	495
100	1150	1.7[a]	375	320	2.3	2.2	7.5	6.1	1.7	11.1	8.0	11.0	4.0	660
150	1300	2.4[a]	485	410	3.0	3.0	10.6	8.7	2.4	16.0	11.0	16.0	6.0	990
Maintenance of Mature Breeding Bulls														
500	–	8.3	640	300	9.5	–	20.3	16.6	4.6	20	15	53	21	–
600	–	9.6	735	345	10.8	–	23.8	19.5	5.4	22	17	64	26	–
700	–	10.9	830	390	12.3	–	26.9	22.1	6.1	25	19	74	30	–
800	–	12.0	915	430	13.9	–	29.5	24.2	6.7	27	21	85	34	–
900	–	13.1	1000	470	15.2	–	32.2	26.4	7.3	30	23	95	38	–
1000	–	14.1	1075	505	16.9	–	34.8	28.6	7.9	32	25	106	42	–
1100	–	15.1	1160	545	18.2	–	37.0	30.4	8.4	35	27	117	47	–
1200	–	16.1	1235	580	19.5	–	39.7	32.5	9.0	38	29	127	51	–
1300	–	17.1	1310	615	20.7	–	42.3	34.7	9.6	40	31	138	55	–
1400	–	18.1	1380	650	21.9	–	44.5	39.8	10.1	43	33	148	59	–

[a]Based on milk replacer.

[b]Weeks of age.

Appendix Table 9. Daily nutrient requirements of lactating dairy cattle (adapted from NRC, 1971).

Body weight, kg	Dry feed, kg	Protein: Total, g	Protein: Digestible, g	Energy: NE_{milk}[b] Mcal	Energy: DE, Mcal	Energy: ME, Mcal	TDN, kg	Ca, g	P, g	Carotene, mg	Vitamin A, 100 IU
Maintenance of Mature Lactating Cows[c]											
350	5.0	468	220	6.9	12.3	10.1	2.8	14	11	37	15
400	5.5	521	245	7.6	13.6	11.2	3.1	17	13	42	17
450	6.0	585	275	8.3	15.0	12.3	3.4	18	14	48	19
500	6.5	638	300	9.0	16.3	13.4	3.7	20	15	53	21
550	7.0	691	325	9.6	17.6	14.4	4.0	21	16	58	23
600	7.5	734	345	10.3	18.9	15.5	4.2	22	17	64	26
650	8.0	776	365	10.9	19.8	16.2	4.5	23	18	69	28
700	8.5	830	390	11.6	21.1	17.3	4.8	25	19	74	30
750	9.0	872	410	12.2	22.0	18.0	5.0	26	20	79	32
800	9.5	915	430	12.8	23.3	19.1	5.3	27	21	85	34
Maintenance and Pregnancy (Last 2 Months of Gestation)											
350	6.4	570	315	8.7	15.8	13.0	3.6	21	16	67	27
400	7.2	650	355	9.7	17.2	14.1	4.0	23	18	76	30
450	7.9	730	400	10.7	19.4	15.9	4.4	26	20	86	34
500	8.6	780	430	11.6	21.1	17.3	4.8	29	22	95	38
550	9.3	850	465	12.6	22.9	18.8	5.2	31	24	105	42
600	10.0	910	500	13.5	24.6	20.2	5.6	34	26	114	46
650	10.6	960	530	14.4	26.4	21.6	6.0	36	28	124	50
700	11.3	1000	555	15.3	27.7	22.7	6.3	39	30	133	53
750	12.0	1080	595	16.2	29.5	24.2	6.7	42	32	143	57
800	12.6	1150	630	17.0	31.2	25.6	7.1	44	34	152	61
Milk Production (Nutrients Required per kg of Milk)[d]											
% Fat											
2.5		66	42	0.59	1.10	0.91	0.255	2.4	1.7		
3.0		70	45	0.64	1.23	0.99	0.280	2.5	1.8		
3.5		74	48	0.69	1.34	1.06	0.305	2.6	1.9		
4.0		78	51	0.74	1.46	1.13	0.330	2.7	2.0		
4.5		82	54	0.78	1.57	1.21	0.355	2.8	2.1		
5.0		86	56	0.83	1.68	1.28	0.380	2.9	2.2		
5.5		90	58	0.88	1.79	1.36	0.405	3.0	2.3		
6.0		94	60	0.93	1.90	1.43	0.430	3.1	2.4		

[a] Adapted from NRC (1971).

[b] The energy requirements for maintenance, reproduction, and milk production of lactating cows are expressed in terms of NE_{milk}.

[c] Maintenance of lactating cows = 0.085 Mcal $NE_{milk}/kg^{3/4}$. To allow for growth, add 20 percent to the maintenance allowance during the first lactation and 10 percent during the second lactation.

[d] The energy requirement is presented as the actual amount required with no adjustment to compensate for any reduction in feed value at high levels of feed intake. To account for depressions in digestibility, which occur at high planes of nutrition with certain types of rations, such as corn silage, coarse textures grains or forages with high cell-wall content (e.g. Bermuda grass, sorghum, etc.), an increase of 3 percent feed should be allowed for each 10 kg of milk produced above 20 kg/day.

Appendix Table 10. Nutrient content of rations for dairy cattle (NRC, 1971).

Nutrients	Quantity per kg of dry matter															
	Calf milk Replacer		Calf starter		Heifer grower ration		Dry cow ration		Lactating cow rations daily milk production						Mature bull ration	
									< 20 kg		20-30 kg		> 30 kg			
	Mn	Mx	Mn	Mx	Mn	Mx	Mn	Mx	Mn	Mx	Mn	Mx	Mn	Mx	Mn	Mx
Protein, g	220.0		160.0		100.0		85		140		150		160		77	
Digestible, g	200.0		120.0		62.0		51		105		114		123		36	
Energy, Mcal																
Digestible (DE)	4.2		3.2		2.9		2.3		2.7		2.9		3.1		2.5	
Metabolizable (ME)	3.4		2.6		2.4		1.9		2.1		2.3		2.5		2.0	
NE_m	1.7		0.8		0.8		1.1								1.2	
NE_{gain}	0.8		0.7		0.4											
$NE_{lactation}$									1.4		1.6		1.8			
TDN, g	950		720		660		530		600		650		700		560	
Ether extract, g	100.0		25		20		20		20		20		20		20	
Crude fiber, g	0	30		150	150		150		130		130		130		150	
Calcium, g	5.5		4.1		3.4		3.4		4.3		4.7		5.3		2.4	
Phosphorus, g	4.2		3.2		2.6		2.6		3.3		3.5		3.9		1.8	
Magnesium, g	0.6		0.7		0.8		0.8		1.0		1.0		1.0		0.8	
Potassium, g	7.0		7.0		7.0		7.0		7.0		7.0		7.0		7.0	
Sodium, g	1.0		1.0		1.0		1.0		1.8		1.8		1.8		1.0	
Sodium chloride, g	2.5		2.5		2.5		2.5		4.5		4.5		4.5		2.5	
Sulfur, g	2.0		2.0		2.0		2.0		2.0		2.0		2.0		2.0	
Iron, mg	100.0		100.0		100.0		100.0		100.0		100.0		100.0		100.0	
Cobalt, mg	0.1	10	0.1	10	0.1	10	0.1	10	0.1	10	0.1	10	0.1	10	0.1	10
Copper, mg	10.0	100	10.0	100	10.0	100	10.0	100	10.0	100	10.0	100	10.0	100	10	100
Manganese, mg	20.0		20.0		20.0		20.0		20.0		20.0		20.0		20	
Zinc, mg	40.0	500	40.0	500	40.0	500	40.0	1000	40.0	1000	40.0	1000	40.0	1000	40	1000
Iodine, mg	0.1		0.1		0.1		0.6		0.6		0.6		0.6		0.1	
Molybdenum, mg		6		6		6		6		6		6		6		6
Fluorine, mg		40		30		30				40		40		40		40
Selenium, mg	0.1	5	0.1	5	0.1	5	0.1	5	0.1	5	0.1	5	0.1	5	0.1	5
Carotene, mg	9.5		4.2		4.0		8.0		8.0		8.0		8.0		8.0	
Vitamin A equiv., IU	3800		1600		1500		3200		3200		3200		3200		3200	
Vitamin D, IU	600		250		250		300		300		300		300		300	
Vitamin E, mg	300															

[a] The following minimum quantities of B-complex vitamins are suggested for milk replacers: niacin, 2.6 mg; pantothenic acid, 13 mg; riboflavin, 6.5 mg; pyridoxine, 6.5 mg; thiamine, 6.5 mg; folic acid, 0.5 mg; biotin, 0.1 mg; vitamin B_{12}, 0.07 mg; choline, 2.6 g. It appears that adequate amounts of these vitamins are furnished when calves have functional rumens (usually at 6 weeks of age) by a combination of rumen synthesis and natural feedstuffs.

Appendix Table 11 Nutrient requirements of sheep.[a]

Body wt.		Daily gain or loss		Daily dry matter[1] per animal		Daily dry matter[1]	Energy			Total protein	DP,[3]	Ca,	P,
kg	lb	g	lb	kg	lb	% of live wt.	TDN, %	DE,[2] Mcal/kg	ME, Mcal/kg	%	%	%	%
Ewes[4]													
Maintenance													
50	110	10	.02	1.0	2.2	2.0	55	2.42	1.98	8.9	4.8	.30	.28
60	132	10	.02	1.1	2.4	1.8	55	2.42	1.98	8.9	4.8	.28	.26
70	154	10	.02	1.2	2.6	1.7	55	2.42	1.98	8.9	4.8	.27	.25
80	176	10	.02	1.3	2.9	1.6	55	2.42	1.98	8.9	4.8	.25	.24
Non-lactating; First 15 Weeks Gestation													
50	110	30	.07	1.1	2.4	2.2	55	2.42	1.98	9.0	4.9	.27	.25
60	132	30	.07	1.3	2.9	2.1	55	2.42	1.98	9.0	4.9	.24	.22
70	154	30	.07	1.4	3.1	2.0	55	2.42	1.98	9.0	4.9	.23	.21
80	176	30	.07	1.5	3.3	1.9	55	2.42	1.98	9.0	4.9	.22	.21
Last 6 Weeks Gestation or Last 8 Weeks Lactation Suckling Singles[5]													
50	110	175(+45)	.39	1.7	3.7	3.3	58	2.55	2.09	9.3	5.2	.24	.23
60	132	180(+45)	.40	1.9	4.2	3.2	58	2.55	2.09	9.3	5.2	.23	.22
70	154	185(+45)	.41	2.1	4.6	3.0	58	2.55	2.09	9.3	5.2	.21	.20
80	176	190(+45)	.42	2.2	4.8	2.8	58	2.55	2.09	9.3	5.2	.21	.20
90	198	195(+45)	.43	2.3	5.1	2.6	58	2.55	2.09	9.3	5.2	.21	.20
First 8 Weeks Lactation Suckling Singles or Last 8 Weeks Lactation Suckling Twins[6]													
50	110	-25(+80)	-.06	2.1	4.6	4.2	65	2.86	2.35	10.4	6.2	.52	.37
60	132	-25(+80)	-.06	2.3	5.1	3.9	65	2.86	2.35	10.4	6.2	.50	.36
70	154	-25(+80)	-.06	2.5	5.5	3.6	65	2.86	2.35	10.4	6.2	.48	.34
80	176	-25(+80)	-.06	2.6	5.7	3.2	65	2.86	2.35	10.4	6.2	.48	.34
First 8 Weeks Lactation Suckling Twins													
50	110	-60	-.13	2.4	5.3	4.8	65	2.86	2.35	11.5	7.2	.45	.33
60	132	-60	-.13	2.6	5.7	4.3	65	2.86	2.35	11.5	7.2	.44	.32
70	154	-60	-.13	2.8	6.2	4.0	65	2.86	2.35	11.5	7.2	.43	.31
80	176	-60	-.13	3.0	6.6	3.7	65	2.86	2.35	11.5	7.2	.42	.30
Replacement Lambs and Yearlings[7]													
30	66	180	.40	1.3	2.9	4.3	62	2.73	2.24	10.0	5.8	.45	.25
40	88	120	.26	1.4	3.1	3.5	60	2.65	2.17	9.5	5.3	.44	.24
50	110	80	.18	1.5	3.3	3.0	55	2.42	1.98	8.9	4.8	.42	.23
60	132	40	.09	1.5	3.3	2.5	55	2.42	1.98	8.9	4.8	.43	.24
70	154	10	.02	1.4	3.1	2.0	55	2.42	1.98	8.9	4.8	.46	.26

Appendix Table 11. — continued.

Rams													
Replacement Lambs and Yearlings[7]													
40	88	250	.55	1.8	4.0	4.5	65	2.86	2.35	10.2	6.0	.35	.19
60	132	200	.44	2.3	5.1	3.8	60	2.65	2.17	9.5	5.3	.31	.17
80	176	150	.33	2.8	6.2	3.5	55	2.42	1.98	8.9	4.8	.28	.16
100	220	100	.22	2.8	6.2	2.8	55	2.42	1.98	8.9	4.8	.30	.17
120	265	50	.11	2.6	5.7	2.2	55	2.42	1.98	8.9	4.8	.33	.18
Lambs													
Fattening[8]													
30	66	200	.44	1.3	2.9	4.3	64	2.81	2.30	11.0	6.7	.37	.23
35	77	220	.48	1.4	3.1	4.0	67	2.95	2.42	11.0	6.7	.34	.21
40	88	250	.55	1.6	3.5	4.0	70	3.08	2.53	11.0	6.7	.31	.19
45	99	250	.55	1.7	3.7	3.8	70	3.08	2.53	11.0	6.7	.29	.18
50	110	220	.48	1.8	4.0	3.6	70	3.08	2.53	11.0	6.7	.28	.17
55	121	200	.44	1.9	4.2	3.5	70	3.08	2.53	11.0	6.7	.26	.16
Early-weaned[9]													
10	22	250	.55	0.6	1.3	6.0	73	3.21	2.63	16.0	11.5	.47	.28
20	44	275	.60	1.0	2.2	5.0	73	3.21	2.63	16.0	11.5	.50	.30
30	66	300	.66	1.4	3.1	4.7	73	3.21	2.63	17.0	9.5	.51	.31

[a]From NRC (1974)

[1]To convert dry matter to an as-fed feed basis, divide dry matter by percentage dry matter.

[2]1 kg TDN = 4.4 Mcal DE (digestible energy). DE may be converted to ME (metabolizable energy) by multiplying by 82 percent.

[3]DP = digestible protein.

[4]Values are for ewes in moderate condition, not excessively fat or thin. For fat ewes feed at next lower weight; for thin ewes feed at next highest weight.

[5]Values in parentheses for ewes suckling singles last 8 weeks of lactation.

[6]Values in parentheses for ewes suckling twins last 8 weeks of lactation.

[7]Replacement lambs (ewe & ram) requirements start at time they are weaned.

[8]Maximum gains expected—if lambs are held for later market, they should be fed similar to replacement ewe lambs. Lambs capable of gaining faster than indicated need to be fed at higher level; self-feeding permits lambs to finish most rapidly.

[9]40 kg early-weaned lamb fed same as finishing lamb of equal weight.

Appendix Table 12. Nutrient requirements of breeding swine: % or amount/kg of diet.

Nutrients		Bred gilts and sows[a]	Lactating gilts and sows[b]	Boars (young and adult)[c]
Protein and energy				
Crude protein	%	14	15	14
Digestible energy	kcal	3,300	3,300	3,300
Inorganic nutrients				
Calcium	%	0.75	0.6	0.75
Phosphorus	%	0.50	0.4	0.50
NaCl (salt)	%	0.5	0.5	0.5
Vitamins				
β-Carotene	mg	8.2	6.6	8.2
Vitamin A	IU	4,100	3,300	4,100
Vitamin D	IU	275	220	275
Thiamin	mg	1.4	1.1	1.4
Riboflavin	mg	4.1	3.3	4.1
Niacin	mg	22.0	17.6	22.0
Pantothenic acid	mg	16.5	13.2	16.5
Vitamin B_{12}	μg	13.8	11.0	13.8

[a]Liveweight range (kg): 110 to 160.
[b]Liveweight range (kg): 140 to 200.
[c]Liveweight range (kg): 110 to 180.

Appendix Table 13. Nutrient requirements of growing and finishing swine: % or amount/kg of diet.

		Diet, liveweight class (kg)[a]						
		Full-fed on cereal grains			Full-fed on corn		Full-fed on wheat, barley, oats	
		5–10	10–20	20–35	35–60	60–100	35–60	60–100
Nutrients		Requirements						
Protein and energy								
Crude protein	%	22	18	16	14	13	15	14
Digestible energy	kcal	3,500	3,500	3,300	3,300	3,300	3,100	3,100
Inorganic nutrients								
Calcium	%	0.80	0.65	0.65	0.50	0.50	0.50	0.50
Phosphorus	%	0.60	0.50	0.50	0.40	0.40	0.40	0.40
Sodium	%	–––	0.10	0.10	–––	–––	–––	–––
Chlorine	%	–––	0.13	0.13	–––	–––	–––	–––
Vitamins								
β-Carotene	mg	4.4	3.5	2.6	2.6	2.6	2.6	2.6
Vitamin A	IU	2,200	1,750	1,300	1,300	1,300	1,300	1,300
Vitamin D	IU	220	200	200	125	125	125	125
Thiamin	mg	1.3	1.1	1.1	1.1	1.1	1.1	1.1
Riboflavin	mg	3.0	3.0	2.6	2.2	2.2	2.2	2.2
Niacin[b]	mg	22.0	18.0	14.0	10.0	10.0	10.0	10.0
Pantothenic acid	mg	13.0	11.0	11.0	11.0	11.0	11.0	11.0
Vitamin B_6	mg	1.5	1.5	1.1	–––	–––	–––	–––
Choline	mg	1,100	900	–––	–––	–––	–––	–––
Vitamin B_{12}	μg	22	15	11	11	11	11	11

[a]The following shows expected daily gain (in kilograms) for each of the liveweight classes (left to right): 0.30, 0.50, 0.60, 0.75, 0.90, 0.70, and 0.80.

[b]It is assumed that all the niacin in the cereal grains and their by-products is in a bound form and thus is largely unavailable.

Appendix Table 14. Nutrient requirements of mature horses, pregnant mares, and lactating mares (NRC, 1973).

	Daily feed		Percentage of ration or amount per kg of feed				
Body weight, kg	Per animal, kg	Percentage of live weight	Digestible energy, Mcal	Protein, %	Digestible protein, %	Ca, %	P, %
Mature Horses at Rest (maintenance)							
200	3.00	1.5	2.75	10.0	5.3	0.26	0.20
400	5.04	1.3	2.75	10.0	5.3	0.31	0.24
500	5.96	1.2	2.75	10.0	5.3	0.33	0.25
600	6.83	1.1	2.75	10.0	5.3	0.35	0.26
Mature Horses at Light Work (2 hr/day)							
200	3.80	1.9	2.75	10.0	5.3	0.21	0.15
400	6.68	1.7	2.75	10.0	5.3	0.24	0.18
500	7.96	1.6	2.75	10.0	5.3	0.25	0.18
600	9.23	1.5	2.75	10.0	5.3	0.26	0.19
Mature Horses at Medium Work (2 hr/day)							
200	4.79	2.4	2.75	10.0	5.3	0.19	0.14
400	8.65	2.2	2.75	10.0	5.3	0.20	0.15
500	10.43	2.1	2.75	10.0	5.3	0.20	0.15
600	12.22	2.0	2.75	10.0	5.3	0.20	0.15
Mares, Last 90 Days of Pregnancy							
200	3.16	1.6	2.75	11.5	6.9	0.33	0.25
400	5.41	1.4	2.75	11.5	6.9	0.36	0.28
500	5.31	1.3	2.75	11.5	6.9	0.38	0.29
600	7.25	1.2	2.75	11.5	6.9	0.39	0.29
Mares, Peak of Lactation							
200	5.54	2.8	2.75	13.5	8.7	0.61	0.41
400	8.91	2.2	2.75	13.3	8.4	0.47	0.40
500	10.04	2.0	2.75	13.1	8.3	0.47	0.37
600	10.92	1.8	2.75	12.9	8.0	0.58	0.39

Appendix Table 15. Nutrient requirements of growing horses (NRC, 1973).

				Daily Feed[a]		Percentage of ration or amount per kg of feed				
Age, Mo.	Body weight, kg	Percentage of mature weight	Daily gain, kg	Per animal, kg	Percentage of live weight	Digestible energy, Mcal	Protein, %	Digestible protein, %	Ca, %	P, %
200-kg Mature Weight										
3	60	25.0	0.70	2.94	5.9	7.43	17.9	13.0	0.59	0.37
6	90	45.0	0.50	3.10	3.4	8.53	14.9	10.2	0.53	0.34
12	135	67.5	0.20	2.89	2.1	7.95	11.7	7.1	0.41	0.25
18	165	82.5	0.10	2.94	1.8	8.08	10.7	6.2	0.35	0.22
42	200	100	0	3.00	1.5	8.24	10.0	5.3	0.29	0.20
400-kg Mature Weight										
3	85	21.3	1.00	3.80	4.5	10.44	19.5	14.6	0.68	0.43
6	170	42.5	0.65	4.51	2.7	12.41	14.2	9.5	0.68	0.48
12	260	65.0	0.40	4.96	1.9	13.63	12.1	7.5	0.45	0.30
18	330	82.5	0.25	5.13	1.6	14.10	11.2	6.6	0.37	0.27
42	400	100	0	5.04	1.3	13.86	10.0	5.3	0.32	0.24
500-kg Mature Weight										
3	110	22.0	1.10	4.39	4.0	12.07	19.0	14.1	0.69	0.44
6	225	45.0	0.80	5.60	2.5	15.40	13.4	9.6	0.82	0.51
12	324	65.0	0.55	6.11	1.9	16.81	12.3	7.7	0.43	0.28
18	400	80.0	0.35	6.24	1.6	17.16	11.3	6.7	0.37	0.26
42	500	100	0	5.96	1.2	16.39	10.0	5.3	0.34	0.25
600-kg Mature Weight										
3	140	23.3	1.25	5.15	3.7	14.16	18.6	13.7	1.01	0.63
6	265	44.2	0.85	6.26	2.4	17.21	13.9	9.2	0.81	0.51
12	385	64.1	0.60	6.86	1.8	18.86	12.2	7.6	0.48	0.30
18	480	80.0	0.35	6.98	1.5	19.20	11.1	6.6	0.45	0.28
42	600	100	0	6.83	1.1	18.79	10.0	5.3	0.35	0.26

[a]Assume 2.75 Mcal of digestible energy per kg of 100 percent dry feed.

Appendix Table 16. Energy, protein, and amino acid requirements of chickens (NRC, 1971).

	Broilers		Replacement pullets (egg or meat type)			Laying and Breeding hens (egg or meat type)
	0–6 weeks	6–9 weeks	0–6 weeks	6–14 weeks	14–20 weeks	
Metabolizable energy, Kcal/kg	3,200	3,200	2,900	2,900	2,900	2,850
Protein, %	23	20	20	16	12	15
Arginine, %	1.4	1.2	1.2	0.95	0.72	0.8
Glycine and/or serine, %	1.15	1.0	1.0	0.8	0.6	?
Histidine, %	0.46	0.4	0.4	0.32	0.24	?
Isoleucine, %	0.86	0.75	0.75	0.6	0.45	0.5
Leucine, %	1.6	1.4	1.4	1.1	0.84	1.2
Lysine, %	1.25	1.1	1.1	0.9	0.66	0.5
Methionine, %	0.86	0.75	0.75	0.6	0.45	0.53
or						
Methionine, %	0.46	0.4	0.4	0.32	0.24	0.28
Cystine, %	0.40	0.35	0.35	0.28	0.21	0.25
Phenylalanine, %	1.5	1.3	1.3	1.05	0.78	?
or						
Phenylalanine, %	0.8	0.7	0.7	0.55	0.42	?
Tyrosine, %	0.7	0.6	0.6	0.5	0.36	?
Threonine, %	0.8	0.7	0.7	0.55	0.42	0.4
Tryptophan, %	0.23	0.2	0.2	0.16	0.12	0.11
Valine, %	1.0	0.85	0.85	0.7	0.5	?

Appendix Table 17. Vitamin, mineral, and linoleic acid requirements of chickens[a] (in % or amount/kg of feed) (NRC, 1971).

	Starting chickens (0-8 weeks)	Growing chickens (8-18 weeks)	Laying hens	Breeding hens
Vitamin A activity, IU	*1,500*	*1,500*	4,000	4,000
Vitamin D, ICU	200	200	500	500
Vitamin E, IU	*10*	?	?	?
Vitamin K_1, mg	0.53	?	?	?
Thiamin, mg	1.8	?	?	0.8
Riboflavin, mg	3.6	1.8	2.2	3.8
Panthothenic acid, mg	10	10	2.2	10
Niacin, mg	27	11	*10*	*10*
Pyridoxine, mg	3	?	3	4.5
Biotin, mg	0.09	?	?	0.15
Choline, mg	1,300	?	?	?
Folacin (starch diet), mg	0.55	?	0.25	0.35
Folacin (sugar diet), mg	1.2	?	?	?
Vitamin B_{12}, mg	0.009	?	?	0.003
Linoleic acid, %	?	?	1.0	1.0
Calcium, %	1.0	*0.8*	2.75	2.75
Phosphorus, %	0.7	*0.4*	0.6	0.6
Sodium, %	0.15	0.15	0.15	0.15
Potassium, %	0.2	0.16	?	?
Manganese, mg	55	?	?	33
Iodine, mg	0.35	0.35	0.30	0.30
Magnesium, mg	500	?	?	?
Iron, mg	*80*	?	?	?
Copper, mg	*4*	?	?	?
Zinc, mg	*50*	?	?	*65*
Selenium, mg	*0.1*	?	?	?

[a]These figures are estimates of requirements and include no margins of safety. Italicized figures are tentative.

SUBJECT MATTER INDEX